W9-AGD-326

Atmospheric Science

AN INTRODUCTORY SURVEY

Atmospheric Science

AN INTRODUCTORY SURVEY

John M. Wallace Peter V. Hobbs
University of Washington

Academic Press
San Diego New York Boston
London Sydney Tokyo Toronto

This book is printed on acid-free paper. 

Academic Press
An Elsevier Science Imprint
525 B Street, Suite 1900, San Diego, California 92101-4495, USA
http://www.academicpress.com

Academic Press
32 Jamestown Road, London NW1 7BY, UK
http://www.academicpress.com

Library of Congress Cataloging-in-Publication Data

Wallace, John M.
Atmospheric science.
Includes bibliographical references.
1. Atmosphere. 2. Meteorology. I. Hobbs, Peter Victor, Date joint author. II. Title.
QC861.2.W34 551.5 76-19493
International Standard Book Number: 0-12-732950-1

PRINTED IN THE UNITED STATES OF AMERICA
03 04 05 06 07 MM 20 19 18

I am the daughter of Earth and Water,
   And the nursling of the Sky;
I pass through the pores of the ocean and shores;
   I change, but I cannot die.
For after the rain when with never a stain
   The pavilion of Heaven is bare,
And the winds and sunbeams with their convex gleams
   Build up the blue dome of air,
I silently laugh at my own cenotaph,
   And out of the caverns of rain,
Like a child from the womb, like a ghost from the tomb,
   I arise and unbuild it again.

Percy Bysshe Shelley
"The Cloud"

Contents

Chapter 3 EXTRATROPICAL SYNOPTIC-SCALE DISTURBANCES

Chapter 4 ATMOSPHERIC AEROSOL AND CLOUD MICROPHYSICAL PROCESSES

Chapter 5 CLOUDS AND STORMS

Chapter 6 RADIATIVE TRANSFER

Chapter 7 THE GLOBAL ENERGY BALANCE

Preface

This book has been written in response to a need for a text to support several of the introductory courses in atmospheric sciences commonly taught in universities; namely, introductory survey courses at the junior or senior undergraduate level and beginning graduate level, the undergraduate physical meteorology course, and the undergraduate synoptic laboratory. These courses serve to introduce the student to the fundamental physical principles upon which the atmospheric sciences are based and to provide an elementary description and interpretation of the wide range of atmospheric phenomena dealt with in detail in more advanced courses. In planning the book we have assumed that students enrolled in such courses have already had some exposure to calculus and physics at the first-year college level and to chemistry at the high school level.

The subject material is almost evenly divided between physical and dynamical meteorology. In the general area of physical meteorology we have introduced the basic principles of atmospheric hydrostatics and thermodynamics, cloud physics, and radiative transfer (Chapters 2, 4, and 6, respectively). In addition, we have covered selected topics in atmospheric chemistry, aerosol physics, atmospheric electricity, aeronomy, and physical climatology. Coverage of dynamical meteorology consists of a description of large-scale atmospheric motions (with emphasis on middle latitudes), an introduction to the primitive equations, and an elementary interpretation of the general circulation (Chapters 3, 8, and 9, respectively). In the discussion of clouds and storms (Chapter 5) we have attempted to integrate material from physical and dynamical meteorology. In arranging the chapters we have purposely placed the material on synoptic meteorology near the beginning of the book (Chapter 3) in order to have it available as an introduction to the daily weather map discussions which are an integral part of many introductory survey courses.

The book is divided into nine chapters. Most of the basic theoretical material is covered in the even-numbered chapters (2, 4, 6, and 8). Chapters 1 and 3 are almost entirely descriptive, while Chapters 5, 7, and 9 are mainly interpretive in character. Much of the material in the odd-numbered chapters is straightforward enough to be covered by means of reading assignments, especially in graduate courses. However, even with extensive use of reading assignments we recognize that it may not be possible to completely cover a book of this length in a one-semester undergraduate course. In order to facilitate the use of the book for such courses, we have purposely arranged the theoretical chapters in such a way that certain of the more difficult sections can be omitted without serious loss of continuity. These sections are indicated by means of footnotes. Descriptive and interpretive material in the other chapters can be omitted at the option of the instructor.

The book contains 150 numerical problems and 208 qualitative problems which illustrate the application of basic physical principles to problems in the atmospheric sciences. In addition, the solutions of 48 of the numerical problems are incorporated into the text. We have purposely designed problems that require a minimum amount of mathematical manipulation in order to place primary emphasis on the proper application of physical principles. Universal constants and other data needed for the solution of quantitative problems are given on pages xvi–xvii.

It should be noted that many of the qualitative problems at the ends of the chapters require some original thinking on the part of the student. We have found such questions useful as a means of stimulating classroom discussion and helping the students to prepare for examinations.

Throughout the book we have consistently used SI units which are rapidly gaining acceptance within the atmospheric sciences community. A list of units and symbols is given on pages xv–xvi.

The book contains biographical footnotes which summarize the lives and work of scientists mentioned in the text who have made major contributions to the atmospheric sciences. Brief as these are, we hope that they will give the student a sense of the long history of meteorology and its firm foundations in the physical sciences. As a matter of policy we have included footnotes only for individuals who are deceased or retired.

We wish to express our gratitude to the University of Washington and the National Science Foundation for their support of our teaching, research, and other scholarly activities which contributed to this book. While working on the book, one of us (J.M.W.) was privileged to spend six months on an exchange visit to the Computer Center of the Siberian Branch of the Soviet Academy of

Sciences, Novosibirsk, USSR and a year at the U.S. National Center for Atmospheric Research under the auspices of the Advanced Study Program. The staff members and visitors at both of these institutions made many important contributions to the scientific content of the book. Thanks go also to many other individuals in the scientific community who provided help and guidance.

We wish especially to express our gratitude to colleagues in our own department who provided a continuous source of moral support, constructive criticism and stimulating ideas. Finally, we wish to acknowledge the help received from many individuals who aided in the preparation of the final manuscript, as well as the many interim manuscripts that preceded it.

J. M. W.
P. V. H.

Atmospheric Sciences Department
University of Washington,
Seattle, Washington
December, 1976

Units and Numerical Values

The units used in this book conform to the Système International d'Unités (i.e., the SI system) which is the internationally accepted form of the metric system.

Quantity	*Name of unit*	*Symbol*	*Definition*
Basic units			
Length	meter	m	
Mass	kilogram	kg	
Time	second	s	
Electrical current	ampere	A	
Temperature	degree Kelvin	°K	
Derived units			
Force	newton	N	$kg\,m\,s^{-2}$
Pressure	pascal	Pa	$N\,m^{-2} = kg\,m^{-1}\,s^{-2}$
Energy	joule	J	$kg\,m^2\,s^{-2}$
Power	watt	W	$J\,s^{-1} = kg\,m^2\,s^{-3}$
Electrical potential difference	volt	V	$W\,A^{-1} = kg\,m^2\,s^{-3}\,A^{-1}$
Electrical charge	coulomb	C	$A\,s$
Electrical resistance	ohm	Ω	$V\,A^{-1} = kg\,m^2\,s^{-3}\,A^{-2}$
Electrical capacitance	farad	F	$A\,s\,V^{-1} = kg^{-1}\,m^{-2}\,s^4\,A^2$

Quantity	*Name of unit*	*Symbol*	*Definition*
Frequency	hertz	Hz	s^{-1}
Celsius temperature	degree Celsius	°C	°K − 273.15
Temperature interval	degree	deg or °	K or C need not be specified
Acceptable			
Volume	liter	l	10^{-3} m^3
Pressure	millibar	mbar (abbreviated to mb in this book)	10^2 Pa

The following prefixes are used to construct decimal multiples of units.

Multiple	*Prefix*	*Symbol*	*Multiple*	*Prefix*	*Symbol*
10^{-1}	deci	d	10	deca	da
10^{-2}	centi	c	10^2	hecto	h
10^{-3}	milli	m	10^3	kilo	k
10^{-6}	micro	μ	10^6	mega	M
10^{-9}	nano	n	10^9	giga	G
10^{-12}	pico	p	10^{12}	tera	T
10^{-15}	femto	f			
10^{-18}	atto	a			

SOME USEFUL NUMERICAL VALUES

Universal constants	
Universal gas constant (R^*)	8.3143×10^3 J deg^{-1} kmol^{-1}
Boltzmann's constant (k)	1.381×10^{-23} J deg^{-1} molecule^{-1}
Avogadro's number (N_A)	6.022×10^{26} kmol^{-1}
Stefan–Boltzmann constant (σ)	5.6696×10^{-8} W m^{-2} deg^{-4}
Planck's constant (h)	6.6262×10^{-34} J s
Velocity of light (c^*)	2.998×10^8 m s^{-1}
Permittivity of a vacuum (ε_0)	8.85×10^{-12} C^2 N^{-1} m^{-2}
The earth	
Average radius (R_E)	6.37×10^6 m
Acceleration due to gravity at surface of earth (g_0)	9.81 m s^{-2}
Angular velocity of rotation (Ω)	7.292×10^{-5} rad s^{-1}
Average distance from sun to surface of earth (d)	1.50×10^{11} m
Solar irradiance on a perpendicular plane at distance d from sun	1.38×10^3 W m^{-2}

Dry air

Apparent molecular weight (M_d)	28.97
Gas constant (R_d)	287 J deg^{-1} kg^{-1}
Density at 0°C and 1000 mb pressure (varies as p/T)	1.275 kg m^{-3}
Specific heat at constant pressure (c_p)	1004 J deg^{-1} kg^{-1}
Specific heat at constant volume (c_v)	717 J deg^{-1} kg^{-1}
Thermal conductivity at 0°C (independent of pressure)	2.40×10^{-2} J m^{-1} s^{-1} deg^{-1}

Water substance

Molecular weight (M_w)	18.016
Gas constant for water vapor (R_v)	461 J deg^{-1} kg^{-1}
Density of liquid water at 0°C	1.000×10^3 kg m^{-3}
Density of ice at 0°C	0.917×10^3 kg m^{-3}
Specific heat of water vapor at constant pressure	1952 J deg^{-1} kg^{-1}
Specific heat of water vapor at constant volume	1463 J deg^{-1} kg^{-1}
Specific heat of liquid water at 0°C	4218 J deg^{-1} kg^{-1}
Specific heat of ice at 0°C	2106 J deg^{-1} kg^{-1}
Latent heat of vaporization at 0°C	2.500×10^6 J kg^{-1}
Latent heat of vaporization at 100°C	2.25×10^6 J kg^{-1}
Latent heat of fusion at 0°C	3.34×10^5 J kg^{-1}

Chapter

1

A Brief Survey of the Atmosphere

1.1 INTRODUCTION

The atmospheric sciences comprise a number of related and overlapping disciplines devoted to the description and understanding of phenomena in the atmospheres of the earth and the other planets. Traditionally, the atmospheric sciences have been divided into two disciplines: *meteorology* (from the Greek *meteoros*, meaning lofty, and *logos*, meaning discourse) which is concerned with atmospheric phenomena and their time-dependent behavior, and *climatology* which is concerned with the long-term statistical properties of the atmosphere that constitute *climate* (for example, mean values and range of variability of various measurable quantities, such as temperature, and the frequencies of various events, such as rain or high winds, as a function of geographical location, season, and time of day).

Meteorology, in turn, has traditionally been divided into three main subdisciplines: physical, synoptic, and dynamic meteorology. *Physical meteorology* is concerned with atmospheric structure and composition, the transfer of electromagnetic radiation and acoustic waves through the atmosphere, the physical processes involved in the formation of clouds and precipitation, atmospheric electricity, and a wide range of other problems that are closely related to the disciplines of physics and chemistry. Recently, within physical meteorology, there has developed the field of *aeronomy*, which deals exclusively with phenomena in the upper atmosphere. *Synoptic meteorology* is concerned with the description, analysis, and forecasting of large-scale atmospheric motions. It is rooted in empirical approaches to weather analysis and forecasting which were developed around the turn of the century, following the establishment of the first station networks that provided simultaneous (that is, *synoptic*) weather data over large areas. *Dynamic meteorology* is also concerned with atmospheric motions and their time evolution, but, in contrast to synoptic meteorology, it employs analytical approaches based upon the principles of fluid dynamics. With the increasing sophistication of the methods of weather analysis and forecasting, the distinction between synoptic and dynamic meteorology is rapidly diminishing. The only really pure dynamic meteorologist is one who does not know what a weather map looks like, and the only really pure synoptic meteorologist is one who does not make use of any of the equations that govern atmospheric motions! Both species are approaching extinction.

Climatology can also be subdivided into a number of subdisciplines. Perhaps the most fundamental breakdown is between *physical climatology*, which is concerned with the underlying causes of climate; *climatography*, which deals with the formulation and presentation of climatic statistics on the global, regional, local, and microscales; and *applied climatology*, which deals with the application of climatic statistics to the solution of practical problems. It is clear that climate, on all scales, is determined by meteorological processes. Thus, the distinction between physical climatology and meteorology is not indicative of any real separation between the two disciplines as they exist today; rather, it is a reflection of the largely independent historical development of the two fields. The distinction between meteorology and climatology is being further diminished by the increasing recognition of the fact that climate is continually changing; it can no longer be completely characterized by a single collection of statistics, but rather, like most meteorological phenomena, it must be treated as a time-dependent problem.

Just as the boundaries between various subdisciplines within the atmospheric sciences are gradually becoming less distinct, the whole of atmospheric sciences is becoming less isolated from other scientific endeavors. To cite a few examples:

- Aeronomers, solar physicists, and space physicists are studying the mechanisms by which disturbances on the sun produce a wide variety of phenomena in the earth's upper atmosphere.
- Reconstruction of the climatic history of the earth is involving cooperative efforts among atmospheric scientists, geochemists, geologists, oceanographers, and glaciologists.
- Interactions between the atmosphere and the oceans are receiving increasing attention from both atmospheric scientists and oceanographers.
- Increasing numbers of applied mathematicians and computer scientists are working on atmospheric problems involving numerical modeling.

- The problems of air pollution require specialists with a knowledge of the atmospheric sciences, aerosol physics, and chemistry.
- The atmospheres of other planets, previously the subject of speculation, are now being studied experimentally, using remote sensing techniques and space probes, as well as theoretically.

Much of the current strength and vitality of the atmospheric sciences is due to the fact that the increasing emphasis upon more complex and interdisciplinary types of problems comes at a time when increasingly sophisticated satellite technology is providing scientists with their first opportunity to observe and monitor the atmosphere in a global context from vantage points in space (see Fig. 1.1), and high-speed computers are making it possible to deal with

Fig. 1.1 Image of the earth as seen in reflected visible radiation, 1700 Greenwich Civil Time, 12 June 1975. (NOAA photo.)

complex physical problems in terms of numerical models. These new observational and computational tools provide a basis for attacking complex, interdisciplinary problems that were considered out of reach even a decade ago.

The growing body of knowledge within the atmospheric sciences is being applied to a wide range of practical problems, including

- forecasting of atmospheric phenomena that influence man's activities (for example, day to day weather, hazards to aircraft, droughts, severe storms, events in the upper atmosphere that disrupt radio communications),
- assessment of the impact of man's activities upon the atmospheric environment (for example, local air pollution, inadvertent modification of atmospheric composition, weather, and climate),
- beneficial modification of certain selected physical processes which act on localized scales (for example, fog dispersal, hail suppression, enhancement and redistribution of precipitation), and
- provision of basic atmospheric statistics needed for long-range planning purposes (for example, land use, building design, specifications for aircraft and space vehicles).

In this book we try to capture some of the broad scope and the interdisciplinary character of the work that is going on today within the atmospheric sciences while at the same time presenting in some depth the fundamental concepts that underlie the traditional disciplines upon which the atmospheric sciences are based.

1.2 ORIGIN AND COMPOSITION OF THE ATMOSPHERE

In comparison to the sun, the atmosphere of the earth is remarkably deficient in the noble gases (helium, neon, argon, xenon, and krypton). Various theories have been proposed to account for the virtual absence of these elements in the earth's atmosphere. The prevailing belief is that either the earth formed in a way which systematically excluded gases (for example, by the agglomeration of pieces of solid material similar to that in meteorites),[†] or else the gaseous materials that formed the original atmosphere were lost soon after the earth formed. In either case it is probable that the earth had no atmosphere at, or shortly after, the time of its formation some 4.5×10^9 years ago. The atmosphere that we observe today is believed to have come into existence as a result of the expulsion of volatile substances from the interior in association with volcanic activity as depicted in Fig. 1.2.

Anyone who has been within smelling distance of an active volcano might have difficulty in seeing much resemblance between the atmosphere, as we

[†] Such material presumably included small amounts of volatile substances (materials capable of existing in gaseous form within the range of temperatures found on the surface of the earth) such as water, which could have been present as ice or in chemical combination with solid substances.

Fig. 1.2 Emissions from St. Augustine Volcano, Alaska, following a major eruption, February 1976. (Photo: L. F. Radke.)

know it, and the "raw material" from which it is believed to have evolved. The present atmosphere is roughly 76% nitrogen and 23% oxygen, by mass, as indicated in Table 1.1. By contrast, the gaseous emissions from volcanoes are a mixture of roughly 85% water vapor, 10% carbon dioxide, and up to a few percent nitrogen and sulfur or sulfur compounds (sulfur dioxide and hydrogen sulfide). Free oxygen is notably absent in volcanic emissions.

Table 1.1

Composition of the earth's atmosphere below 100 km

Constituent	Molecular weight	Content (fraction of total molecules)
Nitrogen (N_2)	28.016	0.7808 (75.51% by mass)
Oxygen (O_2)	32.00	0.2095 (23.14% by mass)
Argon (A)	39.94	0.0093 (1.28% by mass)
Water vapor (H_2O)	18.02	0–0.04
Carbon dioxide (CO_2)	44.01	325 parts per million
Neon (Ne)	20.18	18 parts per million
Helium (He)	4.00	5 parts per million
Krypton (Kr)	83.7	1 parts per million
Hydrogen (H)	2.02	0.5 parts per million
Ozone (O_3)	48.00	0–12 parts per million

In order to understand how the present atmosphere might have formed from the volatile substances expelled from the interior of the earth it is necessary to see the atmosphere not as an isolated entity but as a component of a coupled system that also comprises the *hydrosphere* (the total mass of water substance on or above the earth's surface), the *biosphere* (all animal and plant life), and the sedimentary part of the *lithosphere* (the earth's crust). The total mass of volatile material contained in this coupled system is on the order of 0.025% of the mass of the earth. It is also important to bear in mind that the mass of the atmosphere is very small in comparison to some of the other components of this coupled system; for example, it represents only about 1/300th of the mass of the hydrosphere.

1.2.1 Evolution of the hydrosphere

As will be shown in Section 2.6, the atmosphere is capable of holding only a minute fraction of the mass of water vapor that has been injected into it during volcanic eruptions. Thus the earliest volcanic activity on the earth's surface must have given rise to clouds and rain, and with it the formation of bodies of water on the earth's surface. The present "inventory" of the hydrosphere is shown in Table 1.2.

Table 1.2

An inventory of the hydrosphere[a,b]

Component	Percentage of mass of hydrosphere
Oceans	97.
Ice	2.4
Fresh water (underground)	0.6
Fresh water in lakes, rivers, etc.	0.02
Atmosphere	0.001

[a] Total mass $= 1.36 \times 10^{21}$ kg $= 2.66 \times 10^{6}$ kg m^{-2} over surface of earth.

[b] Based on data given in H. H. Lamb, "Climate: Present, Past and Future," Methuen Co. Ltd., London, 1972, p. 482.

If the rate of release of steam by volcanoes over the past century is representative of the average rate over the lifetime of the earth, then the present mass of the hydrosphere is smaller, by as much as two orders of magnitude, than the amount of water that has been injected into the atmosphere. A possible explanation of this discrepancy is leakage from the hydrosphere in the deep oceans along the seams in the earth's crust. Another possibility is that large amounts of water have been destroyed as a result of photodissociation by ultraviolet radiation.

If we accept the hypothesis that Venus, Earth, and Mars were formed in a similar manner, from similar materials, it is reasonable to ask why oceans exist on Earth, but not on Mars or Venus. The absence of oceans on Mars is easily explained: the surface of the planet is too cold to sustain water in liquid form. If water vapor has been the principal component of volcanic emissions on Mars, most of it should be stored in the form of ice in the polar cap regions. Data obtained from recent Mariner missions to Mars suggest that ice may indeed be present in these regions. In view of the high surface temperature of Venus ($\simeq 700°K$) it seems unlikely that any appreciable amounts of liquid water, ice, or hydrated compounds can be present. Thus either the solid material from which Venus was formed contained far less water than that from which the earth was formed or Venus has lost nearly all of its water.[†]

1.2.2 Atmospheric oxygen and life

There are at least two possible sources of atmospheric oxygen: the dissociation of water,

$$2H_2O \rightarrow 2H_2 + O_2 \tag{1.1}$$

and the photosynthesis reaction,

$$H_2O + CO_2 \rightarrow \{CH_2O\} + O_2 \tag{1.2}$$

Both reactions involve the absorption of solar radiation; (1.1) requires ultraviolet radiation and (1.2) requires visible radiation.

It is well established that the photosynthesis reaction (1.2) has produced significant amounts of oxygen on the earth (much more, in fact, than the amount now present in the atmosphere, as will be shown in Section 1.2.3). However, it is not known whether the amount of oxygen produced by photosynthesis is sufficient to account for the present state of oxidation of the materials in the earth's crust, relative to their state of oxidation at the time of the earth's formation.

The possible role of the photodissociation reaction (1.1) as a source of oxygen is a matter of some controversy. There is considerable uncertainty regarding the reaction rate for (1.1), which depends upon other photochemical reactions that compete for the same ultraviolet radiation. Furthermore, the rate of production of oxygen in (1.1) is crucially dependent upon the rate at which the hydrogen produced in the reaction escapes to space through the mechanism discussed in Section 1.3.3. If the escape rate is much slower than the production rate (which is a distinct possibility) then most of the oxygen produced in (1.1) will recombine with hydrogen to form water again.

The production of oxygen by photosynthesis is closely linked with biological processes, since the $\{CH_2O\}$ monomer produced in (1.2) is the basic building

[†] A possible loss mechanism is discussed in R. M. Goody and J. E. G. Walker, "Atmospheres," Prentice-Hall, Inc., Englewood Cliffs, New Jersey, 1972, pp. 132–139.

block for the carbohydrate molecules that form the cells in plant life. In view of the abundance of oxygen in the earth's atmosphere and the nearly complete absence of oxygen in the atmospheres of Venus and Mars, which are lifeless, or nearly lifeless, it is tempting to conclude that most of the oxygen in the earth's atmosphere was produced by photosynthesis. In the remainder of this subsection we will proceed under this supposition.

It is presently believed that the crucial first stage in the evolution of single-celled organisms, which took place over 4×10^9 years ago, required an oxygen-free (reducing) environment. There is geological evidence that indicates that primitive forms of plant life had evolved to the point where they began to release very small amounts of oxygen through the photosynthesis reaction (1.2) between 2 and 3×10^9 years ago. It is believed that these early life forms developed in an aqueous environment, far enough below the surface to be protected from the sun's lethal ultraviolet radiation, but close enough to the surface to have access to visible radiation needed for photosynthesis.[†]

By means of processes discussed in Section 7.2.4, the buildup of oxygen in the atmosphere led to the formation of the ozone (O_3) layer in the upper atmosphere, which filters out incoming solar radiation in the ultraviolet part of the spectrum. With the development of the ozone layer, less and less ultraviolet radiation penetrated to the earth's surface. In this increasingly favorable environment, plant life was able to spread upward into the uppermost layers of the ocean, thereby gaining access to increasing amounts of visible radiation, an essential ingredient in the photosynthesis reaction. More oxygen—less ultraviolet radiation—more access to visible radiation—more abundant plant life—still more oxygen production: through this bootstrap process, life may have slowly but inexorably worked its way upward toward the surface until it finally emerged onto land some 400 million years ago.

1.2.3 The oxygen and carbon budgets

For every molecule of diatomic oxygen produced in (1.2) one molecule of carbon is incorporated into an organic compound. Most of these carbon atoms are oxidized again in respiration or in the decay of organic matter

$$\{CH_2O\} + O_2 \rightarrow H_2O + CO_2 \tag{1.3}$$

However for every few tens of thousands of carbon molecules photosynthesized, one molecule escapes oxidation by being buried or "fossilized." Most of the

[†] It seems likely that the single-celled organisms originated sufficiently near the surface to receive some ultraviolet energy, since laboratory experiments have shown that when a mixture of methane, ammonia, water, and hydrogen is irradiated with ultraviolet light (or sparked by a corona discharge) amino acids, which are basic to life, are formed. Some of the newly formed molecules may then have been transported to lower depths before photodissociation by ultraviolet light occurred.

Another way in which life might have originated is during freezing. Recent laboratory experiments have shown that when ice forms from certain dilute, aqueous solutions, some organic molecules are formed, including adenine which is one of the four components of DNA (which carries the genetic code).

earth's unoxidized carbon is contained in shales, while lesser amounts are stored in more concentrated form in fossil fuels (coal, oil, and natural gas). The relatively "short-term" storage of organic carbon in the biosphere represents a minute fraction of the total storage. More quantitative information on the relative amounts of carbon stored in various forms is given in Table 1.3.

Table 1.3

Inventory of carbon near the earth's surface[a]

Biosphere	marine	1
	nonmarine	1
Atmosphere (in CO_2)		70
Ocean (in dissolved CO_2)		4000
Fossil fuels		800
Shales		800,000
Carbonate rocks		2,000,000

[a] Given in relative units. After P. K. Weyl, "Oceanography," John Wiley & Sons, New York, 1970.

The burning of fossil fuels undoes the work of photosynthesis. At the present rate of fuel consumption, man burns in one year what it took photosynthesis about a thousand years to produce. This rate of consumption seems less alarming when one bears in mind that photosynthesis has been at work for hundreds of millions of years. One can take further comfort from the fact that the bulk of the organic carbon in the earth's crust is stored in a form that is far too dilute for man to exploit.

Of the net amount of oxygen that has been produced by plant life (that is, production by photosynthesis minus destruction through decay) during the earth's history, only about 10% is presently stored in the atmosphere. Most of the oxygen has found its way into oxides such as Fe_2O_3 and carbonate compounds ($CaCO_3$ and $MgCO_3$) in the earth's crust. The formation of carbonate compounds is of particular interest since it is the major sink for the vast amounts of carbon dioxide that have been released in volcanic activity.

Carbonates are formed by means of ion exchange reactions that take place within certain marine organisms, the most important being the one-celled foraminifera. The dissolved carbon dioxide forms a weak solution of carbonic acid

$$H_2O + CO_2 \rightarrow H_2CO_3 \tag{1.4}$$

Then there follows a sequence of reactions the net result of which is

$$H_2CO_3 + Ca^{++} \rightarrow CaCO_3 + 2H^+ \tag{1.5}$$

The $CaCO_3$ enters into the shells of animals, which are eventually compressed into limestone in the earth's crust. Magnesium carbonate is produced by a similar sequence of reactions. The hydrogen ions released in (1.5) react with metallic oxides in the earth's crust, from which they steal an oxygen atom to form another water molecule. The stolen oxygen atom is eventually replaced by one from the atmosphere. Thus oxygen is removed from the atmosphere during the formation of carbonates and it is given back to the atmosphere when carbonates dissolve. It has been proposed that foraminifera and other carbonate-producing sea animals, by virtue of their role as mediators in the process of carbonate formation, have the ability to regulate the amount of oxygen present in the atmosphere, which has been remarkably constant over the past few million years.

The widespread occurrence of the fossils of sea animals in limestone deposits suggests that ion exchange reactions in sea water have played an important role in the removal of carbon dioxide from the earth's atmosphere. Thus, the dominance of carbon dioxide in the Martian atmosphere may be due, at least in part, to the absence of liquid water on the surface. In contrast to the situation on Mars, the massive carbon dioxide atmosphere of Venus may be a consequence of the high surface temperatures on that planet. At such temperatures there should exist an approximate state of equilibrium between the amount of carbon dioxide in the atmosphere and the carbonate deposits in rocks on the surface, as expressed by the reaction

$$CaCO_3 \rightleftarrows CaO + CO_2 \tag{1.6}$$

There are indications that the rate of removal of carbon dioxide from the earth's atmosphere is not large enough to keep pace with the ever increasing rate of input due to the burning of fossil fuels. The concentration of carbon dioxide in the atmosphere has been rising steadily since the early part of this century, as indicated in Fig. 1.3. The present rate of increase is roughly half as large as the rate at which carbon dioxide is being added to the atmosphere by the burning of fossil fuels.

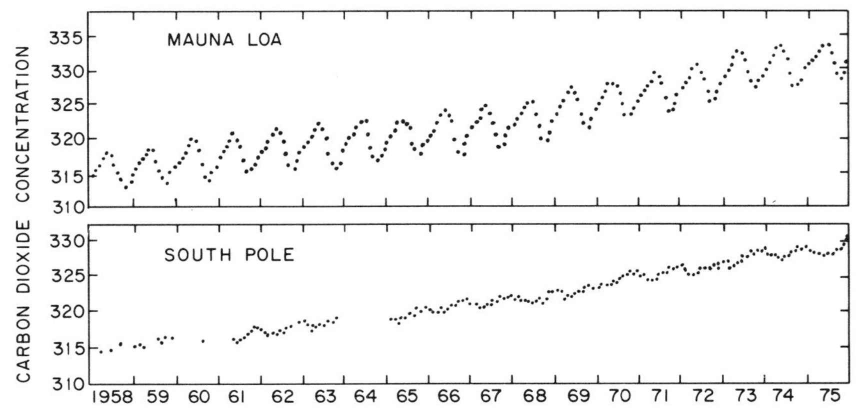

Fig. 1.3 Carbon dioxide concentrations at Mauna Loa, Hawaii, and at the South Pole in units of parts per million. (Courtesy of C. D. Keeling.)

1.2.4 Some other atmospheric constituents

By means of ion exchange reactions analogous to (1.5) and fixation by soil microorganisms a small fraction (perhaps 20%) of the nitrogen released into the atmosphere has entered into nitrates in the earth's crust. However, by virtue of its chemical inertness and low solubility in water ($\simeq 1/70$ that of carbon dioxide), most of the nitrogen released by volcanoes has remained in the atmosphere. Because of the nearly complete removal of water and carbon dioxide by the processes described above, nitrogen has become the dominant gaseous constituent of the earth's atmosphere.

Sulfur and its compounds hydrogen sulfide and sulfur dioxide, which are released into the earth's atmosphere by volcanic emissions, are quickly oxidized to sulfur trioxide which dissolves in cloud droplets to form a dilute solution of sulfuric acid. ("Acid rain" is observed downwind of major industrial centers where sulfur-bearing coal and oil are burned.) After being "scavenged" out of the atmosphere by precipitation particles, the sulfate ions combine with metal ions to form sulfates within the earth's crust. Sulfur dioxide may also react with ammonia in the presence of liquid water to produce ammonium sulfate (see Section 4.1.3).

Argon is far more abundant in the earth's atmosphere than any of the other noble gases. This apparent discrepancy is a reflection of the fact that about 99.7% of the atmospheric argon is ^{40}A, which is a by-product of the radioactive decay of ^{40}K in the solid earth. Helium in the earth's atmosphere is also mainly a by-product of radioactive decay.

1.3 THE DISTRIBUTION OF ATMOSPHERIC MASS AND GASEOUS CONSTITUENTS

Because of the earth's gravitational field, the atmosphere exerts a downward force on the earth's surface. The force per unit area due to the weight of the atmosphere can be expressed as a pressure. The mean atmospheric pressure averaged over the earth's surface can be closely approximated by $M_A g_0/4\pi R_E^2$, where M_A is the total mass of the atmosphere (5.14×10^{18} kg), g_0 is the mean gravitational acceleration (9.8 m s^{-2}), and R_E is the mean radius of the earth (6.37×10^6 m). Substituting these numerical values into the expression it is readily verified that the mean atmospheric pressure at the earth's surface is roughly 10^5 Pa.†

† In this book we will adhere rather strictly to the use of the SI units as outlined on page xv. However, in view of the fact that, as of the time of this writing, relatively few meteorologists have even heard of the term *pascal* (Pa), we will make a reluctant exception to our general policy and express pressure in the more familiar units of *millibars* (mb). (The *bar* is a remnant of cgs units, where 1 bar $\equiv 10^6$ dyn cm^{-2}. Therefore, 1 mb $= 10^2$ Pa, so that the mean pressure at the earth's surface is roughly 1000 mb.) In making quantitative calculations, the problem of keeping track of units is greatly simplified if all quantities, including pressure, are expressed in SI units. Thus, although pressure will usually be expressed in terms of millibars in the text, it should always be converted to pascals when working quantitative problems.

1.3.1 Vertical profiles of pressure and density

The vertical variability of pressure and density is much larger than the horizontal and temporal variability of these quantities. Therefore it is useful to define a "standard atmosphere" which represents the horizontal and time-averaged structure of the atmosphere as a function of height only, as displayed in Fig. 1.4. At any given level, up to about 100 km, the atmospheric pressure and density are nearly always within 30% of the corresponding "standard atmosphere" values. It may be noted that within the lowest 100 km, the logarithm of the pressure drops off almost linearly with height; that is to say,

$$\log[p(z)] \simeq \log[p(0)] - Bz \qquad (1.7)$$

where $p(z)$ is pressure at height z above sea level, $p(0)$ is the pressure at sea level, and B is a constant which is related to the average slope of the pressure curve. Making use of the identity

$$\ln x = 2.3 \log x$$

where ln indicates a logarithm to the base e (= 2.718), (1.7) can be rewritten in the form

$$\ln \frac{p(z)}{p(0)} \simeq -\frac{z}{H}$$

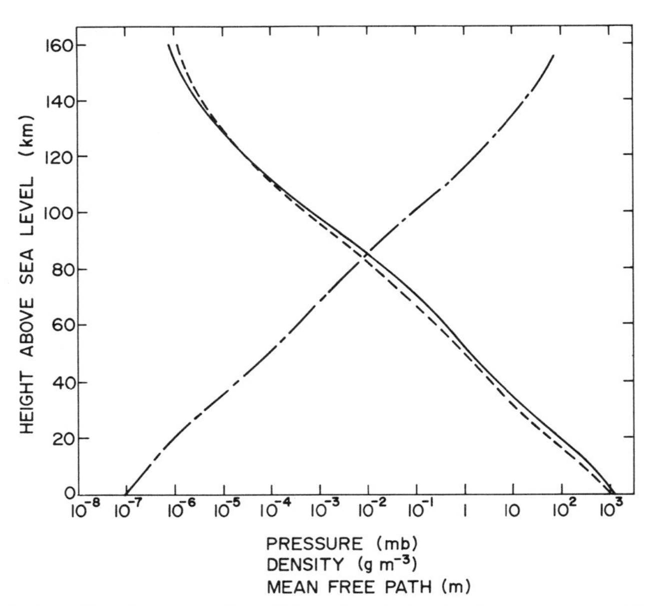

Fig. 1.4 Vertical profile of pressure in millibars (---), density in grams per cubic meter (—), and mean free path in meters (— - —) for U.S. Extension to International Civil Aviation Organization Standard Atmosphere. (Adapted from "CRC Handbook of Chemistry and Physics," 54th Edition, © CRC Press, Cleveland, 1973, pp. F186–190. Used by permission of CRC Press, Inc.)

where $H \equiv 1/(2.3B)$. Taking the antilog of both sides, we obtain

$$p(z) \simeq p(0)\exp(-z/H) \tag{1.8}$$

which states that pressure drops off by a factor e in passing upward through a layer of depth H. H is called the *scale height* of the atmosphere. Since the density curve is nearly straight and has almost the same slope as the pressure curve, it follows that the density $\rho(z)$ at height z is also given approximately by

$$\rho(z) \simeq \rho(0)\exp(-z/H) \tag{1.9}$$

The approximate value of the scale height H can be determined by noting that $\log(p)$ drops off by about 6.3 in the lowest 100 km. Substituting into (1.7) we obtain $B = 0.063$, from which it follows that $H \simeq 7$ km. A physical interpretation of the pressure and density profiles is given in Section 2.2.

It should be emphasized that the atmosphere is remarkably thin in comparison to the dimensions of the earth. Half the mass of the atmosphere lies below the 500-mb level, which has a mean height of roughly $5\frac{1}{2}$ km (less than 0.001 earth radius) above sea level, and about 99% of the mass of the atmosphere lies within the lowest 30 km above sea level.

1.3.2 Atmospheric composition as a function of height

In the absence of sources and sinks, the ratios of various gaseous constituents at any level in the atmosphere are determined by two competing physical processes: molecular diffusion and mixing due to fluid motions.

Diffusion by random molecular motions tends to produce an atmosphere in which the mean molecular weight of the mixture of gases gradually decreases with height to the point where only the lightest gases (hydrogen and helium) are present at the highest levels. In effect, each gaseous constituent behaves as if it alone were present. The density of each gas drops off exponentially with height, as indicated by (1.9), but the scale height H is different for each gas. The density of the lighter gases drops off more slowly than that of the heavier gases, with scale height being inversely proportional to molecular weight. The physical basis for this relationship is discussed in Section 2.2.

In contrast to molecular diffusion, the mixing due to the motions of macroscale air parcels does not discriminate on the basis of molecular weight. Within the range of levels where this process predominates, atmospheric composition tends to be independent of height.

The relative effectiveness of molecular diffusion increases in proportion to the root mean square velocity of the random molecular motions, and the mean free path between collisions. The dependence of the rate of diffusion upon mean free path is also implicit in the process of separation of the sexes as they seek out their respective restrooms during the intermission at a concert. The more crowded the lobby (that is, the smaller the "mean free path"), the slower the rate at which the separation takes place. In the mixing by fluid motions the analog of the mean free path is the "mixing length" which depends upon the spectrum of scales of motion present in the atmosphere.

Of the various factors that influence the relative effectiveness of molecular diffusion and mixing by fluid motions, by far the most important is the increase in mean free path with height, which is illustrated in Fig. 1.4. In the lower atmosphere the mean free path is so short that the time required for the vertical separation of the heavier and lighter constituents by molecular diffusion is many orders of magnitude longer than the time required for turbulent fluid motions to homogenize them. Near an altitude of 100 km the two competing processes are of roughly comparable importance, while well above 100 km the vertical mixing of atmospheric constituents is essentially controlled by molecular diffusion. The level of transition from turbulent mixing to molecular diffusion is called the *turbopause*. For the purpose of describing atmospheric composition, the well-mixed region below the turbopause is called the *homosphere*; the region above is called the *heterosphere*.

The composition of the lower part of the heterosphere is strongly influenced by the photodissociation of diatomic oxygen, which gives rise to large numbers of free oxygen atoms. Above about 120 km most of the atmospheric oxygen is in the atomic form. The production of atomic oxygen is discussed further in Section 7.2.3.

At higher levels there is a noticeable increase in the relative abundance of the lighter constituents, due to the effects of molecular diffusion. The heaviest major constituent, diatomic nitrogen, drops off most rapidly with height. Around 500 km the atmosphere is predominantly atomic oxygen, with only traces of diatomic nitrogen and the very light constituents, helium and hydrogen. Above 1000 km helium and hydrogen are the dominant species.

The structure of the heterosphere is strongly dependent upon temperature, which varies by a factor of three or more in response to solar activity as discussed in Section 7.2.1. At low temperatures the transition to lighter species takes place at relatively low levels, whereas at high temperatures it takes place at higher levels. Above 300 km, the pressure and density at any given level vary by an order of magnitude or more in response to changes in solar activity.

1.3.3 Escape of the lighter constituents

Above about 500 km the mean free path between collisions is so long that individual molecules follow ballistic trajectories, like rockets. For all species of molecules there exists a single *escape velocity* V_e for which the kinetic energy of the molecule is equivalent to the potential energy that needs to be supplied to lift it out of the earth's gravitational field. Escape velocity is a function of height only: in the earth's atmosphere at a level of 500 km it is on the order of 11 km s^{-1}.

The most probable velocity of any molecular species is given by

$$V_0 = \sqrt{\frac{2kT}{Mm_H}} \tag{1.10}$$

where k is Boltzmann's† constant (1.38×10^{-23} J deg^{-1}), T is the absolute temperature, M is the molecular weight, and m_H is the mass of a hydrogen atom (1.67×10^{-27} kg).‡ Note that (1.10) implies an equipartition of kinetic energy among the various constituents such that the mean velocity of the lighter molecules is higher than that of the heavier ones. The individual molecules within a gas exhibit a range of velocities scattered about V_0. The kinetic theory of gases predicts that only about 2% of the molecules have velocities greater than $2V_0$, and only one molecule in 10^4 has a velocity greater than $3V_0$. Additional examples are given in Table 1.4.

In the earth's atmosphere, temperatures at the base of the "escape region" (*or exosphere*, as it is sometimes called) are on the order of 600°K. Substituting this value into (1.10) it is readily verified that, for hydrogen atoms ($M = 1$), $V_0 \simeq 3$ km s^{-1}. Therefore, according to Table 1.4, for each collision near 500 km, the probability of escape (that is, $V > V_e$) is slightly greater than 10^{-6}. The corresponding time period required for the escape of all the hydrogen from the earth's atmosphere turns out to be much less than the lifetime of the earth; hence, the relative absence of free hydrogen in the atmosphere despite its continual production due to the dissociation of water. For atomic oxygen ($M = 16$), $V_0 \simeq 0.8$ km s^{-1} and the probability of escape is $\simeq 10^{-84}$. The rate of escape of atomic oxygen is so slow that the cumulative loss over the lifetime of the earth is negligible.

Table 1.4

Fraction of gas molecules with velocities V greater than various multiples of the most probable velocity V_0

V/V_0	Fraction	V/V_0	Fraction
1	0.5	6	10^{-20}
2	0.02	10	10^{-50}
3	10^{-4}	15	10^{-90}
4	10^{-6}		

1.3.4 Variable constituents

In contrast to the other gases listed in Table 1.1 the concentrations of water vapor and ozone are highly variable in space and time. Although they are present only as trace constituents, both gases play extremely important roles

† **Ludwig Boltzmann** (1844–1906) Austrian physicist. Made fundamental contributions to the kinetic theory of gases. Adhered to the view that atoms and molecules are real at a time when this concept was in dispute. Committed suicide.

‡ For a derivation of this relationship and the basis for Table 1.4, see F. W. Sears, "Thermodynamics, The Kinetic Theory of Gases and Statistical Mechanics," Addison-Wesley, Reading, Massachusetts, 2nd Ed., 1953, pp. 232–239.

in the atmospheric energy balance and in the absorption of radiation passing through the atmosphere, as discussed in Chapter 7.

The main source of atmospheric water vapor is evaporation from the earth's surface and the main sink is condensation which takes place in clouds. The typical "lifetime" of a molecule of water vapor in the atmosphere is only on the order of a week. The concentration of water vapor is largest near the ground and it drops to very small values above 10 km.

Ozone is generated by photochemical reactions in the layer between 20 and 60 km, as discussed in Section 7.2.4, and (in much smaller amounts) in polluted air over cities. At the earth's surface, ozone is rapidly destroyed by reacting with plants and dissolving in water, whereas, at levels between 10 and 25 km, it is a rather stable chemical species with a lifetime on the order of months. In the long-term statistical average, there is a slow downward flux of ozone from the high level source region to the sink at the earth's surface.

1.4 CHARGED PARTICLES IN THE ATMOSPHERE

Although charged particles account for only a minute fraction of the mass of the atmosphere, they play a crucial role in a wide range of geophysical phenomena including lightning, auroras, the reflection of radio waves, fluctuations in the geomagnetic field, and the maintenance of the fair weather electric field between the earth's surface and the upper atmosphere.

A number of distinctly different physical processes contribute to the production of charged particles:

- X-ray and ultraviolet radiation from the sun ionizes air molecules. Virtually all the sun's ionizing radiation is absorbed at levels above 60 km. (For further details, see Section 7.2.2.)
- High-energy cosmic rays emanating from the sun and from sources outside the solar system penetrate into the lower atmosphere where they leave trails of ionized air molecules.
- Over land areas, the radioactive decay of substances within the earth's crust provides an additional source of ions.
- Electric charges may be separated within clouds (see Section 4.6.1).

1.4.1 The ionosphere

Because of the rapid increase in the mean free path of particles with height and the transition to more stable species of ions at higher levels, the free electrons produced by the sun's ionizing radiation in the upper atmosphere have much longer lifetimes than those resulting from the various sources at lower levels. Hence, most of the atmosphere's free electrons are located at levels above 60 km, where they exist in sufficient numbers to affect radio wave propagation.

This region of the atmosphere is sometimes called the *ionosphere* or, sometimes, the *Heaviside*† *Layer.*

A vertical profile of the number density of free electrons is shown in Fig. 1.5, together with the total particle density. It can be seen that the concentration of free electrons increases monotonically with height from very small values below 60 km to a maximum value near 300 km. There are some small undulations labeled D and E in the figure. These irregularities in the vertical gradient of electron concentration have a profound effect upon the propagation of radio waves. Before the results shown in Fig. 1.5 were established by means of *in situ* measurements from rockets and satellites, it was widely believed, on the basis of radio wave propagation experiments, that the bump at E (~ 110 km) corresponded to a distinct maximum in electron density. At that time, the terms *E-layer* and *F-layer* came into use as a means of distinguishing between the lower and upper maxima. The term *D-region* was used to denote a lower layer in which strong absorption of radio waves takes place due to collisions between electrons and neutral particles. Later it was discovered that the D-region contains a separate bump in the electron density profile.

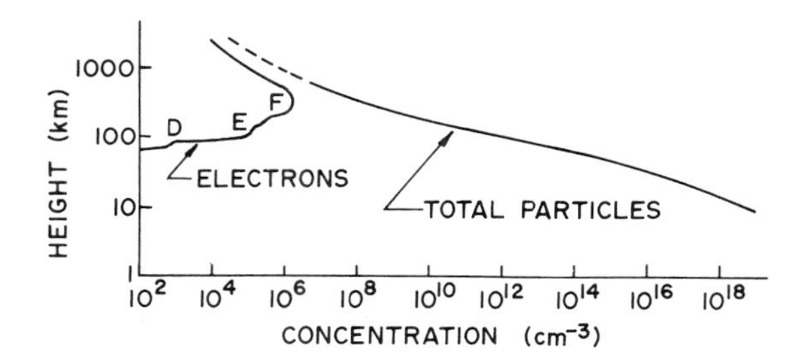

Fig. 1.5 Total particle density and number density of electrons as a function of height. (After R. G. Fleagle and J. A. Businger, "An Introduction to Atmospheric Physics," Academic Press, New York, 1963, p. 235.)

Electron densities in the ionosphere decline in the hours after sunset, in response to the interruption of the ionizing flux of solar radiation. The extent of the decrease varies with height. Most of the electrons in the D- and E-regions recombine with positive ions during the night. With the virtual disappearance of the D-region, the absorption of radio waves is reduced to the point where signals reflected from higher layers of the ionosphere interfere with reception in the AM band. The F-region exhibits a smaller diurnal variability, which is often masked by irregular fluctuations in electron density.

† **Oliver Heaviside** (1850–1925) Self-taught English mathematician and physicist. Developed duplex telegraphy at age 23. Predicted the existence of an ionized air layer (the ionosphere) and its effects on radio transmission in 1902.

Electron density also varies in response to events on the sun which modulate the X rays reaching the ionosphere. For example, strong solar flares are accompanied by bursts of X rays which increase the electron densities within the D-region on the daylight side of the earth. Enhanced electron densities give rise to increased absorption of radio waves, which sometimes causes fadeouts in long-range communications that depend upon the reflection of radio waves from the ionosphere. Coupling between solar activity and the earth's ionosphere is discussed further in Section 7.2. Still another source of variability within the D-region is large-scale motions in the polar regions which occur in association with the "sudden warming" phenomenon described in Section 1.5.1.

Within the E-region the mean free path is sufficiently short that the movement of positive ions is controlled, to a large extent, by the drift of the neutral constituents, which account for an overwhelming fraction of the mass. However, within this same region, the free electrons are constrained to move along the magnetic field lines. Hence, whenever the neutral atmosphere within the E-region moves across the earth's magnetic field lines, charge separation takes place, currents flow, and voltages are induced; the effect is analogous to the generation of electrical power in a dynamo. Currents generated in the E-region are responsible for variations in geomagnetism and in the structure of the F-region. An important input of energy into the "dynamo" is associated with atmospheric tidal motions driven not by gravitational effects but by the diurnal variation in solar heating.[†]

1.4.2 The fair weather atmospheric electric field

Below an altitude of a few tens of kilometers there is a downward-directed electric field in the atmosphere during fair weather. Above this layer of relatively strong electric field is a region extending upward to the top of the ionosphere in which the electrical conductivity is so high that it is essentially at a constant electric potential. This region of constant electric potential is called the *electrosphere*. Since the electrosphere is a good electrical conductor, it serves as an almost perfect electrostatic shield. Thus charged particles from outside the electrosphere (for example, those associated with auroral displays) rarely penetrate below the electrosphere and the effects of thunderstorms in the lower atmosphere do not extend above the electrosphere.

The magnitude of the fair weather electric field near the surface of the earth, averaged over the whole globe, is about 120 V m^{-1}; averaged over the ocean it is closer to 130 V m^{-1} and in industrial regions it is as high as 360 V m^{-1}. The high value in the latter case is due to the fact that industrial pollutants decrease the electrical conductivity of the air because large, slow-moving

[†] An elementary and very readable discussion of the ionosphere and magnetosphere is given in J. A. Ratcliffe, "Sun, Earth and Radio," McGraw-Hill, New York, 1970.

particles tend to replace ions of higher mobility. Since the vertical current density (which is equal to the product of the electric field and electrical conductivity) must be the same at all levels, the electric field must increase if the conductivity decreases. At heights above about 100 m, the conductivity of the air increases with height and therefore the fair weather electric field decreases with height. The increase in electrical conductivity with height is due to the greater ionization by cosmic rays and diminishing concentrations of large particles with increasing altitude. Thus at 10 km above the earth's surface the fair weather electric field is only 3% of its surface value. The average potential of the electrosphere with respect to the earth is about 300 kV.

The presence of the downward-directed fair weather electric field implies that the electrosphere carries a net positive charge and the earth's surface a net negative charge. Lord Kelvin,[†] who in 1860 first suggested the existence of a conducting layer in the upper atmosphere in relation to atmospheric electricity, also suggested that the earth and the electrosphere act as a gigantic spherical capacitor, the inner conductor of which is the earth, the outer conductor the electrosphere, and the (leaky) dielectric the air. The electric field is nearly constant despite the fact that the current flowing in the air (which averages about 2 to 4×10^{-12} A m^{-2}) would be large enough to discharge the capacitor in a matter of minutes. Thus, there must be an electrical generator in the system. In 1920 Wilson[‡] proposed that the generators are thunderstorms, and this idea is now almost universally accepted. Thunderstorms separate electric charges (by mechanisms to be discussed in Section 4.6.1) in such a way that their upper regions become positively charged and their bases negatively charged. The upper positive charges are leaked to the base of the electrosphere through the relatively highly conducting atmosphere at these levels. Below a thunderstorm the electrical conductivity of the air is low. However, under the influence of the very large electric fields, a current of positive charges flows upwards from the earth (through trees and other pointed obstacles). This is called the *point discharge current*.[§] Precipitation particles are polarized by the fair weather electric field, and the field beneath thunderstorms, in such a way that they tend to preferentially collect positive ions as they fall to the ground. A positive charge equivalent to about 30% of that from point discharges is returned to the earth in this way. Finally, lightning flashes (which will be discussed in Section 4.6.2) transport negative charges from the bases of thunderstorms to the ground.

[†] **Lord Kelvin 1st Baron (William Thomson)** (1824–1907) Scottish mathematician and physicist. Entered Glasgow University at age 11. At 22 became Professor of Natural Philosophy at the same university. Carried out incomparable work in thermodynamics, electricity, and hydrodynamics.

[‡] **C. T. R. Wilson** (1869–1959) Scottish physicist. Invented the Wilson cloud chamber used in studies of ionizing radiation and charged particles. Carried out important studies on condensation nuclei and atmospheric electricity. Awarded Nobel Prize in physics in 1927.

[§] Point discharges at mastheads, etc., are known as *St. Elmo's fire*.

A schematic of the complete global electrical circuit is shown in Fig. 1.6. A rough electrical budget for the earth (in units of C km^{-2} yr^{-1}) is 90 units gained from the fair weather conductivity, 100 units lost through point discharges, 30 units gained from precipitation, and 20 units lost due to the transfer of negative charges to the earth by lightning.

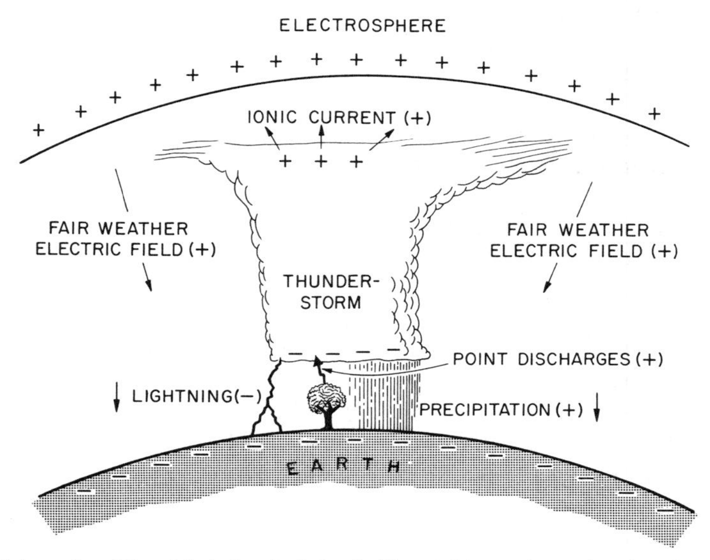

Fig. 1.6 Schematic of the global electrical circuit. The positive and negative signs in parentheses indicate the signs of the charges transported in the direction of the arrows.

1.4.3 The magnetosphere

Above 500 km, collisions between individual particles are so infrequent that there is relatively little interaction between charged particles and the neutral constituents of the atmosphere. At these levels the motions of charged particles are strongly constrained by the presence of the earth's magnetic field lines; hence, the term *magnetosphere.*

To a first approximation, the earth's magnetic field is equivalent to that produced by a central dipole inclined at an angle of 11° to the axis of planetary rotation. However, the field lines are somewhat distorted from an idealized dipole shape due to the influence of the *solar wind,*† a stream of ionized particles emanating outward from the sun with velocities on the order of 500 km s^{-1}. On the daylight side of the earth the field lines are compressed slightly due to the pressure of the solar wind, while on the night side they are stretched outward

† The solar wind is discussed further in Section 7.2.1.

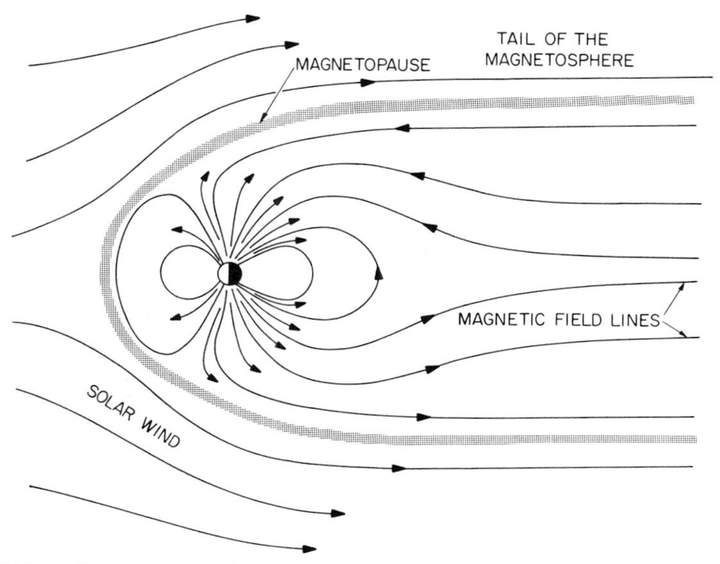

Fig. 1.7 Schematic cross section through the earth's magnetosphere. (Adapted from P. M. Banks and G. Kockarts, "Aeronomy," Academic Press, New York, 1973, Part A, p. 115.)

in a long tail as shown in Fig. 1.7. The earth's *magnetosphere* acts as an obstacle to the solar wind, with the *magnetopause* separating solar wind particles, which flow around the magnetosphere, from charged particles trapped within the magnetosphere.

Disturbances on the sun spew out bursts of particles which have the effect of increasing the speed and particle density of the solar wind. When these clouds of solar particles reach the earth they produce a wide variety of effects, including changes in the magnetic field, anomalous currents in the ionosphere, and the injection of high-energy particles into the lower polar ionosphere. These high-energy particles, which are somehow generated within the earth's magnetosphere, are responsible for the brilliant auroral displays that often occur several days after strong outbursts of solar activity. Disturbances within the earth's magnetosphere are also closely related to the structure of the interplanetary magnetic field, which rotates with the sun.

1.5 THE TEMPERATURE DISTRIBUTION

The early exploration of the vertical temperature structure of the atmosphere was carried out by means of *in situ* measurements made during manned balloon flights. These heroic and sometimes tragic expeditions into the atmosphere provided the first conclusive evidence that within the lowest 10 km temperature usually decreases with altitude at a rate of about 7 deg km^{-1}. This so-called "lapse rate" is highly variable from place to place, but it never exceeds 10 deg km^{-1} except near the ground. In 1902, Teisserenc de Bort[†] and Assmann[‡] independently discovered the existence of a higher layer, above about 10 km, in which the temperature usually remains constant or even increases with further increase in altitude. De Bort called the lowest layer above the ground the *troposphere* and the newly discovered upper layer the *stratosphere.*

Up until the 1920's the monitoring of the temperature structure of the upper atmosphere was an expensive venture that could only be undertaken on an occasional basis. The invention of the radiosonde by Molchanov[§] in 1927 introduced a new era of atmospheric exploration in which routine temperature soundings could be taken simultaneously over a whole synoptic network, using unmanned balloons carrying relatively inexpensive instrument packages which radioed their observations back to earth. Modern radiosonde systems are capable of making observations up to about 40 km. Similar instrument packages borne aloft by rockets have been used, together with acoustic sounding techniques, to probe the layer of the atmosphere between 40 and 80 km. At higher levels much of the information on temperature structure has been deduced indirectly from measurements of satellite drag, based upon observations of orbital changes.

The vertical distribution of temperature for the "standard atmosphere" is shown in Fig. 1.8. This profile is representative of typical conditions in middle latitudes. As indicated in the figure, the vertical profile can be divided into four distinct layers: *troposphere*, *stratosphere*, *mesosphere*, and *thermosphere*. The tops of these layers are called the *tropopause*, *stratopause*, *mesopause*, and *thermopause*, respectively.

The troposphere (literally, the turning or changing sphere) accounts for more than 80% of the mass and virtually all of the water vapor, clouds, and precipitation in the earth's atmosphere. It is characterized by rather strong vertical mixing; for example, in clear air it is not unusual for an air molecule to traverse the entire depth of the troposphere over the course of a few days. In the updrafts of large thunderstorms, particles can travel from the vicinity of the ground up

[†] **Teisserenc de Bort** (1855–1913) French meteorologist. Founded one of the first aerological observatories at Trappes, near Paris, spending nearly his entire private fortune on it. Began kite observations in 1897 (although American meteorologists were the first to use kites systematically for meteorological purposes). One of the pioneers in the use of balloons (originally paper) for atmospheric soundings (1898). Extended the use of kites and balloons to on-board ships and to the tropics. Detected trade-wind inversion. Began the technique of radio probing the atmosphere with transmitters on balloons. Also privately financed the first *International Cloud Atlas* (1896).

[‡] **Richard Assmann** (1845–1918) German meteorologist. First to suggest the use of rubber balloons for probing the atmosphere. Developed a method for measuring wind velocity with balloons followed by theodolite. Invented the aspiration psychrometer named after him, which is still widely used.

[§] **Pavel A. Molchanov** (1893–1941) Soviet meteorologist who devoted most of his scientific career to exploring and documenting the vertical structure of the atmosphere. Made numerous contributions to the development of meteorological instrumentation used on kites and balloons. Director of USSR Main Geophysical Observatory, 1936–1939.

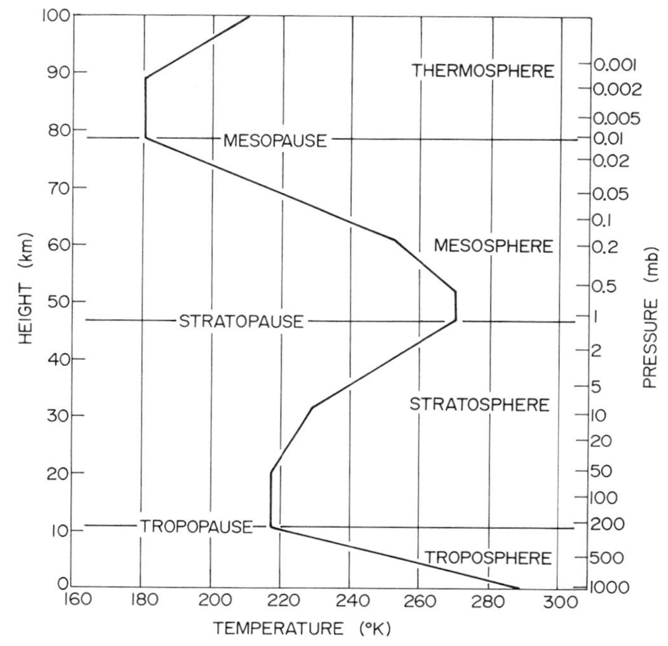

Fig. 1.8 Vertical temperature profile for the U.S. Standard Atmosphere.

to the tropopause in a matter of a few minutes. As a result of the rapid vertical mixing and the "scavenging" of aerosol (small solid and liquid particles) by precipitation, the "mean residence time" for the tropospheric aerosol is rather short, ranging from a few days to a few weeks.

The transition from the troposphere to the stratosphere is usually marked by an abrupt change in the concentrations of some of the variable trace constituents; water vapor decreases sharply while ozone concentration often increases by an order of magnitude within the first few kilometers above the tropopause. These strong concentration gradients just above the tropopause are a reflection of the fact that there is relatively little mixing between dry, ozone-rich stratospheric air and relatively moist, ozone-poor tropospheric air. Another indication of the lack of mixing is the fact that debris from past nuclear explosions and dust from large volcanic eruptions are present in much higher concentrations in stratospheric air than in tropospheric air. In contrast to the situation in the troposphere, substantial numbers of particles remain in the stratosphere for periods of a year or longer after the event that originally produced them. Because of these extremely long residence times, the stratosphere behaves as a "reservoir" for certain types of atmospheric pollution.

The stratosphere (literally, the layered sphere) is characterized by very small vertical mixing. Thin layers of aerosol are observed to persist for long periods of time within certain height ranges. Even the most vigorous thunderstorm updrafts are unable to penetrate more than a few kilometers into the lower stratosphere before they lose their buoyancy, flatten out, and turn into layered clouds. The strongly layered structure of the stratosphere is related to the vertical temperature gradient, as explained in Section 2.7.1.

It can be seen in Fig. 1.8 that the pressure at the stratopause level is about 1 mb, compared to about 1000 mb at the earth's surface. Thus the troposphere and stratosphere together account for about 99.9% of the atmospheric mass. Of the remaining mass, about 99% is contained within the mesosphere and the remaining 1% in the thermosphere.

The mesosphere (literally, the middle sphere) overlaps with the lower part of the ionosphere and the lower part of the region in which auroras are sometimes observed. Like the troposphere it is a region in which temperature decreases with height, and vertical air movements are not strongly inhibited. During summer there is sometimes enough lifting to produce thin cloud layers in the upper mesosphere over parts of the polar regions. Under ordinary conditions the concentrations of particles in these clouds are far too small to render them visible from the ground. However, at twilight mesospheric clouds are sometimes still in sunlight while the lower atmosphere is in the earth's shadow. Under such conditions such clouds are visible from the ground as *noctilucent clouds.*

The thermosphere extends upward to an altitude of several hundred kilometers where temperatures range from 500°K to as much as 2000°K, depending upon the amount of solar activity, as indicated in Fig. 1.9. The *thermopause* is the level of transition to a more or less isothermal (constant temperature) profile. Above 500 km molecular collisions are so infrequent that temperature, as such, is difficult to define. At these levels neutral particles and charged particles move more or less independently and thus there is no reason that their temperatures should be the same. Outside the magnetosphere, the environmental temperature is determined by the solar wind.

It is interesting to note that the earth is the only one of the inner planets that has an intermediate temperature maximum (that is, a stratopause). Both Mars and Venus have a well-defined troposphere and thermosphere, separated by a more or less isothermal layer. The physical processes responsible for the observed vertical temperature structure in the earth's atmosphere are discussed in Section 7.2.

From a comparison of Figs. 1.4 and 1.8 two relationships are apparent between temperature, pressure, and density:

- The ratio of pressure to density (as measured by the horizontal displacement between the two curves in Fig. 1.4) is larger in warm layers than in cold layers.
- Below the turbopause, pressure and density decrease with height more rapidly in cold layers than in warm layers.

These relationships are a consequence of the ideal gas equation and the hypsometric equation, which are discussed in Sections 2.1 and 2.2.2, respectively.

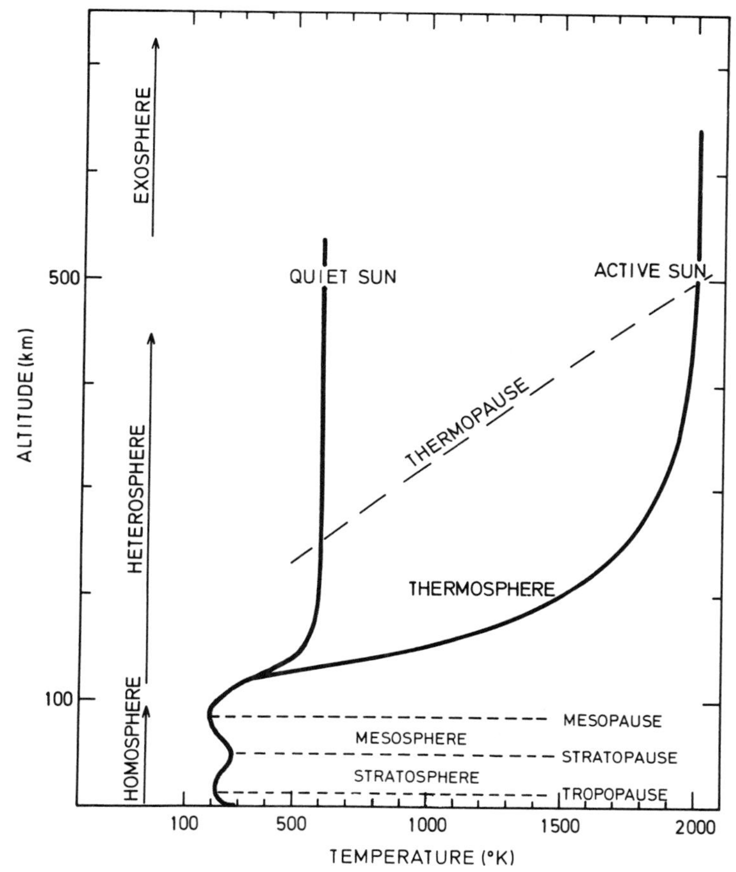

Fig. 1.9 Vertical temperature distribution in the earth's atmosphere with emphasis on the thermosphere. (After P. M. Banks and G. Kockarts, "Aeronomy," Academic Press, New York, 1973, Part A, p. 3.)

1.5.1 Climatological variability

The layered temperature profile described in connection with Fig. 1.8 is present at nearly all latitudes and in all seasons. However, there is considerable latitudinal and seasonal variability in the details of the vertical temperature profile, as seen in Fig. 1.10.

Within the troposphere temperature decreases with latitude, with the latitudinal gradient being about twice as steep in the winter hemisphere as in the summer hemisphere. The tropopause is considerably higher and colder over the tropics than over the polar regions. (Note the extremely low temperatures at the equatorial tropopause.) In the lower stratosphere the latitudinal temperature distribution is rather complicated: The summer hemisphere is characterized by a cold equator and warm pole, while the winter hemisphere displays a

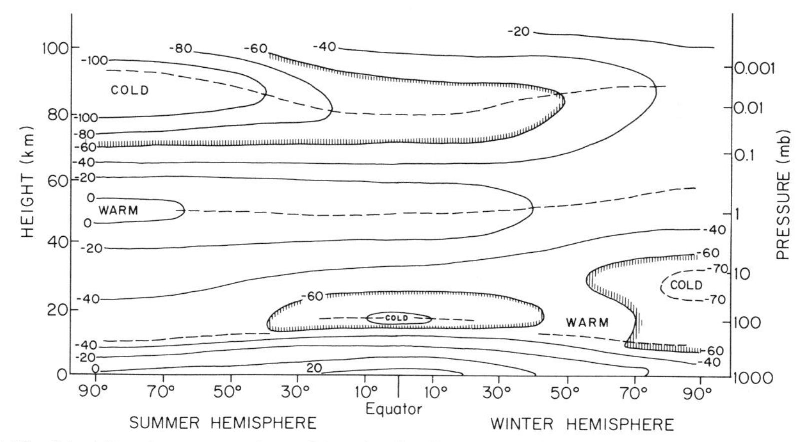

Fig. 1.10 Meridional cross section of longitudinally averaged temperature in degrees Celsius at the time of the solstices. Dashed lines indicate tropopause, stratopause, and mesopause. (Courtesy of R. J. Reed.)

distinct temperature maximum over middle latitudes. The cold pool of stratospheric air over the winter pole is highly variable, particularly in the northern hemisphere where it has been known to disappear completely for periods of a few weeks during midwinter. During the onset of these warm intervals, the stratospheric temperatures over individual stations have been observed to rise by as much as 70 deg over the course of a week! The term *sudden warming* is often used to describe this phenomenon. At the stratopause there is a monotonic temperature gradient between the warm summer pole and the cold winter pole, while at the mesopause level the situation is just the opposite: the summer pole is cold and the winter pole is warm.

The existence of temperature minima at the equatorial tropopause and at the mesopause in the summer hemisphere may seem somewhat surprising. A partial explanation of these features is given in Section 9.10.

A more detailed view of the temperature distribution in the troposphere and lower stratosphere is given in Fig. 1.11. Note the prominent break in the tropopause over middle latitudes.

The seasonal variation of temperature is strongly influenced by the distribution of land and sea. During summer the air over the continents tends to be warmer than the air over the surrounding oceanic regions, while during the winter it tends to be colder. The temperature contrasts between continental and oceanic regions tend to be most pronounced within the lowest few kilometers above the earth's surface, but they are present throughout the depth of the troposphere.

Temperature displays a pronounced diurnal variability in certain regions of the atmosphere. The strongest variability is observed in the upper thermosphere, where day to night temperature differences are on the order of hundreds of degrees. There are also important (but much smaller) diurnal temperature

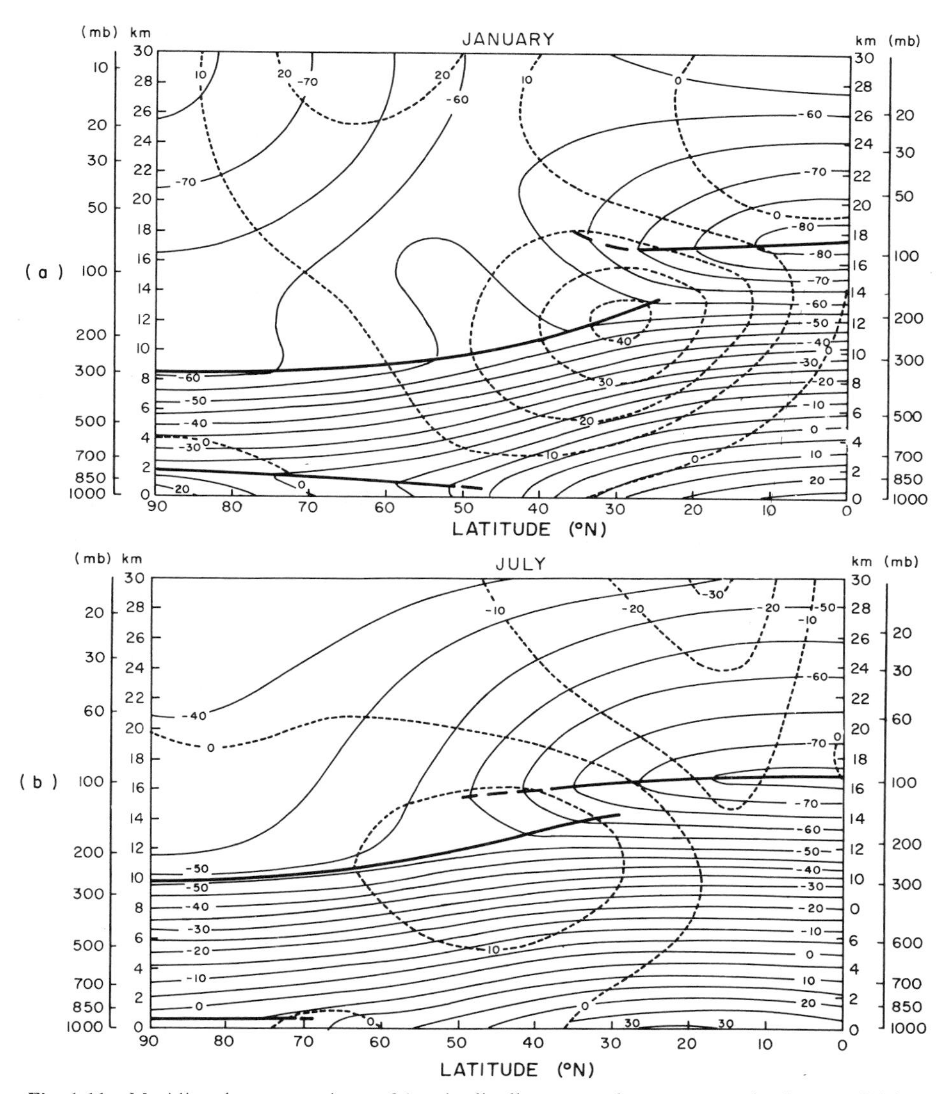

Fig. 1.11 Meridional cross sections of longitudinally averaged temperature in degrees Celsius (——) and zonal wind in meters per second (– – –) for the northern hemisphere in January (a) and July (b). Positive zonal winds indicate flow from west to east. Heavy lines denote the tropopause and the Arctic inversion. (After *Arctic Forecast Guide*, U.S. Navy Weather Research Facility, 1962.)

changes around the stratopause level, which give rise to strong tidal motions in the earth's upper atmosphere. In the troposphere the day to night temperature changes are generally less than a degree except within the lowest few kilometers over land. At the earth's surface, over land, the diurnal temperature range is on the order of 10 deg at most locations, but it reaches values of more than 20 deg over high-altitude desert regions.

At any particular instant in time the distribution of temperature within the troposphere exhibits large deviations from the seasonally and longitudinally averaged cross section shown in Fig. 1.10. Much of the equator to pole temperature gradient tends to be concentrated in narrow bands called *frontal zones* such as the one described in Chapter 3. Such zones may be marked by strong east–west temperature contrasts as well as north–south contrasts.

1.6 WINDS IN THE EARTH'S ATMOSPHERE

Atmospheric fluid motions exhibit a wide spectrum of horizontal scales,[†] ranging from the dimensions of the earth itself down to scales comparable to the mean free path of the individual molecules. For the purposes of discussing atmospheric motions it is convenient to divide the spectrum as follows:

- *planetary-scale motions:* the broadest features of the global circulation, including the longitudinally averaged wind field and longitudinally dependent features with horizontal dimensions comparable to the scales of the major continents and oceans.
- *synoptic-scale motions:* waves or eddies with horizontal dimensions large enough to be resolved by a conventional observing network with a station spacing on the order of a few hundred kilometers,[‡] but smaller than planetary-scale motions defined above. (Since most day to day weather changes are associated with this range of scales, we will devote the whole of Chapter 3 to describing synoptic-scale motions.)
- *mesoscale motions:* waves, eddies, or jetlike features with horizontal dimensions ranging from a few tens of kilometers up to a few hundred kilometers. Within this category fall some of the narrower jet streams, circulations associated with frontal zones, rain bands and squall lines in hurricanes and middle latitude storms, mountain lee waves, and a wide variety of other phenomena. Mesoscale motions are inherently difficult to study without special networks, since they are too large to be within the field of view of an individual station, but too small to be resolved by the conventional synoptic network. Some important mesoscale features associated with cloud and precipitation systems will be described in Chapter 5.

[†] For the purposes of this discussion, horizontal *scale* can be defined as the distance over which the wind changes by an amount comparable to the magnitude of the wind itself. For example, if the flow consists of closed circulations or eddies, the horizontal scale could be defined as the radius of a typical eddy.

[‡] The conventional upper air synoptic network consists of roughly 1000 stations at which radiosondes are released at 12- or 24-h intervals (usually midnight and/or noon Greenwich Civil Time). Winds are obtained by tracking the balloons as they ascend. The spacing between stations in the network ranges from a few hundred kilometers over some land areas to more than a thousand kilometers over large areas of the oceans. Additional wind data are provided by commercial aircraft which monitor flight-level winds, and the tracking of cloud motions by satellites.

- *small-scale motions:* the remaining part of the spectrum. Included in this category are a wide range of diverse phenomena, some of which will be discussed in Chapters 5, 7, and 9.

A common property of planetary-, synoptic-, and mesoscale motions is the fact that they are quasi-horizontal (that is to say, the vertical motion component is more than an order of magnitude smaller than the horizontal wind component). Motions smaller than the mesoscale exhibit roughly comparable horizontal and vertical motion components.

1.6.1 Longitudinally averaged zonal winds

When the atmospheric circulation is averaged with respect to longitude, the zonal (east–west) wind component turns out to be about an order of magnitude larger than the meridional (north–south) component at most locations. A latitude–height cross section of longitudinally averaged zonal wind at the time of the solstices is shown in Fig. 1.12. In both summer and winter hemispheres there are westerly (west to east) jets in middle latitudes at an altitude of 10 km. Note that this position corresponds closely to the "tropopause break" in Fig. 1.11. The tropospheric jet in the winter hemisphere is considerably stronger than the one in the summer hemisphere. The strongest zonal winds in the section are associated with the matched pair of *mesospheric jets* at the 60-km level. Over middle latitudes the wind at these levels undergoes a dramatic seasonal reversal between winter westerlies and summer easterlies.

There are a number of features in the longitudinally averaged zonal wind field which do not appear explicitly in Fig. 1.12. The "sudden warming" phenomenon described in the previous section is accompanied by large changes in

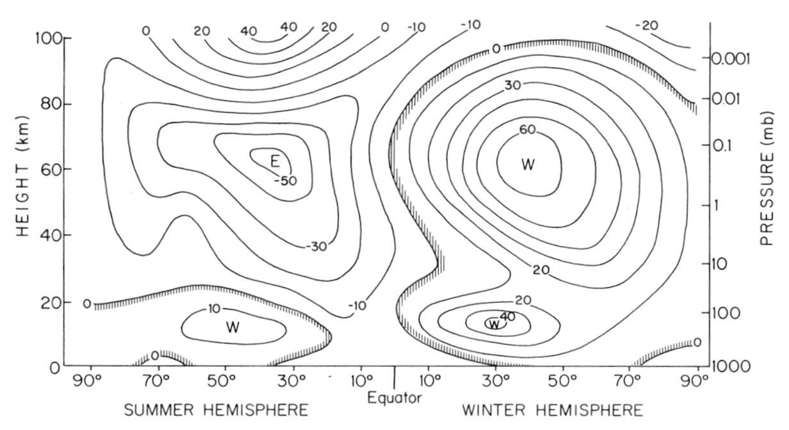

Fig. 1.12 Meridional cross section of longitudinally averaged zonal wind in meters per second at the time of the solstices. Positive zonal winds denote flow from west to east. (Courtesy of R. J. Reed.)

the longitudinally averaged zonal wind at high latitudes in the winter stratosphere. During the winters without warmings the westerlies in this region are considerably stronger than those indicated in Fig. 1.12. The midwinter warmings are accompanied by a pronounced weakening and sometimes even a complete disappearance of the westerlies at stratospheric levels. However, they have little effect upon the tropospheric circulation. The longitudinally averaged zonal circulation of the tropical stratosphere is characterized by a strong semiannual (half yearly) oscillation at upper levels and a more irregular quasi-biennial (nearly, but not quite, two yearly) oscillation at lower levels. Both phenomena are evident in Fig. 1.13.

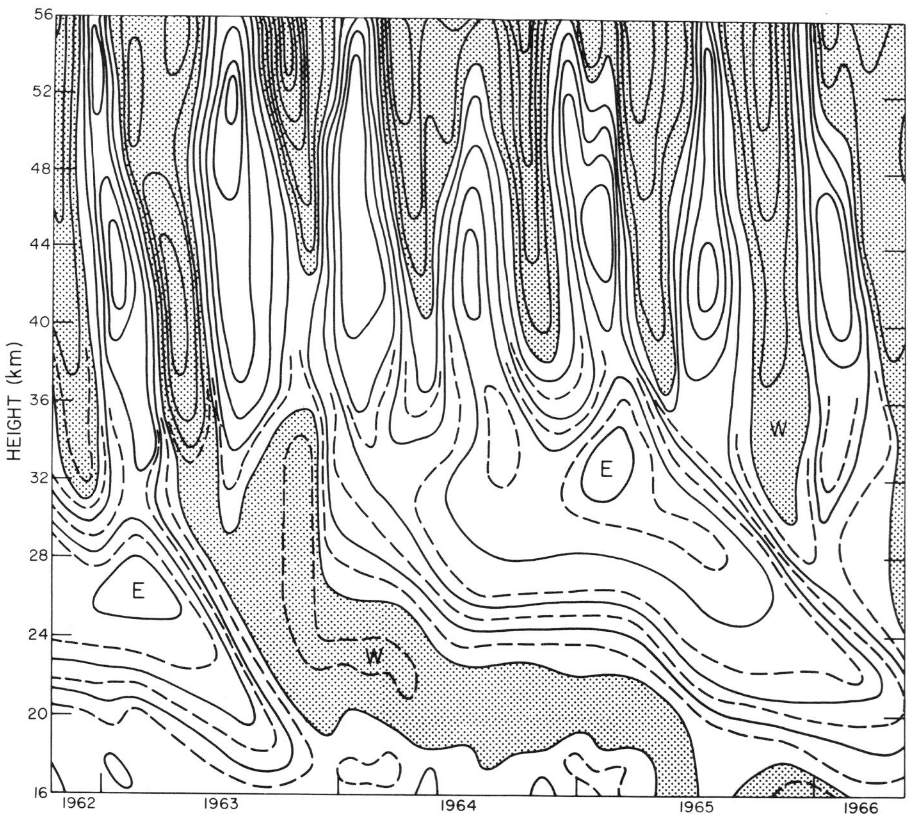

Fig. 1.13 Time–height section of zonal wind in the vicinity of the equator based on monthly averaged data. Solid lines have been drawn at intervals of 10 $\mathrm{m\,s^{-1}}$. Shaded regions denote flow from west to east, and the letters W denote maxima of such *westerly* flow. E's denote maxima of easterly (east to west) flow. At the higher levels there is a prominent semiannual oscillation while below 35 km there is a longer-term variability associated with the quasi-biennial oscillation. [From Wallace, *Rev. Geophys. and Space Phys.*, **11**, 196 (1973), copyrighted by American Geophysical Union.]

1.6.2 Tropospheric winds at middle and high latitudes

The winds at middle and high latitudes tend to blow parallel to the isobars[†] or height contours[‡] leaving low pressure to the left in the northern hemisphere (to the right in the southern hemisphere).[§¶] Furthermore, at any given latitude the speed of the wind tends to be inversely proportional to the spacing of the isobars or height contours. Thus, as a "channel" between two adjacent contours narrows, the wind speed increases and vice versa. Because of this so-called *geostrophic relationship* between wind and pressure, it is possible to deduce the gross features of the wind field from an inspection of maps showing the distribution of pressure at various levels.

Figure 1.14 shows the mean height contours of the 500-mb surface over the northern hemisphere for January. (The 500-mb level is usually located between 5 and 6 km above sea level, in the middle troposphere.) In both seasons the 500-mb surface is highest in the tropics and lowest in polar regions. Thus the prevailing winds are from the west throughout middle and high latitudes. It may be noted that the flow pattern is not quite circular about the north pole. There are regions of relatively low heights over the east coasts of North America and Asia. The prevailing winds sweep southward around these "troughs" in the pressure field, and northward around the "ridges" located near the west coasts of the continents. The corresponding 500-mb height contours for the southern hemisphere tend to be more circularly symmetric about the pole, and they do not change as much between summer and winter.*

The annual average sea level pressure pattern for the northern hemisphere is shown in Fig. 1.15. In contrast to the rather simple circumpolar vortex at the 500-mb level, the flow at these low levels displays a number of closed circulations. (Note that, in accordance with the geostrophic relationship described above, the flow around centers of low pressure is counterclockwise and the flow around centers of high pressure is clockwise.) The prominent low-pressure areas over the northern Atlantic and Pacific Oceans (referred to as the "*Icelandic*" and "*Aleutian*" *lows*, respectively) are characterized by frequent storms. Further

[†] *Isobars* are lines lying within a horizontal plane (for example, sea level) which connect points of equal air pressure.

[‡] *Height contours* of a particular pressure surface (for example, 500 mb) describe the "topography" of that surface just as the contours on a geodetic survey map describe the topography of the land surface. To a very close approximation isobars on a horizontal surface and height contours of a nearby pressure surface have the same shape.

[§] This relationship was first noted by Professor Buys Ballot of Utrecht in 1857, who stated: "If in the northern hemisphere you stand with your back to the wind, pressure is lower on your left hand than on your right." In the southern hemisphere the reverse is true.

[¶] **Christopher H. D. Buys-Ballot** (1817–1890) Dutch meteorologist. Became professor of mathematics at University of Utrecht at age 30. Director of Dutch Meteorological Institute (1854–87). Labored unceasingly for the widest possible network of synoptic observations.

* See E. Palmén and C. W. Newton, "Atmospheric Circulation Systems," Academic Press, New York, 1969, pp. 72–73.

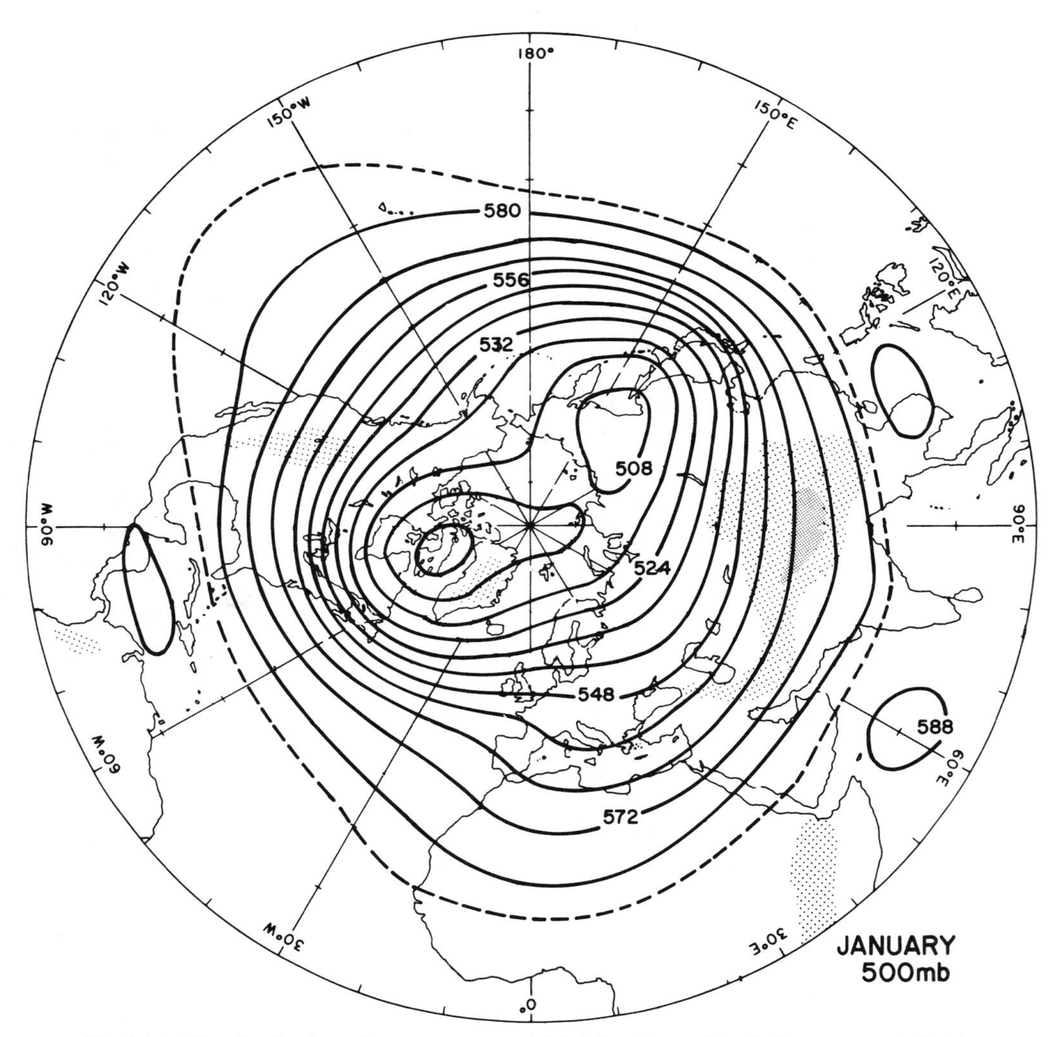

Fig. 1.14 The distribution of mean geopotential height (almost identical to geometric height above sea level) on the 500-mb surface for January. Contours are labeled in tens of meters. [Based on data in *Meteorol. Abhandl.*, **4**(2), Part II, 32 (1958) as adapted by E. Palmén and C. W. Newton, "Atmospheric Circulation Systems," Academic Press, New York, 1969, p. 68.]

to the south, over both oceans, near 30°N, lies a belt of high pressure, which separates the predominently easterly flow in the subtropics from the southwesterly flow over the higher-latitude oceans. These *subtropical high-pressure belts* are characterized by light winds and an absence of storms.

The sea level pressure pattern over land is dominated by a belt of high pressure centered around 50°N. The highest values are observed over Mongolia and southern Siberia. Pressures are generally low over the subtropical land masses; particularly over India and the Arabian peninsula.

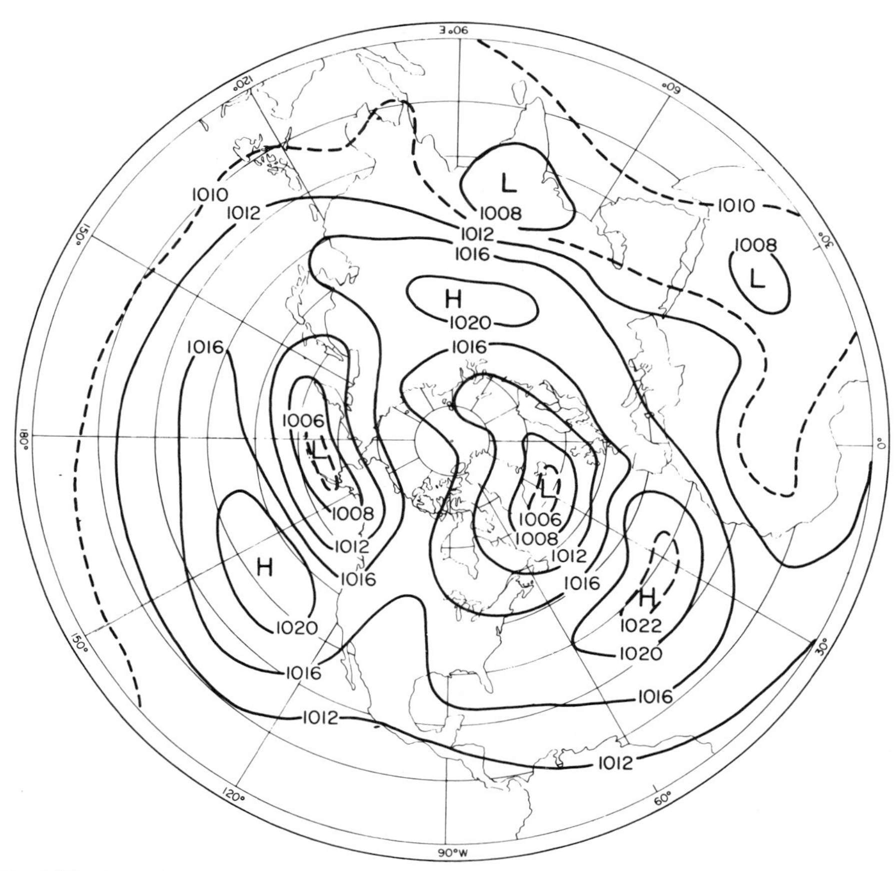

Fig. 1.15 Annual average sea level pressure in millibars. [From *Mon. Wea. Rev.*, **104**(12) (1976).]

Figure 1.16 shows the amplitude and phase of the annual cycle in sea level pressure. Generally speaking, land areas are characterized by higher pressure during winter and lower pressure during summer, while ocean areas display an annual cycle with the opposite phase. Hence, during winter, the Siberian high and the Aleutian and Icelandic lows are much more pronounced than in the annual average. In contrast, the highs over the subtropical oceans and the lows over the subtropical land masses tend to be most pronounced during summer, and weak or nonexistent during winter.

It should be noted that Figs. 1.14 and 1.15 refer to climatological averages. The wind and pressure fields in middle and high latitudes display a large amount of day to day variability in association with the evolution and movement of synoptic-scale disturbances. Hence the surface and upper air charts on individual days are characterized by prominent synoptic-scale features that do not appear in Figs. 1.14 and 1.15. For examples of the structure of the wind field on a typical day, the reader is referred to Chapter 3.

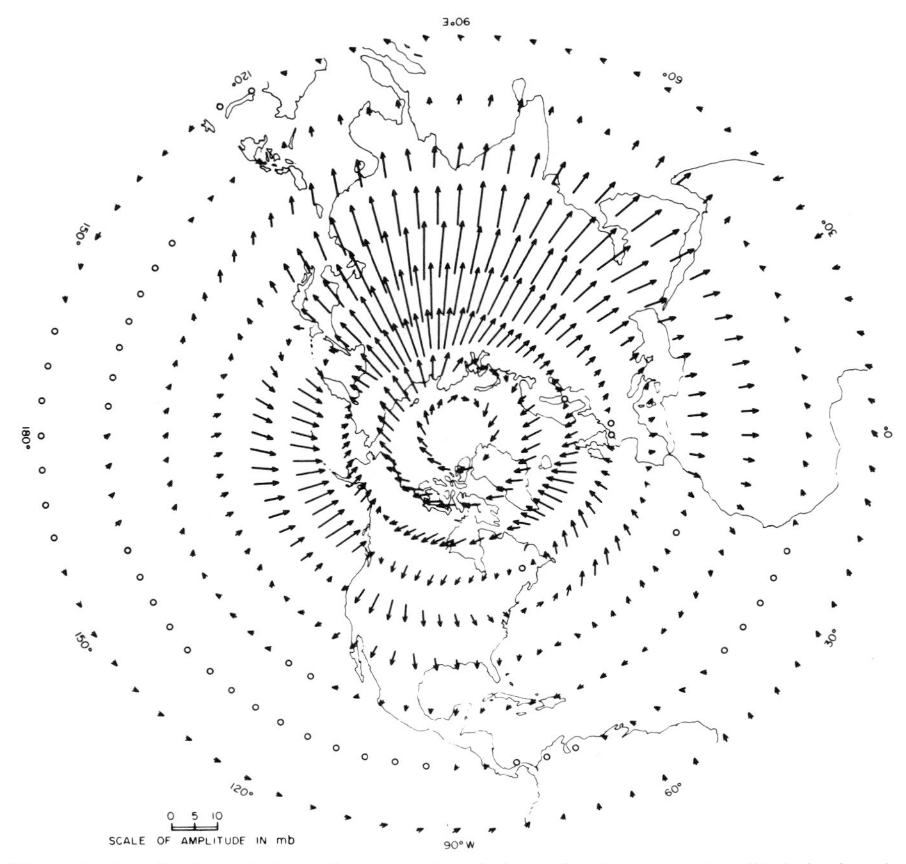

Fig. 1.16 Amplitude and phase of the annual cycle in sea level pressure. Amplitude is given by the length of the arrows in relation to the scale at the lower left. Phase (time of maximum sea level pressure) can be determined from the orientation of the arrows. Arrows directed from pole toward equator denote a January 1 maximum; those directed from east to west denote an April 1 maximum; those directed from equator to pole denote a July 1 maximum, etc. Open circles denote amplitudes too small to plot. [From *Mon. Wea. Rev.*, **104**(12) (1976).]

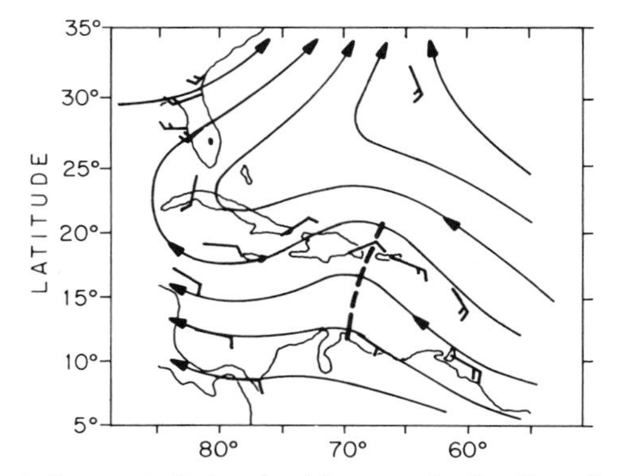

Fig. 1.17 A typical easterly wave in the low-level flow over the Carribean Sea. Axis of the "trough" of the wave (---), direction of the flow at about 3 km above sea level (——); the arrows denote actual wind observations coded according to the conventions shown in Table 3.1. (From *Proc. Symp. Tropical Meteorol.*, Amer. Meteorol. Soc., Boston, 1970, p. G-IV-3.)

1.6.3 Tropospheric winds at low latitudes

In contrast to the highly transient character of the middle- and high-latitude circulation, the winds in the tropics tend to blow somewhat more steadily from one day to the next.

The low-level circulation over the Atlantic and Pacific Oceans is dominated by the easterly flow around the equatorward flank of the subtropical high-pressure belts near 30°N and 30°S. Superimposed upon this planetary-scale easterly flow are weak synoptic-scale disturbances which cause the wind to fluctuate between northeasterly and southeasterly as shown in Fig. 1.17. These disturbances are called *easterly waves* because they are observed to move from east to west, in contrast to middle-latitude disturbances which usually move from west to east. Also endemic to the tropical oceans is a class of extremely intense, circular vortices called *tropical cyclones*, which account for the strongest sustained surface winds observed anywhere in the earth's atmosphere. Most of these storms develop during the warm season, over a few well-defined areas, as shown in Fig. 1.18.

Within the lowest kilometer above the sea surface the easterly flow is particularly steady and has a distinct equatorward component. Hence, the prevailing winds in the tropical North Atlantic and North Pacific are from the northeast, and those in the tropical South Atlantic and South Pacific are from the southeast. In the early days of sailing ships these wind regimes came to be known as the *northeast* and *southeast trades*, respectively. Until relatively recently, it was widely believed that the tradewind belts in the northern and southern hemispheres were separated by a region of calm winds along the equator called the "doldrums." More recent evidence indicates that, with the exception of the extreme western Pacific, the transition between the northeast and southeast trades usually takes place within a very narrow belt located several degrees north of the equator. This so-called *intertropical convergence zone* (*ITCZ*), where the northeast and southeast trades flow together, is characterized by strong upward motion and heavy rainfall, as will be shown in the next section.

At the longitudes of the major continents the low-level tropical wind field exhibits a strong seasonal dependence, with a tendency toward onshore (sea to land) flow during summer and offshore flow during winter. The seasonal reversal is particularly pronounced over southeast Asia and adjacent regions of the Indian Ocean where the prevailing winds blow from the southwest during summer and northeast during winter. These seasonal wind regimes are known as *monsoons* (from the Arabic word *mausin*—a season). Over most of India, the summer (southwest) monsoon is characterized by heavy rainfall while the winter (northeast) monsoon is extremely dry.[†]

[†] More comprehensive surveys of tropical wind systems are given in H. Riehl, "Tropical Meteorology," McGraw-Hill, New York, 1954; E. Palmén and C. W. Newton, "Atmospheric Circulation Systems," Academic Press, New York, 1969; C. S. Ramage, "Monsoon Meteorology," Academic Press, New York, 1971.

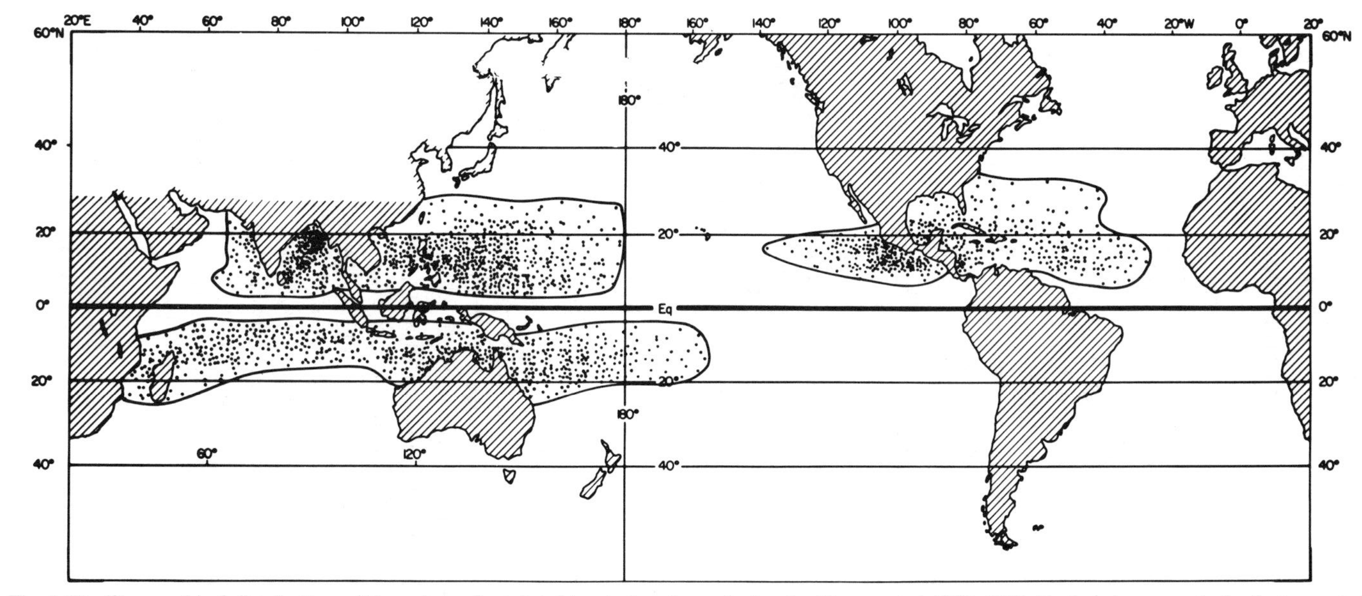

Fig. 1.18 Geographical distribution of the points of origin of tropical cyclones during the 20-yr period 1952–1971. Each dot represents the first reported location of a tropical cyclone which subsequently developed sustained winds in excess of 20 m s^{-1}. About two-thirds of these storms eventually developed winds of "hurricane force" (> 33 m s^{-1}). (Courtesy of W. M. Gray.)

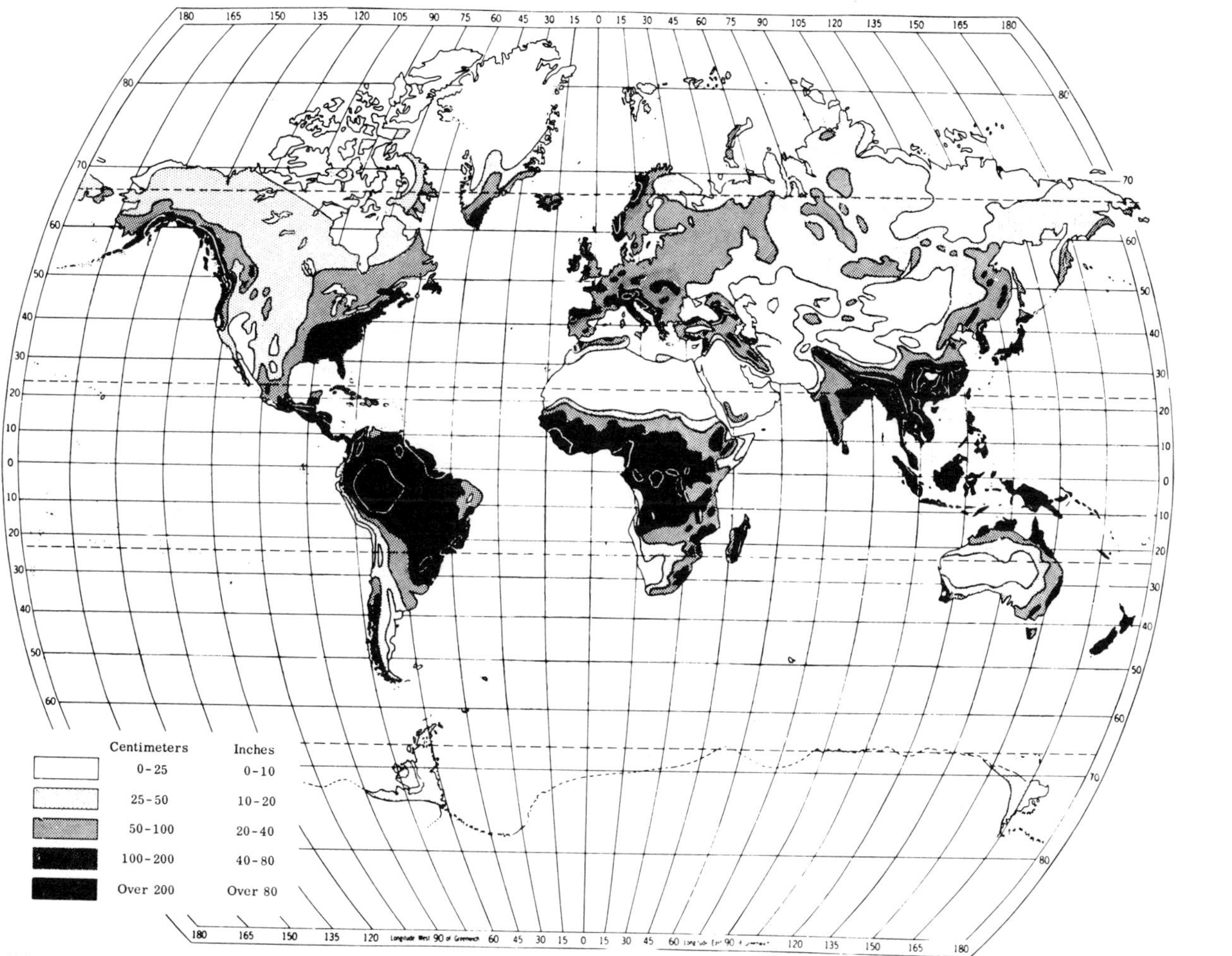

Fig. 1.19 Distribution of average annual precipitation over the continents. (After H. J. Critchfield, "General Climatology," Prentice-Hall, Englewood Cliffs, New Jersey, 2nd Edition, 1966, p. 65.)

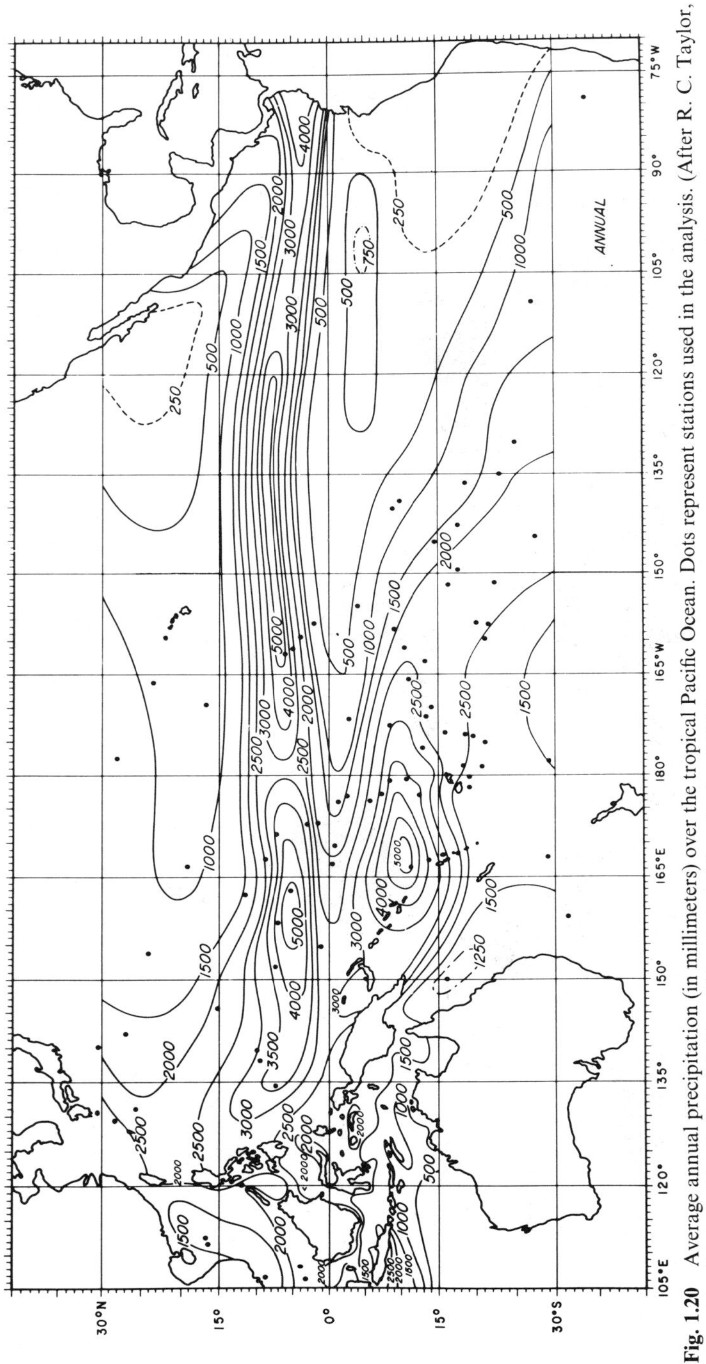

Fig. 1.20 Average annual precipitation (in millimeters) over the tropical Pacific Ocean. Dots represent stations used in the analysis. (After R. C. Taylor, *An Atlas of Pacific Islands Rainfall*, Hawaii Institute of Geophysics, 1973.)

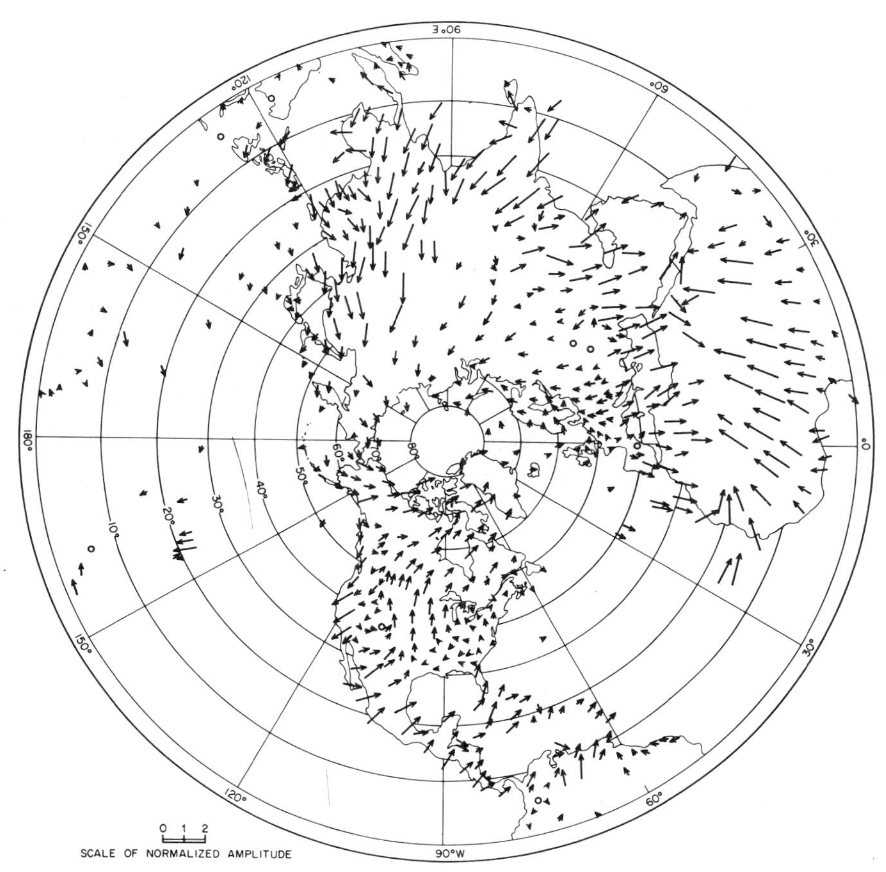

Fig. 1.21 Normalized amplitude and phase of the annual cycle in precipitation. Amplitudes are normalized by dividing by the mean monthly precipitation averaged over the year at each station. Normalized amplitude is given by the length of the arrows in relation to the scale at the lower left. Phase (time of maximum precipitation) can be determined from the orientation of the arrows. Arrows directed from pole toward equator denote a January 1 maximum; those directed from east to west denote an April 1 maximum; those directed from equator to pole denote a July 1 maximum, etc. Open circles denote amplitudes too small to plot. [From *Mon. Wea. Rev.*, **104**, 1095 (1976).]

1.7 PRECIPITATION

The geographical distribution of annual average precipitation is shown in Fig. 1.19. It should be noted that the analysis has been highly smoothed in mountain areas, where there exist very large local gradients. Heavy precipitation is observed over large areas of the tropical continents and on the windward slopes of mountain ranges at higher latitudes. The major desert regions of the world correspond to subtropical latitudes, particularly along the western coasts of continents and the regions north of the major east–west mountain ranges in Asia. Precipitation is also very light over much of the polar regions.

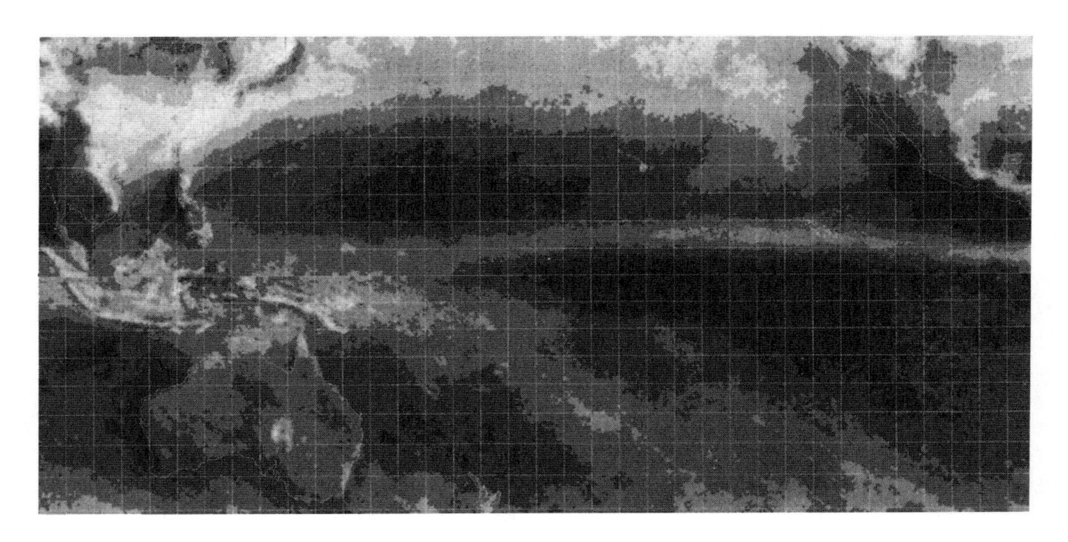

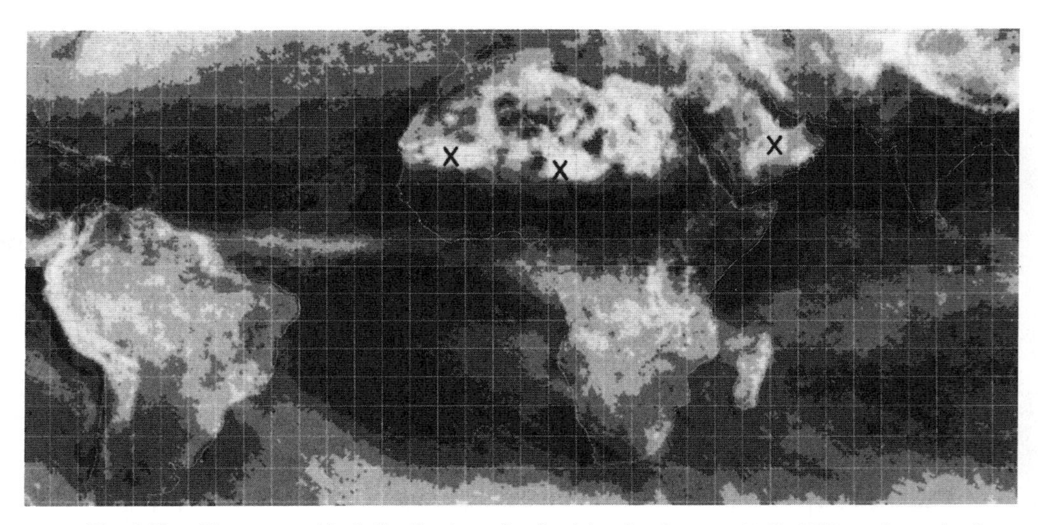

Fig. 1.22a The geographical distribution of reflectivity for January 1967–1970 as determined from satellite observations. Most of the bright areas in the figure are characterized by persistent cloudiness and relatively heavy precipitation. However, the following exceptions should be noted: areas indicated by X's denote desert regions where the earth's surface is highly reflective and areas indicated by Y's denote regions of persistent low, nonprecipitating cloud decks. Tick marks along the side denote the position of the equator; the Mercator grid lines are spaced at intervals of 5 degrees of latitude and longitude. (From U.S. Air Force and U.S. Department of Commerce, *Global Atlas of Relative Cloud Cover*, 1967–1970, Washington, 1971.)

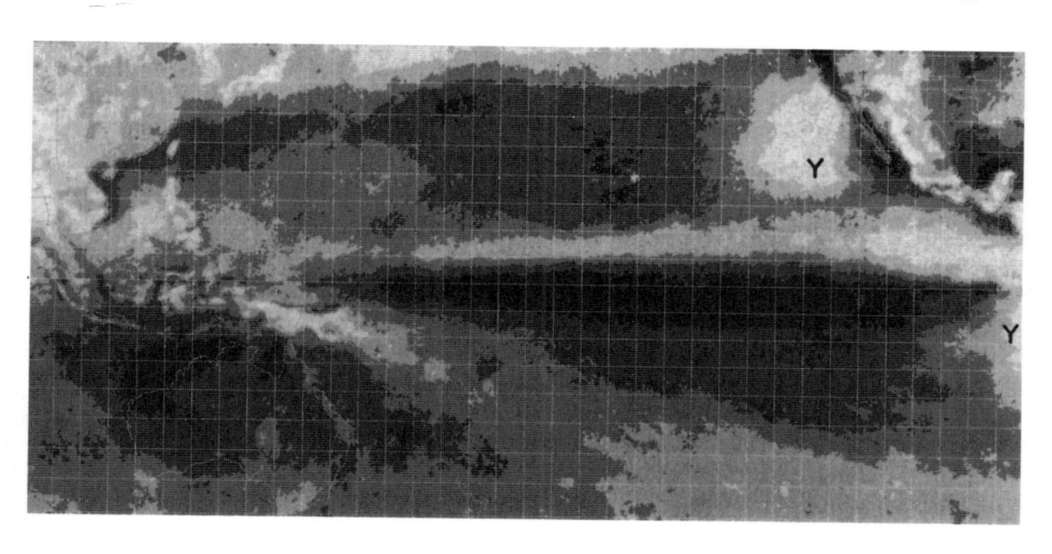

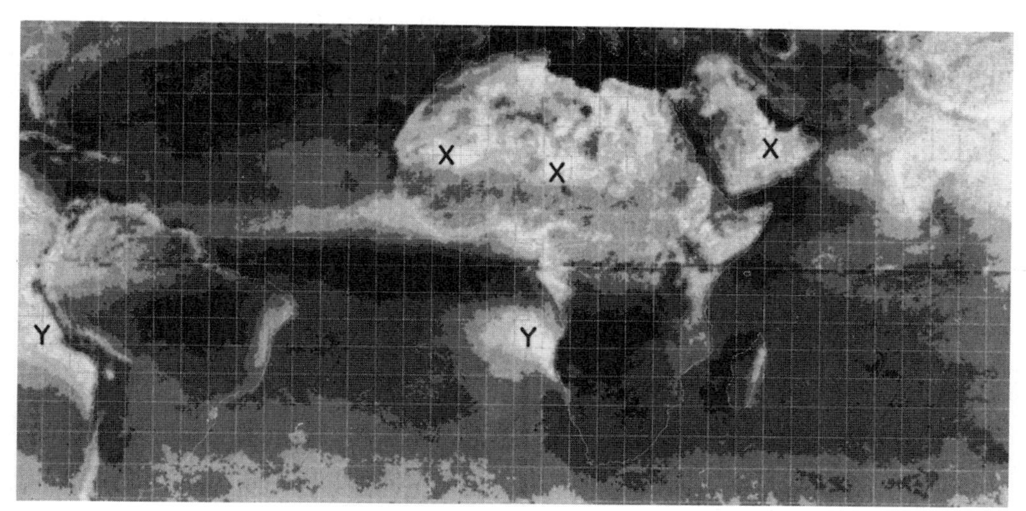

Fig. 1.22b Same as Fig. 1.22a but for July 1967–1970.

The distribution of precipitation over the oceans is subject to considerable uncertainty because of the problem of obtaining representative observations from widely spaced island stations. Definitive data are available only for parts of the tropical Pacific region where there exist numerous small atolls with long records of precipitation observations. The distribution over the tropical Pacific is shown in Fig. 1.20. Note the remarkably large spatial variability over this relatively homogeneous oceanic region. The prominent precipitation maximum around 7°N is associated with the intertropical convergence zone. Tropical islands located near the center of the mean position of the ITCZ have lush vegetation and rain forest climates while those located just a few hundred kilometers north or south may be desert islands. At most longitudes there exists a dry belt along the equator and a secondary maximum in the southern hemisphere near the other convergence line in the wind field.

Over much of the world most of the average annual precipitation falls during a well-defined "wet season." The timing and relative importance of the "wet season" can be represented in terms of the phase and amplitude of the annual cycle in mean monthly precipitation, which is displayed vectorially in Fig. 1.21 for the northern hemisphere. The marked predominance of summer rainfall over much of Asia is associated with the summer monsoon circulation which carries moisture northward from the Indian Ocean. Relatively wet summers and dry winters are also observed over much of Europe north of the Alps and North America east of the Rockies. Winter is the wet season over the Mediterranean and adjacent land areas, central Asia, and western North America.

Figures 1.22a and 1.22b show the distribution of sunlight reflected from the earth and its atmosphere for the months of January and July, respectively, based upon images obtained from satellites. With the few exceptions noted in the figure caption, the bright regions are indicative of frequent cloudiness and high monthly average precipitation. The intertropical convergence zone shows up quite clearly over both the Atlantic and Pacific Oceans during both months. The other cloud band in the southern hemisphere, northeast of Australia, corresponds closely to position of the secondary rainfall maximum in Fig. 1.20. Most tropical land masses show large seasonal differences in brightness that are consistent with the concept of a wet summer monsoon and a dry winter monsoon, as described in the previous section. The seasonal differences are particularly pronounced over most of tropical Asia, South America (south of the equator), Africa (south of 15°N), extreme northern Australia, Florida, and the highlands of Mexico.

It is interesting to note that in the equatorial Pacific the year to year variability in precipitation is much larger than the normal seasonal variability. Some islands within this zone have wet climates as defined by their average annual rainfall, and yet they may experience intervals as long as two or three years with little or no precipitation.

Over much of the world, precipitation also exhibits a preference for certain times of day. Figure 1.23 shows the amplitude and phase of the diurnal cycle in the frequency of summer thunderstorms over the conterminous United States. Most areas are characterized by an afternoon maximum, but a broad area over the central part of the country exhibits a preference for nocturnal thunderstorms. Summer rainfall displays a similar pattern of diurnal variability.

Even in regions with pronounced wet seasons, most of the precipitation falls in association with the passage of transient, synoptic-scale disturbances. Although the vertical motions that accompany these disturbances are only on the order of a few centimeters per second, they exert a strong control on cloudiness and precipitation, which tend to be confined to regions of synoptic-scale ascent. An example of the modulation of cloudiness by transient, synoptic-scale disturbances can be seen in Fig. 1.24, which shows a time–longitude section of cloudiness along the ITCZ, as viewed once a day from polar-orbiting satellites. The diagonal "cloud lines" in the section are indicative of rain areas that drift westward at a rate of about 6° of longitude per day. These rain areas correspond to the regions of ascent in easterly waves similar to the ones shown in Fig. 1.17. The distribution of precipitation in middle-latitude disturbances will be discussed in Section 5.5.

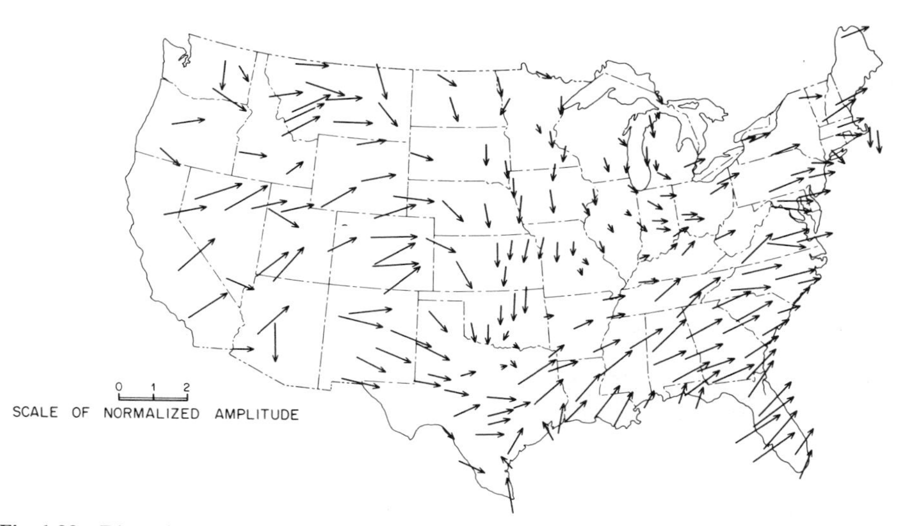

Fig. 1.23 Diurnal variation of hourly thunderstorm frequency over the United States. Normalized amplitude of the diurnal cycle is given by the length of the arrows in relation to the scale at bottom left. (Amplitudes are normalized by dividing by the mean hourly thunderstorm frequency averaged over the 24 hr of the day at each station.) Phase (time of maximum thunderstorm frequency) is indicated by the orientation of the arrows. Arrows directed from north to south denote a midnight maximum, arrows directed from east to west denote a 6 a.m. maximum, those from south to north denote a midday maximum, etc. [Based on data in *Mon. Wea. Rev.*, **103**, 409 (1975).]

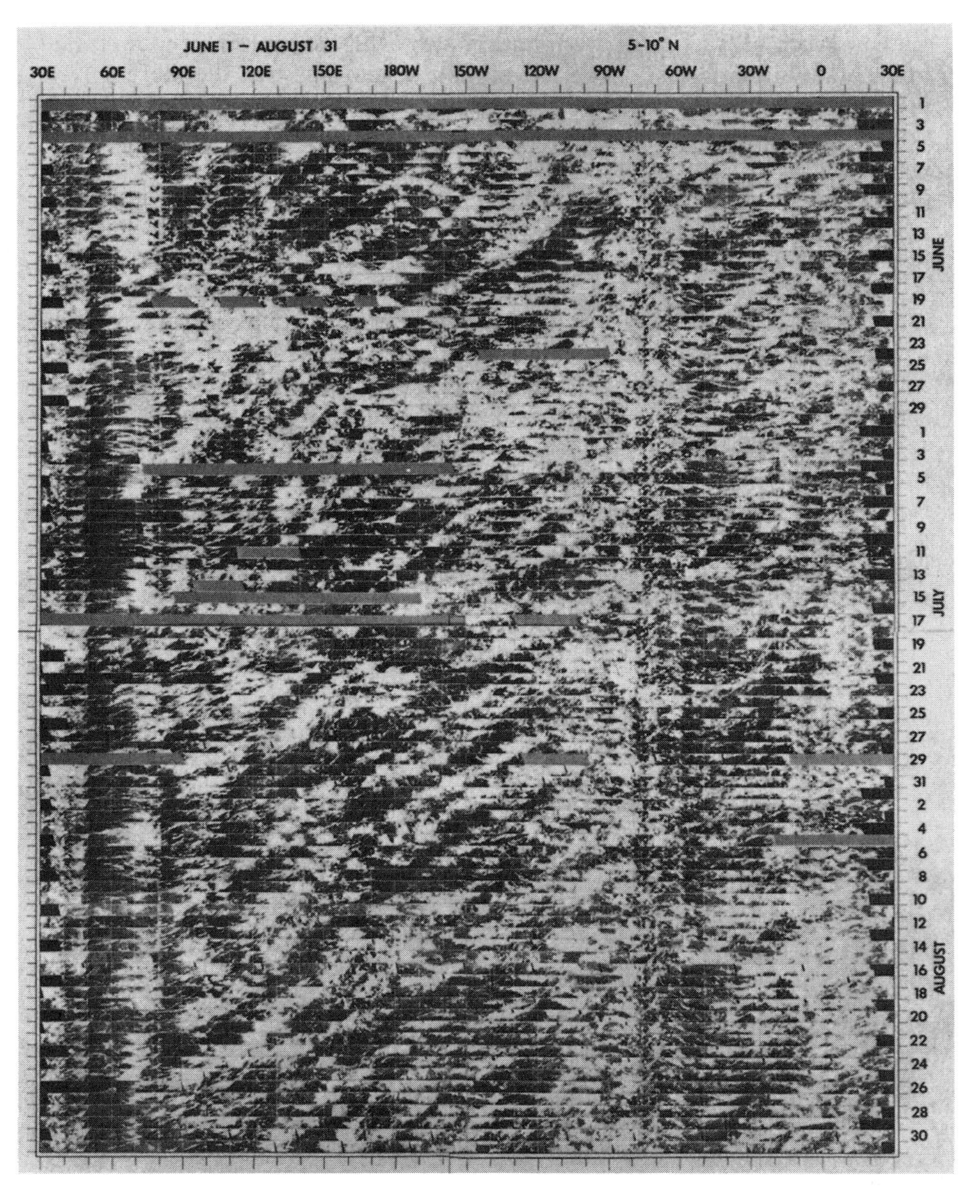

Fig. 1.24 Time longitude section of satellite viewed cloudiness, June–August, 1967, in the latitude band 5°N–10°N. Gray areas denote missing data. (From ESSA Technical Report NESC 56, U.S. Department of Commerce, Washington, D.C., 1970, p. 24.)

PROBLEMS†

1.1 Explain or interpret the following on the basis of the principles discussed in this chapter:

(a) It is much easier to establish the amount of oxygen produced by photosynthesis than the amount produced by the dissociation of water.

(b) Nitrogen is the dominant constituent in the earth's atmosphere, yet it is only a minor constituent in the atmospheres of Venus and Mars.

(c) The ratio of argon to neon is much larger in the earth's atmosphere than in the sun's atmosphere.

(d) The atmosphere of Venus is almost a hundred times more massive than the atmosphere of Earth.

(e) The evolution of the atmosphere of any planet is crucially dependent upon its distance from the sun and the strength of its gravitational field.

(f) Atmospheric pressure decreases with height.

(g) Below 100 km vapor trails become distorted and break up into puffs, whereas above 100 km they simply disappear.

(h) Most of the ozone in the earth's atmosphere is confined to the stratosphere.

(i) Other conditions being the same, the more massive the atmosphere of a planet the more rapid the escape of lighter constituents.

(j) The escape of hydrogen from the earth's atmosphere is more rapid during periods of strong solar activity.

(k) The terms *E-layer* and *F-layer* have been replaced by *E-region* and *F-region*.

(l) Many AM radio stations broadcast only during the daylight hours.

(m) Much more charge separation takes place within the E-region than within the D- or F-regions.

(n) Auroral displays are largely confined to high latitudes.

(o) Auroral displays can usually be forecast a few days in advance.

(p) Temperature exhibits a discontinuity at the magnetopause.

(q) The distribution of temperature is not quite the same in the northern and southern hemispheres.

(r) The 500-mb height contours are more circularly symmetric about the pole in the southern hemisphere than in the northern hemisphere.

1.2 It has been estimated that, if global energy consumption continues to expand in a manner consistent with prevailing trends of the past few decades, about 3×10^{13} kg of fossil fuels will be consumed annually by the year 2000. If half the resulting carbon dioxide remained in the atmosphere, what would be the annual rate of increase in atmospheric carbon dioxide around the year 2000? Assume that the fuels are 80% carbon by Mass.

Answer 5.6 parts per million per year by volume

1.3 Consider a perfectly elastic ball of mass m bouncing up and down above a horizontal surface under the action of a downward gravitational acceleration g. Prove that in the time average over an integral number of bounces the downward force exerted by the ball upon the surface is equal to the weight of the ball. [Hint: The downward force is equal to the downward momentum imparted to the surface with each bounce, divided by the time interval between successive bounces.] Does this result suggest anything about the "weight" of an atmosphere composed of gas molecules?

† Numerical values of constants and other parameters required to solve some of the problems in this and other chapters can be found on page 00.

1.4 If the earth's atmosphere consisted of an incompressible fluid whose density was everywhere equal to that observed at sea level (about 1.25 kg m^{-3}), how deep would it have to be in order to account for the observed sea level pressure (1013 mb)?

Answer 8.27 km

1.5 If the density of air decreases exponentially with height from a value of 1.25 kg m^{-3} at sea level, calculate the *scale height* H that is consistent with the observed sea level pressure of 1013 mb. [Integrate (1.9) from sea level to infinity to obtain the atmospheric mass per unit area.]

Answer 8.27 km

1.6 Calculate the most probable velocity of hydrogen atoms at a temperature of 600°K.

Answer 3.15 km s^{-1}

1.7 On the basis of the sea level pressure data shown in Figs. 1.15 and 1.16 determine whether the surface westerly winds over the North Atlantic and Pacific are stronger during summer or winter.

Answer winter

1.8 On the basis of the data shown in Figs. 1.10 and 1.12 verify that, in those regions in which temperature decreases with increasing latitude, the zonal wind is becoming more westerly with increasing height, and vice versa.

1.9 On the basis of the data shown in Figs. 1.15 and 1.16 locate the region of the northern hemisphere that experiences the highest monthly sea level pressure. How high is the pressure and during what month is it observed? Check your result by consulting a climatological reference book that shows the climatological pressure patterns for the northern hemisphere during individual months, or at least during January and July.

1.10 During September, October, and November the average surface pressure over the northern hemisphere rises at a rate of about 1 mb per (30-day) month. Calculate the (mass) average northward velocity across the equator that is required to account for this pressure rise. Why do you suppose that the mass of the atmosphere over the northern hemisphere increases during these months?

Answer 2.5 mm s^{-1}

1.11 If the electric field in the lower atmosphere is 100 V m^{-1}, determine the density of charge on the earth's surface. (The permittivity ε_0 of free space is 8.85×10^{-12} C^2 N^{-1} m^{-2}.)

Answer -8.85×10^{-10} C m^{-2}

Chapter

2

Atmospheric Thermodynamics

Thermodynamics plays an important role in our quantitative understanding of atmospheric phenomena, ranging from the smallest cloud microphysical processes to the general circulation of the atmosphere. The purpose of this chapter is to introduce some fundamental ideas and relationships in thermodynamics and to apply them to a number of simple, but important, atmospheric situations. Further applications of the concepts developed in this chapter occur throughout the book.

In the first section we consider the equation of state for an ideal gas and its application to dry air, water vapor, and moist air. In Section 2.2 an important meteorological relationship, known as the hydrostatic equation, is derived and interpreted. The next section is concerned with the relationship between the mechanical work done by a system and the heat the system receives, as expressed by the First Law of Thermodynamics. There follows several sections concerned with applications of the foregoing to the atmosphere. Finally, in Sections 2.8 and 2.9, which are intended for the more advanced student, the Second Law of Thermodynamics and the concept of entropy are introduced and criteria for thermodynamic equilibrium are discussed.

2.1 THE GAS LAWS

Laboratory experiments have shown that the pressure, volume, and temperature of any material can be related by an *equation of state*. All gases are found to follow approximately the same equation of state over a wide range of conditions. This equation of state is referred to as the *ideal* (*or perfect*) *gas equation*. For most purposes we may assume that atmospheric gases, whether considered individually or as a mixture, obey the ideal gas equation exactly. In this section we consider various forms of the ideal gas equation and its application to dry and moist air.

The ideal gas equation may be written as

$$pV = mRT \tag{2.1}$$

where p, V, m, and T are the pressure, volume, mass, and temperature (in degrees Kelvin) of the gas, respectively, and R is a constant (called the *gas constant*) for 1 kg of a gas, the value of which depends on the particular gas under consideration. Since $m/V = \rho$, where ρ is the density of the gas, the ideal gas equation may also be written in the form

$$p = R\rho T \tag{2.2}$$

For a unit mass (1 kg) of gas $m = 1$ and we may write (2.1) as

$$p\alpha = RT \tag{2.3}$$

where α is the specific volume of the gas; that is, the volume occupied by a unit mass. It should be noted that $\alpha = 1/\rho$.

If the temperature is constant, (2.1) reduces to *Boyle's*[†] *law*, which states that, if the temperature of a fixed mass of gas is constant, the volume of the gas is inversely proportional to its pressure. Changes in the physical state of a body which occur at constant temperature are termed *isothermal*. Also implicit in (2.1) are *Charles'*[‡] *two laws*. The first of these laws states that, for a fixed mass of gas at constant pressure, the volume of the gas is directly proportional to its absolute temperature. The second of Charles' laws states that, for a fixed mass of gas held within a fixed volume, the pressure of the gas is proportional to its absolute temperature.

We define now a *kilogram-molecular weight* or *kilomole* (abbreviated to kmol) of a material as its molecular weight M expressed in kilograms. For example, the molecular weight of water is 18.016; therefore, one kilomole of water is

[†] The Hon. **Sir Robert Boyle** (1627–1691) Fourteenth child of the first Earl of Cork. Physicist and chemist, often called the "father of modern chemistry." Discovered Boyle's law in 1662. One of the founders of the Royal Society of London. Boyle declared: "The Royal Society values no knowledge but as it has a tendency to use it"!

[‡] **Jacques A. C. Charles** (1746–1823) French physical chemist and inventor. Pioneer in the use of hydrogen in man-carrying balloons. When Benjamin Franklin's experiments with lightning became known, Charles repeated them with his own innovations. Franklin visited Charles and congratulated him on his work.

18.016 kg of water. The number of kilomoles n in mass m (in kilograms) of material is given by

$$n = \frac{m}{M} \tag{2.4}$$

Since the masses contained in 1 kmol of different materials bear the same ratios to each other as the molecular weights of the materials, a kilomole of one material must contain the same number of molecules as a kilomole of any other material. Therefore, the number of molecules in a kilomole of any material is a universal constant, which is called *Avogadro's*† *number* N_A. The value‡ of N_A is 6.022×10^{26}.

According to Avogadro's hypothesis, gases containing the same number of molecules occupy the same volumes at the same temperature and pressure. Therefore, for a kilomole of any gas the value of the gas constant is the same and is referred to as the *universal gas constant* R^*. The value of R^* is 8314.3 J deg^{-1} kmol^{-1}. Since

$$R^* = MR \tag{2.5}$$

it follows from (2.1), (2.4), and (2.5) that the ideal gas equation for n kilomoles of any gas is

$$pV = nR^*T \tag{2.6}$$

The gas constant for one molecule of any gas is also a universal constant, known as *Boltzmann's constant* k. Since the gas constant for N_A molecules is R^*, we have

$$k = \frac{R^*}{N_A} \tag{2.7}$$

Hence for a gas containing n_0 molecules per unit volume, the ideal gas equation is

$$p = n_0 kT \tag{2.8}$$

If the pressure and specific volume of dry air are p_d and α_d, respectively, the ideal gas equation in the form of (2.3) becomes for dry air:

$$p_d \alpha_d = R_d T \tag{2.9}$$

where R_d is the gas constant for 1 kg of dry air. For a mixture of gases, such as air, we can define an *apparent molecular weight* M_d as the total mass (in kilograms) of the constituent gases divided by the total number of kilomoles in

† **Amedeo Avogadro, Count of Quaregna** (1776–1856) Practiced law before turning to science at age 23. Became a professor of physics at the University of Turin later in life. His famous hypothesis was published in 1811 but was not generally accepted until a half century later. Introduced the term "molecule."

‡ In the cgs system, Avogadro's number is the number of molecules in a gram-molecular weight of any material and its value is therefore one-thousandth of that given here.

the mixture; that is,

$$M_d = \sum_i m_i \Big/ \sum_i \frac{m_i}{M_i} \tag{2.10}$$

where m_i and M_i represent the mass and molecular weight of the ith constituent in the mixture. The apparent molecular weight of dry air is 28.97. The gas constant for 1 kg of dry air is therefore given by

$$R_d = \frac{R^*}{M_d} = 287 \quad \text{J deg}^{-1}\ \text{kg}^{-1} \tag{2.11}$$

The ideal gas equation may also be applied to the individual gaseous components of air. For example, for water vapor,

$$e\alpha_v = R_v T \tag{2.12}$$

where e and α_v are the vapor pressure and specific volume of water vapor, respectively, and R_v the gas constant for 1 kg of water vapor. Since the molecular weight of water is M_w (= 18.016), and the gas constant for M_w kg of water vapor is R^*, we have

$$R_v = \frac{R^*}{M_w} = 461 \quad \text{J deg}^{-1}\ \text{kg}^{-1} \tag{2.13}$$

From (2.11) and (2.13),

$$\frac{R_d}{R_v} = \frac{M_w}{M_d} \equiv \varepsilon = 0.622 \tag{2.14}$$

It should be noted that since air is a mixture of gases it obeys Dalton's† law of partial pressures which states that the total pressure exerted by a mixture of gases which do not interact chemically is equal to the sum of the partial pressures of the gases. The *partial pressure* of a gas is the pressure it would exert at the same temperature as the mixture if it alone occupied the volume that the mixture occupies.

Problem 2.1 Calculate the density of water vapor which exerts a pressure of 9 mb at 20°C.

Solution From (2.12),

$$e\alpha_v = R_v T$$

where e is the water vapor pressure (in pascals), α_v the specific volume of the water vapor (in $\text{m}^3\ \text{kg}^{-1}$), R_v the gas constant for 1 kg of water vapor (in $\text{J deg}^{-1}\ \text{kg}^{-1}$), and T the

† **John Dalton** (1766–1844) English chemist. Initiated modern atomic theory. In 1787 he commenced a meteorological diary which he continued all his life, recording 200,000 observations. First to describe color blindness. He "never found time to marry"!

temperature in degrees Kelvin. From (2.13),

$$R_v = \frac{R^*}{M_w} = \frac{8314.3}{18.016} \quad \text{J deg}^{-1}\,\text{kg}^{-1} = 461 \quad \text{J deg}^{-1}\,\text{kg}^{-1}$$

Since 1 mb = 100 Pa, 9 mb = 900 Pa. Therefore, at 20°C or 293°K,

$$900\,\alpha_v = 461 \times 293$$

and

$$\alpha_v = 150 \quad \text{m}^3\,\text{kg}^{-1}$$

therefore, the density ρ_v of the water vapor is given by

$$\rho_v = \frac{1}{\alpha_v} = 6.67 \times 10^{-3} \quad \text{kg m}^{-3}$$

2.1.1 Virtual temperature

Moist air has a lower apparent molecular weight than dry air. Therefore, the gas constant for 1 kg of moist air is larger than that for 1 kg of dry air. However, rather than use a gas constant for moist air, the exact value of which would depend on the amount of water vapor in the air, it is more convenient to retain the gas constant for dry air and use a fictitious temperature (called the *virtual temperature*) in the ideal gas equation. We can derive an expression for the virtual temperature in the following way.

Consider a volume V of moist air at temperature T and total pressure p which contains mass m_d of dry air and mass m_v of water vapor. The density ρ of the moist air is given by

$$\rho = \frac{m_d + m_v}{V} = \rho_d{}' + \rho_v{}' \tag{2.15}$$

where $\rho_d{}'$ is the density which the same mass of dry air would have if it *alone* occupied the volume V and $\rho_v{}'$ the density which the same mass of water vapor would have if it *alone* occupied the volume V. We may call these *partial densities.* At first glance it might appear from (2.15) that the density of moist air is greater than that of dry air. This is not the case since the partial density $\rho_d{}'$ is less than the true density of dry air.† Applying the ideal gas equation in the form of (2.2) to the water vapor and dry air in turn, we have

$$e = R_v \rho_v{}' T$$

and

$$p_d{}' = R_d \rho_d{}' T$$

† The fact that moist air is less dense than dry air was first clearly stated by Sir Isaac Newton in his "Opticks" 1717. However, this idea was not generally understood until the latter half of the eighteenth century.

where e and p_d are the partial pressures exerted by the water vapor and the dry air, respectively. Also, from Dalton's law of partial pressures,

$$p = p_d' + e$$

Combining the last three equations,

$$\rho = \frac{p - e}{R_d T} + \frac{e}{R_v T}$$

or

$$\rho = \frac{p}{R_d T}\left[1 - \frac{e}{p}(1 - \varepsilon)\right]$$

where ε is defined by (2.14). The last equation may be written as

$$p = R_d \rho T_v \tag{2.16}$$

where

$$T_v = \frac{T}{1 - (e/p)(1 - \varepsilon)} \tag{2.17}$$

T_v is called the *virtual temperature*. If this fictitious temperature, rather than the actual temperature, is used for moist air, the total pressure p and density ρ of the moist air are related by the ideal gas equation with the gas constant the same as that for a unit mass of *dry air* (R_d). It follows that the virtual temperature is the temperature that dry air must have in order to have the same density as the moist air at the same pressure. Moist air is less dense than dry air; therefore, the virtual temperature is always greater than the actual temperature. However, even for very warm, moist air the virtual temperature exceeds the actual temperature by only a few degrees (see Problem 2.8).

2.2 THE HYDROSTATIC EQUATION AND ITS APPLICATIONS

The atmosphere is in motion at all times. However, the upward force acting on a thin slab of air, due to the decrease in pressure with height, is generally very closely in balance with the downward force due to gravitational attraction. We will now derive an important equation assuming that these forces are exactly in balance.

Let us consider a vertical column of air with unit cross-sectional area (Fig. 2.1). The mass of air between heights z and $z + dz$ in the column is $\rho\, dz$, where ρ is the density of the air at height z. The force acting on this column due to the weight of the air is $g\rho\, dz$, where g is the acceleration due to gravity at height z. Now let us consider the net vertical force on the block due to the pressure of the surrounding air. We will assume that in going from height z to height

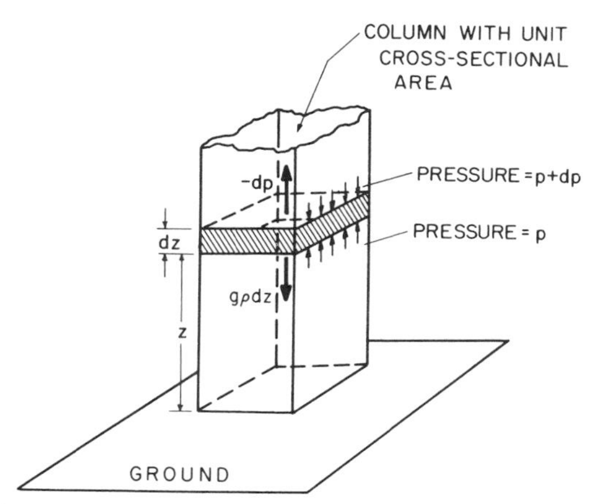

Fig. 2.1 Balance of vertical forces in an atmosphere with no vertical acceleration. The small arrows on the right side of the shaded block indicate the pressure exerted on the air in the block by the air above and below it in the column. The heavy arrows represent the vertical forces: the downward weight of the air and the upward force due to the vertical pressure gradient. Note that the incremental pressure change dp is a negative quantity, since pressure decreases with height.

$z + dz$ the pressure changes by the amount dp as indicated in Fig. 2.1. Since we know that pressure decreases with height, dp must be a negative quantity, and the upward pressure on the lower face of the shaded block must be slightly greater than the downward pressure on the upper face of the block. Thus the net vertical force on the block due to the vertical gradient of pressure is upward and given by the positive quantity $-dp$ as indicated in the figure. The balance of forces in the vertical requires that

$$-dp = g\rho \, dz$$

or

$$\frac{dp}{dz} = -g\rho \tag{2.18}$$

Equation (2.18) is termed the *hydrostatic equation.*† It should be noted that the negative sign in (2.18) ensures that the pressure decreases with increasing height. Since $\rho = 1/\alpha$, (2.18) can be rearranged to give

$$g \, dz = -\alpha \, dp$$

† Strictly speaking, the left-hand side of (2.18) should be written in partial notation as $\partial p/\partial z$, since the variation of pressure is taken with the other independent variables held constant. However, in the remainder of this chapter the distinction between total and partial derivatives will not be important, since (2.18) will only be used in situations where the dependent variables are treated as functions of height only. The partial notation will be introduced in Chapter 8.

If the pressure at height z is $p(z)$, we have, from (2.18),

$$-\int_{p(z)}^{p(\infty)} dp = \int_z^\infty g\rho \, dz$$

or, since $p(\infty) = 0$,

$$p(z) = \int_z^\infty g\rho \, dz \tag{2.19}$$

That is, the pressure at level z is equal to the weight of the air in the vertical column of unit cross-sectional area lying above that level. If the mass of earth's atmosphere were uniformly distributed over the globe, the pressure at sea level would be 1013 mb, or 1.013×10^5 Pa, which is referred to as *normal atmospheric pressure* and abbreviated as 1 atm.

2.2.1 Geopotential

The *geopotential* Φ at any point in the atmosphere is defined as the work that must be done against the earth's gravitational field in order to raise a mass of 1 kg from sea level to that point. In other words, Φ is the gravitational potential for unit mass. The units of geopotential are J kg^{-1} or $\text{m}^2\ \text{s}^{-2}$. The force (in newtons) acting on 1 kg at height z above sea level is numerically equal to g. The work (in joules) in raising 1 kg from z to $z + dz$ is $g\,dz$; therefore,

$$d\Phi = g\,dz = -\alpha\,dp \tag{2.20}$$

The geopotential $\Phi(z)$ at height z is thus given by

$$\Phi(z) = \int_0^z g\,dz \tag{2.21}$$

where the geopotential $\Phi(0)$ at sea level ($z = 0$) has, by convention, been taken as zero. It should be emphasized that the geopotential at a particular point in the atmosphere depends only on the height of that point and not on the path through which the unit mass is taken in reaching that point. The work done in taking a mass of 1 kg from point A with geopotential Φ_A to point B with geopotential Φ_B is $\Phi_B - \Phi_A$.

We can also define a quantity called the *geopotential height* Z as

$$Z \equiv \frac{\Phi(z)}{g_0} = \frac{1}{g_0}\int_0^z g\,dz \tag{2.22}$$

where g_0 is the globally averaged acceleration due to gravity at the earth's surface (taken as 9.8 m s^{-2}). Geopotential height is used as the vertical coordinate in most atmospheric applications in which energy plays an important role (for example, the large-scale motions discussed in Chapter 8). It can be seen from Table 2.1 that the values of z and Z are almost the same in the lower atmosphere where $g_0 \simeq g$.

Table 2.1

Values of the geometric height (z), geopotential height (Z), and acceleration due to gravity (g) at 40° latitude

z (km)	Z (km)	g (m s^{-2})
0	0	9.802
1	1.000	9.798
10	9.986	9.771
20	19.941	9.741
30	29.864	9.710
60	59.449	9.620
90	88.758	9.531
120	117.795	9.443
160	156.096	9.327
200	193.928	9.214
300	286.520	8.940
400	376.370	8.677
500	463.597	8.427
600	548.314	8.186

In meteorological practice it is not convenient to deal with density ρ, which cannot be measured directly. By making use of (2.2) or (2.16) to eliminate ρ in (2.18), we obtain

$$\frac{dp}{dz} = -\frac{pg}{RT} = -\frac{pg}{R_d T_v}$$

Rearranging the last expression and using (2.20),

$$d\Phi = -RT\frac{dp}{p} = -R_d T_v \frac{dp}{p} \tag{2.23}$$

If we now integrate between pressure levels p_1 and p_2, with geopotentials Φ_1 and Φ_2, respectively,

$$\Phi_2 - \Phi_1 = -R_d \int_{p_1}^{p_2} T_v \frac{dp}{p}$$

Dividing both sides of the last equation by g_0 and reversing the limits of integration yields

$$Z_2 - Z_1 = \frac{R_d}{g_0}\int_{p_2}^{p_1} T_v \frac{dp}{p} \tag{2.24}$$

2.2.2 Scale height and the hypsometric equation

For an isothermal (temperature constant with height) and dry atmosphere, (2.24) becomes

$$Z_2 - Z_1 = H \ln(p_1/p_2) \tag{2.25}$$

or

$$p_2 = p_1 \exp\left[-\frac{(Z_2 - Z_1)}{H}\right] \tag{2.26}$$

where

$$H = \frac{RT}{g_0} = \frac{R_d T_v}{g_0} = 29.3 T_v \tag{2.27}$$

H is called the *scale height*. If $Z_2 - Z_1$ is set successively equal to 0, H, $2H$, $3H, \ldots, p_2/p_1$ is equal to 1, exp(−1), exp(−2), exp(−3), ... That is, the pressure decreases by a factor e (=2.718) for each increase H in geopotential height. It should be noted that (2.26) is equivalent to (1.8) which was derived empirically in Section 1.3.1.

Since the atmosphere is well mixed below the turbopause, the pressure and densities of the individual gases decrease with altitude at the same rate and with a scale height corresponding to the apparent molecular weight of the mixture. If we take a value for T_v of 288°K near the earth's surface, the scale height H for air in the atmosphere is found from (2.27) to be $8\frac{1}{2}$ km. Above the turbopause (about 105 km) the vertical distribution of gases is largely controlled by molecular diffusion and a scale height may then be defined for each of the individual gases in air. Since for each gas the scale height is proportional to the gas constant for a unit mass of a gas, which in turn is inversely proportional to the molecular weight of that gas, the pressures (and densities) of heavier gases fall off more rapidly with height above the turbopause than do those of lighter gases.

Problem 2.2 If the ratio of the number density of oxygen atoms to the number density of hydrogen atoms at a geopotential height of 200 km above the earth's surface is 10^5, calculate the ratio of the number densities of these two constituents at a geopotential height of 1400 km assuming an isothermal atmosphere with a temperature of 2000°K.

Solution At these altitudes the distribution of the individual gases is determined by diffusion and therefore by (2.26). Also, at constant temperature the ratio of the number densities of two gases is equal to the ratio of their pressures. From (2.26),

$$\frac{(p_{1400\text{km}})_\text{O}}{(p_{1400\text{km}})_\text{H}} = \frac{(p_{200\text{km}})_\text{O} \exp[-1200 \text{ km}/H_\text{O}(\text{km})]}{(p_{200\text{km}})_\text{H} \exp[-1200 \text{ km}/H_\text{H}(\text{km})]}$$

$$= 10^5 \exp\left[-1200 \text{ km}\left(\frac{1}{H_\text{O}} - \frac{1}{H_\text{H}}\right)\right]$$

From the definition of scale height we have, at 2000°K,

$$H_\text{O} = \frac{R^*}{16}\frac{2000}{9.8} \text{ m} \quad \text{and} \quad H_\text{H} = \frac{R^*}{1}\frac{2000}{9.8} \text{ m}$$

therefore,

$$\frac{1}{H_\text{O}} - \frac{1}{H_\text{H}} = 8.84 \times 10^{-6} \quad \text{m}^{-1}$$

and

$$\frac{(p_{1400\text{km}})_{\text{O}}}{(p_{1400\text{km}})_{\text{H}}} = 10^5 \exp(-10.6) = 2.47$$

Hence, the ratio of the number densities of oxygen to hydrogen atoms at a geopotential height of 1400 km is 2.47.

The temperature of the atmosphere generally varies with height. In this case (2.24) may be integrated if we define a mean virtual temperature $\bar{T}_v$ with respect to ln p as shown in Fig. 2.2. That is,

$$\bar{T}_v \equiv \int_{\ln p_2}^{\ln p_1} T_v\, d(\ln p) \Big/ \int_{\ln p_2}^{\ln p_1} d(\ln p) = \int_{p_2}^{p_1} T_v \frac{dp}{p} \Big/ \ln\left(\frac{p_1}{p_2}\right) \tag{2.28}$$

Then, from (2.24) and (2.28),

$$Z_2 - Z_1 = \frac{R_d \bar{T}_v}{g_0} \ln\left(\frac{p_1}{p_2}\right) = \bar{H} \ln\left(\frac{p_1}{p_2}\right) \tag{2.29}$$

where the scale height $\bar{H}$ is now defined as

$$\bar{H} \equiv \frac{R_d \bar{T}_v}{g_0} = 29.3\bar{T}_v \tag{2.30}$$

Equation (2.29) is called the *hypsometric equation.*

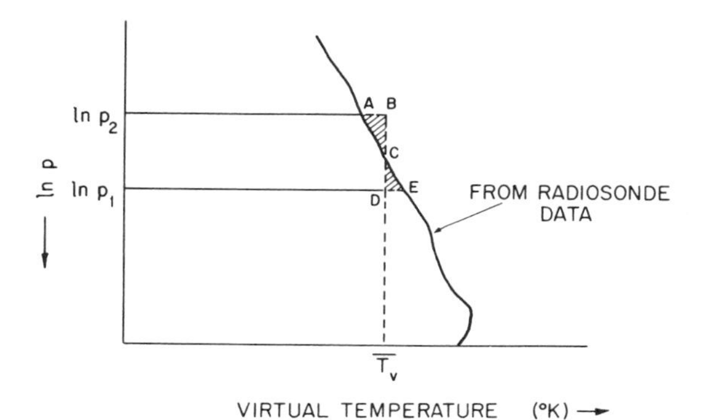

Fig. 2.2 ln p versus T_v diagram. If area ABC = area CDE, $\bar{T}_v$ is the mean virtual temperature with respect to ln p between the pressure levels p_1 and p_2.

2.2.3 Thickness and heights of constant pressure surfaces

The difference in geopotential height $Z_2 - Z_1$ between any two levels in the atmosphere is called the *thickness* of the intervening layer. It can be seen from (2.29) and (2.30) that the thickness of the layer between any two pressure levels p_2 and p_1 is proportional to the mean virtual temperature of the layer.

We can visualize that as $\bar{T}_v$ increases, the air between the two pressure levels expands so that the layer becomes thicker.

Problem 2.3 Calculate the thickness of the layer between the 1000- and 500-mb pressure surfaces (a) at a point in the tropics where the mean virtual temperature of the layer is 9°C, and (b) at a point in the polar regions where the corresponding mean virtual temperature is -40°C.

Solution From (2.29)

$$\Delta Z = Z_{500\text{mb}} - Z_{1000\text{mb}} = \frac{R_d \bar{T}_v}{g_0} \ln\left(\frac{1000}{500}\right) = 20.3\bar{T}_v$$

Therefore, for the tropics with $\bar{T}_v = 282$°K, $\Delta Z = 5725$ m. For polar regions with $\bar{T}_v = 233$°K, $\Delta Z = 4730$ m.

Thickness may readily be evaluated from radiosonde data which provide measurements of the pressure, temperature, and humidity at various levels in the atmosphere. The virtual temperature T_v at each level may be found from the measurements of temperature and humidity and these values plotted against the pressure on a $\ln p$ versus T_v diagram (see Fig. 2.2). The mean virtual temperature $\bar{T}_v$ for the layer can be computed on an equal area basis (that is, area ABC = area CDE in Fig. 2.2) and the thickness evaluated using (2.29). If p_1 is the pressure at ground level, then the value of Z_2 determined from (2.29) is the geopotential height at which the air pressure is p_2. Given sounding data from a network of stations it is possible to construct topographical maps of the distribution of geopotential height on selected pressure surfaces (see Chapter 3). Calculations such as this are carried out routinely by weather services throughout the world.

In moving from a given pressure surface to another located above or below it, the change in the geopotential height is geometrically related to the thickness of the intervening layer which, in turn, is directly proportional to the mean virtual temperature of the layer. Thus, if the three-dimensional distribution of virtual temperature is known together with the distribution of geopotential height on one pressure surface, it is possible to infer the distribution of geopotential height of any other pressure surface. The same hypsometric relationship between the three-dimensional temperature field and the shape of pressure surface can be used in a qualitative way to gain some useful insights into the three-dimensional structure of atmospheric disturbances. Let us consider the following examples:

- From the earth's surface up to the tropopause the core of a hurricane is warmer than its surroundings. Consequently, the intensity of the storm (as measured by the depression of the isobaric surfaces) must decrease with height (Fig. 2.3a). Such *warm core lows* always exhibit their greatest intensity near the ground and diminish with increasing height above the ground.

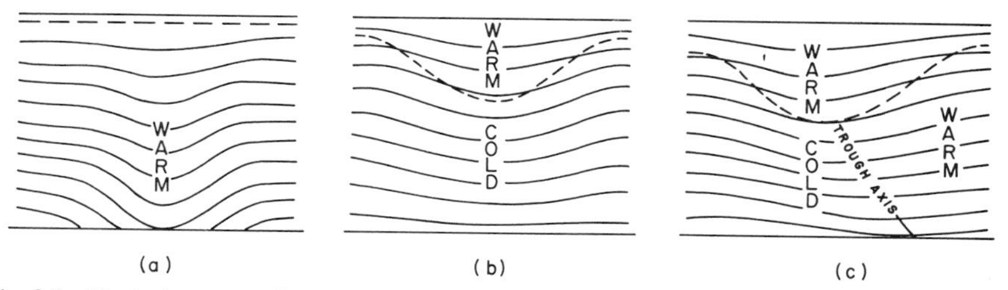

Fig. 2.3 Vertical cross sections through (a) a hurricane, (b) a "cold core" upper tropospheric low in middle latitudes, and (c) a middle-latitude disturbance which tilts westward with increasing height. The solid lines indicate various constant pressure surfaces and the dashed lines represent the tropopause. The sections are drawn such that the thickness between adjacent pressure surfaces is smaller in regions labeled *cold* and larger in regions labeled *warm*.

- Some upper air lows do not extend downward to the ground as indicated in Fig. 2.3b. Such lows are referred to as being *cold core* because they exhibit negative thickness anomalies relative to their surroundings.
- Most middle-latitude disturbances have vertical structure similar to that shown in Fig. 2.3c with the coldest air and the lowest thickness to the west of the surface low. Such systems slope westward with height in agreement with the hypsometric equation.
- The disturbances depicted in Fig. 2.3b and c exhibit amplitude maxima at the tropopause level and damp out rapidly with height in the lower stratosphere. This behavior is consistent with a phase reversal in the temperature field at the tropopause level; for example, warm air in the lower stratosphere is situated over the cold low in the upper troposphere. These concepts are illustrated further in Chapter 3.

2.2.4 Reduction of pressure to sea level

In mountainous regions the difference in surface pressure from one station to another is largely due to differences in elevation. In order to isolate that part of the pressure field which is due to the passage of weather systems, it is necessary to reduce the pressures to a common reference level. For this purpose, sea level is normally used.

Let the subscripts g and 0 refer to conditions at the ground and at sea level ($Z = 0$), respectively. For the layer between the earth's surface and sea level the hypsometric equation (2.29) assumes the form

$$Z_g = \bar{H} \ln \frac{p_0}{p_g}$$

which can be solved to obtain the sea level pressure

$$p_0 = p_g \exp\left(\frac{Z_g}{\bar{H}}\right) = p_g \exp\left(\frac{g_0 Z_g}{R_d \bar{T}_v}\right) \tag{2.31}$$

where (2.30) has been used to obtain the last expression which shows how the sea level pressure depends on the mean virtual temperature between ground and sea level. If Z_g is small, the scale height $\bar{H}$ can be evaluated from the ground temperature. Also, if $Z_g/\bar{H} \ll 1$, the exponential in (2.31) can be approximated by $1 + Z_g/\bar{H}$. Since $\bar{H} \simeq 8$ km for the observed range of ground temperatures on earth, this approximation is satisfactory provided that Z_g is less than a few hundred meters. With this approximation, (2.31) becomes

$$p_0 - p_g \simeq p_g \frac{Z_g}{\bar{H}} = p_g \left(\frac{g_0 Z_g}{R_d \bar{T}_v} \right) \tag{2.32}$$

Since $p_g \simeq 1000$ mb and $\bar{H} \simeq 8$ km, the pressure correction (in millibars) is roughly equal to Z_g (in meters) divided by 8. In other words, near sea level the pressure falls about 1 mb for every 8 m of vertical ascent.

When Z_g is on the order of 1 km or greater, there is difficulty in estimating what the mean virtual temperature of the layer would be in the absence of topography. In practice, a number of empirical corrections are applied to the surface temperature in the estimation of the scale height $\bar{H}$. These procedures are not entirely satisfactory in eliminating the effects of topography. Therefore, sea level pressure analysis in mountainous regions still leaves much to be desired.

Problem 2.4 Calculate the geopotential height of the 1000-mb pressure surface when the pressure at sea level is 1014 mb. The scale height of the atmosphere may be taken as 8 km.

Solution From the hypsometric equation (2.29),

$$Z_{1000\text{mb}} = \bar{H} \ln\left(\frac{p_0}{1000}\right) = \bar{H} \ln\left(1 + \frac{p_0 - 1000}{1000}\right) \simeq \bar{H}\left(\frac{p_0 - 1000}{1000}\right)$$

where p_0 is the sea level pressure and the relationship $\ln(1 + x) \simeq x$ for $x \ll 1$ has been used. Substituting $\bar{H} \simeq 8000$ m into this expression gives

$$Z_{1000\text{mb}} \simeq 8(p_0 - 1000)$$

Therefore, with $p_0 = 1014$ mb, the geopotential height Z_{1000} of the 1000-mb pressure surface is found to be 112 m above sea level. [Note: similar expressions can be derived to relate the height of other pressure surfaces to the distribution of pressure on a nearby constant height level; the constant of proportionality will, in general, depend on the pressure levels.]

Problem 2.5 Derive a relationship for the height of a given pressure surface (p) in terms of the pressure p_0 and temperature T_0 at sea level assuming that the temperature decreases uniformly with height at a rate Γ deg km^{-1}.

Solution Let the height of the pressure surface be z; then its temperature T is given by

$$T = T_0 - \Gamma z \tag{2.33}$$

Combining the hydrostatic equation (2.18) with the ideal gas equation (2.2) yields

$$\frac{dp}{p} = -\frac{g}{RT} dz \tag{2.34}$$

From (2.33) and (2.34)

$$\frac{dp}{p} = -\frac{g}{R(T_0 - \Gamma z)} dz$$

Integrating this equation between pressure levels p_0 and p and corresponding heights 0 and z and neglecting the variation of g with z, we obtain

$$\int_{p_0}^{p} \frac{dp}{p} = -\frac{g}{R} \int_0^z \frac{dz}{(T_0 - \Gamma z)}$$

or

$$\ln \frac{p}{p_0} = \frac{g}{R\Gamma} \ln\left(\frac{T_0 - \Gamma z}{T_0}\right)$$

Therefore,

$$z = \frac{T_0}{\Gamma}\left[1 - \left(\frac{p}{p_0}\right)^{R\Gamma/g}\right] \tag{2.35}$$

(This equation forms the basis for the calibration of aircraft altimeters. An altimeter is simply an aneroid barometer which measures the air pressure p. However the scale of the altimeter is expressed as the height z of the aircraft, where z is related to p by (2.35) with values for T_0, p_0, and Γ appropriate to the U.S. Standard Atmosphere, namely, $T_0 = 288°\text{K}$, $p_0 = 1013.25$ mb, and $\Gamma = 6.50$ deg km^{-1}.)

2.3 THE FIRST LAW OF THERMODYNAMICS

In addition to the macroscopic kinetic and potential energy that a body as a whole may possess, it also contains *internal energy* due to the kinetic and potential energy of its molecules or atoms. Increases in internal kinetic energy in the form of molecular motions are manifested as increases in the temperature of the body, while changes in the potential energy of the molecules are caused by changes in their relative configurations.

Let us suppose that a body of unit mass takes in a certain quantity of heat energy q (measured in joules), which it can receive by either thermal conduction or radiation. As a result the body may do a certain amount of *external* work w (also measured in joules). The excess of the energy supplied to the body over and above the external work done by the body is $q - w$. Therefore, if there is no change in the macroscopic kinetic and potential energy of the body, it follows from the principle of conservation of energy that the internal energy of the body must increase by $q - w$. That is,

$$q - w = u_2 - u_1 \tag{2.36}$$

where u_1 and u_2 are the internal energies of the body before and after the change. In differential form (2.36) becomes

$$dq - dw = du \tag{2.37}$$

where, dq is the differential increment of heat added to the body, dw the differential element of work done by the body, and du the differential increase in internal energy of the body.[†] Equations (2.36) and (2.37) are statements of the *First Law of Thermodynamics*. In fact, (2.37) provides a definition of du. It should be noted that the change in internal energy du is a function only of the initial and final states of the body and is therefore independent of the manner by which the body is transferred between these two states.

In order to visualize the work term dw in (2.37) in a simple case, consider a substance, often called the *working substance*, contained in a cylinder of fixed cross-sectional area which is fitted with a movable, frictionless piston (Fig. 2.4). The volume of the material is then proportional to the distance from the base of the cylinder to the face of the piston, and can be represented on the horizontal line of the graph shown in Fig. 2.4. The pressure of the substance in the cylinder can be represented on the vertical line of this graph. Therefore, every state of the substance, corresponding to a given position of the cylinder, is represented by a point of the graph. When the substance is in equilibrium at a state represented by the point P on this graph its pressure is p and its volume V. If the piston moves outwards through an incremental distance dx, while the pressure remains essentially constant at p, the work dW done by the substance in expanding is equal to the force exerted on the piston (this force is equal to pA where A is the cross sectional area of the piston) multiplied by the distance

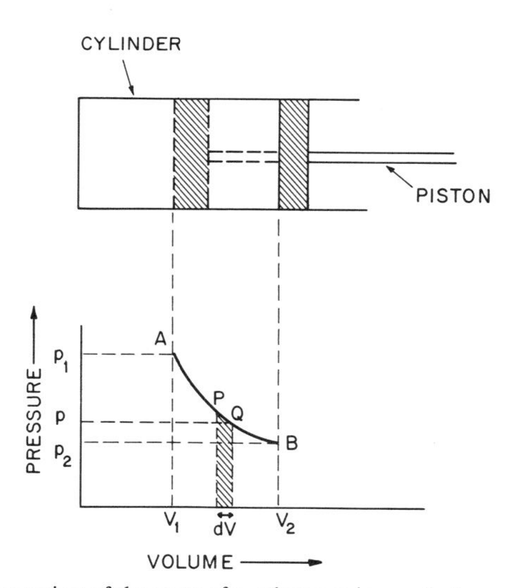

Fig. 2.4 Representation of the state of a substance in a cylinder on a $p-V$ diagram.

[†] It should be noted that neither the heat q nor the work w are functions of state. Therefore, dq and dw are not *perfect differentials*. The internal energy u is a function of state; therefore, du is a perfect differential. If $z = f(x, y)$, dz is a perfect differential provided that

$$\frac{\partial^2 f}{\partial y\,\partial x} = \frac{\partial^2 f}{\partial x\,\partial y}$$

dx through which the piston moves. That is,

$$dW = pA\,dx = p\,dV \tag{2.38}$$

In other words, the work done by the substance when its volume increases by a small amount is equal to the pressure of the substance multiplied by its increase in volume. It should be noted that $dW = p\,dV$ is equal to the shaded area in the graph shown in Fig. 2.4; that is, it is equal to the area under the curve PQ. When the substance passes from state A with volume V_1 to state B with volume V_2 (Fig. 2.4), during which its pressure p changes, the work W done by the material is equal to the area under the curve AB. That is,

$$W = \int_{V_1}^{V_2} p\,dV \tag{2.39}$$

Equations (2.38) and (2.39) are quite general and represent the work done by any material due to a change in its volume.

The pressure–volume (or p–V) diagram shown in Fig. 2.4 is an example of a *thermodynamic diagram* in which the physical state of a substance is represented by two thermodynamic variables. Such diagrams are very useful in meteorology; we will discuss other examples later in this chapter.

If we are dealing with a unit mass of material, the volume V is replaced by the specific volume α and the work dw which is done when the specific volume increases by $d\alpha$ is

$$dw = p\,d\alpha \tag{2.40}$$

Combination of (2.37) and (2.40) yields

$$dq = du + p\,d\alpha \tag{2.41}$$

which is an alternative statement of the First Law of Thermodynamics.[†] It should be noted that the First Law of Thermodynamics can be applied to any system; however, here we shall generally apply it only to gases.

2.3.1 Joule's law

Following a series of laboratory experiments on air, Joule[‡] concluded in 1848 that when a gas expands without doing external work (for example, by expanding into a chamber which has been evacuated), and without taking

[†] We have assumed here that the only external work done by the body is due to its volume changing. We will see in Section 2.9 that there are other ways in which a body may do external work. We have also assumed that the macroscopic kinetic and potential energy remain constant. However, it can be shown (see, for example, R. G. Fleagle and J. A. Businger, "An Introduction to Atmospheric Physics," Academic Press, New York, 1963, p. 37) that even if the macroscopic kinetic and potential energies of a parcel of air in the atmosphere are changing, the First Law still takes the form of (2.41).

[‡] **James Prescott Joule** (1818–1889) English physicist, one of the great experimentalists of the nineteenth century. Started his scientific work (carried out in laboratories in his home and at his own expense) at age 19. Measured the mechanical equivalent of heat. Recognized the dynamical nature of heat, and developed the principle of conservation of energy. Derived the relationship for the heat produced by an electric current.

in or giving out heat, the temperature of the gas does not change. This statement, which is known as *Joule's law*, is strictly true only for an ideal gas.

Joule's law leads to an important conclusion concerning the internal energy of any ideal gas. If the gas neither does external work nor takes in or gives out heat, $dw = 0$ and $dq = 0$ in (2.37), so that $du = 0$. Also, according to Joule's law, under these conditions the temperature of the gas does not change, which implies that the kinetic energy of the molecules remains constant. Therefore, since the internal energy of the gas is also constant, that part of the internal energy due to the potential energy must also remain unchanged even though the volume of the gas changes. In other words, the internal energy of an ideal gas is independent of volume if the temperature is constant. This implies that the molecules of an ideal gas do not exert any attractive or repulsive forces on each other, so that the internal energy depends only upon the temperature.

2.3.2 Specific heats

Suppose a small quantity of heat dq is given to a unit mass of material and, as a consequence, its temperature increases from T to $T + dT$ without a phase change occurring. The ratio dq/dT is called the *specific heat* of the material. However, the specific heat defined in this way can have any number of values, depending on how the material changes as it receives the heat. If the volume of the material is kept constant, *a specific heat at constant volume* c_v is defined which is given by

$$c_v \equiv \left(\frac{dq}{dT}\right)_{\alpha \mathrm{const}} \tag{2.42}$$

But if the specific volume is constant (2.41) becomes $dq = du$; therefore,

$$c_v = \left(\frac{du}{dT}\right)_{\alpha \mathrm{const}}$$

But for an ideal gas Joule's law applies and therefore u depends upon temperature alone and we may write

$$c_v = \frac{du}{dT} \tag{2.43}$$

Therefore, from (2.41) and (2.43), the First Law of Thermodynamics for an ideal gas can be written in the form[†]

$$dq = c_v\, dT + p\, d\alpha \tag{2.44}$$

[†] The term dq is sometimes called the *diabatic* (or nonadiabatic) *heating* or *cooling*, where "diabatic" means involving the transfer of heat. The term "diabatic" would be redundant if "heating" and "cooling" were always taken to mean "the addition or removal of heat." However, "heating" and "cooling" are often used in the sense of "to raise or lower the temperature of," in which case it is meaningful to distinguish between that part of the temperature change dT due to diabatic effects (dq) and that part due to adiabatic effects ($p\, d\alpha$).

We may also define a *specific heat at constant pressure* c_p as

$$c_p \equiv \left(\frac{dq}{dT}\right)_{p\text{const}} \tag{2.45}$$

where the material is allowed to expand as the heat is added and its temperature rises but its pressure is kept constant. In this case, a certain amount of the heat added will have to be expended to do work as the material expands against the constant pressure of the environment. Therefore, a larger quantity of heat must be added to the material to raise its temperature by a given amount than if the volume of the material were kept constant. This can be seen mathematically in the following way. Equation (2.44) can be rewritten in the form

$$dq = c_v\, dT + d(p\alpha) - \alpha\, dp \tag{2.46}$$

Making use of the ideal gas equation (2.3), this becomes

$$dq = (c_v + R)\, dT - \alpha\, dp \tag{2.47}$$

At constant pressure, the last term in (2.47) vanishes; therefore, from (2.45),

$$c_p = c_v + R \tag{2.48}$$

The specific heats at constant volume and at constant pressure for dry air are 717 and 1004 J $\text{deg}^{-1}\ \text{kg}^{-1}$, respectively, and the difference between them is 287 J $\text{deg}^{-1}\ \text{kg}^{-1}$, which is numerically equal to the gas constant for dry air. For an ideal monatomic gas $c_p : c_v : R = 5:3:2$ and for an ideal diatomic gas $c_p : c_v : R = 7:5:2$.[†]

By combining (2.47) and (2.48) we obtain an alternate form of the First Law of Thermodynamics:

$$dq = c_p\, dT - \alpha\, dp \tag{2.49}$$

2.3.3 Enthalpy

If heat is added to a material at constant pressure, so that the specific volume of the material increases from α_1 to α_2, the work done by a unit mass of the material is $p(\alpha_2 - \alpha_1)$. Therefore, from (2.41), the heat dq added to a unit mass of the material at constant pressure is given by

$$dq = (u_2 - u_1) + p(\alpha_2 - \alpha_1) = (u_2 + p\alpha_2) - (u_1 + p\alpha_1)$$

where u_1 and u_2 are, respectively, the initial and final internal energies for unit mass. Therefore, at constant pressure,

$$dq = h_2 - h_1$$

[†]See, for example, F. W. Sears, "Thermodynamics, the Kinetic Theory of Gases and Statistical Mechanics," Addison-Wesley Publishing Co., 1953, p. 246–50.

where h is the *enthalpy* of a unit mass of material which is defined by

$$h \equiv u + p\alpha \tag{2.50}$$

Differentiating this expression, we obtain

$$dh = du + d(p\alpha)$$

Substituting for du from (2.43) and combining with (2.46), we obtain

$$dq = dh - \alpha\, dp \tag{2.51}$$

which is yet another form of the First Law of Thermodynamics.

By comparing (2.49) and (2.51) we note that

$$dh = c_p\, dT \tag{2.52}$$

or, in integrated form,

$$h = c_p T \tag{2.53}$$

where h is taken as zero when $T = 0$.

By combining (2.20), (2.51), and (2.53) we obtain

$$dq = d(h + \Phi) = d(c_p T + \Phi) \tag{2.54}$$

Hence, for an air parcel moving about in an hydrostatic atmosphere, the quantity $(h + \Phi)$ is constant if the gas under consideration neither gains nor loses heat (that is, $dq = 0$).[†]

2.4 LATENT HEATS

Under certain conditions heat may be supplied to a substance without its temperature changing. In this case the increase in internal energy is associated entirely with a change in molecular configurations produced by a change of phase. For example, if heat is supplied to ice at normal atmospheric pressure and 0°C, the temperature remains constant until all of the ice has melted. The *latent heat of melting* is defined as the heat required to convert a unit mass of a material from the solid to the liquid phase without a change in temperature. The temperature at which this phase change occurs is called the *melting point*. At normal atmospheric pressure and temperature the latent heat of melting of the water substance is 3.34×10^5 J kg^{-1}; the *latent heat of fusion* has the same value. Similarly, the *latent heat of vaporization* is the heat required to convert a unit mass of material from the liquid to the vapor phase without a change in temperature. For the water substance at normal atmospheric pressure and 0°C, the latent heat of vaporization is 2.500×10^6 J kg^{-1}; the *latent heat of condensation* has the same value.

[†] Strictly speaking, (2.54) holds only for an atmosphere in which there are no fluid motions. However, it is correct to within a few percent for the earth's atmosphere where the kinetic energy of fluid motions represents only a very small fraction of the total energy (see Problem 7.7). An exact relationship can be obtained by using Newton's Second Law of Motion and the continuity equation in place of (2.20) in the derivation. See E. Lorenz, "On the Nature and Theory of the General Circulation of the Atmosphere," Unipub, 1967, p. 15.

We will see in Section 2.8.2 that the melting and boiling points of a material depend on the pressure. Also, the latent heats of fusion and vaporization vary with temperature (see Problem 2.16).

2.5 ADIABATIC PROCESSES

If a material changes its physical state (that is, if its pressure, volume, or temperature change) without any heat being either added to it or withdrawn from it, the change is said to be *adiabatic*. Suppose that the initial state of a material is represented by the point A on the p–V diagram in Fig. 2.5 and that when it undergoes an isothermal transformation it moves along the line AB. If the same body underwent a similar change in volume but under adiabatic conditions, the transformation would be represented by a curve such as AC which is called an *adiabat*. The reason why the adiabat AC is steeper than the isotherm AB can be seen as follows: During the adiabatic compression the internal energy increases (since $dq = 0$ and $p\,d\alpha$ is negative in 2.41) and therefore the temperature of the material rises. However, for the isothermal compression, the temperature remains constant. Hence, $T_C > T_B$ and therefore $p_C > p_B$.

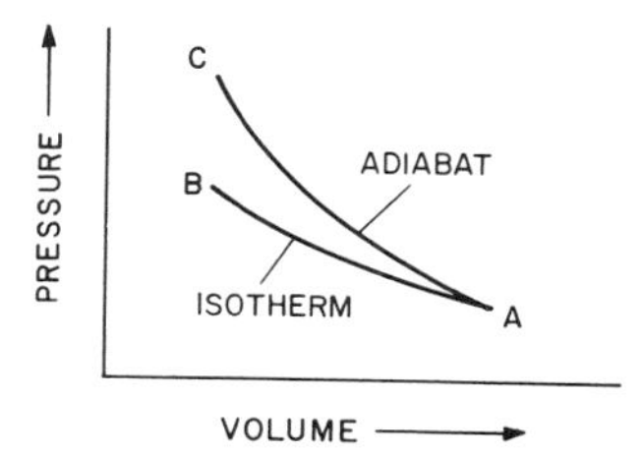

Fig. 2.5 Representation of an isothermal (AB) and an adiabatic (AC) transformation on a p–V diagram.

2.5.1 Concept of an air parcel

In many fluid mechanics problems, mixing is viewed as a result of the random motions of individual molecules. In the atmosphere molecular mixing is important only within a centimeter of the earth's surface and at levels above the turbopause (~105 km). At intermediate levels virtually all the vertical mixing is accomplished by the exchange of well-defined air parcels with horizontal dimensions ranging from a few centimeters to the scale of the earth itself.

In order to gain some insights into the nature of vertical mixing in the atmosphere it is useful to consider the behavior of an air parcel of infinitesimal dimensions that is assumed to be

- thermally insulated from its environment so that its temperature changes adiabatically as it rises or sinks,

- always at exactly the same pressure as the environmental air at the same level,† which is assumed to be in hydrostatic equilibrium, and
- moving slowly enough that its kinetic energy is a negligible fraction of its total energy.

Although in the case of real air parcels one or more of these assumptions is nearly always violated to some extent, this simple, idealized model is helpful in understanding some of the physical processes that influence the distribution of vertical motions and vertical mixing in the atmosphere.

2.5.2 The adiabatic lapse rate

We will now derive an expression for the rate of change of temperature with height of a parcel of dry air which moves about in the earth's atmosphere while always satisfying the conditions listed above. Since the air parcel undergoes only adiabatic transformations ($dq = 0$), and the atmosphere is in hydrostatic equilibrium, for a unit mass of air in the parcel we have, from (2.54),

$$d(c_p T + \Phi) = 0 \tag{2.55}$$

Dividing through by dz and making use of (2.20) we obtain

$$-\left(\frac{dT}{dz}\right)_{\text{dry parcel}} = \frac{g}{c_p} \equiv \Gamma_d \tag{2.56}$$

where Γ_d is called the *dry adiabatic lapse rate*. Since an air parcel expands as it rises in the atmosphere, its temperature will decrease with height so that Γ_d as defined by (2.56) is a positive quantity. Substituting $g = 9.81$ m s^{-2} and $c_p = 1004$ J kg^{-1} deg^{-1} into (2.56) gives $\Gamma_d = 0.0098$ deg m^{-1} or 9.8 deg km^{-1}.

It should be emphasized again that Γ_d is the rate of change of temperature following a parcel of dry air which is being raised or lowered adiabatically. The actual lapse rate of temperature (which we will indicate by Γ) in the atmosphere, as measured by a radiosonde, averages 6–7 deg km^{-1} in the troposphere but it takes on a wide range of values at individual locations.

2.5.3 Potential temperature

The *potential temperature* θ of an air parcel is defined as the temperature which the parcel of air would have if it were expanded or compressed adiabatically from its existing pressure and temperature to a standard pressure p_0 (generally taken as 1000 mb).

We can derive an expression for the potential temperature of air in terms of its pressure p, temperature T, and the standard pressure p_0 as follows. For an

† Any pressure differences between the parcel and its environment give rise to sound waves which produce a rapid adjustment. Temperature differences, on the other hand, are eliminated by much slower processes.

adiabatic transformation ($dq = 0$), (2.49) becomes

$$c_p\,dT - \alpha\,dp = 0$$

Combining this with (2.3) and rearranging terms yields

$$\frac{c_p}{R}\frac{dT}{T} - \frac{dp}{p} = 0$$

Integrating upward from p_0 (where $T = \theta$) to p, we obtain

$$\frac{c_p}{R}\int_\theta^T \frac{dT}{T} = \int_{p_0}^{p} \frac{dp}{p}$$

or

$$\frac{c_p}{R}\ln\frac{T}{\theta} = \ln\frac{p}{p_0}$$

Taking the antilog of both sides,

$$\left(\frac{T}{\theta}\right)^{c_p/R} = \frac{p}{p_0}$$

or

$$\theta = T\left(\frac{p_0}{p}\right)^{R/c_p} \tag{2.57}$$

Equation (2.57) is called *Poisson's*† *equation.* For dry air, $R = R_d = 287$ J deg^{-1} kg^{-1} and $c_p = 1004$ J deg^{-1} kg^{-1}; therefore, $R/c_p = 0.286$.

If an air parcel is subjected to only adiabatic transformations as it moves through the atmosphere, its potential temperature remains constant. Parameters which remain constant during certain transformations are said to be *conserved.* Thus, potential temperature is a conservative quantity for adiabatic transformations. Potential temperature is an extremely useful parameter in atmospheric thermodynamics, since atmospheric processes are often close to adiabatic, and therefore θ remains essentially constant (see Sections 8.1.2 and 8.7.1).

2.5.4 The pseudoadiabatic chart

Poisson's equation may be conveniently solved in graphical form. If pressure is plotted on a distorted scale, in which the distance from the origin is proportional to p^{R_d/c_p} or $p^{0.286}$, (2.57) becomes

$$T = (\text{const})\theta p^{0.286} \tag{2.58}$$

Each value of θ represents a dry adiabat which is defined by a straight line with a particular slope which passes through the point $p = 0$, $T = 0$. If the pressure

† **Simeon Denis Poisson** (1781–1840) French mathematician. Studied medicine but turned to applied mathematics and became the first professor of mechanics at the Sorbonne.

scale is inverted so that p increases downward, the relation takes the form shown in Fig. 2.6, which is the basis for the *pseudoadiabatic chart* frequently used in meteorological computations. The region of the chart of greatest interest in the atmosphere is the portion shown within the dotted lines in Fig. 2.6, and this is generally the only portion printed. A copy of the normally printed pseudo-adiabatic chart (but extending only to 100 mb) is shown on the back endpapers of this book. The sloping heavy black lines are the *dry adiabats* (that is, lines of constant potential temperature), some of which have the potential temperatures (in degrees Kelvin) printed over them. It should be noted that, as required by the definition of potential temperature, the actual temperature of the air (given on the abscissa) at 1000 mb is equal to its potential temperature. Several other sets of lines are printed on the pseudoadiabatic chart, some of which will be discussed later in this chapter.

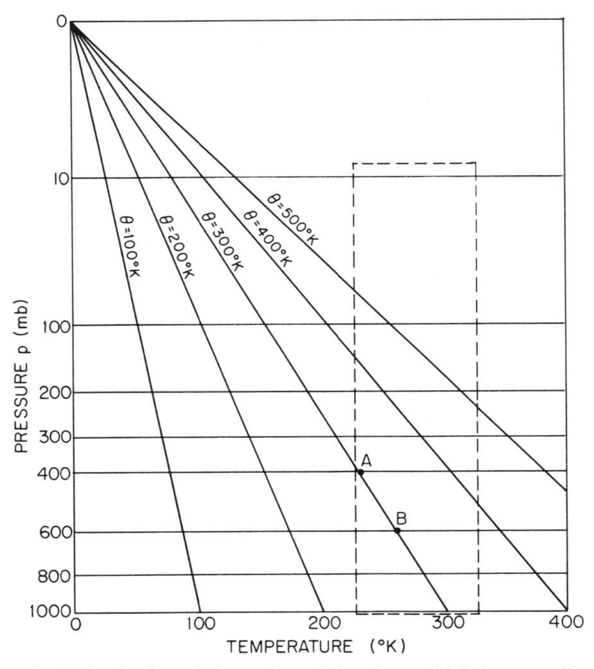

Fig. 2.6 The pseudoadiabatic chart. The region of the chart which is generally printed and used in meteorological computations is enclosed by the dotted lines. A portion of this chart is shown on the back endpapers of this book.

Problem 2.6 A parcel of air has a temperature of 230°K at the 400-mb level. What is its potential temperature? What temperature will the parcel have if it descends adiabatically to the 600-mb level?

Solution This problem is solved with reference to Fig. 2.6 but the student should follow the same steps on the pseudoadiabatic chart (back endpapers).

The original state of the air parcel is indicated by point A on Fig. 2.6 where its pressure is 400 mb and temperature (on the abscissa) 230°K or −43°C. The value of the dry adiabat which passes through point A is 300°K, which is therefore the potential temperature of the air. If an adiabatic transformation occurs to a pressure of 600 mb, the air along the dry adiabat with potential temperature 300°K descends to point B on Fig. 2.6 where the pressure is 600 mb and its new temperature (read off the abscissa) is −13°C or 260°K.

2.6 WATER VAPOR IN THE AIR

So far we have indicated the presence of water vapor in the air through the vapor pressure e which it exerts and we have allowed for its effect on the density of air by introducing the concept of a virtual temperature. However, the amount of water vapor present in a certain quantity of air may be expressed in many different ways; some of the more important of these are considered below.

2.6.1 Moisture parameters

Mixing ratio

The amount of water vapor in a certain volume of air may be defined as the ratio of the mass m_v of water vapor to the mass m_d of dry air; this is called the *mixing ratio* w. That is,

$$w \equiv \frac{m_v}{m_d} \tag{2.59}$$

The mixing ratio is generally expressed in grams of water vapor per kilogram of air. In the atmosphere the magnitude of w is typically a few grams per kilogram in middle latitudes but in the tropics it can reach values of 20 g kg^{-1}. It should be noted that if neither condensation nor evaporation take place, the mixing ratio of an air parcel is a conservative quantity.

Problem 2.7 If air contains water vapor with a mixing ratio of 5.5 g kg^{-1} and the total pressure is 1026.8 mb, calculate the vapor pressure e.

Solution The partial pressure exerted by any constituent in a mixture of gases is proportional to the number of kilomoles of the constituent in the mixture. Therefore, the pressure e due to water vapor in the air is given by

$$e = \frac{m_v/M_w}{m_d/M_d + m_v/M_w} p \tag{2.60}$$

where m_v and m_d are the masses of water vapor and dry air in the mixture, M_w the molecular weight of water, M_d the apparent molecular weight of dry air, and p the total pressure of the moist air. From (2.59) and (2.60) we obtain

$$e = \frac{w}{w + \varepsilon} p \tag{2.61}$$

where $\varepsilon = 0.622$ as defined in (2.14). Substituting $p = 1026.8$ mb and $w = 5.5 \times 10^{-3}$ kg kg^{-1} into (2.61), we obtain $e = 9$ mb.

Problem 2.8 Calculate the virtual temperature of moist air at 30°C which has a mixing ratio of 20 g kg^{-1}.

Solution From (2.17) and (2.61)

$$T_v = T\frac{\varepsilon + w}{\varepsilon(1 + w)}$$

Dividing the denominator into the numerator in this expression and neglecting terms in w^2 and higher orders, we obtain

$$T_v = T\left(1 + \frac{1 - \varepsilon}{\varepsilon} w\right)$$

or, substituting $\varepsilon = 0.622$,

$$T_v = T(1 + 0.61w) \tag{2.62}$$

With $T = 303°K$ and $w = 20 \times 10^{-3}$ kg kg^{-1}, (2.62) gives $T_v \simeq 306.69°K$ or 33.69°C.

Saturation vapor pressures

Consider a small box containing air at temperature T°K and let the floor of the box be covered with pure water. If the air in the box is initially dry, the water will evaporate and the water vapor pressure in the air will increase. Eventually an equilibrium state will be reached where the rate of evaporation of molecules from the water will be equal to their rate of condensation on the water from the moist air. When this condition is realized the air is said to be *saturated with respect to a plane surface of pure water*, and the pressure e_s exerted by the water vapor is called the *saturation vapor pressure over a plane surface of pure water*. Similarly, air which is in an equilibrium state with respect to a plane surface of ice is said to be saturated with respect to ice and the pressure e_{si} exerted by the water vapor is called *the saturation vapor pressure over a plane ice surface.*

The magnitudes of the saturation vapor pressures depend only on temperature and they both increase rapidly with increasing temperature. The variations with temperature of e_s and $e_s - e_{si}$ are shown in Fig. 2.7. It can be seen that $e_s > e_{si}$ at all temperatures (because water evaporates more readily than ice) and that the magnitude of $e_s - e_{si}$ reaches a peak value at about -12°C. It follows that if an ice particle is in water-saturated air it will grow due to the deposition of water vapor upon it. In Chapter 4 we will see that this mechanism is important in the growth of precipitable particles in clouds.

Saturation mixing ratios

The *saturation mixing ratio* w_s with respect to water is defined as the ratio of the mass m_{vs} of water vapor in a given volume of air saturated with respect to a plane surface of water to the mass m_d of the dry air. That is,

$$w_s = \frac{m_{vs}}{m_d} \tag{2.63}$$

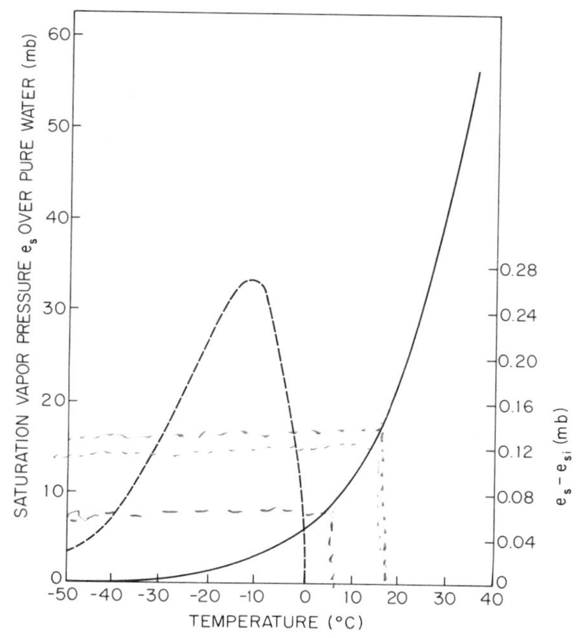

Fig. 2.7 Variations with temperature of the saturation vapor pressure e_s over a plane surface of pure water (——) and the difference between e_s and the saturation vapor pressure over a plane surface of ice e_{si} (– – –).

Since water vapor and dry air both obey the ideal gas equation,

$$w_s = \frac{\rho'_{vs}}{\rho_d'} = \frac{e_s/(R_v T)}{(p - e_s)/(R_d T)} \tag{2.64}$$

where ρ'_{vs} is the partial density of water vapor required to saturate air with respect to water at temperature T, ρ_d' the partial density of the dry air (see Section 2.1.1), and p the total pressure. Combining (2.64) with (2.14), we obtain

$$w_s = 0.622 \frac{e_s}{p - e_s}$$

For the range of temperatures observed in the earth's atmosphere, $p \gg e_s$; therefore,

$$w_s \simeq 0.622 \frac{e_s}{p} \tag{2.65}$$

That is, at a given temperature the saturation mixing ratio is inversely proportional to the total pressure.

Since e_s depends only on temperature, it follows from (2.65) that w_s is a function of temperature and pressure. Hence, w_s can be plotted as a function of state on a pseudoadiabatic chart. Lines of constant saturation mixing ratio with respect to water are printed on the pseudoadiabatic chart and are labeled with the value of w_s in grams of water vapor per kilogram of dry air. These are the sloping, unbroken, overprinted, red lines on the pseudoadiabatic chart (back endpapers). It can be seen that at constant pressure, w_s increases with increasing temperature and at constant temperature it increases with decreasing pressure.

Relative humidity; dew point and frost point

The relative humidity (RH) with respect to water is the ratio (expressed as a percentage) of the actual mixing ratio w to the saturation mixing ratio w_s with respect to water at the same temperature and pressure. That is,

$$\mathrm{RH} \equiv 100 \frac{w}{w_s} \tag{2.66}$$

The *dew point* T_d is the temperature to which air must be cooled at constant pressure in order for it to become saturated with respect to a plane surface of water. In other words, the dew point is the temperature at which the saturation mixing ratio w_s with respect to water becomes equal to the actual mixing ratio w. It follows that the humidity at temperature T and pressure p is given by

$$\mathrm{RH} = 100 \frac{w_s \text{ (at temperature } T_d \text{ and pressure } p)}{w_s \text{ (at temperature } T \text{ and pressure } p)} \tag{2.67}$$

The *frost point* is defined as the temperature to which air must be cooled at constant pressure in order to saturate it with respect to a plane surface of ice. Saturation mixing ratios and relative humidities with respect to ice may be defined in analogous ways to the corresponding definitions with respect to water. When the terms mixing ratio and relative humidity are used without qualification they are with respect to water.

Problem 2.9 Air at 1000 mb and 18°C has a mixing ratio of 6 g kg^{-1}. What is its relative humidity and dew point?

Solution This problem may be solved using the pseudoadiabatic chart (back endpapers). The students should duplicate the following steps. First locate the point with pressure 1000 mb and temperature 18°C. We see from the chart that the saturation mixing ratio for this state is about 12.9 g kg^{-1}. Since the air specified in the problem only has a mixing ratio of 6 g kg^{-1}, in this state it is unsaturated and its relative humidity is, from (2.66), 100 × 6/12.9 = 46.5%. To find the dew point we move from right to left along the 1000-mb ordinate until we intercept the saturation mixing ratio line of magnitude 6 g kg^{-1}; this occurs at an abscissa value of 6.4°C. Therefore, if the air is cooled at constant pressure, the water vapor it contains will just saturate it with respect to water at a temperature of 6.4°C. Therefore, by definition, the dew point of the air is 6.4°C.

At the earth's surface, the pressure varies only slightly. Therefore, the dew point is a good indicator of the moisture content of the air. In warm, humid weather it is also a convenient indicator of the level of human discomfort. For example, most people begin to feel uncomfortable when the dew point rises above 20°C, and air with a dew point above about 24°C is generally regarded as extremely humid or "sticky." Fortunately, dew points much above this temperature are rarely observed even in the tropics. In contrast to the dew point, relative humidity depends as much upon the temperature of the air as upon its moisture content. On a sunny day the relative humidity may drop by as much as 50% from morning to afternoon, just because of the rise in air temperature. Neither is relative humidity a good indicator of the level of human discomfort. For example, a relative humidity of 70% may feel quite comfortable at a temperature of 20°C, but it would cause considerable discomfort to most people at a temperature of 30°C.

Lifting condensation level

The *lifting condensation level* is defined as the level to which a parcel of moist air can be lifted adiabatically before it becomes saturated with respect to a plane surface of water. During lifting the mixing ratio w of the air and its potential temperature θ remain constant, but the saturation mixing ratio w_s decreases until it becomes equal to w at the lifting condensation level. Therefore, the lifting condensation level is located at the intersection of the potential temperature line passing through the temperature T and pressure p of the parcel of air and the w_s line which passes through the pressure p and dew point T_d of the air parcel (Fig. 2.8). Since the dew point and lifting condensation level are related in the manner indicated in Fig. 2.8, knowledge of either one is sufficient to determine the other. Similarly, a knowledge of T and p and any one moisture parameter is sufficient to determine all the other moisture parameters that we have defined.

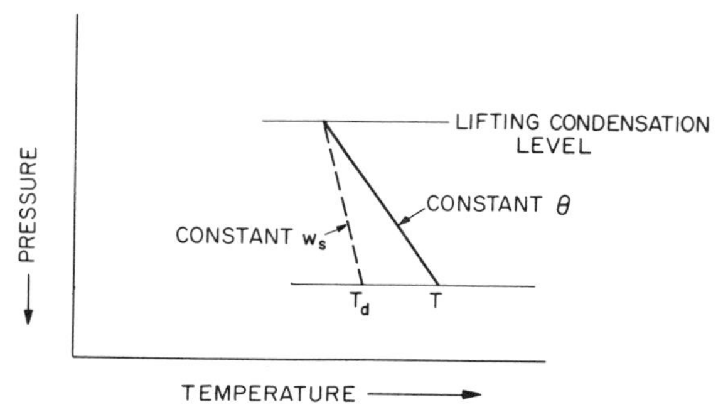

Fig. 2.8 Location of the lifting condensation level of a parcel of air at pressure p and temperature T and dew point T_d on a pseudoadiabatic chart.

Wet-bulb temperature

Another important moisture parameter is the wet-bulb temperature T_w which is defined as the temperature to which a parcel of air is cooled by evaporating water into it at constant pressure until the air is saturated with respect to a plane surface of liquid water. The wet-bulb temperature is measured directly with a thermometer, the bulb of which is covered with a moist cloth over which air is drawn. The heat required to evaporate water from the bulb is supplied by the cooling of the air which comes into contact with it. When this air becomes saturated with water, the temperature of the wet bulb reaches a steady value and the reading is taken. It should be noted that an evaporating cloud droplet or raindrop will be at the wet-bulb temperature.

The definition of the wet-bulb temperature is rather similar to that of the dew point but there is a distinct difference. If the unsaturated air approaching the wet-bulb has a mixing ratio w, the dew point T_d is the temperature to which the air must be cooled at constant pressure in order to become saturated. The air which leaves the wet-bulb has a mixing ratio w' which saturates it at temperature T_w. If the air approaching the wet-bulb is unsaturated, w' is greater then w; therefore, T_w is greater than T_d. Therefore, $T_d \leq T_w \leq T$, where the equality signs apply only under saturated conditions.

2.6.2 Saturated-adiabatic and pseudoadiabatic processes

When a parcel of air rises in the atmosphere its temperature decreases with altitude at the dry adiabatic lapse rate (see Section 2.5) until the air becomes saturated with water vapor. Further lifting results in the condensation of liquid water (or deposition of ice) which releases latent heat. Consequently, the rate of decrease in the temperature of the rising parcel becomes less. If all of the condensation products remain in the rising parcel, the process may still be considered to be adiabatic (and reversible), even though latent heat is released in the system, provided that heat does not pass through the boundaries of the parcel. The air is then said to undergo a *saturated-adiabatic process*. If, on the other hand, all of the condensation products immediately fall out of the parcel of air, the process is irreversible and also not strictly adiabatic; it is referred to as *pseudoadiabatic*. However, as the reader is asked to show in Problem 2.45, the amount of heat carried by the condensation products is small compared to that carried by the air. Therefore, the saturated-adiabatic lapse rate is essentially the same as the pseudoadiabatic lapse rate.

2.6.3 The saturated-adiabatic lapse rate

We will now derive an expression for the rate of change in temperature with height of a parcel of air undergoing a saturated adiabatic process.

Substitution of (2.20) into (2.54) yields

$$dq = c_p\, dT + g\, dz \tag{2.68}$$

Now, if the saturation mixing ratio of the air with respect to water is w_s, the quantity of heat dq released into (or absorbed from) a unit mass of dry air due to condensation (or evaporation) of liquid water is $-L\, dw_s$, where L is the latent heat of condensation. Therefore,

$$-L\,dw_s = c_p\,dT + g\,dz \tag{2.69}$$

If we neglect the small amounts of water vapor associated with a unit mass of dry air which are also warmed (or cooled) by the release (or absorption) of the latent heat, then c_p in (2.69) is the specific heat at constant pressure of dry air. Dividing both sides of (2.69) by $c_p\,dz$ and rearranging terms, we obtain

$$\frac{dT}{dz} = -\frac{L}{c_p}\frac{dw_s}{dz} - \frac{g}{c_p}$$

or

$$\frac{dT}{dz} = -\frac{L}{c_p}\frac{dw_s}{dT}\frac{dT}{dz} - \frac{g}{c_p}$$

Therefore, the *saturated adiabatic lapse rate* Γ_s is defined by

$$\Gamma_s \equiv -\frac{dT}{dz} = \frac{\Gamma_d}{1 + (L/c_p)(dw_s/dT)} \tag{2.70}$$

where Γ_d is the dry adiabatic lapse rate which, from (2.56), is equal to g/c_p. It should be noted that all derivatives in the above three equations refer to rates of change following a parcel of saturated air which is being raised or lowered in the atmosphere under adiabatic or pseudoadiabatic conditions.†

The magnitude of Γ_s is not constant but depends on the pressure and temperature. Since dw_s/dT is always positive, it follows from (2.70) that $\Gamma_s < \Gamma_d$. Actual values of Γ_s range from about 4 deg km^{-1} near the ground in warm, humid air masses where dw_s/dT is very large, to typical values of 6–7 deg km^{-1} in the middle troposphere. For typical temperatures near the tropopause, Γ_s is only slightly less than Γ_d because the moisture capacity is so small that the effect of condensation is negligible.

Lines which show the decrease in temperature with height of a parcel of air which is rising or sinking in the atmosphere under saturated adiabatic (or pseudoadiabatic) conditions are called *saturated adiabats* (or pseudoadiabats). On the pseudoadiabatic chart (back endpaper) these are slightly curved, broken, red lines.

Problem 2.10 A parcel of air with an initial temperature of 15°C and dew point 2°C is lifted adiabatically from the 1000-mb level. Determine its lifting condensation level and temperature at that level. If the air parcel is lifted a further 200 mb above its lifting condensation level, what is its final temperature and how much liquid water is condensed during this rise?

† William Thomson (later Lord Kelvin) was the first (in 1862) to derive quantitative estimates of the dry and saturated adiabatic lapse rates based on theoretical arguments. He, in turn, was indebted to James Joule for the key idea that latent heat release must be important in the saturated case.

Solution We solve this problem using the pseudoadiabatic chart (back endpapers). The student should duplicate the following steps on this chart. First locate the initial state of the air on the chart at the intersection of the 15°C abscissa with the 1000-mb ordinate. Since the dew point of the air is 2°C, the magnitude of the saturation mixing ratio line which passes through the 1000-mb pressure level at 2°C is the actual mixing ratio of the air at 15°C and 1000 mb. From the chart this is found to be 4.4 g kg^{-1}. Since the saturation mixing ratio at 1000 mb and 15°C is about 10.7 g kg^{-1}, the air is initially unsaturated. Therefore, when it is lifted it will follow a dry adiabat (that is, a line of constant potential temperature) until it intercepts the saturation mixing ratio line of magnitude 4.4 g kg^{-1}. Following upwards along the dry adiabat (θ = 288°K) which passes through 1000 mb and 15°C, the saturation mixing ratio line of 4.4 g kg^{-1} is intercepted at the 830-mb level. This is the lifting condensation level of the air parcel. The temperature of the air at this point is −0.5°C. For lifting above this level the air parcel will follow a saturated adiabat. Following the saturated adiabat which passes through 830 mb and −0.5°C up to the 630-mb level, the final temperature of the air is found to be about −15°C. The saturation mixing ratio at 630 mb and −15°C is 1.8 g kg^{-1}. Therefore, 4.4 − 1.8 = 2.6 g of water must have condensed out of each kilogram of air during the rise from 830 to 630 mb.

2.6.4 Equivalent potential temperature and wet-bulb potential temperature

Substituting (2.3) into (2.49),

$$\frac{dq}{T} = c_p \frac{dT}{T} - R \frac{dp}{p} \tag{2.71}$$

From (2.57) the potential temperature θ is given by

$$\ln \theta = \ln T - \frac{R}{c_p} \ln p + \text{const}$$

or, differentiating,

$$c_p \frac{d\theta}{\theta} = c_p \frac{dT}{T} - R \frac{dp}{p} \tag{2.72}$$

Combining (2.71) and (2.72) and substituting $dq = -L\, dw_s$, we obtain

$$-\frac{L}{c_p T} dw_s = \frac{d\theta}{\theta} \tag{2.73}$$

The student is asked to show in Problem 2.33 that

$$\frac{L}{c_p T} dw_s \simeq d\left(\frac{Lw_s}{c_p T}\right) \tag{2.74}$$

From (2.73) and (2.74)

$$-\frac{Lw_s}{c_p T} = \ln \theta + \text{const} \tag{2.75}$$

We will define the constant of integration by requiring that at low temperatures, as $w_s/T \to 0$, $\theta \to \theta_e$. Then

$$-\frac{Lw_s}{c_pT} = \ln\left(\frac{\theta}{\theta_e}\right)$$

or

$$\theta_e = \theta \exp\left(\frac{Lw_s}{c_pT}\right) \tag{2.76}$$

The quantity θ_e defined by (2.76) is called the *equivalent potential temperature.* It can be seen that θ_e is the potential temperature θ of a parcel of air when its saturation mixing ratio w_s is zero. Therefore, the equivalent potential temperature may be found as follows. The air is expanded pseudoadiabatically until all the vapor has condensed, released its latent heat, and fallen out. The air is then compressed dry adiabatically to the standard pressure of 1000 mb when it will attain the temperature θ_e. (If the air is initially unsaturated, w_s and T are the saturation mixing ratio and temperature at the point where the air first becomes saturated after being lifted dry adiabatically.) We have seen in Section 2.5 that potential temperature is a conservative quantity for adiabatic transformations. The equivalent potential temperature, however, is conserved during both dry- and saturated-adiabatic processes. Therefore, on pseudoadiabatic charts, saturated adiabats are labeled with a number that gives the equivalent potential temperature (in degrees Kelvin) of air which rises along that adiabat (see chart, back endpapers).

If the line of constant equivalent potential temperature (that is, the pseudoadiabat) that passes through the wet-bulb temperature of a parcel of air is traced back on a pseudoadiabatic chart to the point where it intersects the 1000-mb isobar, the temperature at this intersection is called the *wet-bulb potential temperature* θ_w of the air parcel. Like the equivalent potential temperature, the wet-bulb potential temperature is conserved during both dry- and saturated-adiabatic processes. The wet-bulb potential temperature is therefore valuable as a tracer of air masses (see below).

2.6.5 Normand's rule

The following theorem, known as Normand's[†] rule, is extremely helpful in many computations involving the pseudoadiabatic chart. Normand's rule states that on a pseudoadiabatic chart the lifting condensational level of an air parcel is located at the intersection of the potential temperature line which passes through the point located by the temperature and pressure of the air parcel, the equivalent potential temperature line (that is, pseudoadiabat) which passes through the point located by the wet-bulb temperature and pressure of

[†] **Sir Charles William Blyth Normand** (1889–) British meteorologist. Director-General of Indian Meteorological Service, 1927–1944.

the air parcel, and the saturation mixing ratio line which passes through the point determined by the dew point and pressure of the air. This rule is illustrated in Fig. 2.9 for the case of an air parcel with temperature T, pressure p, dew point T_d, and wet-bulb temperature T_w. It can be seen that, if T, p, and T_d are known, T_w may be readily determined using Normand's rule. Also by extrapolating the θ_e line that passes through T_w back to the 1000-mb level, the wet-bulb potential temperature θ_w may be found (Fig. 2.9).

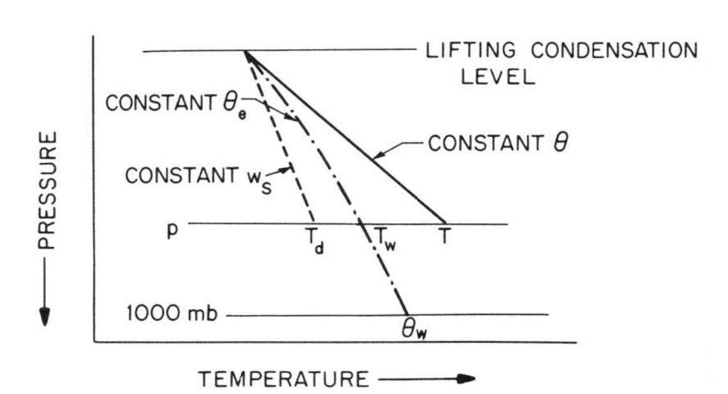

Fig. 2.9 Illustration of Normand's rule. The method for determining the wet-bulb potential temperature θ_w of the air is also illustrated.

2.6.6 Some effects of irreversible condensation processes

When a parcel of air is lifted above its lifting condensation level so that condensation occurs, and if the products of the condensation (that is, precipitation particles) fall out, the latent heat gained by the air during this process will be retained if the parcel returns to its original level. In this sense, such a process is irreversible. The effects of the saturated ascent coupled with the adiabatic descent are

- net increases in the temperature and potential temperature of the parcel,
- a decrease in moisture content (as indicated by changes in the mixing ratio, relative humidity, dew point, or wet-bulb temperature),
- no change in the equivalent potential temperature or wet-bulb potential temperature.

The following problem illustrates these points.

Problem 2.11 An air parcel at 950 mb has a temperature of 14°C and a mixing ratio of 8 g kg^{-1}. What is the wet-bulb potential temperature of the air? The air parcel is lifted to the 700-mb level by passing over a mountain, and 70% of the water vapor that is condensed out by the ascent is removed by precipitation. Determine the temperature, potential temperature, mixing ratio, and wet-bulb potential temperature of the air after it has returned to the 950-mb level on the other side of the mountain.

Solution The problem can be solved by carrying out the following steps on a pseudoadiabatic chart (back endpapers). Locate the initial state of the air at 950 mb and 14°C. The saturation mixing ratio for this state is found from the chart to be 10.6 g kg^{-1}. Therefore, since the air has a mixing ratio of only 8 g kg^{-1}, it is unsaturated. The wet-bulb potential temperature can be determined using the method indicated schematically in Fig. 2.9 which is as follows. Trace the constant potential temperature line (291.5°K) that passes through the initial state of the air up to the point where it intersects the saturation mixing ratio line with value 8 g kg^{-1}. This occurs at a pressure of about 890 mb which is the lifting condensation level (LCL) of the air. Now follow the equivalent potential temperature line that passes through the point back down to the 1000-mb level and read off the temperature at this point on the abscissa—it is 14°C. This is in the wet-bulb potential temperature of the air.

When the air is lifted over the mountain, its temperature and pressure on the pseudo-adiabatic chart are given by points on the potential temperature line with value 291.5°K up to the LCL at 890 mb. With the further ascent to the 700-mb level, the air follows the saturated adiabat that passes through the LCL (it has an equivalent potential temperature of 315°K). This saturated adiabat intersects the 700-mb level at a point where the saturation mixing ratio is 4.7 g kg^{-1}. Therefore, $8 - 4.7 = 3.3$ g kg^{-1} of water vapor has to condense out between the LCL and the 700-mb level, and 70% of this, or 2.3 g kg^{-1}, is precipitated out. Therefore, at the 700-mb level there is 1.0 g kg^{-1} of liquid water in the air. The air descends on the other side of the mountain at the saturated adiabatic lapse rate until it evaporates all of the liquid water, at which point the saturation mixing ratio will have risen to $4.7 + 1.0 = 5.7$ g kg^{-1}. At this point the air is at 760 mb and 1.8°C (with potential temperature 297.5K). Thereafter, it descends along a dry adiabat to the 950-mb level where its temperature is 20°C and the mixing ratio is still 5.7 g kg^{-1}. If the method indicated in Fig. 2.9 is applied again, the wet-bulb potential temperature of the air will be found to be unchanged at 14°C. [Note: The heating of air during its passage over a mountain, 6 deg in this example, is responsible for the remarkable warmth of *Föhn* or *Chinook* winds, which often blow downward along the lee side of mountain ranges.†‡]

2.7 THE CONCEPT OF STATIC STABILITY

2.7.1 Unsaturated air

Consider a layer of the atmosphere in which the actual lapse rate Γ (as measured by a radiosonde) is less than the dry adiabatic lapse rate Γ_d (Fig. 2.10a). If a parcel of unsaturated air originally located at level O is raised slightly to the level defined by points A and B, its temperature will fall to T_A, which is lower than the environmental temperature T_B at this level. Since the parcel immediately adjusts to the pressure of its environment, it is clear from the ideal gas equation that the colder parcel must be denser than the warmer environmental air. Therefore, the parcel tends to return to its original level. If the parcel is displaced downwards from O it becomes warmer than its environment and

† The first person to explain the Föhn wind in this way appears to have been J. von Hann in his classic book "Lehrbuch der Meteorologie," Willibald Keller, Leipzig, 1901.

‡ **Julius F. von Hann** (1839–1921) Austrian meteorologist. Introduced thermodynamic principles into meteorology. Developed theories for mountain and valley winds. Published the first comprehensive treatise on climatology (1883).

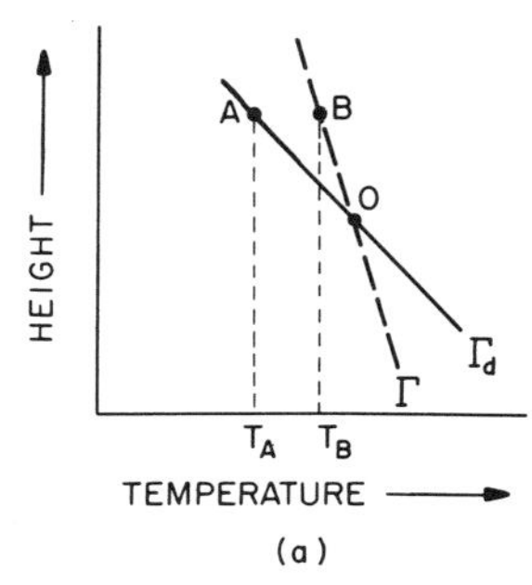

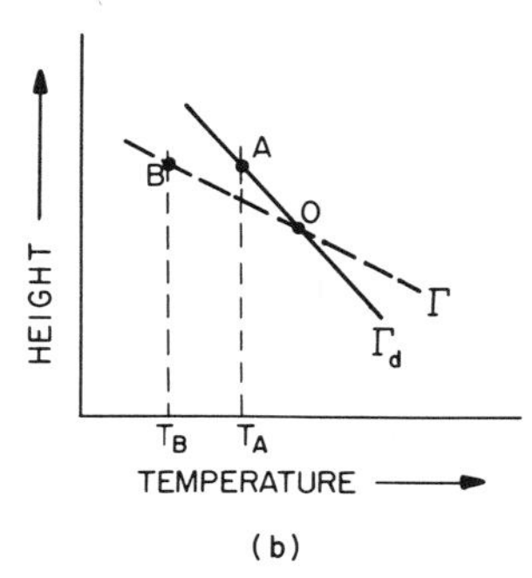

Fig. 2.10 Conditions for (a) positive static stability and (b) negative static stability for unsaturated air. Negative static stability can only exist very close to the ground.

tends to rise back to its original level. In both cases, the parcel of air encounters a restoring force and vertical mixing is inhibited. Thus, the condition $\Gamma < \Gamma_d$ is one of *positive static stability* for unsaturated air. In general, the larger the difference $\Gamma_d - \Gamma$ the greater the restoring force for a given displacement and the greater the static stability.

Problem 2.12 An unsaturated air parcel of unit mass is given a small upward displacement δz from its equilibrium level in a stably stratified atmosphere. Derive an expression for the restoring force that the parcel experiences.

Solution Let ρ' and T' be the density and temperature of the displaced parcel and ρ and T be the density and temperature of the ambient air at the same level. For the environmental air

$$\frac{dp}{dz} = -\rho g$$

Now since dp/dz is the same for the parcel as it is for the environmental air, it follows that if $\rho' > \rho$ the parcel experiences a net downward force

$$(\rho' - \rho)g$$

per unit volume, or

$$\left(\frac{\rho' - \rho}{\rho}\right)g$$

per unit mass. Making use of (2.2) the force can be expressed in the form

$$g\left(\frac{1/T' - 1/T}{1/T}\right) = g\left(\frac{T - T'}{T'}\right) \simeq g\left(\frac{T - T'}{T}\right)$$

provided that δz is small. But $(T - T') = (\Gamma_d - \Gamma)\,\delta z$; therefore, the restoring force can be expressed in the form

$$\frac{g(\Gamma_d - \Gamma)\,\delta z}{T}$$

Layers of air with negative lapse rates (that is, temperatures increasing with height) are called *inversions*. It is clear from the above discussion that these layers are marked by very strong static stability. A low-level inversion can act as a "lid" which traps pollution laden air beneath it (Fig. 2.11). The fact that the stratosphere corresponds to a deep inversion layer explains its strongly layered structure. The effects of stability on clouds will be discussed in Chapter 5.

The condition $\Gamma = \Gamma_d$ corresponds to zero or *neutral stability*. When unsaturated air parcels are displaced vertically in such a layer, they warm (or cool) at exactly the same rate as the environment and therefore they do not encounter any restoring force.

If $\Gamma > \Gamma_d$ (Fig. 2.10b), a parcel of unsaturated air displaced upward from O will find itself at A with a temperature greater than that of its environment.

Fig. 2.11 Photograph of Mt. Rainier (altitude 4.4 km) taken from an aircraft. A temperature inversion has confined most of the natural and man-made aerosol from the earth's surface to below the level of the inversion. Consequently, the visibility below the inversion is poor while above the inversion it is good. The accumulation of water at the inversion has also produced a layer of stratocumulus cloud at this level. (Photo: L. F. Radke.)

Therefore, it will be less dense than the environmental air and will continue to rise. Similarly, if the parcel is displaced downwards it will be cooler than its environment and continue to sink. Such *unstable* situations generally do not persist in the free atmosphere, since the instability is eliminated by strong vertical mixing as fast as it forms. The only exception is in the layer just above the ground under conditions of very strong heating from below (see Section 7.3.2).

Problem 2.13 Show that if the potential temperature θ increases with increasing altitude the atmosphere has positive static stability with respect to dry air.

Solution Combining (2.69) and (2.73), we obtain

$$c_p T \frac{d\theta}{\theta} = c_p \, dT + g \, dz$$

Dividing through by $c_p T \, dz$ yields

$$\frac{1}{\theta} \frac{d\theta}{dz} = \frac{1}{T} \left(\frac{dT}{dz} + \frac{g}{c_p} \right) \tag{2.77}$$

Now, $-dT/dz$ is the actual lapse rate Γ of the air and the dry adiabatic lapse rate Γ_d is g/c_p; therefore, (2.77) may be written as

$$\frac{1}{\theta} \frac{d\theta}{dz} = \frac{1}{T} (\Gamma_d - \Gamma) \tag{2.78}$$

However, we have shown above that when $\Gamma < \Gamma_d$ the air is characterized by positive static stability. It follows that under these same conditions $d\theta/dz$ is positive; that is, the potential temperature increases with height.

2.7.2 Saturated air

If a parcel of air is saturated its temperature will decrease with height at the saturated adiabatic lapse rate Γ_s. It follows from arguments similar to those given above that if Γ is the actual lapse rate of temperature in the atmosphere, saturated air will be stable, neutral, or unstable with respect to vertical displacements, depending on whether $\Gamma < \Gamma_s$, $\Gamma = \Gamma_s$, or $\Gamma > \Gamma_s$, respectively. It should be noted that when a radiosonde sounding is plotted on a pseudo-adiabatic chart the relative magnitudes of Γ, Γ_d, and Γ_s can be seen quite readily (see Problem 2.51).

2.7.3 Conditional and convective instability

If the actual lapse rate Γ of the atmosphere lies between the saturated adiabatic lapse rate Γ_s and the dry adiabatic lapse rate Γ_d, a parcel of air that is lifted sufficiently far above its equilibrium level will become warmer than its environment. This situation is illustrated in Fig. 2.12, where a parcel lifted from its equilibrium level at O cools dry adiabatically until it reaches its lifting condensation level at A. At this level it is colder than the environmental temperature.

Further lifting produces cooling at the moist adiabatic lapse rate, so the temperature of the parcel of air follows the moist adiabat ABC. If the air parcel is sufficiently moist, the moist adiabat through A will cross the environmental sounding; the point of intersection is shown as B in Fig. 2.12. Up to this point the parcel was colder and denser than its environment, and an expenditure of energy was required in order to lift it. If lifting had stopped prior to this point, the parcel would have returned to its equilibrium level. Once above point B, the parcel develops a positive buoyancy which carries it upward even in the absence of further lifting. For this reason, B is referred to as the *level of free convection.* The level of free convection depends on the amount of moisture in the rising parcel of air as well as the magnitude of the lapse rate Γ.

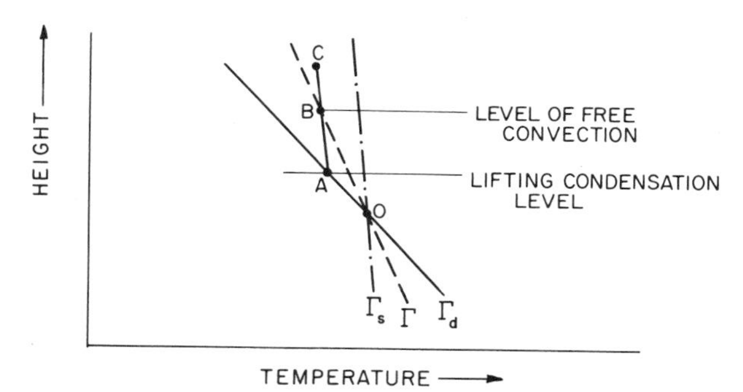

Fig. 2.12 Conditions for conditional instability. Γ_s and Γ_d are the saturated and dry adiabatic lapse rates and Γ the actual lapse rate of temperature of the air.

From the above discussion it is clear that, when $\Gamma_s < \Gamma < \Gamma_d$, there is the possibility of gravitational instability if vertical motions are large enough to lift air parcels beyond their level of free convection. Such an atmosphere is said to be *conditionally unstable.* If vertical motions are weak, this type of stratification can be maintained indefinitely.

The potential for gravitational instability is also related to the vertical stratification of water vapor. In the profile shown in Fig. 2.13, the mixing ratio decreases rapidly with height within the inversion layer AB which marks the top of a moist layer. Now suppose that this layer is lifted. An air parcel at A will reach its lifting condensation level almost immediately, and beyond that point it will cool moist adiabatically. On the other hand, an air parcel starting at point B will cool dry adiabatically through a deep layer before it reaches its lifting condensation level. Therefore, as the inversion layer is lifted, the top part of it cools much more rapidly than the bottom part and the lapse rate quickly becomes destabilized. Sufficient lifting may cause the layer to become conditionally unstable, even if the entire sounding is absolutely stable to begin with. The criterion for this so-called *convective* (or *potential*) *instability* is that $d\theta_e/dz$ be negative within the layer.

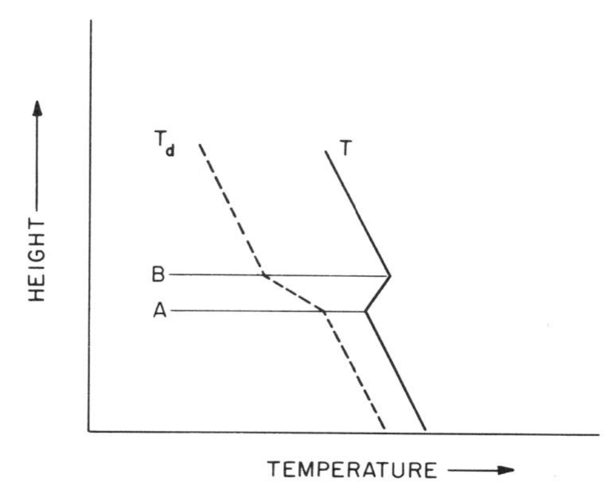

Fig. 2.13 Conditions for convective instability. T and T_d are the temperature and dew point of the air.

Throughout virtually the entire tropics the lapse rate is conditionally unstable up to about 15 km, and convectively unstable up to about 6 km as well. Yet deep convection breaks out only within a few percent of the area where there is sufficient lifting to release the instability. The greatest potential for violent convection develops in geographical regions which are subject to incursions of warm, humid air at low levels while dry, conditionally unstable air prevails aloft. For example, in the southern Great Plains of the United States during summer, a southerly flow of warm, very moist air from the Gulf of Mexico is often capped by a sharp inversion which marks the transition to a dry, westerly current aloft. With sufficient lifting, this situation can result in violent convective activity which manifests itself in the form of severe thunderstorms which are often accompanied by hail, damaging winds, and tornadoes (see Section 5.3).

2.8 THE SECOND LAW OF THERMODYNAMICS AND ENTROPY†

The First Law of Thermodynamics (Section 2.3) is a statement of the principle of conservation of energy. The Second Law of Thermodynamics, which was deduced in various forms by Carnot,‡ Clausius,§ and Kelvin, is concerned with the maximum fraction of a quantity of heat that can be converted into useful work. The fact that for any given system there is a theoretical limit to this conversion factor was first clearly demonstrated by Carnot, who also introduced the important concepts of cyclic and reversible processes.

† The remaining sections of this chapter can be omitted in an introductory undergraduate course.

‡ **Nicholas Leonard Sadi Carnot** (1796–1832) Born in Luxenbourg. Admitted to the Ecole Polytechnique at age 16. Became a captain in the Corps of Engineers. Founded the science of thermodynamics. Died of cholera at age 36.

§ **Rudolf Clausius** (1822–1888) German physicist. Contributed to the sciences of thermodynamics, optics, and electricity.

2.8.1 The Carnot cycle

A *cyclic* process is a series of changes in the state of a substance (called the *working substance*) in which its volume changes and it does external work, subject to the conditions that the working substance returns to its initial conditions. It should be noted that in a cyclic process the initial and final states of the working substance are the same. Therefore, the internal energy of the working substance remains unchanged and, from (2.36), the *net* heat absorbed by the working substance is equal to the external work that it does in the cycle. A working substance is said to undergo a *reversible* transformation if each state of the system is in equilibrium, so that a reversal in the direction of an infinitesimal change returns the working substance and the environment to their original states. If during one cycle of an engine a quantity of heat Q_1 is absorbed and heat Q_2 is rejected, the amount of mechanical work done by the engine is $Q_1 - Q_2$ and its *efficiency* η is defined as

$$\eta \equiv \frac{\text{Mechanical work done by engine}}{\text{Heat absorbed by the working substance}} = \frac{Q_1 - Q_2}{Q_1} \tag{2.79}$$

The efficiency of an engine is also given by

$$\eta = \frac{q_1 - q_2}{q_1} \tag{2.80}$$

where q_1 and q_2 are the heats absorbed and rejected by a unit mass of the working substance during one cycle of the engine.

Carnot was concerned with the important practical problem of the efficiency with which heat engines can do mechanical work. He envisaged an ideal heat engine (Fig. 2.14) consisting of working substance contained in a cylinder (Y) with insulating walls and a conducting base (B) which is fitted with an insulated, frictionless piston (P) to which a variable load may be applied, a nonconducting stand (S) on which the cylinder may be placed in order to insulate its base, an infinite warm reservoir of heat (H) at constant temperature T_1, and an infinite

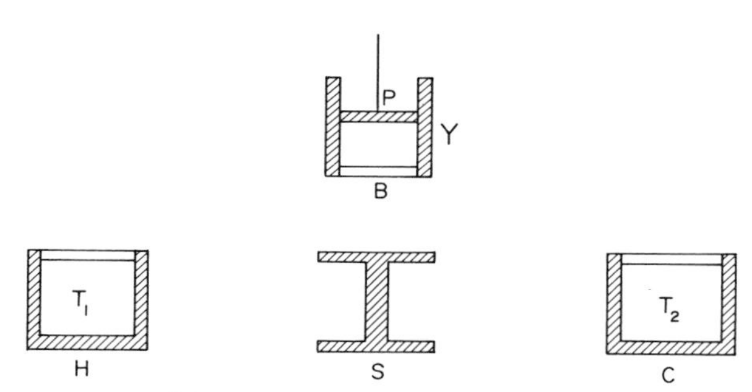

Fig. 2.14 The components of Carnot's ideal heat engine. Shaded and unshaded areas indicate insulating and conducting material, respectively.

cold reservoir for heat (C) at constant temperature T_2 (where $T_1 > T_2$). Heat is supplied to the working substance contained in the cylinder by the warm reservoir, and extracted from it by the cold reservoir. As the working substance expands (or contracts) the cylinder moves outward (or inward) and external work is done by (or on) the engine.

Carnot's cycle consists of taking the working substance in the cylinder through the following four operations which together constitute a reversible, cyclic transformation:

(i) The substance starts at a condition represented by A on the $p-V$ diagram in Fig. 2.15 with temperature T_2. The cylinder is placed on the stand S and the piston pushed in. Since heat can neither enter nor leave the working substance in the cylinder when it is on the stand, the working substance undergoes an adiabatic compression to the state represented by B in Fig. 2.15 in which its temperature has risen to T_1.

(ii) The cylinder is now placed on the warm reservoir H from which it extracts a quantity of heat Q_1. During this process the working substance expands isothermally at temperature T_1 to point C in Fig. 2.15.

(iii) The cylinder is returned to the nonconducting stand and the working substance undergoes an adiabatic expansion along CD in Fig. 2.15 until its temperature falls to T_2.

(iv) Finally, the cylinder is placed on the cold reservoir and the working substance is compressed isothermally along DA back to its original state A. In this transformation the working substance gives up a quantity of heat Q_2 to the cold reservoir.

It follows from (2.39) that the net amount of work done by the working substance during the Carnot cycle is equal to the area contained within the figure ABCD in Fig. 2.15. Also, since the working substance is returned to its original state, the net work done is equal to $Q_1 - Q_2$ and the efficiency of the engine is given by (2.79). In this cyclic operation the engine has done work by transferring a certain quantity of heat from a warmer (H) to a cooler (C) body. The Second

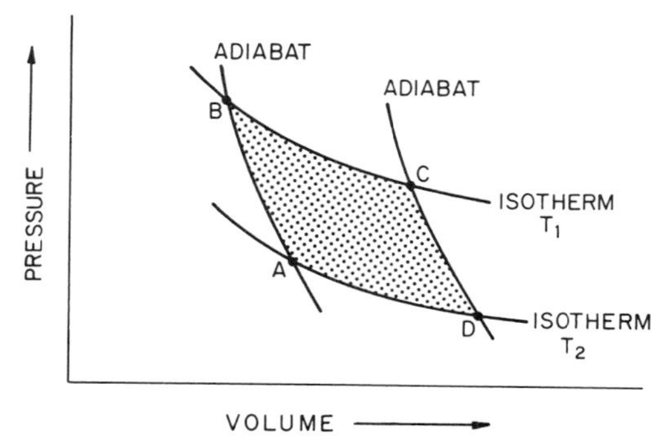

Fig. 2.15 Representation of a Carnot cycle on a $p-V$ diagram.

Law of Thermodynamics states that it is only by transferring heat from a warmer to a colder body that heat can be converted into work in a cyclic process. In Problem 2.36 the student is invited to prove that *no engine can be more efficient than a reversible engine working between the same limits of temperature and all reversible engines working between the same temperature limits have the same efficiency*. The validity of these two statements, which are known as *Carnot's Theorems*, depends on the truth of the Second Law of Thermodynamics.

Problem 2.14 Show that in a Carnot cycle the ratio of heat Q_1 absorbed from the warm reservoir at temperature T_1°K to the heat Q_2 rejected to the cold reservoir at temperature T_2°K is equal to T_1/T_2.

Solution To prove this important relationship we let the substance in the Carnot engine be 1 kmol of an ideal gas and we take it through the Carnot cycle indicated in Fig. 2.15.

For the adiabatic transformation from A to B we have (using the adiabatic equation which the reader is asked to prove in Problem 2.23)

$$p_A V_A^{\gamma} = p_B V_B^{\gamma}$$

where γ is the ratio of the specific heat at constant pressure to the specific heat at constant volume. For the isothermal transformation from B to C

$$p_B V_B = p_C V_C$$

The transformation from C to D is adiabatic; therefore, from the adiabatic equation,

$$p_C V_C^{\gamma} = p_D V_D^{\gamma}$$

and, for the isothermal change from D to A,

$$p_D V_D = p_A V_A$$

Combining these four equations gives

$$\frac{V_C}{V_B} = \frac{V_D}{V_A} \qquad (2.81)$$

Consider now the heats absorbed and rejected by the ideal gas. In passing from state B to C, heat Q_1 is absorbed from the warm reservoir and

$$Q_1 = \int_B^C p\, dV$$

or

$$Q_1 = \int_B^C \frac{R^* T_1}{V}\, dV$$

therefore,

$$Q_1 = R^* T_1 \ln\left(\frac{V_C}{V_B}\right) \qquad (2.82)$$

Similarly, the heat Q_2 rejected to the cold reservoir in the isothermal transformation from from D to A is given by

$$Q_2 = R^* T_2 \ln\left(\frac{V_D}{V_A}\right) \qquad (2.83)$$

From (2.82) and (2.83)

$$\frac{Q_1}{Q_2} = \frac{T_1 \ln(V_C/V_B)}{T_2 \ln(V_D/V_A)} \tag{2.84}$$

Therefore, from (2.81) and (2.84),

$$\frac{Q_1}{Q_2} = \frac{T_1}{T_2} \tag{2.85}$$

A practical example of a heat engine is the steam engine in which the warm reservoir is the boiler, the cold reservoir the condenser, and water in liquid and vapor forms is the working substance which expands after it has absorbed heat and thereby moves pistons and/or turbine blades and does work.

Carnot's cycle can be reversed in the following way. Starting from point A in Fig. 2.15, the material in the cylinder may be expanded at constant temperature until the state represented by point D is reached. During this process a quantity of heat Q_2 is *absorbed* from the cold reservoir. An adiabatic expansion takes the substance from state D to C. The substance is then compressed from state C to state B during which transformation a quantity of heat Q_1 is *rejected* to the warm reservoir. Finally, the substance is expanded adiabatically from state B to state A.

In this reversed cycle Carnot's ideal engine serves as a *refrigerating machine*, for a quantity of heat Q_2 is taken from a cold body (the cold reservoir) and heat $Q_1 (Q_1 > Q_2)$ is given to a hot body (the warm reservoir). In order to accomplish this transfer of heat a quantity of mechanical work equivalent to $Q_1 - Q_2$ has been expended by some outside agency (for example, an electric motor) to drive the refrigerator. We are therefore led to an alternative statement of the Second Law of Thermodynamics: namely, "heat cannot of itself (that is, without the performance of work by some external agency) pass from a cold to a warm body."

2.8.2 Entropy

We have seen that isotherms are distinguished from each other by differences in temperature and that a dry adiabat can be characterized by its potential temperature. However, there is another way of measuring the differences between adiabats. Consider the three adiabats A_1, A_2, and A_3 on the p–V diagram shown in Fig. 2.16. In passing reversibly from one adiabat to another along an isotherm (for example, in one operation of a Carnot cycle) heat is absorbed or rejected, where the amount of heat Q_{rev} depends on the temperature T of the isotherm (the subscript rev indicates that the heat is exchanged reversibly). Moreover, it follows from (2.85) that the ratio Q_{rev}/T is the same no matter which isotherm is chosen in passing from one adiabat to another. Therefore, the ratio Q_{rev}/T is a measure of the difference between the two adiabats; it is called the difference in *entropy* (S). More precisely, we may define

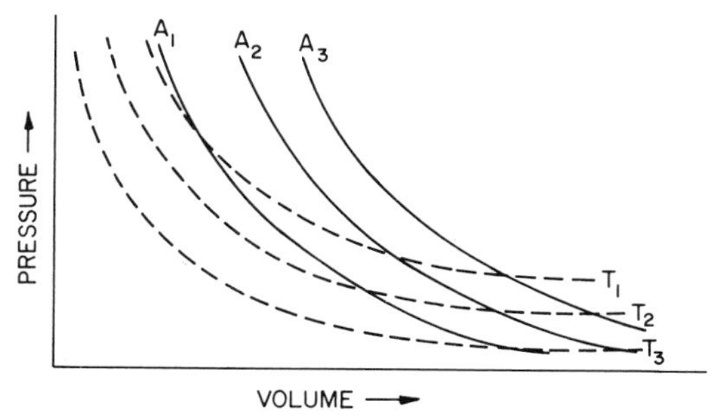

Fig. 2.16 Isotherms (---) and adiabats (——) on a p–V diagram.

the increase in the entropy dS of a substance as

$$dS = \frac{dQ_{\text{rev}}}{T} \tag{2.86}$$

where dQ_{rev} is the quantity of heat which is added reversibly to the substance at temperature T. For a unit mass of the substance,

$$ds = \frac{dq_{\text{rev}}}{T} \tag{2.87}$$

Entropy is a function only of the state of a substance and not the path by which the substance is brought to that state. We see from (2.41) and (2.87) that the First Law of Thermodynamics for a reversible transformation may be written as

$$T\,ds = du + p\,d\alpha \tag{2.88}$$

It should be noted that in this form the First Law contains only functions of state; therefore, all three differentials are perfect.

When a substance passes from state 1 to state 2, the change in entropy of a unit mass is

$$s_2 - s_1 = \int_1^2 \frac{dq_{\text{rev}}}{T} \tag{2.89}$$

Combining (2.71) and (2.72) we obtain

$$\frac{dq}{T} = c_p \frac{d\theta}{\theta}$$

Therefore, since the processes leading to (2.71) and (2.72) were reversible, we have

$$ds = c_p \frac{d\theta}{\theta} \tag{2.90}$$

Integrating (2.90) we obtain a relationship between entropy and potential temperature:

$$s = c_p \ln \theta + \text{const} \tag{2.91}$$

Transformations in which the entropy remains constant (and therefore the potential temperature is also constant) are called *isentropic*.

Problem 2.15 Calculate the change in entropy when 5 g of water at 0°C are raised to 100°C and converted into steam at that temperature. (Latent heat of vaporization of water at 100°C is 2.253×10^6 J kg^{-1}.)

Solution The increase in entropy raising the water from 0°C (273°K) to 100°C (373°K) is given by

$$S_{373} - S_{273} = \int_{273}^{373} \frac{dQ_{\text{rev}}}{T}$$

and

$$dQ_{\text{rev}} = mc\, dT$$

where m is the mass and c the specific heat of the material. Taking c to be constant and equal to 4.18×10^3 J kg^{-1} deg^{-1},

$$S_{373} - S_{273} = 20.9 \int_{273}^{373} \frac{dT}{T} = 20.9 \ln \frac{373}{273}$$

$$= 20.9 \times 0.312 = 6.52 \text{ J deg}^{-1}$$

The increase in entropy in converting 5 g of water to steam at 100°C is, from (2.86), $5 \times 2.253 \times 10^3/373$ or 30.2 J deg^{-1}.

The total increase in entropy is therefore 36.72 J deg^{-1}.

Let us consider now the change in entropy in the Carnot cycle shown in Fig. 2.15. The transformations from A to B and from C to D are both adiabatic and reversible; therefore, in these two transformations there can be no changes in entropy. In passing from state B to state C, the substance takes in reversibly a quantity of heat Q_1 from the source at temperature T_1; therefore, the entropy of the source decreases by an amount Q_1/T_1. In passing from state D to state A, a quantity of heat Q_2 is rejected reversibly to the sink at temperature T_2; therefore, the entropy of the sink increases by Q_2/T_2. Since the working substance is taken in a cycle, and therefore returned to its original state, it cannot undergo any net change in entropy. Therefore, the net increase in entropy in the complete Carnot cycle is $Q_2/T_2 - Q_1/T_1$. However, we have shown in Problem 2.14 that $Q_1/T_1 = Q_2/T_2$. Hence, there is no change in entropy in a Carnot cycle.

Since *any* reversible cycle may be divided up into an infinite number of adiabatic and isothermal transformations, and therefore an infinite number of Carnot cycles, it follows that in any reversible cycle the total change in entropy is zero. This important result is yet another way of stating the Second Law of Thermodynamics.

It is interesting to note that if temperature (in degrees Kelvin) is taken as the ordinate and entropy as the abscissa, the Carnot cycle assumes a rectangular

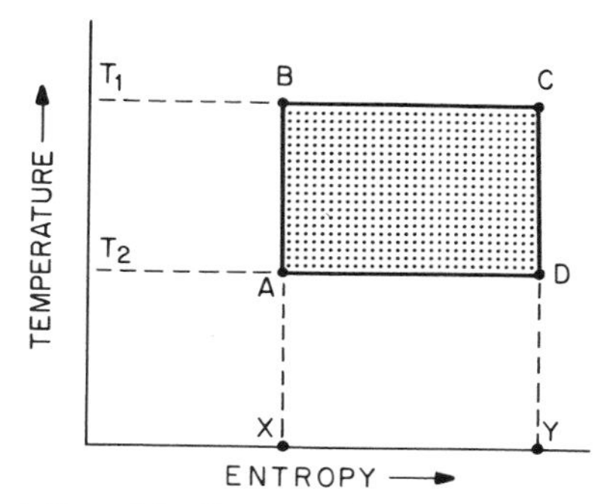

Fig. 2.17 Representation of the Carnot cycle on a temperature–entropy diagram.

shape, as shown in Fig. 2.17, where the letters A, B, C, and D correspond to the state points in the previous discussion. Adiabatic processes (AB and CD) are represented by vertical lines (that is, lines of constant entropy) and isothermal processes (BC and DA) by horizontal lines. From (2.86) it is evident that in a cyclic transformation ABCDA, the heat Q_1 taken in reversibly by the working substance from the warm reservoir is given by the area XBCY and the heat Q_2 rejected by the working substance to the cold reservoir is given by the area XADY. Therefore, the work $Q_1 - Q_2$ done in the cycle is given by the difference between the two areas, which is equivalent to the shaded area ABCD. Any reversible heat engine or refrigerator can be represented as a closed loop on a temperature–entropy diagram, the area of which is proportional to the work done by or on the system, depending upon whether the loop is traversed clockwise or counterclockwise.

2.8.3 The Clausius–Clapeyron equation

We will now utilize the Carnot cycle to derive an important relationship, known as the *Clausius–Clapeyron equation* (also referred to as the First Latent Heat Equation), for the change in the saturated vapor pressure above a liquid with temperature or the change in the melting point of a solid with pressure.

Let the substance in the cylinder of a Carnot heat engine be a mixture of liquid in equilibrium with its saturated vapor, and let the initial state of the substance be represented by point A in Fig. 2.18 in which the saturated vapor pressure is $e_s - de_s$ at temperature $T - dT$. The adiabatic transformation from state A to state B, where the saturated vapor pressure is e_s at temperature T, can be achieved by placing the cylinder on the nonconducting stand and compressing the piston infinitesimally (Fig. 2.19). Now let the cylinder be placed on the source of heat at temperature T and let the substance expand isothermally until a unit mass of the liquid evaporates (Fig. 2.19). In this transformation the pressure remains constant at e_s and the substance passes from state B to state C (Fig. 2.18). If the specific volumes of liquid and vapor at temperature T are α_1 and α_2, respectively, the increase in the volume of the system in passing

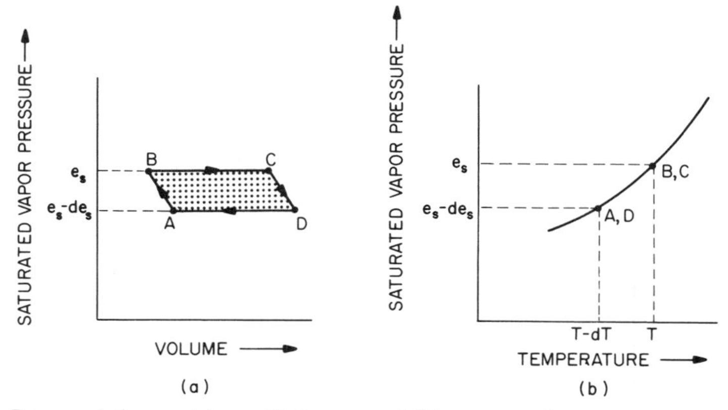

Fig. 2.18 Representation on (a) a p–V diagram and (b) a saturated vapor pressure versus temperature diagram, of the states of a mixture of liquid and its saturated vapor taken through a Carnot cycle.

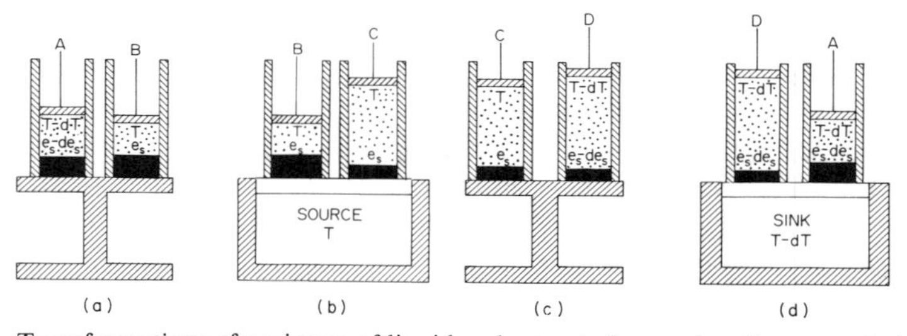

Fig. 2.19 Transformations of a mixture of liquid and saturated vapor in a Carnot cycle. The letters A, B, C, D indicate the states of the mixture shown in Fig. 2.18.

from state B to C is $(\alpha_2 - \alpha_1)$. Also the heat absorbed from the source is L, where L is the latent heat of vaporization. The cylinder is now placed again on the nonconducting stand, and a small adiabatic expansion is carried out from state C to state D in which the temperature falls from T to $T - dT$ and the pressure from e_s to $e_s - de_s$ (Fig. 2.19). Finally, the cylinder is placed on the sink at temperature $T - dT$ and an isothermal and isobaric compression is carried out from state D to state A during which vapor is condensed. All of the transformations are reversible.

From (2.85),

$$\frac{Q_1}{T_1} = \frac{Q_2}{T_2} = \frac{Q_1 - Q_2}{T_1 - T_2} \tag{2.92}$$

where $Q_1 - Q_2$ is the *net* heat absorbed by the substance in the cylinder during a cycle, which is also equal to the work done by the substance in the cycle. However, we have shown in Section 2.3 that the work done during a cycle is equal to the area of the enclosed loop on a p–V diagram. Therefore, from Fig. 2.18, $Q_1 - Q_2 = \text{BC} \times \text{AB} = (\alpha_2 - \alpha_1)\, de_s$. Also, $Q_1 = L$, $T_1 = T$, $T_1 - T_2 =$

dT; therefore, substituting into (2.92),

$$\frac{L}{T} = \frac{(\alpha_2 - \alpha_1)\, de_s}{dT}$$

or

$$\frac{de_s}{dT} = \frac{L}{T(\alpha_2 - \alpha_1)} \tag{2.93}$$

which is the *Clausius–Clapeyron equation* for the variation of the equilibrium vapor pressure e_s with temperature T. The same argument can be applied to an equilibrium mixture of solid and liquid; therefore (2.93) also gives the relationship between the variation with pressure of the melting point T of a solid, where L is the latent heat of fusion and α_1 and α_2 are the specific volumes of the solid and liquid, respectively.

Problem 2.16 Calculate the change in the melting point of ice if the pressure is increased from 1 to 2 atm. (The specific volumes of ice and water at 0°C are 1.0908×10^{-3} and 1.0010×10^{-3} m^3 kg^{-1} and the latent heat of melting is 3.34×10^5 J kg^{-1} at normal atmospheric pressure at 0°C.)

Solution The change dT in the melting point of ice due to a change in pressure dp is, by analogy with (2.93),

$$dT = \frac{T(\alpha_w - \alpha_i)}{L}\, dp$$

where α_i and α_w on the specific volumes of ice and water, respectively, and L the latent heat of melting. Substituting $T = 273$°K, $(\alpha_w - \alpha_i) = (1.0010 - 1.0908)10^{-3}$ m^3 kg^{-1} $= -0.0898 \times 10^{-3}$ m^3 kg^{-1}, $L = 3.34 \times 10^5$ J kg^{-1}, and $dp = 1$ atm $= 1.013 \times 10^5$ Pa, we obtain

$$dT = -\frac{273 \times 0.0898 \times 10^{-3}}{3.34 \times 10^5} 1.013 \times 10^5 \quad \text{deg}$$

or

$$dT = -0.00744 \quad \text{deg}$$

Therefore, an increase in pressure of 1 atm decreases the melting point of ice by about 0.007 deg. (Ice is unusual in this respect; the specific volume of the liquid form of most materials is greater than the specific volume of the solid; therefore, the melting point increases with increasing pressure.)

Problem 2.17 Derive an expression for the variation of the latent heat of vaporization L with temperature T in term of specific heats of the vapor and liquid.

Solution The increase in entropy when a unit mass of liquid vaporizes at temperature T is given by

$$s_v - s_l = \frac{L}{T}$$

where s_v and s_l are the entropies of a unit mass of vapor and liquid, respectively. Differentiating this expression with respect to temperature gives

$$\frac{ds_v}{dT} - \frac{ds_l}{dT} = \frac{1}{T}\frac{dL}{dT} - \frac{L}{T^2}$$

or

$$T\frac{ds_v}{dT} - T\frac{ds_l}{dT} = \frac{dL}{dT} - \frac{L}{T}$$

Now, $ds_v = dq_v/T$ and $ds_l = dq_l/T$; therefore, the two terms on the left-hand side of the last equation are specific heats (evaluated under the conditions that the vapor and liquid are in equilibrium under saturated conditions). If the specific heats of the vapor and liquid under these conditions are c_v and c_l, respectively, we obtain

$$\frac{dL}{dT} - \frac{L}{T} = c_v - c_l$$

This expression is referred to as the *Second Latent Heat Equation.*

2.8.4 The temperature–entropy and skew T–ln p diagrams

In Section 2.5.4 we described one type of thermodynamic diagram, the pseudoadiabatic chart, which is commonly used in meteorology. Now we will describe two more thermodynamic diagrams which are also in widespread use.

Because of the linear relationship between entropy and the logarithm of potential temperature [see (2.91)], the entropy scale on a *temperature–entropy diagram*, such as that shown in Fig. 2.17, can be labeled in terms of potential temperature, provided the scale is logarithmic, as shown in Fig. 2.20a [in meteorological practice it is customary to plot temperature on the abscissa and entropy (or ln θ) on the ordinate, the reverse of that shown in Fig. 2.17]. For isobaric (constant pressure) processes it follows from (2.57) that ln θ = ln T + const. Therefore, since ln θ is an ordinate on the temperature–entropy diagram but ln T is not the abscissa, isobars on this diagram are curved lines as indicated

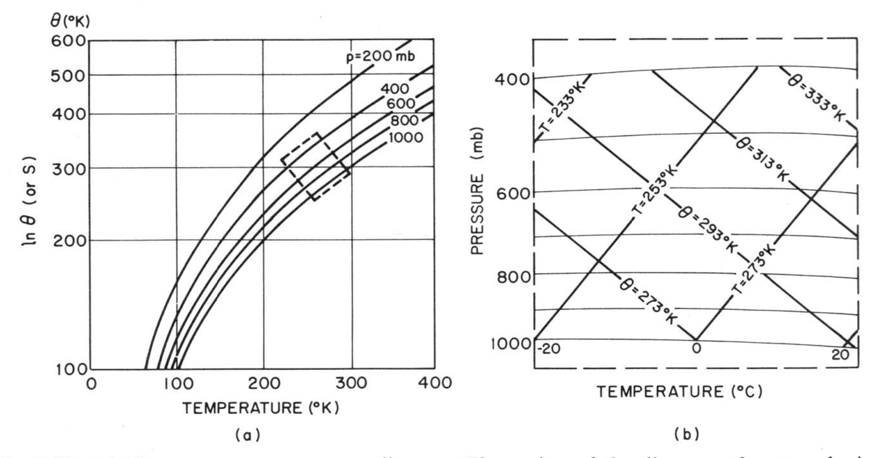

Fig. 2.20 (a) The temperature–entropy diagram. The region of the diagram of meteorological interest is enclosed by the dashed lines and is shown expanded in (b) where the diagram has also been rotated. (b) is the form in which the temperature–entropy diagram is normally encountered.

in Fig. 2.20a. By contrasting the temperature–entropy diagram with the pseudoadiabatic chart discussed previously, it will be seen that one is just a distorted form of the other. In the temperature–entropy diagram adiabats are perpendicular to isotherms and isobars are slightly curved, while in the pseudoadiabatic chart isobars are perpendicular to isotherms and the adiabats are nonparallel lines (see chart, back endpapers). The temperature–entropy[†‡] diagram is often used in place of the pseudoadiabatic chart. One practical advantage of the temperature–entropy diagram is the fact that, when atmospheric temperature soundings are plotted on this diagram, changes in slope are readily discernible because there is a relatively large angle (90°) between isotherms and dry adiabats (the lapse rate in the atmosphere generally lies between isothermal and dry adiabatic). Only a small portion of the entropy–temperature diagram shown in Fig. 2.20a is of interest in meteorology. This portion, which is indicated by dotted lines in Fig. 2.20a, is expanded in Fig. 2.20b; note that this portion has also been rotated as a unit so that the isobars are roughly horizontal. The diagram outlined in Fig. 2.20b is the one that is printed for meteorological use.

Another thermodynamic chart used widely in meteorology is the so-called skew T–ln p diagram in which the ordinate is proportional to ln p and the abscissa is proportional to $(T + \ln p)$. Thus, on this diagram, isotherms are straight, parallel lines (which slope upward to the right at an angle of 45° to the isobars). The equation for a dry adiabat (θ const) is $\ln T = (\text{const}) \ln p + \text{const}$, from (2.57). Since ln p is an ordinate on the skew T–ln p diagram but the abscissa is not ln T, adiabats are not straight lines on this diagram. Instead, adiabats are slightly curved lines which run from the lower right to the upper left of the diagram. The angle between isotherms and dry adiabats on the skew T–ln p diagram is nearly 90°.

2.8.5 Generalized statement of the second law of thermodynamics

So far we have discussed the Second Law of Thermodynamics and entropy in a fairly informal manner and only with respect to ideal reversible processes in which a system moves through a series of equilibrium states. However, all natural processes are spontaneous and irreversible since they move a system from a nonequilibrium state toward a condition of equilibrium.

[†] The temperature–entropy diagram was introduced into meteorology by Shaw. Since entropy is often represented by the symbol ϕ (rather than the S used in this text), the diagram is often referred to as a *tephigram*.

[‡] **Sir (William) Napier Shaw** (1854–1945) English meteorologist. Lecturer in Experimental Physics, Cambridge University, 1877–1899. Director of the British Meteorological Office, 1905–1920. Professor of Meteorology, Imperial College, University of London, 1920–1924 (and founder of that department). Shaw did much to establish the scientific basis of meteorology. His interests ranged from the atmospheric general circulation and forecasting to air pollution. Introduced the millibar into meteorology.

We now state the Second Law more formally in terms of the following four postulates:

- There exists a function of state for a body called the entropy s (per unit mass).
- s may change either because the body comes into thermal contact with its environment ($d_e s$) or as a result of internal changes within the body ($d_i s$). The total change in the entropy of a body ds is given by

$$ds = d_e s + d_i s$$

- The change $d_e s$ is given by

$$d_e s = \frac{dq}{T}$$

where dq is the heat received by a unit mass of the body.
- For reversible changes $d_i s = 0$ and for irreversible changes $d_i s > 0$. Hence,

$$\begin{aligned} ds &= \frac{dq}{T} \quad \text{for reversible changes} \\ ds &> \frac{dq}{T} \quad \text{for irreversible changes} \end{aligned} \tag{2.94}$$

By combining (2.41) with (2.94), we see that

$$T\,ds \geq du + p\,d\alpha \tag{2.95}$$

where the equality sign applies to reversible (equilibrium) transformations and the inequality sign to irreversible (spontaneous) transformations.

2.9 THERMODYNAMIC FUNCTIONS AND EQUILIBRIUM CONDITIONS

In an equilibrium state every possible transformation is reversible and any increase in the entropy of a body is equal in magnitude to the decrease in the entropy of its environment. Hence, a necessary condition for equilibrium is that the total entropy of a body and its environment are constant. This is the most general criterion for an equilibrium state. Indeed, it is so general as to be difficult to apply in practice. We therefore introduce two thermodynamic functions of state, called the Helmholtz[†] and the Gibbs[‡] free energies, which lead to more useful criteria for equilibrium under certain conditions.

[†] **Herman von Helmholtz** (1821–1894) German physiologist and physicist. Investigated vision and hearing. One of three men credited with formulating the law of conservation of energy. Also studied electrical oscillations, indicated the possibility of an electromagnetic theory of light, and studied vortex motions.

[‡] **Josiah Willard Gibbs** (1839–1903) American mathematical physicist. Apart from three years in his youth spent in European universities, he lived all his life in New Haven, Connecticut. His first published paper did not appear until he was 34. Helped to lay the foundations of thermodynamics, statistical mechanics, physical chemistry, and vector analysis. Patented the railroad brake!

The *Helmholtz free energy* f of a unit mass of a body is defined by†

$$f \equiv u - Ts \tag{2.96}$$

Differentiating we obtain

$$df = du - T\,ds - s\,dT \tag{2.97}$$

For a reversible (equilibrium) transformation we may combine (2.95) with (2.97) to yield

$$df = -s\,dT - p\,d\alpha \tag{2.98}$$

Therefore, if a body is in equilibrium and if its temperature and volume are constant,

$$df = 0 \tag{2.99}$$

For a body which undergoes a spontaneous, irreversible transformation, combination of (2.95) and (2.97) yields

$$df < -s\,dT - p\,d\alpha$$

Therefore, for an irreversible transformation at constant temperature and volume,

$$df < 0 \tag{2.100}$$

It follows from (2.99) and (2.100) that a body at constant temperature and volume is in stable equilibrium when its Helmholtz free energy has a minimum value. For this reason the Helmholtz free energy is sometimes called the *thermodynamic potential at constant volume* (constant temperature understood) in order to indicate its analogy with the potential energy of a mechanical system which also has a minimum value under equilibrium conditions.

The *Gibbs free energy* g of a unit mass of a body is defined by†

$$g \equiv u - Ts + p\alpha \tag{2.101}$$

Differentiating we obtain

$$dg = du - T\,ds - s\,dT + p\,d\alpha + \alpha\,dp \tag{2.102}$$

Combining (2.95) and (2.102) for the case of a reversible (equilibrium) transformation, we obtain

$$dg = -s\,dT + \alpha\,dp \tag{2.103}$$

Therefore, if temperature and pressure are constant, for a body in equilibrium,

$$dg = 0 \tag{2.104}$$

† Since we have defined the specific (unit mass) variables we have used lower case letters. In general, the Helmholtz and Gibbs functions are represented by F and G, respectively.

For a body which undergoes a spontaneous (irreversible) transformation, combination of (2.95) and (2.102) yields

$$dg < -s\,dT + \alpha\,dp \tag{2.105}$$

Therefore, for an irreversible transformation at constant temperature and pressure,

$$dg < 0 \tag{2.106}$$

Hence, the criterion for the thermodynamic equilibrium of a body at constant temperature and pressure is that the Gibbs free energy has a minimum value. The Gibbs free energy is sometimes called the *thermodynamic potential at constant pressure* (constant temperature understood).

We now wish to relate the Helmholtz and Gibbs free energies to the external work that a body can do under various conditions. In discussing the First Law of Thermodynamics in Section 2.3 we tacitly assumed that the only external work that a body can do is the work of expansion $p\,d\alpha$. However, a body may also perform external work through other means (for example, electrical or by creating new surface area between two phases—see Section 4.2.1). In general, therefore, we should write the First Law for a unit mass of a body as†

$$dq = du + dw_{\text{tot}} \tag{2.107}$$

or, for a reversible transformation,

$$T\,ds = du + dw_{\text{tot}} \tag{2.108}$$

where dw_{tot} is the total work done by a unit mass of a body. It follows from (2.97) and (2.108) that for a reversible process

$$dw_{\text{tot}} = -df - s\,dT$$

therefore, at constant temperature,

$$dw_{\text{tot}} = -df \tag{2.109}$$

That is, the total external work done by a body in a reversible, isothermal process is equal to the decrease in the Helmholtz free energy of the body.

If da is the external work done by a unit mass of a body over and above any $p\,d\alpha$ work, that is, if

$$da = dw_{\text{tot}} - p\,d\alpha \tag{2.110}$$

by combining (2.102), (2.108), and (2.110) we obtain, for a reversible process,

$$da = -dg - s\,dT - \alpha\,dp$$

Therefore, if both temperature and pressure are constant,

$$da = -dg \tag{2.111}$$

† Here, as in Section 2.3, we are assuming that the kinetic energy and the gravitational potential energy of the body are constant.

That is, the external work done by a body (exclusive of any $p\,d\alpha$ work) in a reversible, isothermal-isobaric process is equal to the decrease in the Gibbs free energy of the body.

If a single molecule is removed from a material in a certain phase, the temperature and pressure remaining constant, the resulting change in the Gibbs free energy of the material is called the *chemical potential* μ of that phase. In other words, the chemical potential is the Gibbs free energy per molecule.

2.9.1 Some applications

The thermodynamic functions and relations discussed above are important in considering phase changes in the atmosphere. In particular, we will need to utilize them in Chapter 4 in discussing the formation of water droplets by condensation. In order to lay the foundation for this discussion, we will now derive two important results.

Problem 2.18 Show that when a plane surface of liquid is in equilibrium with its vapor the chemical potentials in the liquid and vapor phases are the same.

Solution If a liquid is in equilibrium with its vapor, molecules may evaporate (or condense) at constant temperature and pressure. However, under these conditions, (2.111) holds. Now the only external work done during evaporation is the work of expansion $p\,d\alpha$. Hence, $da = 0$ in (2.111) and therefore $dg = 0$. But if dg remains unchanged during evaporation the chemical potential of the molecules in the liquid phase must be equal to their chemical potential in the vapor phase.

Problem 2.19 Derive an expression for the difference in the chemical potentials of the vapor and liquid phases in terms of the actual vapor pressure e and the temperature T and the saturated vapor pressure e_s for a plane surface of liquid at temperature T.

Solution Let μ_v and μ_l be the chemical potentials in the vapor and liquid phases, respectively, when the pressure is e and the temperature T. If the pressure changes (reversibly) by de at a constant temperature, it follows from the definition of chemical potential and (2.103) applied to the case of a single vapor molecule that

$$d\mu_v = v_v\,de$$

where v_v is the volume occupied by a single molecule in the vapor phase at temperature T and pressure e. Similarly, the change in the chemical potential for the liquid phase for a (reversible) change in pressure de at constant temperature is given by

$$d\mu_l = v_l\,de$$

where v_l is the volume occupied by a single molecule in the liquid phase. Combining the last two expressions we obtain

$$d(\mu_v - \mu_l) = (v_v - v_l)\,de$$

or, since $v_v \gg v_l$,

$$d(\mu_v - \mu_l) = v_v\,de$$

Applying the ideal gas equation to one molecule in the vapor phase we have

$$ev_v = kT$$

where k is Boltzmann's constant. From the last two expressions

$$d(\mu_v - \mu_l) = \frac{kT}{e}\, de$$

Now we have shown in Problem (2.18) that $\mu_v = \mu_l$ when $e = e_s$; hence,

$$\int_{0(\mu_v = \mu_l)}^{\mu_v - \mu_l} d(\mu_v - \mu_l) = \int_{e_s}^{e} \frac{kT}{e}\, de$$

or,

$$\mu_v - \mu_l = kT \ln \frac{e}{e_s} \tag{2.112}$$

PROBLEMS

2.20 Answer or explain the following in the light of the principles discussed in this chapter:

(a) Air released from a tire is cooler than its surroundings.

(b) Under certain conditions an ideal gas can undergo a change of state without doing external work.

(c) A parcel of air cools when it is lifted. Dry parcels cool more rapidly than moist parcels.

(d) The gas constant for moist air is greater than that for dry air.

(e) A liquid boils when its saturation vapor pressure is equal to the atmospheric pressure.

(f) Pressure in the atmosphere increases approximately exponentially with depth, whereas the pressure in the ocean increases approximately linearly with depth.

(g) Describe a procedure for converting station pressure to sea level pressure.

(h) Under what condition(s) does the hypsometric equation predict an exponential decrease of pressure with height?

(i) If a low pressure system is colder than its surroundings, the amplitude of the pressure anomaly increases with height.

(j) The 1000–500-mb thickness is predicted to increase from 5280 to 5460 m at a given station. Assuming that the lapse rate remains constant, what change in surface temperature would you predict?

(k) On some occasions low surface temperatures are recorded when the 1000–500-mb thickness is well above normal. Explain this apparent paradox.

(l) In cold climates the air indoors tends to be extremely dry.

(m) Which of the following pairs of quantities are conserved when unsaturated air is lifted: potential temperature and mixing ratio, potential temperature and saturation mixing ratio, equivalent potential temperature and saturating mixing ratio?

(n) Which of the following quantities are conserved during the lifting of saturated air: potential temperature, equivalent potential temperature, mixing ratio, saturation mixing ratio?

(o) Dew points in excess of 25°C are rarely observed.

2.21 Determine the "apparent molecular weight" of the Venusian atmosphere, assuming that it consists of 95% of CO_2 and 5% of N_2 by volume. What is the gas constant for 1 kg of such an atmosphere? (Atomic weights of C, O, and N are 12, 16, and 14, respectively.)

Answer 43.2 and 192.5 J deg^{-1} kg^{-1}

2.22 Assuming an isothermal atmosphere with a temperature of $-33°C$ and a surface pressure of 1000 mb, estimate the levels at which pressure equals 100, 10, and 1 mb, respectively.

Answer 16.2, 32.4, and 48.6 km

2.23 Prove that when an ideal gas undergoes an adiabatic transformation, $pV^\gamma = \text{const}$, where γ is the ratio of the specific heat at constant pressure to the specific heat at constant volume. [Hint: Combine the First Law of Thermodynamics in differential form with the ideal gas equation and then integrate.]

2.24 Calculate the work done in compressing isothermally 2 kg of dry air to one-tenth of its volume at 15°C.

Answer 3.807×10^5 J

2.25 (a) Starting with the Clausius–Clapeyron Equation (2.93), show that

$$\ln \frac{e_s}{6.11 \text{ mb}} = \frac{L}{R_v}\left(\frac{1}{273} - \frac{1}{T}\right)$$

[Hint: Make use of the fact that $\alpha_2 \gg \alpha_1$ and $e_s = 6.11$ mb at 0°C.]

(b) Use the above equation to calculate the saturation vapor pressure of water vapor at 20°C. Compare your result with Fig. 2.7.

Answer 23.37 mb

2.26 20 liters of air at 20°C, having a relative humidity of 60%, are compressed isothermally to a volume of 4 liters. Calculate the mass of water condensed. The saturation vapor pressure of water at 20°C is 23.37 mb.

Answer 0.14 g

2.27 (a) *Potential density* D is defined as the density which dry air would attain if it were transformed reversibly and adiabatically from its existing conditions to a standard pressure p_0 (usually 1000 mb). If the density and pressure of the air are ρ and p, respectively, show that

$$D = \rho\left(\frac{p_0}{p}\right)^{c_v/c_p}$$

where c_p and c_v are the specific heats of air at constant pressure and constant volume, respectively.

(b) Calculate the potential density of a quantity of air at a pressure of 600 mb and temperature $-15°C$.

(c) Show that

$$\frac{1}{D}\frac{dD}{dz} = -\frac{1}{T}(\Gamma_d - \Gamma)$$

where Γ_d is the dry adiabatic lapse rate, Γ the actual lapse rate of the atmosphere, and T the temperature at height z. [Hint: Take logarithms of the expression given in (a) and then differentiate with respect to height.]

(d) Show that the criteria for stable, neutral, and unstable conditions in the atmosphere are that the potential density decreases with height, is constant with height, and increases with height, respectively. [Hint: Use the expression given in (c).]

Answer (b) 1.167 kg m^{-3}

2.28 A necessary condition for the formation of a mirage is that the density of the air increase with height. Show that this condition is realized if the decrease of atmospheric temperature with height exceeds three and a half times the dry adiabatic lapse rate. [Hint: Proceed as in Problem 2.27(c).]

2.29 The *specific humidity* is defined as the ratio of the mass of vapor in a certain volume to the total mass of air and vapor in the same volume. If the specific humidity of a sample of air is 0.0196 at 30°C, find its virtual temperature. If the total pressure of the moist air is 1014 mb, what is its density?

Answer 33.7°C and 1.15 kg m^{-3}

2.30 A meteorological station is located 50 m below sea level. If the surface pressure at this station is 1020 mb, the virtual temperature at the surface 15°C, and the mean virtual temperature for the 1000–500-mb layer 0°C, compute the height of the 500-mb level above sea level at this station.

Answer 5.663 km

2.31 If water vapor comprises 1% of the volume of the air (that is to say, if it accounts for 1% of the molecules), what is the virtual temperature correction?

Answer $T_v - T \simeq 1$ deg

2.32 An isolated raindrop which is evaporating into air at a temperature of 18°C has a temperature of 12°C. Calculate the mixing ratio of the environmental air. (Saturated mixing ratio of air at 12°C = 8.7 g kg^{-1}.)

Answer 6 g kg^{-1}

2.33 Show that

$$\frac{L}{c_p T}\, dw_s \simeq d\left(\frac{Lw_s}{c_p T}\right)$$

[Hint: Differentiate the right-hand side and, assuming L/c_p is independent of temperature, show that the approximation holds provided

$$\frac{dT}{T} \ll \frac{dw_s}{w_s}$$

Verify this inequality by noting the relative changes in T and w_s for small incremental displacements along saturated adiabats on a pseudoadiabatic chart.]

2.34 The 1000–500-mb layer is subjected to a heat source having a magnitude of 5×10^6 J m^{-2}. Assuming that the atmosphere is at rest (apart from the slight vertical motions associated with the expansion of the layer) calculate the resulting increase in the mean temperature and in the thickness of the layer. [Hint: Remember that pressure is weight per unit area.]

Answer 0.975 deg; 19.8 m

2.35 A hiker sets his altimeter to the correct reading at the beginning of a hike during which he climbs from near sea level to an altitude of exactly 1 km over a 3-h period. During this same time interval the sea level pressure drops 8 mb due to the approach of a storm. Estimate the altimeter reading at the end of the hike.

Answer 1064 m

2.36† Assuming the truth of the Second Law of Thermodynamics, prove the following two statements (known as *Carnot's Theorem*):

† Problems 2.36–2.47 are based on topics discussed in Sections 2.8 and 2.9 which may have been omitted in an introductory, undergraduate course.

(a) No engine can be more efficient than a reversible engine working between the same limits of temperature.

(b) All reversible engines working between the same limits of temperature have the same efficiency.

[Hint: The efficiency of any engine is given by (2.80); the distinction between a reversible (R) and an irreversible (I) engine is that R can be driven backward but I cannot. Consider a reversible and an irreversible engine working between the same limits of temperature. Suppose initially that I is more efficient than R and use I to drive R backwards. Show that this leads to a violation of the Second Law and hence prove that I cannot be more efficient than R.]

2.37 Lord Kelvin introduced the concept of *available energy* which is defined as the maximum amount of heat which can be converted into work by using the coldest available body in a system as the sink in a heat energy. Show that the available energy of the universe is tending to zero and that

$$\text{loss of available energy} = T_0 \text{ (increase in entropy)}$$

where T_0 is the temperature of the coldest available body.

2.38 (a) Sketch the Carnot cycle on a pseudoadiabatic chart.

(b) Prove that for any reversible heat engine, the work done in one cycle is proportional to the area enclosed by the cycle when represented on a graph of θ against p^{R_d/c_p}.

2.39 A reversible engine has a source and sink at temperatures of 100°C and 0°C, respectively. If the engine receives 20 J from the source in every cycle, calculate the work done by the engine in ten cycles. How many joules does the engine reject to the sink in ten cycles?

Answer 53.6 J; 146.4 J

2.40 A refrigerator has an internal temperature of 0°C and is situated in a room with a steady temperature of 17°C. If the refrigerator is driven by an electric motor 1 kW in power, calculate the time required to freeze 20 kg of water already cooled to 0°C when placed in the refrigerator. The refrigerator may be considered to act as an ideal heat engine in reverse.

Answer 6.93 min

2.41 Calculate the change in entropy of 2 g of ice initially at −10°C which is converted to steam at 100°C due to heating.

Answer 17.3 J deg^{-1}

2.42 Calculate the change in entropy when 1 kmol of an ideal diatomic gas initially at 13°C and a pressure of 1 atm changes to a temperature 100°C and a pressure of 2 atm.

Answer Increase of 1966 J deg^{-1}

2.43 Calculate the change in pressure necessary to increase the boiling point of water from 99.5 to 100.5°C. (Specific volume of steam at 100°C = 1.66 m^3 kg^{-1}.)

Answer An increase in pressure of 3.64×10^3 Pa

2.44 (a) By differentiating the enthalpy function, defined by (2.50), show that

$$\left(\frac{\partial p}{\partial T}\right)_s = \left(\frac{\partial s}{\partial \alpha}\right)_p$$

where s is entropy. Note that this is equivalent to the Clausius–Clapeyron equation. [Hint:

$$dh = \left(\frac{\partial h}{\partial s}\right)_p ds + \left(\frac{\partial h}{\partial p}\right)_s dp$$

and, since h is a function of state,

$$\frac{\partial}{\partial p}\left(\frac{\partial h}{\partial s}\right) = \frac{\partial}{\partial s}\left(\frac{\partial h}{\partial p}\right).]$$

(b) By differentiating the Helmholtz function f, defined by (2.96), and then proceeding in an analogous way to (a) above, show that

$$\left(\frac{\partial s}{\partial \alpha}\right)_T = \left(\frac{\partial p}{\partial T}\right)_\alpha$$

(c) By differentiating the Gibb's function g, defined by (2.101), and then proceeding in an analogous way to (a) above, show that

$$\left(\frac{\partial \alpha}{\partial T}\right)_p = -\left(\frac{\partial s}{\partial p}\right)_T$$

2.45 Four grams of liquid water condense out of 1 kg of air during a moist-adiabatic expansion. Show that the total internal energy associated with this amount of water is only 2.4% of that associated with the air.

2.46 A fluid (not necessarily a perfect gas) is flowing through a small aperture or nozzle from one chamber whose pressure is maintained at some constant value p_1 into another chamber whose pressure is maintained at a constant value p_2, where $p_2 < p_1$. Prove that the fluid conserves enthalpy.

2.47 If the fluid in Problem 2.46 is an ideal gas, prove that its temperature doesn't change as a result of its passage through the nozzle. Reinterpret Problem 2.20(a) in light of this result.

A thermodynamic diagram (for example, see pseudoadiabatic chart, back endpapers) should be used to solve the remaining problems.

2.48 The pressure and temperature at the levels at which jet aircraft normally cruise are typically 200 mb and −60°C. Estimate from a pseudoadiabatic chart the temperature of this air if it were adiabatically compressed to 1000 mb. Compare your answer with an accurate computation.

Answer 66°C

2.49 An air parcel at 1000 mb has an initial temperature of 15°C and a dew point of 4°C.

(a) Find the mixing ratio, relative humidity, wet-bulb temperature, potential temperature, and wet-bulb potential temperature of the air.

(b) Determine the foregoing if the parcel rises to 900 mb.

(c) Determine the foregoing if the parcel rises to 800 mb.

(d) Where is the lifting condensation level?

Answer (a) 5.1 g kg^{-1}, 47%, 9.3°C, 288°K, 9.3°C; (b) 5.1 g kg^{-1}, 75%, 4.4°C, 288°K, 9.3°C; (c) 4.4 g kg^{-1}, 100%, −1.0°C, 290°K, 9.3°C; (d) 847 mb

2.50 Air at 1000 mb and 25°C has a wet-bulb temperature of 20°C. Find the dew point. If this air were expanded until all the moisture condensed and fell out and then compressed to 1000 mb, what would be the resulting temperature?

Answer 18°C; 62°C

2.51 Plot the following sounding:

	Pressure level (mb)	*Air temperature* (°C)	*Dew point* (°C)
A	1000	30.0	21.5
B	970	25.0	21.0
C	900	18.5	18.5
D	850	16.5	16.5
E	800	20.0	5.0
F	700	11.0	−4.0
G	500	−13.0	−20.0

(a) State whether the layers AB, BC, CD, and so on, are in stable, unstable, or neutral equilibrium.

(b) State which layers are convectively unstable.

Answer (a) AB, unstable; BC, neutral; CD, neutral; DE, stable; EF, stable; FG, stable; (b) All layers are convectively unstable except CD which is convectively neutral

2.52 Air at a temperature of 20°C and a mixing ratio of 10 g kg^{-1} is lifted from 1000 mb to 700 mb by moving over a mountain. What is the initial dew point of the air? Determine the temperature of the air after it has descended to 900 mb on the other side of the mountain if 80% of the condensed water vapor is removed by precipitation during the ascent.

Answer 14°C; 19.8°C

Chapter

3

Extratropical Synoptic-Scale Disturbances

Throughout middle and high latitudes, day to day weather changes are closely linked to the passage of transient, synoptic-scale disturbances in the tropospheric wind field. Through the systematic display and analysis of synoptic (that is, simultaneous) surface and upper air observations, such disturbances can be identified and tracked through the course of their life histories. In this chapter we will examine an individual winter storm system using conventional synoptic analysis techniques that reveal its three-dimensional structure and time evolution. It should be emphasized at the outset that no two of these disturbances are exactly alike. It is only through the examination of a large number of systems (for example, by following the current synoptic charts over a period of months) that it is possible to develop some appreciation for the wide range of structures and time sequences that are possible. For this example we have attempted to select a storm system that is reasonably representative, to the extent that any single example can be representative of a class of diverse phenomena.

3.1 THE 500-mb FLOW

Figure 3.1 shows the hemispheric distribution of the height[†] of the 500-mb pressure surface at or just after midnight (00)[‡] Greenwich Civil Time (GCT) 20 November 1964. These charts are constructed from measurements obtained

[†] Strictly speaking, we refer here to *geopotential height* as defined in Section 2.2.1.

[‡] Unless otherwise noted, times will be expressed in whole hours.

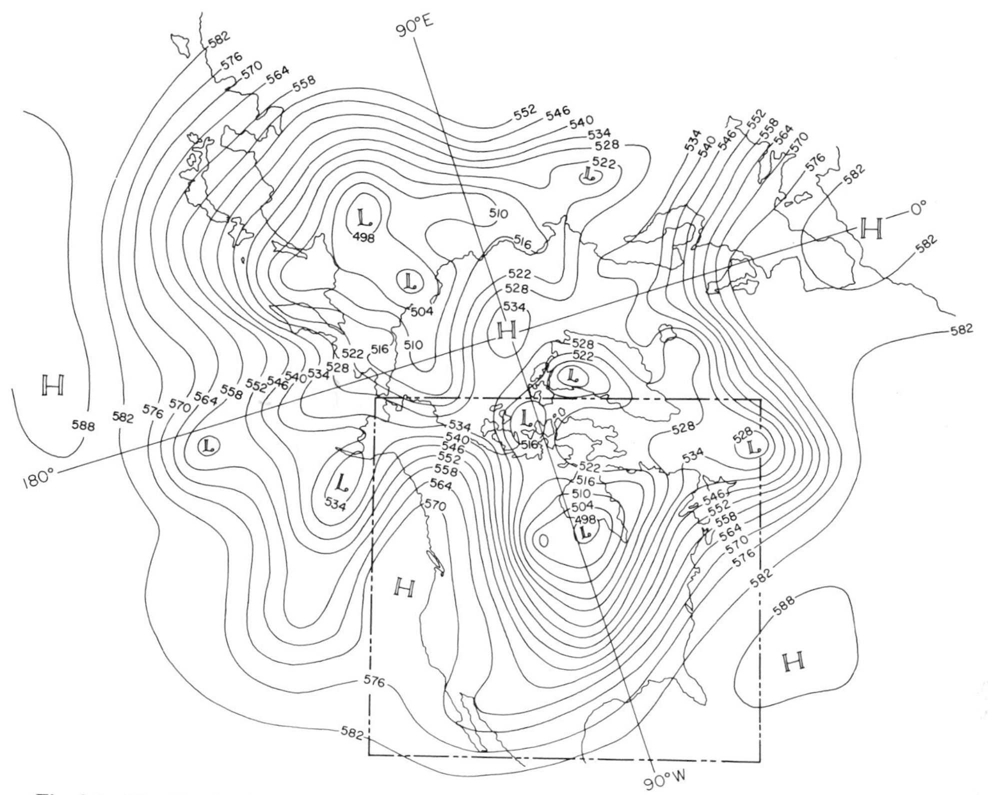

Fig. 3.1 The distribution of geopotential height on the 500-mb surface at 00 GCT 20 November 1964. Labels on contours represent geopotential height, in tens of meters. The letters H and L denote centers of high and low geopotential height, respectively.

from radiosondes, which are launched simultaneously from hundreds of stations scattered over the hemisphere. We recall from Section 1.6 that the winds tend to blow parallel to these height contours, *l*eaving *l*ow heights to the *l*eft in the northern hemisphere. Wind speed tends to be inversely proportional to the spacing between the contours, which are drawn for every 60 m change in height.

At any given instant in time, the hemispheric flow pattern is dominated by large amplitude, synoptic-scale disturbances which vary from day to day in position and intensity. When averaged over a long time period, such as a season, these transient ridges and troughs tend to cancel one another. Hence the climatological 500-mb flow pattern shown in Fig. 1.14 is relatively featureless when compared to a typical instantaneous flow pattern such as the one depicted in Fig. 3.1.

In order to simplify the task of following the time evolution of the 500-mb flow pattern we will concentrate on the limited area enclosed by the rectangle in Fig. 3.1. Four successive 500-mb synoptic charts, spaced at intervals of 12 h,

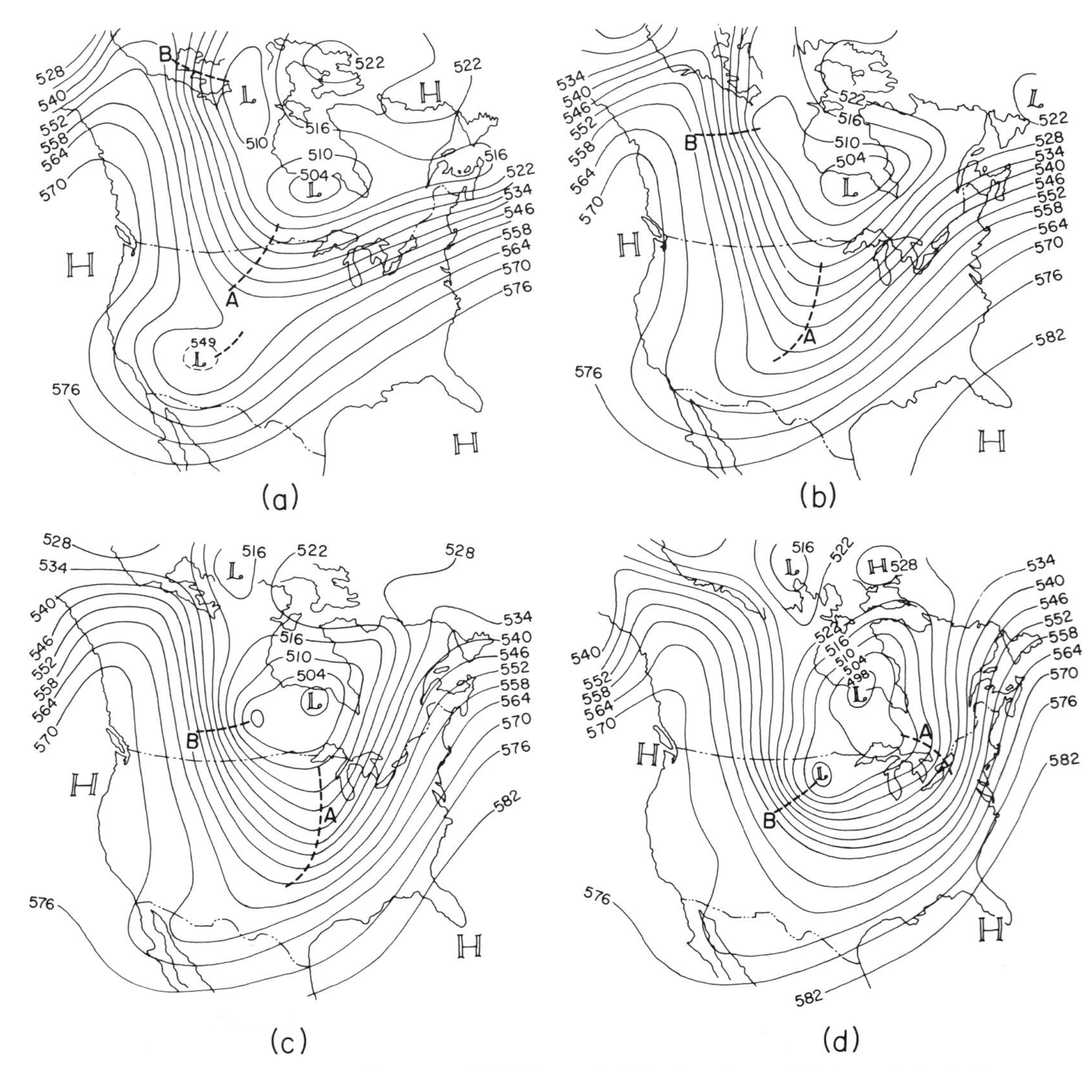

Fig. 3.2 The distribution of geopotential height on the 500-mb surface at 12-h intervals beginning at 00 GCT 19 November 1964 in (a) and ending at 12 GCT 20 November 1964 in (d). Contours are labeled in tens of meters.

are shown in Fig. 3.2. The feature of primary interest is the trough line labeled A in the figure. In chart (a) this feature is in the process of merging with a "cutoff low" that had been situated over the southwestern United States for several days. In charts (b) and (c) the combined trough line is swept eastward toward the Great Lakes. In chart (d), the trough line has come apart again; the northern segment has intensified and moved rapidly northeastward around the low center over central Canada, and the southern segment has been left behind. A second trough line, labeled B in the figure, can be tracked as it moves southward from the Canadian Arctic in (a) into the Northern Plains of the United States in (d). Considerable intensification of this feature can be noted during the 36-h period.

We will examine the upper level structure of this system in more detail in Section 3.4. First, however, we wish to describe the surface weather that occurred in association with trough line A as it moved eastward across the United States.

3.2 SURFACE WEATHER ELEMENTS

In the description of the surface weather, we will begin by confining our attention to the charts for 00 and 12 GCT 19 November and 00 GCT 20 November 1964, the counterparts of (a), (b), and (c) in Fig. 3.2. We will discuss the weather from the standpoint of various observed parameters: wind and pressure, temperature, dew point, precipitation, and pressure tendency. Then, having described the behavior of the various parameters separately, we will show how they are plotted together on a surface synoptic chart, or in the form of time sections at individual stations. In the course of this discussion we will introduce the concept of *fronts*, which is fundamental to the understanding of middle-latitude weather.

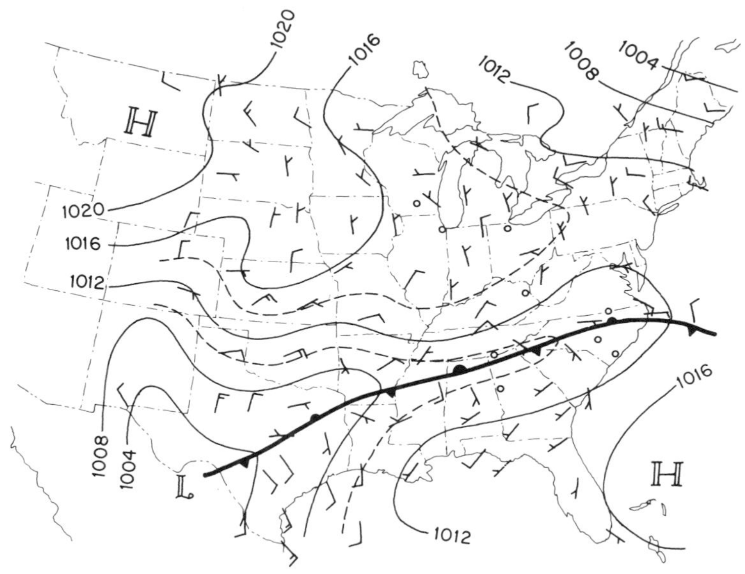

Fig. 3.3a Sea level pressure and surface winds at 00 GCT 19 November 1964. Heavy lines with barbs and half circles denote confluence lines. Lighter lines represent isobars drawn at 4-mb intervals for solid lines, 2-mb for dashed lines. The letters H and L denote centers of high and low pressure, respectively. Winds are plotted using the conventions described in Table 3.1. Small circles denote calm winds.

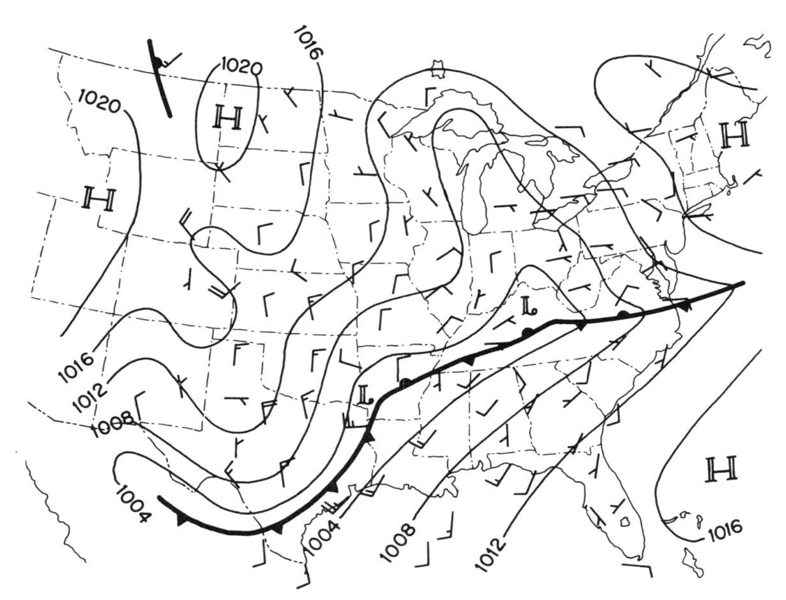

Fig. 3.3b Sea level pressure and surface winds at 12 GCT 19 November 1964. For further details, see caption of Fig. 3.3a.

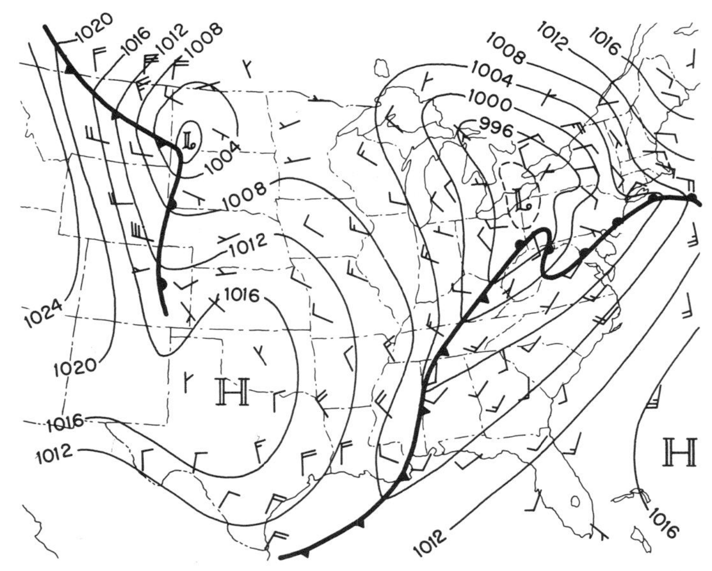

Fig. 3.3c Sea level pressure and surface winds at 00 GCT 20 November 1964. For further details, see caption of Fig. 3.3a.

3.2.1 Wind and pressure

In Fig. 3.3 the surface winds at individual stations are plotted vectorially, using the conventions described in Table 3.1. On the same charts are shown analyses of the sea level pressure, based upon data taken at the same stations as the wind observations. From a cursory inspection of Fig. 3.3 it is apparent

Table 3.1

Abbreviated plotting code for synoptic charts

Wind speed[a]	5	10	15	35	50	120

Wind direction	Northerly (from the north) (0° or 360°)	Northeasterly (45°)	Southeasterly (135°)	Westerly (270°) (southern hemisphere: note orientation of barbs)

Sky cover	Clear	Scattered clouds	Partly cloudy	Cloudy	Sky obscured (outer circle denotes calm wind)

Weather	Light continuous rain	Moderate continuous snow	Rain shower	Past drizzle	Thunderstorm	Dense fog	Sleet or hail	Freezing drizzle

Plotting model	Examples		
TT dd ff PPP ww (N) ±ppa T_dT_d RR	21 247 10 +8	-1 037 ≡ -18 -1 .15	17 936 +30 15 M
ff Wind speed (kt)	15	Calm	5
dd Wind direction	320	—	070°
TT Temperature (°C)	21	−1	17
T_dT_d Dew point (°C)	10	−1	15
PPP Pressure	1024.7	1003.7	993.6
±ppa Pressure tendency	0.8↑ (rising, then steady)	1.8↓	3.0↑ (fall, then larger rise)
N Sky cover	Clear	Obscured	Cloudy
WW Weather	None	Dense fog	Thunderstorm in past hour
RR 6 h precipitation (in.)	None	0.15	Missing

[a] Units: knots (1 knot = 1 nautical mile per hour; 1.95 knots = 1 m s^{-1}). 60 nautical miles = 1 deg of latitude; therefore, knots can easily be related to displacement measured in degrees of latitude over a 12- or 24-h period (for example, 10 knots = 4 deg day^{-1}).

that the surface wind tends to blow parallel to the isobars of sea level pressure, *l*eaving *l*ow pressure to the *l*eft, in accordance with the geostrophic relationship, but there is also some tendency for flow across the isobars from higher toward lower pressure because of the effects of friction.

On the first of the three charts, centers of high pressure are located over the Northern Plains and over the Atlantic Ocean east of Florida. Separating these regions of high pressure is an elongated pressure minimum or "trough" extending from the low center on the Texas–Mexico border, eastward across the Gulf States to North Carolina. This trough in the pressure field coincides with a line of *confluence* (flowing together) in the wind field, which separates a broad southwesterly air current streaming out of the Gulf of Mexico from a northeasterly flow streaming clockwise around the high pressure area over the northern plains. This confluence line is indicated by the heavy line with the pointed and rounded symbols. (The meaning of the symbols will be explained presently.)

On chart (b) in Fig. 3.3, which represents conditions 12 h later, wavelike undulations have developed along the confluence line. The crest of the first "wave" is located over eastern Kentucky, adjacent to the weak pressure minimum designated by the L, and the crest of the second wave is located over southwestern Arkansas, also adjacent to a weak minimum in the pressure field. The northerly flow to the west of the second wave crest appears to be sweeping the confluence line southward through Texas. Further to the east there has been a general northward shift of the confluence line over the 12-h period. The trough along the confluence line has deepened slightly, particularly in the vicinity of the wave crests.

By the time of map (c) in Fig. 3.3, more substantial changes have taken place. The first of the two waves on the confluence line has amplified to become the dominant one. Its crest has moved northeastward into southwestern Pennsylvania, and the associated center of low pressure has deepened by about 10 mb over the 12-h period. A distinct counterclockwise (cyclonic) circulation has developed around the deepening low center. To the east of the wave crest the confluence line is continuing to drift northward, allowing the Gulf air to advance into the Middle Atlantic States, while to the west of the wave crest the cyclonic circulation is sweeping the confluence line southeastward, in advance of the flow of air from the Great Plains. A second confluence line has moved southeastward out of western Canada into the northern plains of the United States. This line is most distinct to the west of the low center, where it separates a northerly flow of Arctic air from a westerly flow of air from the Pacific. The deepening low center over southwestern North Dakota is located near the crest of a wave on this second confluence line.

3.2.2 Temperature; fronts

The distribution of temperature at 00 GCT 19 and 20 November 1964 (which correspond roughly to 6 p.m. local time, 18 and 19 November 1964, respectively) is shown in Fig. 3.4. Here we have chosen to show the raw temperature

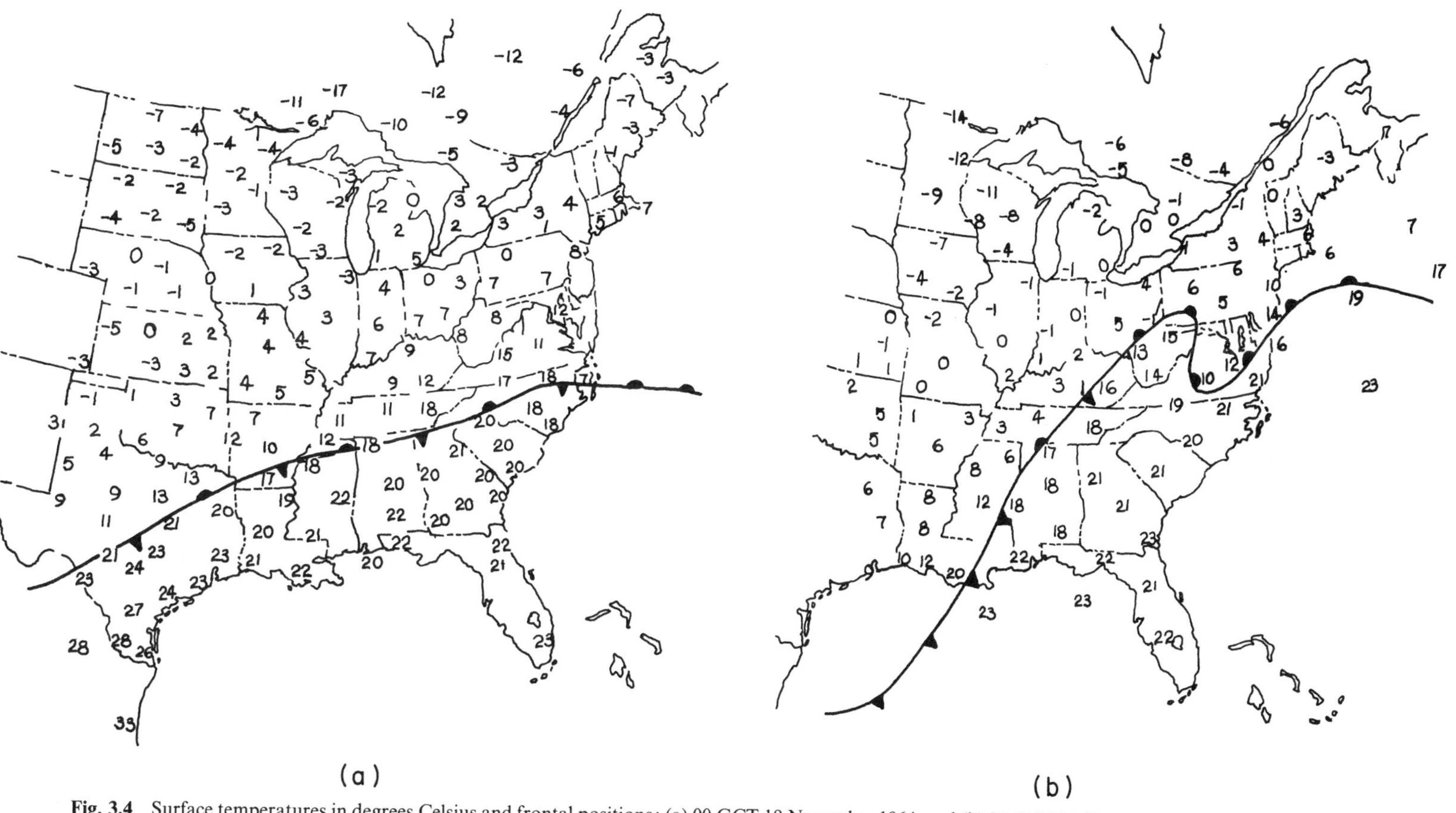

Fig. 3.4 Surface temperatures in degrees Celsius and frontal positions: (a) 00 GCT 19 November 1964, and (b) 00 GCT 20 November 1964. Pointed barbs denote cold fronts, half circles denote warm fronts, and alternating pointed barbs and half circles denote stationary fronts.

data, rather than an analysis of the distribution of surface isotherms. The confluence lines from Fig. 3.3a and c have been transcribed onto these charts.

In Fig. 3.4a, the temperatures to the south of the confluence line are uniformly quite high for the season, which is understandable in view of the fact that the air over this region formerly resided over the warm surface waters of the Gulf of Mexico. Within this warm air mass there is relatively little horizontal temperature gradient; the temperatures along the confluence line are almost as high as they are along the Gulf Coast.

There is no discontinuity in the value of the temperature as one crosses the confluence line. However, within the first 100–200 km to the north of the line there is a band of very strong, horizontal temperature contrast, with temperatures dropping from values of about 20°C along the confluence line itself to about 10°C some 100–200 km to the north of the line. Proceeding further toward the north, the temperatures continue to drop, but at a slower rate. Thus, the confluence line defines the boundary between a rather homogeneous warm air mass and a region of strong thermal contrast between the warm air mass and the colder air to the north. Because of its role as a boundary or dividing line, the confluence line is called a *front*. The region of strong thermal contrast on the "cold air side" of the confluence line is called a *frontal zone* (or sometimes, a *baroclinic zone*).

It should be emphasized that in the context of this definition, it is not quite correct to say that a front marks the boundary between a warm air mass and a colder air mass. The transition between the two air masses takes place within a zone of finite width—the frontal zone. The front is the warm air boundary of the frontal zone and it coincides with the line of confluence in the wind field.

Fronts are labeled in terms of their direction of movement. If the air on the cold side of the frontal zone is advancing into the region formerly occupied by warmer air, the front is called a *cold front*. Cold fronts are denoted on weather charts by triangular shaped "teeth" which point in the direction of movement. For example, the front over the southeastern United States in Figs. 3.3c and 3.4b is moving southeastward as a cold front. Similarly, if the air on the cold side of the front is retreating and being replaced by warmer air, the front is called a *warm front* and denoted on charts by semicircular symbols which point in the direction of frontal movement; in this case, in the direction of the colder air. For example, note the front along the eastern seaboard in Figs. 3.3c and 3.4b. *Stationary fronts* are denoted by alternating cold and warm front symbols on different sides of the line.

To a rather close approximation, fronts behave as *material surfaces* in the atmosphere; that is to say, if one could somehow tag or label the air parcels that lie along a frontal surface at some instant in time and follow them as they move along their respective three-dimensional trajectories through space, these very same air parcels would continue to define the frontal surface at future times. Thus it is almost correct to say air does not move through a frontal surface.

In order to understand why fronts move as they do, it is useful at this point to consider briefly their vertical structure, as depicted schematically in Fig. 3.5, which shows vertical cross sections normal to fronts which are exhibiting various types of movement. All three sections are drawn such that the warm air lies toward the left and the colder air toward the right. Note that regardless of the direction of movement the frontal surface slopes in the direction of the cold air with increasing height. Thus, in all three cases, the frontal zone lies below the frontal surface and the warm air mass lies above it.† Thus it is possible for the warm air to be lifted up and over the frontal surface as shown in Fig. 3.5. In contrast, the air within the frontal zone is "trapped" in the shallow wedge beneath the frontal surface, and thus cannot move relative to the front, or, conversely, the front cannot move relative to it. Hence the direction and speed of movement of the front is determined by the winds within the frontal zone. For example, in Fig. 3.3(c) the front over the southeastern United States is moving southeastward as a cold front, pushed forward by the northwesterly winds within the frontal zone, while the warm front over the eastern seaboard is moving northward, following the retreat of the frontal zone air in that region. It is extremely important to be aware of the need for consistency between frontal movement and the wind field when analyzing sequences of synoptic charts.

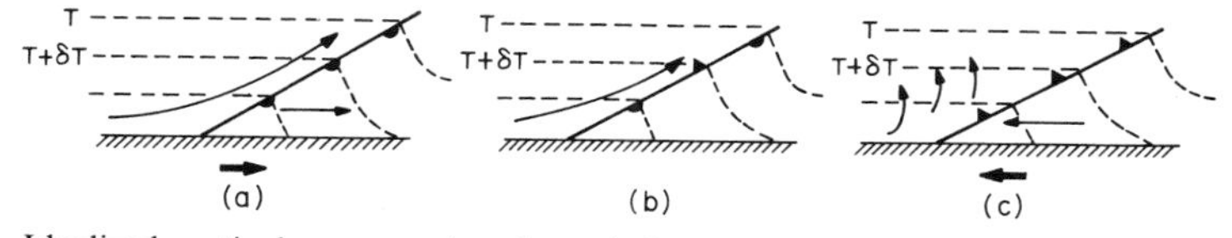

Fig. 3.5 Idealized vertical cross section through frontal zones showing isotherms (---) and air motions relative to the ground (→). (a) Warm front, (b) stationary front with overrunning warm air, and (c) cold front. Heavy arrows at bottom indicate sense of frontal movements.

Lest we overemphasize the role of fronts as a factor in the variability of surface temperature, it should perhaps be mentioned that other factors such as time of day, sky cover, altitude of the station, and proximity to the ocean can be equally, or more, important at times. In fact, there are large regions of the globe in which it is extremely difficult to locate fronts on the basis of gradients of surface temperature:

- over the oceans, where the surface temperature never departs by more than a few degrees from the temperature of the underlying water,
- in mountainous terrain where large differences in station elevation introduce spurious temperature gradients.

† In the case of rapidly moving cold fronts the lowest portion of the front is sometimes retarded by friction so that the frontal zone "overhangs" a narrow strip of warm air at the earth's surface. This effect is usually restricted to the lowest few hundred meters of the atmosphere.

In the analysis of surface charts it is essential to discriminate between the temperature gradients associated with fronts and those due to other influences.

3.2.3 Dew point

Just as frontal zones are characterized by strong horizontal temperature gradients, they also tend to be marked by strong horizontal gradients of dew point, especially when the cold air is of continental origin and the warmer air is of maritime origin, as is often the case over the central and eastern United States. In the 19–20 November 1964 case the distributions of temperature and dew point are so similar that we will not take the space to show the dew points in relation to the front. However, in certain synoptic situations, the dew point gradient is much more reliable than the temperature gradient as an indicator of frontal positions. For example, during summer over land the diurnal temperature range at the ground tends to be considerably larger in cool, dry continental air than in warm, moist air off the Gulf of Mexico, which is often characterized by cloudy skies. Thus, during afternoons, it is not uncommon for surface temperatures in the "cold" air to be just as high as those on the "warm" side of the front, even though there is still considerable thermal contrast 1 km above the ground. Under such conditions the horizontal gradient of dew point is likely to be quite large in the frontal zone, thus providing a clear indication of the frontal position. If such a frontal zone passes a station, moving from northwest to southeast, the observer is likely to notice that the air is becoming much less humid, even though the temperature remains fairly high.

3.2.4 Precipitation

The distribution of precipitation and fog is closely related to the frontal positions as shown in Fig. 3.6. In Fig. 3.6a light rain and snow are falling throughout a broad band to the north of the stationary front as warm air overruns the sloping frontal zone, as indicated in Fig. 3.5b. The light snow extends farther northward along the slopes of the Rockies because of the easterly low-level flow which induces additional lifting in this region where the terrain slopes upward toward the west.

In the 12-h period between (a) and (b) in Fig. 3.6 the precipitation has increased both in intensity and in areal coverage. Snow, sleet, and rain have spread northward in response to increased overrunning above the warm front. Meanwhile an extensive area of fog has developed in the vicinity of the warm front itself. Fog is common wherever warm, moist air passes over a colder, underlying surface. The band of precipitation associated with the cold front is narrow, but rather intense in places, as evidenced by the locally heavy precipitation amounts. Many of the stations in the vicinity of the front reported thunderstorms during the previous 12-h period. In Section 5.5 we will consider in more detail the distribution of precipitation in extratropical cyclonic storms.

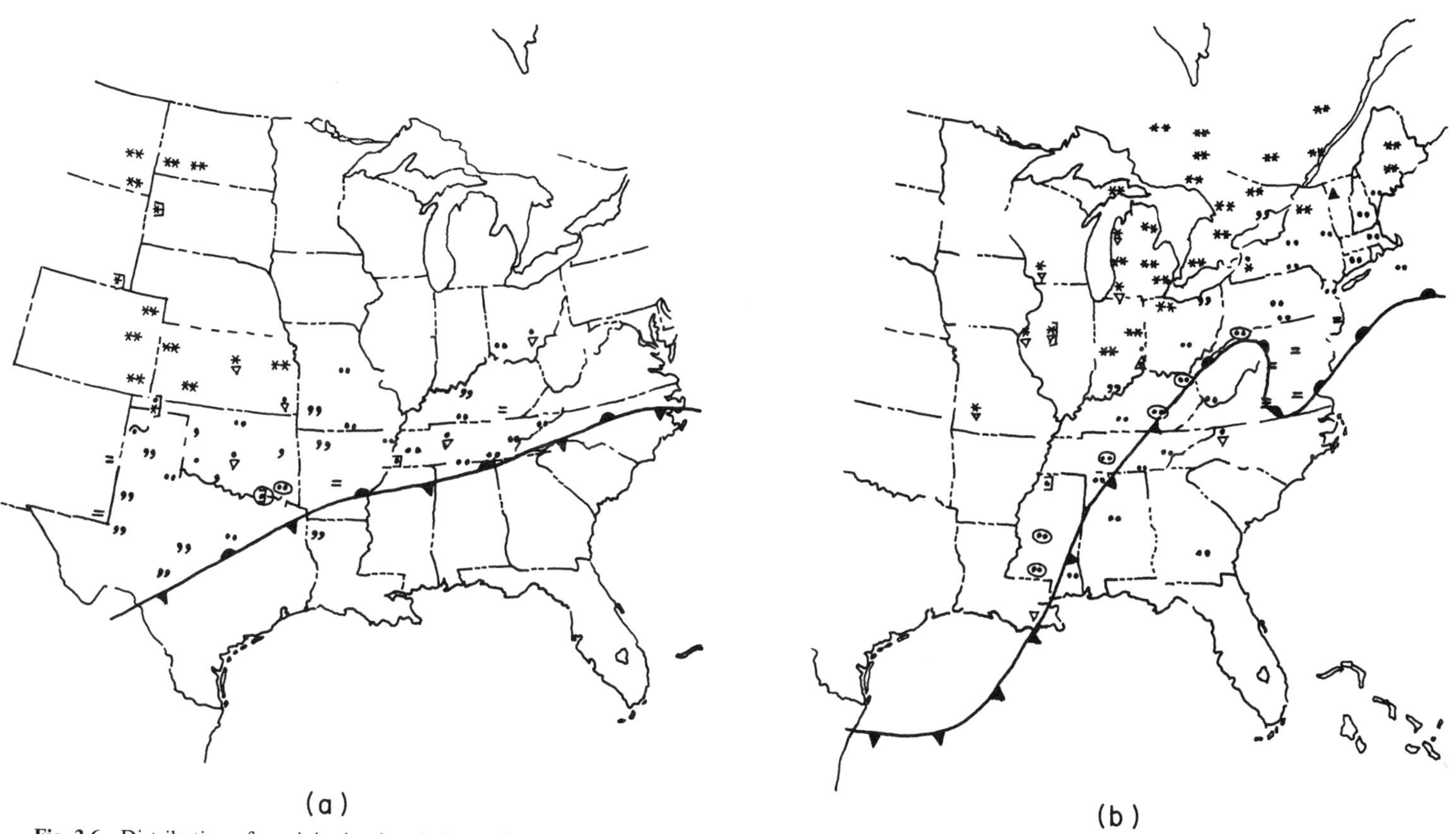

Fig. 3.6 Distribution of precipitation in relation to fronts: (a) 00 GCT 19 November 1964, and (b) 00 GCT 20 November 1964. For an explanation of plotting conventions, see Table 3.1. Symbols with circles around them denote stations that have recorded more than 1 cm of precipitation during the past 6 h.

3.2.5 Pressure tendency

The distribution of pressure change during the past 3 h provides an indication of the direction of motion and the rate of intensification or weakening of various features on the surface chart. The pressure tendency is particularly valuable as an aid in locating fast-moving fronts. For example, as a warm front approaches a station, the layer of relatively dense frontal-zone air adjacent to the ground gradually becomes thinner and warmer. Both effects contribute to a decrease in (hydrostatic) pressure at the ground. The passage of the warm front itself is usually accompanied by a leveling off of the pressure. In a similar manner, the passage of a cold front usually marks the beginning of a period of pronounced pressure rises at the ground. Cold fronts usually slope more steeply, relative to the ground, than warm fronts, and therefore the pressure rises that follow cold fronts tend to be more abrupt than pressure falls that precede warm fronts.

Figure 3.7 shows the distribution of surface pressure tendency at 00 GCT 20 November 1964, the time of the second map in the preceding figures. The lines connecting points at which the same pressure tendency occurs are called *isallobars*. Note the broad area of pronounced pressure falls in advance of the warm front and the narrow band of very strong pressure rises immediately behind the cold front.

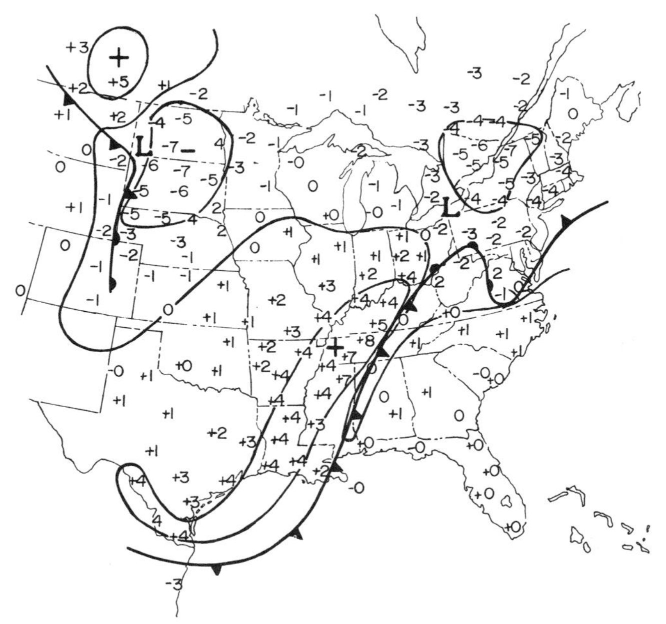

Fig. 3.7 Distribution of pressure change (in whole millibars) during the 3-h period ending at 00 GCT 20 November 1964. Isallobars are drawn at intervals of 4 mb $(3\ h)^{-1}$.

The distribution of pressure tendency also provides an indication of how the intensities of surface cyclones and anticyclones are changing with time. For example, in Fig. 3.7 the pressure is falling in the vicinity of the center of the low pressure area near Buffalo. Thus it is clear that the low is deepening as it moves northeastward and that the associated cyclonic circulation in the wind field is intensifying. In a similar manner the predominance of pressure rises near the center of a high indicates "building up" of the high and an intensification of the related anticyclonic circulation in the wind field.

When interpreting small changes in pressure, it should be borne in mind that the diurnal cycle in solar heating produces small but noticeable pressure fluctuations that have little or nothing to do with the synoptic situation. These "tidal" fluctuations should be subtracted out of the pressure tendencies before trying to infer rates of change that relate to the synoptic scale patterns. Since the tidal pressure changes are geographically and seasonally dependent, these corrections must be made on the basis of climatological data.

3.3 INTERPRETATION OF SYNOPTIC SURFACE REPORTS

An abbreviated version of the plotting model used for surface synoptic observations is shown in Table 3.1. This model incorporates all the meteorological parameters already discussed. The full surface synoptic report contains additional information on visibility, dominant cloud types, and height of the base of the lowest cloud layer, which we will not discuss here. There is also a more extensive set of symbols for describing weather in more detail and for reporting other types of restrictions to visibility such as haze, smoke, dust, and blowing snow. Most of this additional information carried in the surface reports is included primarily because of its applications to aviation. Ship reports include observations of sea surface temperature (or sea–air temperature difference) and a specification of the present rate of movement of the ship.

3.3.1 The synoptic surface chart

The conventional surface chart contains an analysis of the sea level pressure field together with the positions of fronts. In positioning fronts and centers of high and low pressure on the surface chart, the analyst strives to integrate the information plotted for the individual meteorological parameters in order to obtain a representation that is fully consistent with the current synoptic data, and with the previous synoptic chart. A certain amount of subjectivity is inherent in this process, but fortunately there is usually enough redundancy between successive charts, and between the fields of different meteorological parameters, to eliminate most of the ambiguities. In regions of sparse surface synoptic coverage, satellite imagery is the primary basis for positioning the major features on the surface chart.

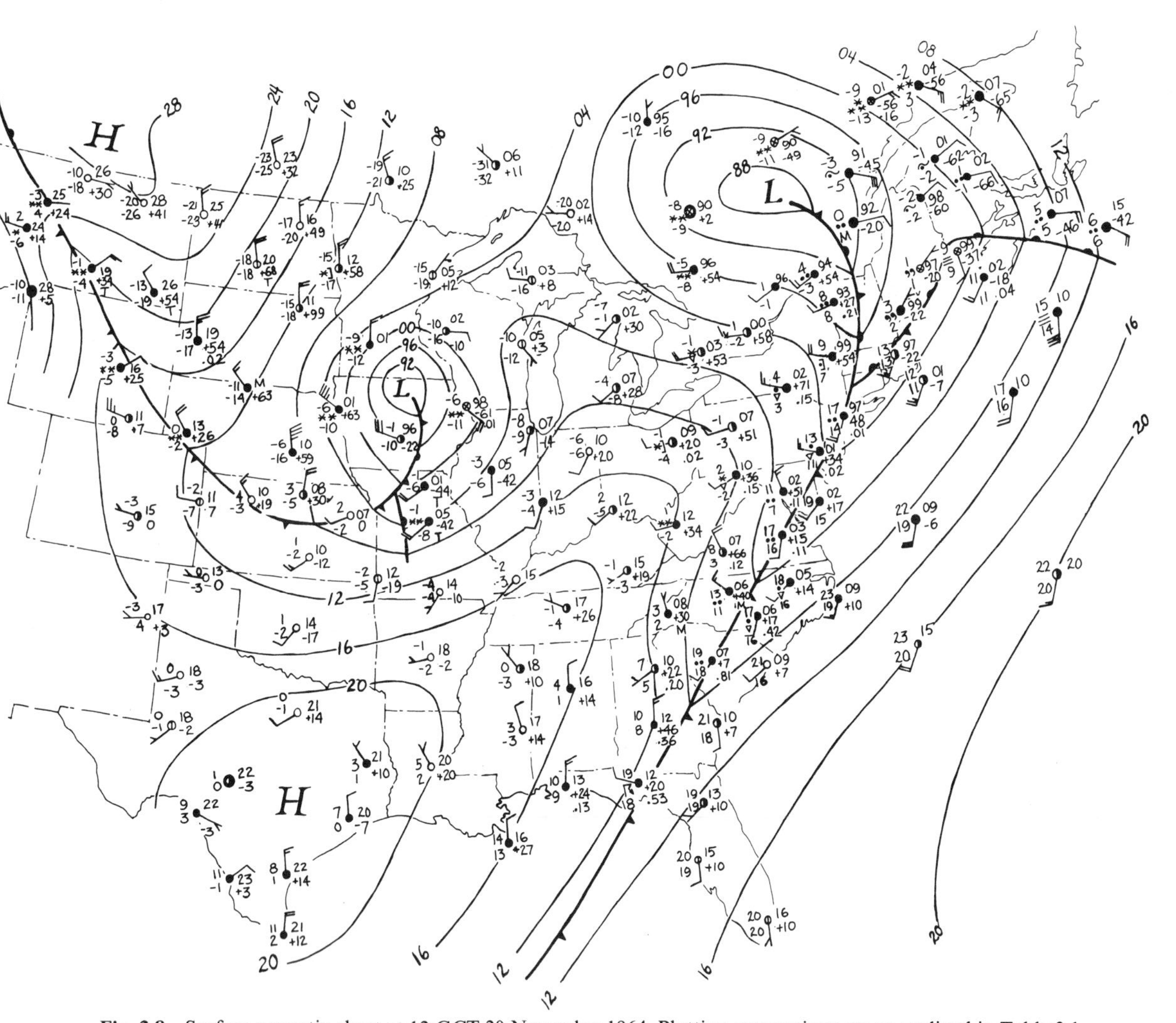

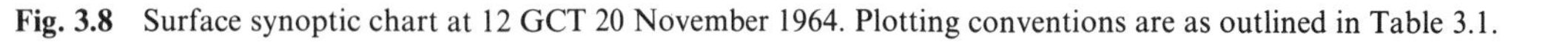

Fig. 3.8 Surface synoptic chart at 12 GCT 20 November 1964. Plotting conventions are as outlined in Table 3.1.

As an example of a surface synoptic chart we show in Fig. 3.8 the analysis for 12 GCT 20 November 1964. Certain familiar features can be identified in this later chart. The area of low pressure that was over the eastern Great Lakes at 00 GCT has moved northeastward into Quebec and deepened slightly, while the wave along the warm front that was located south of Rhode Island has moved northeastward to the coast of New Brunswick. The cold front has progressed rapidly eastward during the 12-h period, and is now approaching the Atlantic coastline.

With the passage of time, the center of lowest pressure is becoming more and more detached from the region of the frontal zone, where it originally formed, as it propagates northward into the cold air. In this latest chart there is evidence of a line of discontinuity connecting the low center over Quebec with the junction of the warm and cold fronts in Connecticut. Stations located ahead (to the east) of the line are experiencing typical pre-warm-front weather (rising temperatures, easterly winds, and falling pressures), while stations located behind (to the west of) the line are experiencing typical post-cold-frontal weather (falling temperatures, decreasing precipitation, gusty westerly winds, and sharply rising pressures). Thus at most stations the passage of this line is marked by a temperature maximum, a pressure minimum, and a windshift. The intensity of the "back to back" frontal zones on either side of the line of discontinuity is directly related to the warmth of the air along the line itself. The thermal contrast is strongest near the junction of the warm and cold fronts, where the air along the line of discontinuity is quite warm, and it becomes progressively weaker as one moves northward along the line, toward the colder air at the center of the low. It is customary to indicate such lines of discontinuity by alternating cold and warm front symbols pointing in the direction of motion, and to refer to them as *occluded fronts*, *occlusions*, or *trowals* (that is, *tro*ugh, *w*arm air *al*oft; a Canadian term).

3.3.2 Time series representation

Figure 3.10 shows time series of surface reports at selected stations in eastern United States and Canada on 19–20 November 1964. (Locations of the stations are shown in Fig. 3.9.) The stations are arranged according to latitude, from north to south. The sequences may be summarized as follows:

- The southernmost stations BR (Brownsville, Texas), LC (Lake Charles, Louisiana), and JA (Jackson, Mississippi) are in the warm air at the beginning of the period. There is a distinct cold front passage accompanied by a distinct windshift and followed by a period of falling temperatures and dew points, and sharply rising pressures. The frontal passage is without precipitation at Brownsville, but it is accompanied by showers and thundershowers at Lake Charles and Jackson.
- NA (Nashville, Tennessee) and HT (a composite of Huntington and Charleston, West Virginia), the next stations to the north, were within the frontal

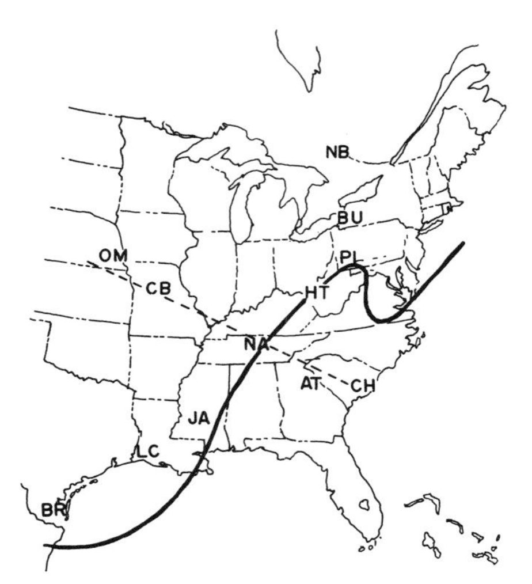

Fig. 3.9 Location of stations used in time sections (Fig. 3.10), soundings (Fig. 3.18), and vertical cross section (Figs. 3.19 and 3.20): (---) orientation of the cross section; (——) the frontal positions at 00 GCT 20 November 1964, the time of the soundings and cross section. Station names are given in the text.

zone at the beginning of the sequence. They experience a period of rising temperatures and falling pressures during the time that the front drifts northward toward them. Then there follows a brief period of southwesterly winds and high temperatures, which begins with the passage of the warm front shortly after 18 GCT 19 November. (The interval of time within the warm sector of the storm is longer for stations located farther to the east.) The brief warm interval is followed by a sharp cold front passage.

- The warm air never reaches PI (Pittsburgh, Pennsylvania) and yet there is evidence of a frontal passage shortly after midnight GCT 20 November. Prior to this time, conditions are similar to those at HT and NA before the warm front passage, and afterwards the sequence is similar to those at HT and NA following the passage of the cold front. Therefore, the front that passed PI has characteristics typical of an occlusion, as described above. Indeed, an occluded front could have been drawn just west of Pittsburgh in Fig. 3.3c.
- The sequence of events at BU (Buffalo, New York) is not as clearly defined as at PI. There is a windshift and a pressure minimum around 03 GCT 20 November, but there is no well-defined temperature maximum at this time. However, prior to the windshift there is evidence of warming aloft, with snow changing to rain, while some time after the windshift the rain changes back to snow. Conditions at BU are indicative of the passage of a weak occluded front, far from the junction of the warm and cold fronts.
- NB (North Bay, Ontario) is situated deep within the cold air throughout the time sequence. There is a pressure minimum and a windshift as the center

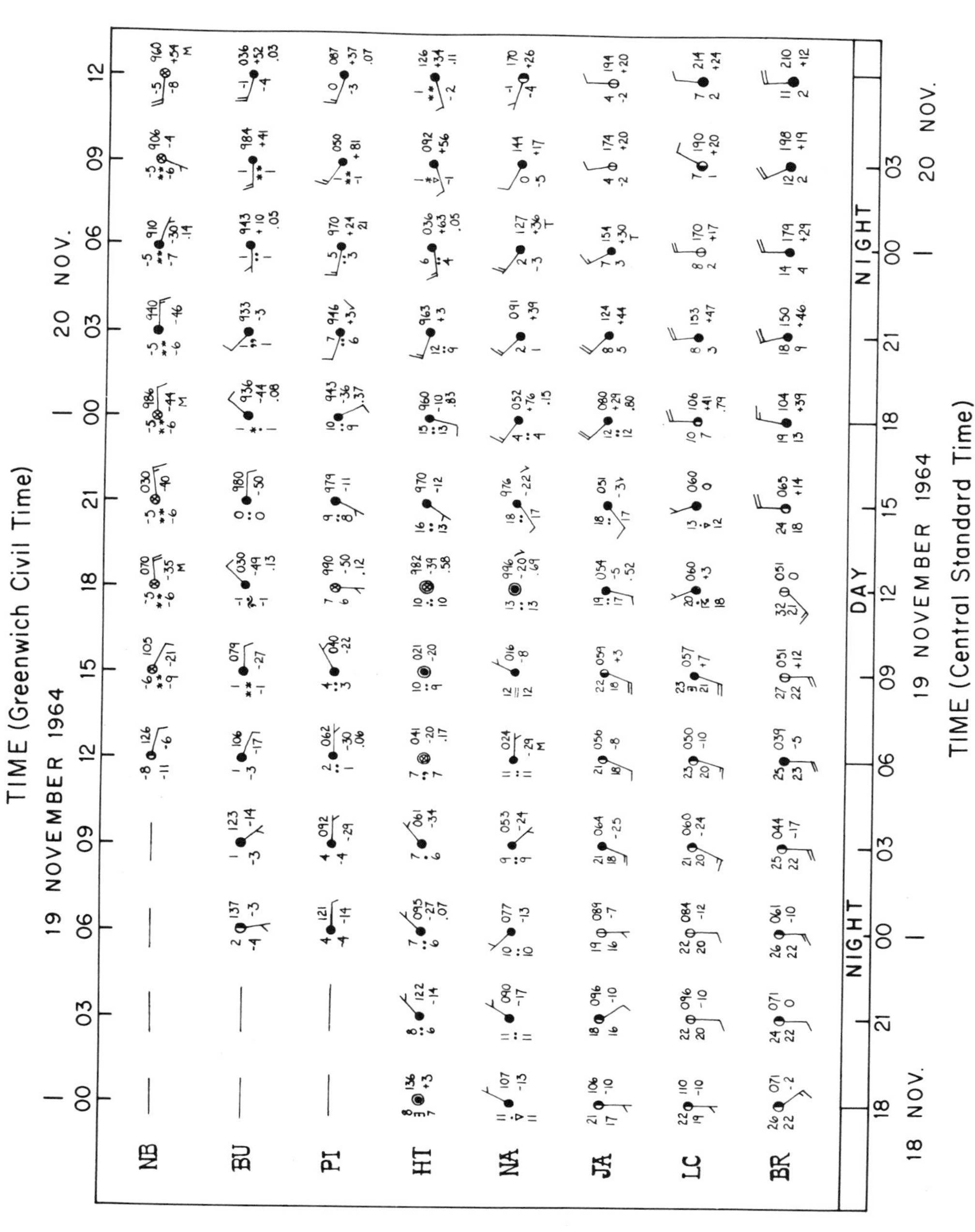

Fig. 3.10 Time series of surface synoptic reports. Station locations are shown in Fig. 3.9, plotting conventions in Table 3.1.

of the low passes the station, but these changes are not associated with any frontal passage.

3.3.3 Models of the life cycle of extratropical cyclones

Most middle-latitude cyclones have much in common with the "classical textbook example" described in this chapter. Important characteristics of such "polar front cyclones" are

- the initial development of the low pressure center along a stationary front; the low center develops on the crest of a wavelike undulation in the shape of the front;
- the ensuing movement of the frontal zone in response to the developing circulation around the deepening low pressure center, with the cold air retreating toward higher latitudes in advance of the surface low, and sweeping equatorward and eastward behind it;
- the propagation of the low pressure center toward the cold air as it deepens, with an occluded front connecting the low center to the junction of the warm and cold fronts; the occlusion process usually marks the end of the period of rapid development (cyclogenesis).

The sequence of events described above is embodied in the idealized model shown in Fig. 3.11, which is very similar to the one first developed by the Bergen school[†] more than 50 yr ago. This model has been widely used by weather forecasters as a basis for interpreting and anticipating changes in the surface synoptic chart.

[†] The school was founded in 1918 by the Norwegian physicist **Vilhelm Bjerknes** (1862–1951), his son **Jacob Bjerknes** (1897–1975), **Halvor Solberg** (1895–), and **Tor Bergeron** (1891–). The elder Bjerknes began his career as a physicist. In the early 1890s he collaborated with Heinrich Hertz and published several fundamental papers in radio science. In the latter part of that decade he turned his attention to the dynamics of atmospheres and oceans. The "circulation theorems," which he developed during this period, provide a theoretical basis for the basic concepts discussed in Section 9.3 of this book. During World War I, when Norway was cut off from most outside weather information, Bjerknes was called upon to found a Geophysical Institute at Bergen. In this role he was successful in convincing the Norwegian government to install a dense network of surface stations which provided data for investigating confluence lines in the surface wind field. These studies led to the concept of fronts and ultimately to models of the life cycle of frontal cyclones. In his characteristically modest manner, Bjerknes credited his younger colleagues with the major scientific achievements of the Bergen school: "During 50 years meteorologists all over the world had looked at weather maps without discovering their most important features. I only gave the right kind of maps to the right young men, and they soon discovered the wrinkles in the face of Weather."

In 1919, J. Bjerknes (aged 22 at the time) published an eight-page paper which introduced the concept of warm, cold, and occluded fronts and correctly explained their relationship to extratropical cyclones. By 1926, in collaboration with Solberg and others, he had described the structure and life cycle of extratropical cyclones. Bergeron made important contributions to the understanding of occluded fronts and the formation of precipitation (see Section 4.5.4).

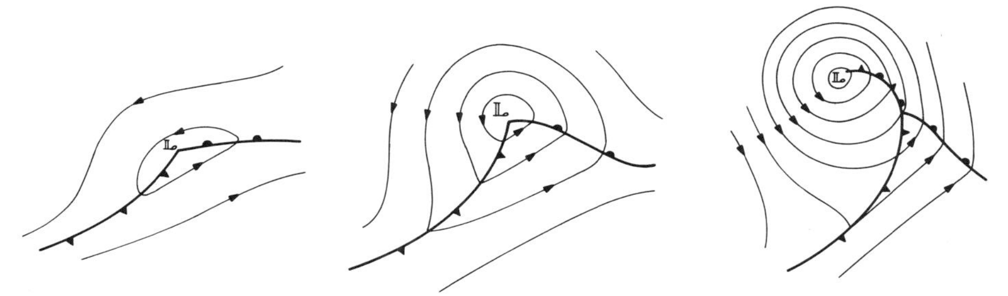

Fig. 3.11 Idealized model of a middle-latitude cyclone in three stages of development showing isobars of sea level pressure and fronts. Arrows indicate the direction of the geostrophic wind. (Adapted from E. Palmén and C. W. Newton, "Atmospheric Circulation Systems," Academic Press, New York, 1969.)

It should be emphasized that relatively few extratropical storms fit the idealized "polar front cyclone" model as well as the one selected for display in this chapter. Clearly defined warm fronts are lacking in many storms,† particularly during the later stages of development, and topographical features often distort or obscure existing fronts,‡ or generate new fronts that would not exist otherwise. Furthermore, there is increasing evidence that some storms develop in isolation from any pre-existing fronts; the fronts that accompany these systems form as part of the process of storm development. Some of these forms of "aberrant behavior" are difficult to understand or predict on the basis of the surface synoptic map alone, but they make sense when viewed as a response to changes that are taking place at upper levels.

3.3.4 Further remarks on occluded fronts

Occluded fronts have been subject to various conflicting interpretations. The term *occluded* means overlapping: it stems from the widely held notion that such fronts form when the cold front "catches up with" part of the warm front during the process of cyclogenesis. When actual frontal movements are carefully examined, there are few, if any, well-documented examples of cold fronts overtaking warm fronts to form occlusions. Rather, it appears that most occluded fronts are essentially new fronts which form as surface lows separate themselves from the junctions of their respective warm and cold fronts and deepen progressively further back into the cold air.

The concept of an occluded front can be understood in terms of the idealized model shown in Fig. 3.12, which stems from the Bergen school. In the cold-type

† Note, for example, the storm over the Northern Plains in Fig. 3.8.

‡ The southward bulge of the warm front in the Middle Atlantic States in Fig. 3.3c is an effect of the Appalachian Mountains. Most warm fronts display a similar distortion when they pass over this region.

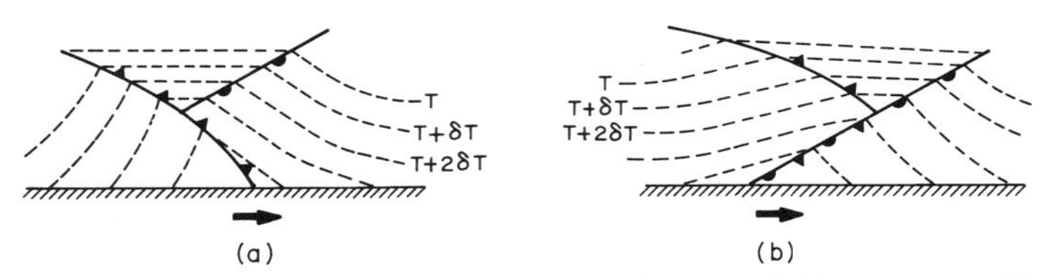

Fig. 3.12 Idealized models of occluded fronts: (a) cold type and (b) warm type. The sketches represent frontal surfaces (——) and isotherms (- - -) in vertical cross sections normal to occluded fronts, which are moving from left to right.

occlusion shown in Fig. 3.12a the cold front extends to the ground and the warm front exists only aloft, whereas in the warm-type occlusion (Fig. 3.12b) the reverse is true. Note that both types of occlusion are consistent with the notion of "back to back" frontal zones at the ground, so that weather conditions prior to the frontal passage are similar to those ahead of a warm front and conditions after the frontal passage are similar to those behind a cold front. It is clear from the diagrams that the passage of a "cold-type occlusion" is marked by an increase in the static stability of the lower troposphere, whereas a warm-type occlusion is marked by a decrease in static stability.

The classical occluded structures shown in Fig. 3.12 are rarely observed in their entirety. One or more of the frontal discontinuities indicated in the figure is often missing or obscured by mesoscale features in the vicinity of the front (as will be discussed in Section 5.5). For example, there may not be a well-defined frontal passage at the ground, or the warm (or cold) frontal zone aloft may lack a well-defined warm air boundary. In such situations it may not be clear whether the occlusion is of the warm or cold type. In view of the wide variety of occluded frontal structures that exist in nature, it is advisable to identify occluded fronts, not in terms of a set of models of frontal configurations but rather in terms of their essential characteristics: namely,

- back to back frontal zones at low levels with the warmest air in the vicinity of the front,
- a trough in sea level pressure.

It will be shown in Section 3.5.1 that an occluded front can also be identified in terms of a ridge in the lower tropospheric thickness field.

3.4 UPPER LEVEL STRUCTURE

The time evolution of the patterns on the surface synoptic chart becomes more understandable when viewed in the context of the synoptic situation in the troposphere as a whole. For example, it is observed that the winds in the middle troposphere (near 500 mb) tend to act as a "steering flow" for features on the surface chart. The upper level patterns also influence the rate of intensification or weakening of surface cyclones and anticyclones, and the amount and type of precipitation that accompanies them.

3.4.1 Upper level synoptic charts

Figures 3.13–3.17 show the distributions of geopotential height and temperature on the 850-, 700-, 500-, 250-, and 100-mb pressure surfaces, respectively, at 00 GCT 20 November 1964 (the same time as the surface charts shown in Figs. 3.3c, 3.4b, and so on).

The 850-mb chart is rather similar to the surface chart discussed previously, but there are some notable differences:

- The closed lows over Buffalo, New York, and western North Dakota at the surface appear as troughs at the 850-mb level.

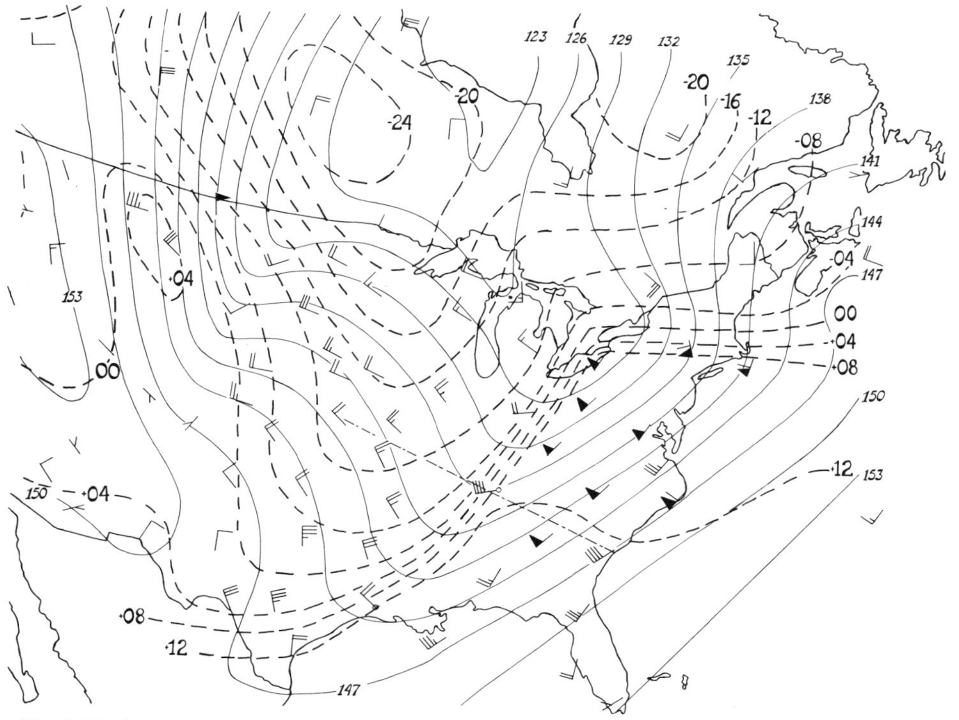

Fig. 3.13 850-mb chart for 00 GCT 20 November 1964: (——) geopotential height contours, drawn at intervals of 30 m and labeled in tens of meters; (– – –) isotherms, labeled in degrees Celsius. Wind speeds are in knots. The stations denoted by the small circles are Lake Charles and Nashville.

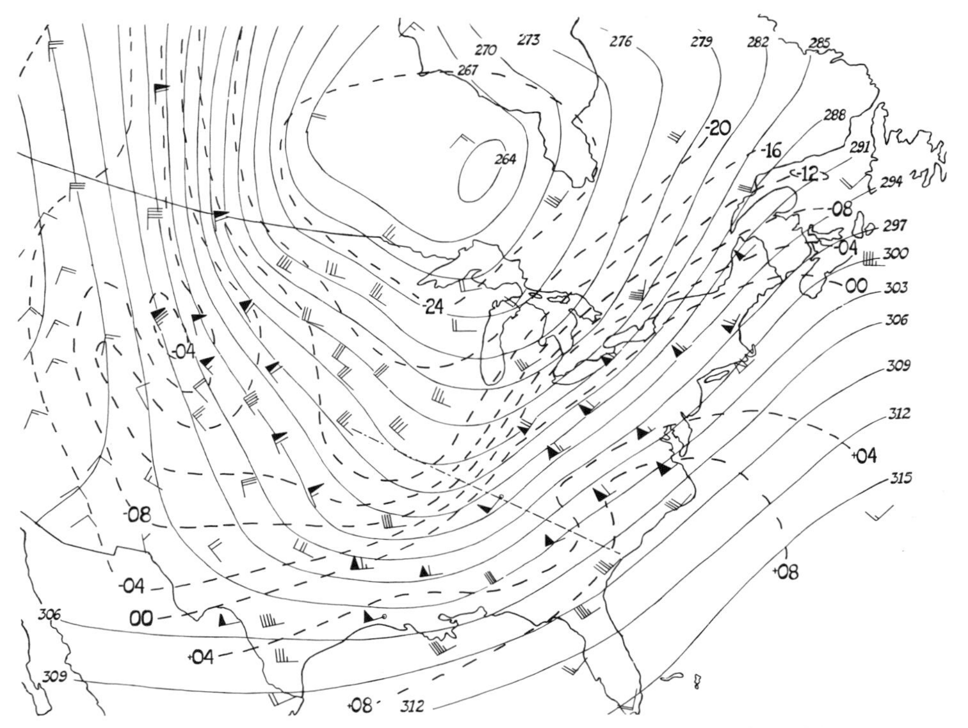

Fig. 3.14 700-mb chart for 00 GCT 20 November 1964: (——) geopotential height contours, drawn at intervals of 30 m and labeled in tens of meters; (---) isotherms, labeled in degrees Celsius. Wind speeds are in knots. The stations denoted by the small circles are Lake Charles and Nashville.

- The winds are generally stronger at 850 mb than at the ground, and there is no evidence of systematic flow across the isobars toward lower pressure as there is on the surface map.
- The cold front (defined as the warm air boundary of the frontal zone) is well past NA (Nashville) and LC (Lake Charles) on the surface chart, but it has just reached them at 850 mb. Thus, the cold front slopes backward toward the cold air with increasing height, in agreement with the model described in Fig. 3.5.
- The warm front is more clearly defined and much farther north at 850 mb than on the surface map. Apparently the "pool" of cold (frontal zone) air over the Middle Atlantic States (see Fig. 3.4b) is very shallow.

Proceeding upward from 850 to 700 mb we note the following:

- The troughs associated with the surface low centers are becoming less distinct and they are displaced upstream relative to their positions on the surface map.

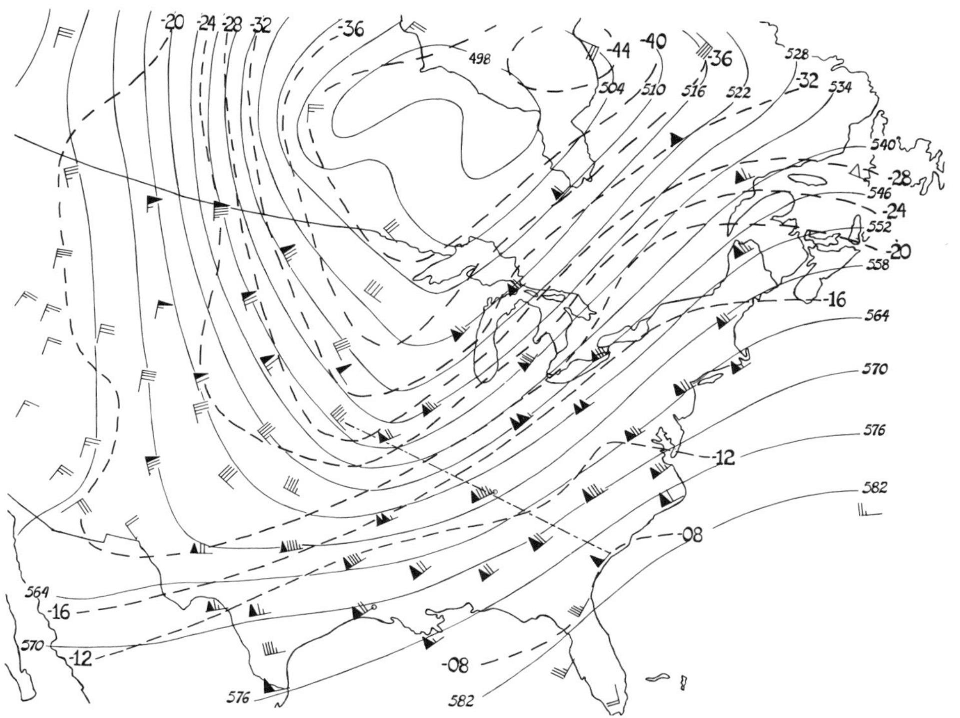

Fig. 3.15 500-mb chart for 00 GCT 20 November 1964: (——) geopotential height contours, drawn at intervals of 60 m and labeled in tens of meters; (– – –) isotherms, labeled in degrees Celsius. Wind speeds are in knots.

- There is a further increase in wind speed with height at most stations.
- The cold front is located still farther toward the northwest of its surface position (note that it has not yet reached NA and LC).
- The warm front is located far to the north of its 850-mb position. Much of the warm front precipitation in Fig. 3.6b coincides with the upper level position of the frontal zone.
- The frontal zones are less clearly defined than at lower levels. For example, it is difficult to identify the southern portion of the cold front at this level.

The transition from the 700-mb chart to the 500-mb chart is marked by a further upstream displacement of the troughs, strengthening of the winds, and sloping of the frontal zones toward the cold air. At the 500-mb level, well-defined frontal zones are rather unusual, but strong thermal contrasts still exist. Note that the gross features of the 500-mb temperature field are similar to those on the charts for lower levels. In fact, throughout the depth of the troposphere, the isotherms have much the same orientation.

We recall from Figs. 1.10 and 1.11 that, in a climatological sense, there is in middle latitudes a distinct break between the high, cold tropical tropopause

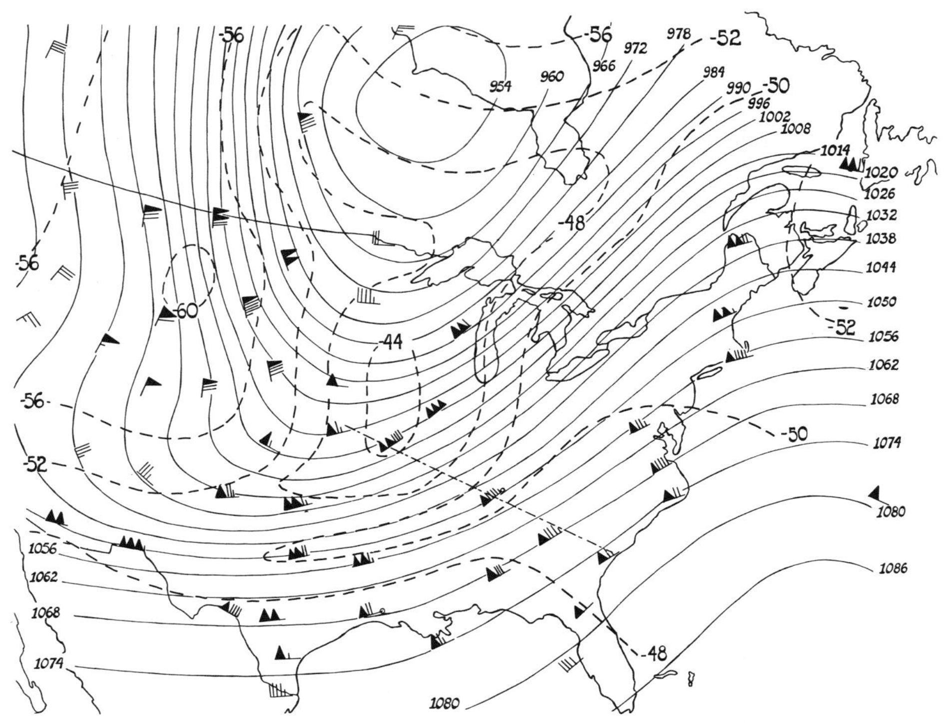

Fig. 3.16 250-mb chart for 00 GCT 20 November 1964: (——) geopotential height contours, drawn at intervals of 60 m and labeled in tens of meters; (– – –) isotherms, labeled in degrees Celsius. Wind speeds are in knots.

and the lower, warmer polar tropopause. It was shown in Fig. 1.11 that this break coincides with the climatological position of the jet stream. The 250-mb chart shown in Fig. 3.16 cuts across this tropopause break at the position of the jet stream. North of the jet stream the 250-mb level is located in the lower stratosphere, while south of the jet stream it is located in the upper troposphere. The warmest air at 250 mb is located on the poleward side of the jet stream in the region of the trough, over the central United States. It will be shown that this warm air coincides with a region of very low tropopause heights.

The 100-mb surface is located well above the tropopause and jet stream. In passing upward into the stratosphere there is a marked decrease in wind speeds and a change in the scale of the circulation patterns: the charts up to the 250-mb level were characterized by a superposition of synoptic-scale and planetary-scale features, whereas only planetary-scale features are present on the 100-mb chart. Note that the temperature field completely reverses between troposphere and stratosphere. For example, at the 100-mb level, the highest temperatures are found over central Canada and the lowest temperatures are found in the vicinity of Florida.

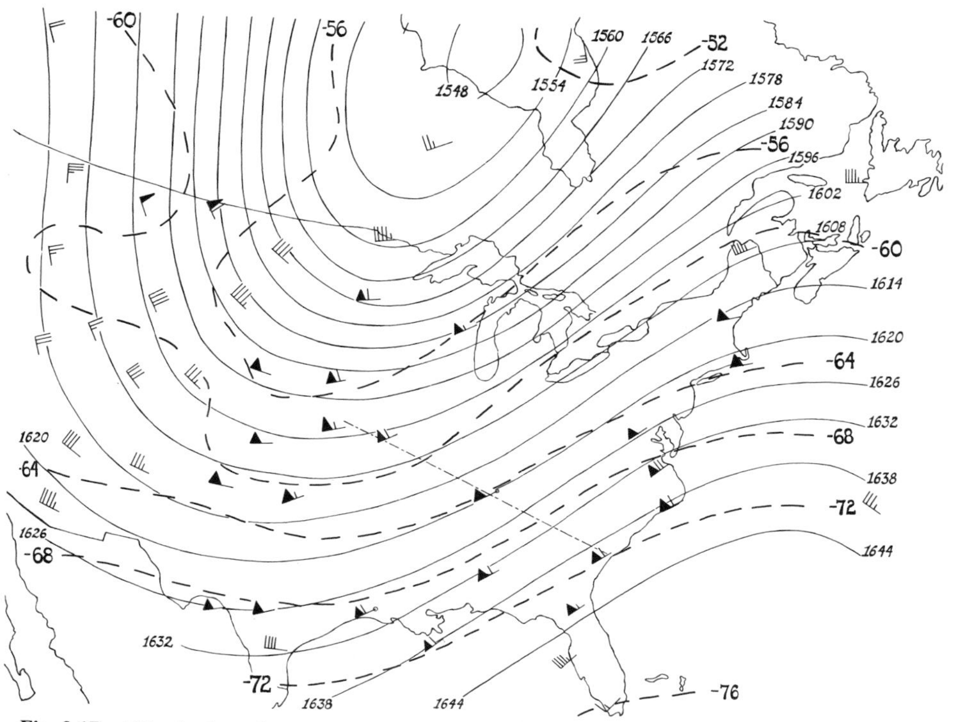

Fig. 3.17 100-mb chart for 00 GCT 20 November 1964: (——) geopotential height contours, drawn at intervals of 60 m and labeled in tens of meters; (- - -) isotherms, labeled in degrees Celsius. Wind speeds are in knots.

3.4.2 Vertical soundings

Vertical temperature soundings for four stations oriented along a line perpendicular to the cold front are shown in Fig. 3.18. The locations of the stations are shown in Fig. 3.9.

The sounding for Athens, Georgia (AT) shows the typical vertical structure of the warm air mass. With the exception of a few minor temperature inversions, the lapse rate is rather uniform and close to moist adiabatic throughout the troposphere. The tropopause is well defined and occurs at a level of about 180 mb or roughly 13 km. At lower latitudes the higher discontinuity in the temperature profile near 100 mb becomes the dominant tropopause.

The sounding for Nashville, Tennessee (NA) is rather similar to the one just described, except for the appearance of the stable frontal-zone air at low levels. The cold front coincides with the top of the inversion, at 850 mb. We recall from Fig. 3.13 that the cold front passes through NA at the 850-mb level, in agreement with the sounding. The increased static stability within the frontal zone is consistent with the thermal structure indicated in Fig. 3.5.

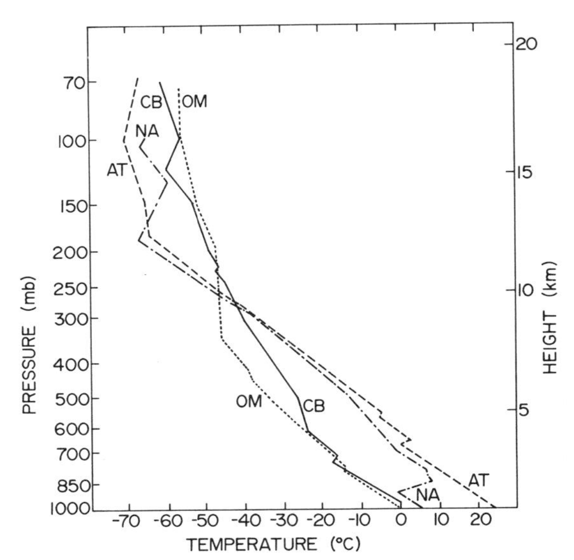

Fig. 3.18 Vertical temperature soundings at 00 GCT 20 November 1964. See Fig. 3.9 for the station locations.

Proceeding farther back into the cold air to Columbia, Missouri (CB) we begin to see more pronounced changes in the vertical temperature profile. The frontal zone is no longer clearly evident, but below the 300-mb level the whole sounding is much colder than the one for NA. The lapse rate gradually decreases from troposphere to stratosphere, but it is impossible to identify any distinct point on the profile at which the transition from troposphere to stratosphere takes place. Above 300 mb the air is warm relative to the first two soundings.

Omaha, Nebraska (OM) is located near the 500-mb trough deep within the cold air mass. On the whole, the sounding for this station is rather similar to the one for CB. The one notable difference is the reappearance of the tropopause, near 350 mb, or roughly 8.5 km. We note that in the climatological cross sections shown in Fig. 1.11 the tropopause does not reach as low as 350 mb, even over the poles. Such extremely low tropopause heights are found only locally in the vicinity of sharp upper tropospheric troughs. In general such regions tend to be characterized by low temperatures throughout the depth of the troposphere, but very high temperatures just above the tropopause. In general, there seems to be a tendency for compensation between tropospheric and lower stratospheric temperatures.

3.4.3 Vertical cross sections

The vertical structure of the atmosphere in the vicinity of the cold front is shown in Fig. 3.19. This cross section was constructed on the basis of temperature and wind soundings at the five stations indicated in the legend and constant pressure level charts such as those shown in Figs. 3.13–3.17. The

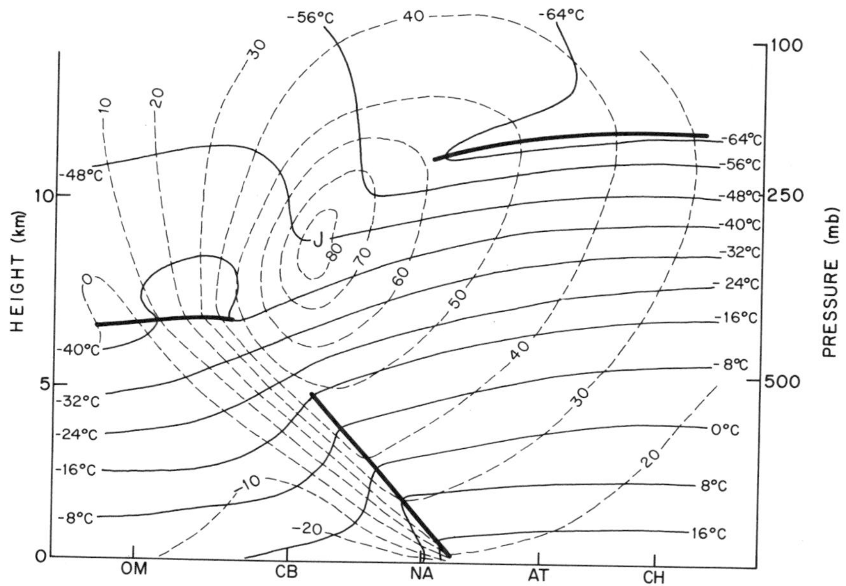

Fig. 3.19 Distribution of isotherms in degrees Celsius (——) and isotachs in meters per second (– – –) in a vertical cross section through the cold front at 00 GCT 20 November 1964. Station locations are indicated in Fig. 3.9. Isotachs refer to the geostrophic wind component normal to the section. Positive values indicate winds directed into the section. The heavy lines indicate the cold front and the tropopause. J refers to the axis of the jet stream.

isotachs (lines of constant wind speed) in the cross section refer to the wind component normal to the section. The sign convention is such that "positive" velocities refer to winds directed into the section, and "negative" velocities refer to winds directed out of the section.† The section is oriented approximately normal to the front and to the jet stream. The following features are evident in the figure:

- There exists a well-defined frontal zone in the lower troposphere, sloping toward the cold air with increasing height. In agreement with Fig. 3.18, the front does not extend as far west as CB.
- Within the frontal zone the isotachs are sloping and very close together, which indicates that the wind component directed into the section is increasing very rapidly with height. Such a region is said to be characterized by strong *vertical wind shear* (literally, a large vertical gradient of the horizontal wind vector).
- The middle-latitude tropospheric jet stream is located within the gap in the tropopause. As noted in the discussion of Fig. 3.16, the 250-mb surface

† The isotach analysis is based not upon actual winds but upon geostrophic winds, as defined in Section 8.4.1. For the purposes of this qualitative discussion the distinction between geostrophic winds and actual winds is unimportant.

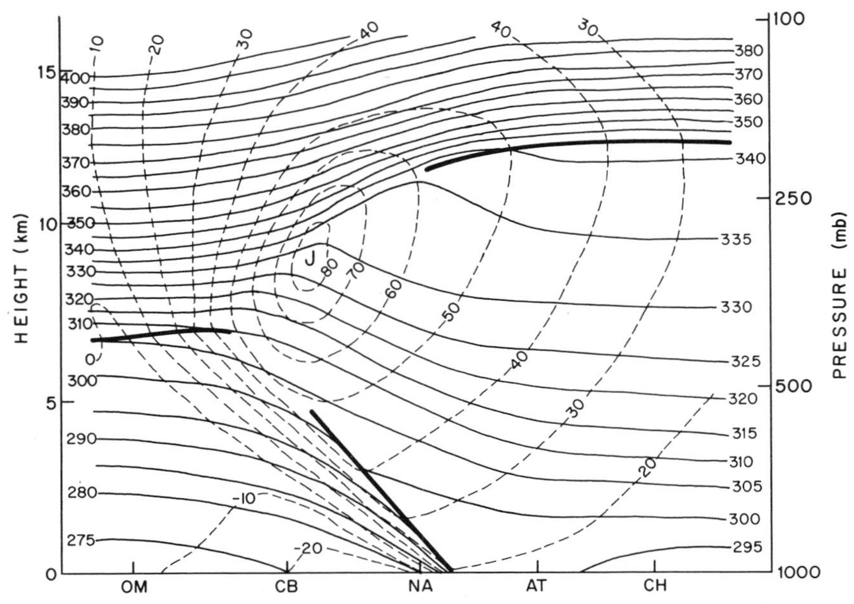

Fig. 3.20 Vertical cross section through frontal zone as shown in the previous figure except that solid lines represent isentropes (lines of constant potential temperature), labeled in degrees Kelvin.

crosses from troposphere to stratosphere at the jet stream, which crosses through the section near CB. From the section it is evident that vertical soundings taken in the vicinity of the jet stream do not show a well-defined tropopause.

- There is a reversal in the horizontal temperature gradient between troposphere and stratosphere.

In vertical cross sections it is sometimes convenient to display the distribution of potential temperature, rather than temperature. Under adiabatic conditions the isentropes (potential temperature lines) in such sections can be closely identified with air motions. In Fig. 3.20 we have reanalyzed the section shown in the previous figure in terms of isentropes and isotachs. In such analyses the stability stratification is directly related to the vertical spacing of the isentropes. Regions of close spacing (for example, the stratosphere and the frontal zone) are characterized by strong static stability.

3.5 THICKNESS AND ITS RELATIONSHIP TO VERTICAL STRUCTURE

Given the distribution of geopotential height on any two pressure surfaces, it is possible to deduce the distribution of thickness for the intervening layer. If the height fields for the two pressure surfaces are analyzed using contour lines that are integral multiples of the same contour interval (for example, 60 m) and superimposed upon the same geographical grid, the corresponding thickness chart can be derived conveniently through the procedure of "graphical subtraction." This procedure makes use of the fact that the thickness is known at each of the intersection points between the two sets of height contours; at each intersection point it is simply the difference between the labels on the two intersecting contour lines. It is clear that the value of the thickness at each of the points where the upper and lower contours intersect must be divisible by the standard contour interval. Thus the family of thickness contours that correspond to integral multiples of the standard contour interval must pass through the intersection points of the upper and lower height contours. It is also clear that one of these thickness contours cannot cross a height contour except at points where the upper and lower height contours intersect. Hence, when the three sets of contours are superimposed, all intersections must be three-way intersections.

3.5.1 The 1000–500-mb layer

Figure 3.21 shows the distribution of thickness for the 1000–500-mb layer at 00 GCT 20 November 1964. Superimposed upon the same analysis are the 500-mb height contours (heavy solid lines) and the 1000-mb height contours (lighter solid lines). The latter set of contours is derived directly from the sea level pressure analysis, using the approximation that the pressure drops 1 mb for each 8 m of vertical ascent. Thus, to within an accuracy of better than 10% the 1000-mb isobar can be relabeled as the zero height contour for the 1000-mb surface, the 1008-mb isobar as the 60-m height contour, the 992-mb isobar as the −60-m height contour, and so on. It is evident that all intersections in Fig. 3.21 are three-way intersections. By referring to the appropriate labels on the lines, it can be readily verified that at any of these intersection points, the 1000-mb height plus the 1000–500-mb thickness, is equal to the 500-mb height. Following along any given 500-mb height contour through several successive intersection points, it can be seen that, if the 1000-mb height increases by 60 m, the thickness decreases by the same amount, and vice versa, so that the sum of the two remains constant. Thus as one approaches a surface low the thickness contours cut across the 500-mb height contours toward lower 500-mb height and vice versa.

In view of the proportionality between thickness and virtual temperature, as discussed in Section 2.2.3, it is reassuring to find that the 1000–500-mb thickness distribution shown in Fig. 3.21 resembles the temperature distributions for the 850-, 700-, and 500-mb levels as displayed in Figs. 3.13–3.15. The vertical averaging inherent in the thickness field tends to smear out the frontal zones somewhat, but they are still evident as regions of strong thickness contrast on the cold side of the surface fronts. Since the 1000-mb surface is rather flat in comparison to the 500-mb surface, there is a strong correspondence

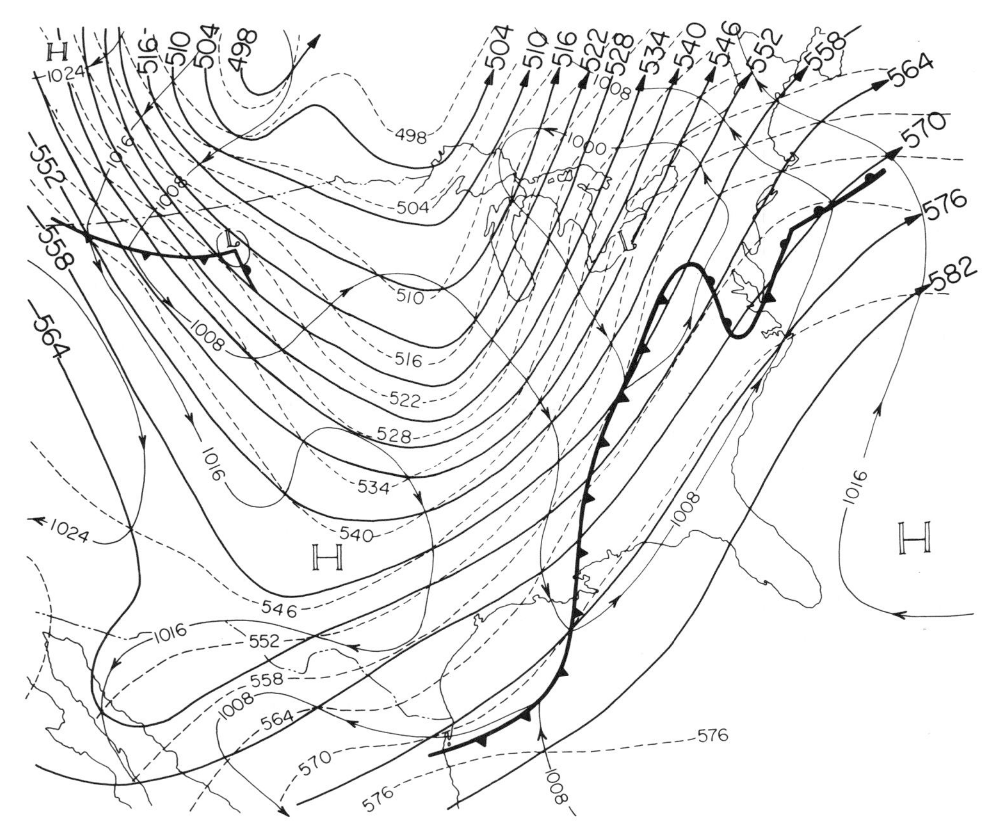

Fig. 3.21 Distribution of sea level pressure (——), 500-mb height (——), and 1000–500-mb thickness (– – –) at 00 GCT 20 November 1964. Thickness and height contours are labeled in tens of meters. Arrows on contours denote the direction of the geostrophic wind. Letters H and L refer to maxima and minima the sea level pressure field.

between the lower tropospheric temperature field and the 500-mb height field. For example, it is evident that the deep 500-mb low over central Canada is largely a reflection of the low 1000–500-mb thicknesses associated with the deep pool of cold air which covers that region.

The northerly flow behind the surface low over the eastern Great Lakes has carried some of the cold Canadian air southward into the central United States, giving rise to a pronounced "trough" in the thickness field over that region. It is because of this trough in the thickness field that the 500-mb trough is located well to the west of the surface low. In a similar manner, the cold air outbreak following behind the surface low over North Dakota shows up as a trough in the thickness field, which is responsible for the 500-mb trough upstream from the position of that surface low. The general tendency for middle-latitude disturbances to slope westward with height in the lower troposphere is thus seen to be a consequence of the distortion of the temperature field by the lower tropospheric flow patterns, which tend to bring colder air equatorward behind the surface lows and warm air poleward ahead of them.

Figure 3.22 shows a model of an idealized middle-latitude (northern hemisphere) disturbance in three stages of development, with the 1000- and 500-mb height fields and the thickness field superimposed as in the previous figure. In the initial chart the surface low is just beginning to form as a wave along a front on the warm side of the region of strong thickness contrast. In the second chart cold air is streaming southward behind the surface low and warm air is advancing northward ahead of it. These distortions in the temperature field are reflected in the growing amplitude of the "wave" in the thickness pattern. Note also how the thickness pattern is closely related to the position of the warm and cold fronts. In this stage of development the surface low is in the process of passing under the jet stream, from the warm to the cold side. In the third chart the occlusion process has begun and the surface low has begun to move across the thickness contours toward lower values as it deepens progressively farther back into the cold air. The junction of the warm and cold fronts remains on the warm side of the region of strong thickness contrast and the occluded front coincides with a warm "ridge" in the thickness field. As the disturbance continues to amplify, the positions of the surface low and the 500-mb trough (or closed low) gradually begin to come into vertical alignment. In fully developed systems the vertical tilt completely disappears and all three sets of contours become mutually parallel.

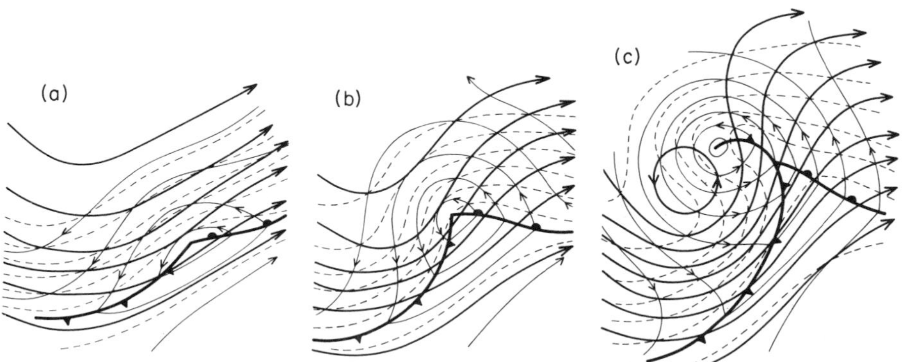

Fig. 3.22 Idealized model of a middle-latitude storm in three stages of development: (a) initial stage, (b) developing stage, and (c) occluded stage. (——) isobars of sea level pressure, (——) contours of 500-mb height, (---) contours of 1000–500-mb thickness. (From E. Palmén and C. W. Newton, "Atmospheric Circulation Systems," Academic Press, New York, 1969, p. 326.)

3.5.2 The 250–100-mb layer

The distribution of temperature in the 250–100-mb layer at 00 GCT 20 November 1964 is displayed in Fig. 3.23. The thickness pattern qualitatively resembles the 100-mb temperature field shown in Fig. 3.17, and it is remarkably similar to the 1000–500-mb thickness pattern except for the fact that the

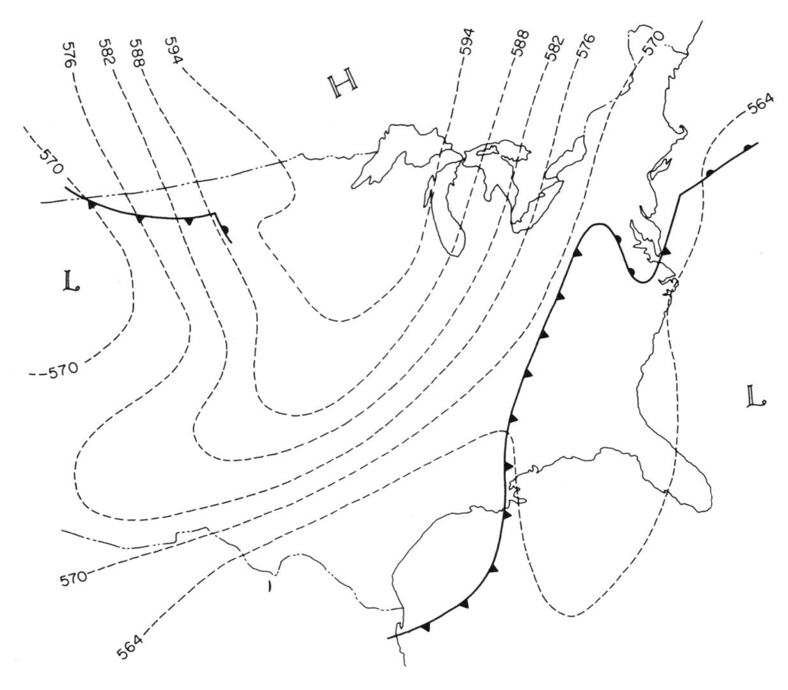

Fig. 3.23 Distribution of 250–100-mb thickness at 00 GCT 20 November 1964. Thickness contours are labeled in tens of meters. Frontal positions are transcribed from the surface chart.

gradients are reversed. The tendency for compensation between tropospheric and lower stratospheric temperature fields was remarked upon in the previous section. We may interpret this tendency for compensation as follows. Immediately above the tropopause the temperature field reverses so that troughs (or closed lows) become warm relative to their surroundings and ridges become cold. The resulting out of phase relationship between thickness and geopotential height is consistent with a reduction in the amplitude of synoptic-scale features as we proceed upward to pressure levels higher in the stratosphere. Troughs tend to be filled in by the positive thickness anomalies and ridges tend to be flattened out by the negative thickness anomalies. As the pressure surfaces become progressively flatter, the wind speeds decrease with height. By the time we reach the 100-mb level, there is little remaining evidence of synoptic-scale features in the geopotential height field. The planetary-scale features present at that level and above have a vertical structure much different from the disturbance that we have considered in this section.

Thus the tendency for compensation between tropospheric and lower stratospheric temperature anomalies is seen to be intimately related to the characteristic vertical structure of a broad class of middle-latitude synoptic-scale disturbances, which exhibit their maximum amplitudes at the tropopause level, and damp out rapidly with height in the lower stratosphere. Unlike the tropospheric temperature gradients, which own their existence to strong latitudinal and geographical contrasts in diabatic heating near the ground, the stratospheric temperature gradients are induced adiabatically by vertical motions in the vicinity of the tropopause. Sinking motion induces warming and suppresses the height of the tropopause above domes of cold tropospheric air, while rising motion induces cooling and lifts the tropopause above the warm tropospheric air masses. We will consider the role of vertical motions in middle-latitude disturbances in more detail in Section 9.5.

PROBLEMS

3.1 Explain or interpret the following:

(a) The temperatures observed at cold fronts are often as high as those observed anywhere within the warm sector of a storm.

(b) Temperature falls that follow the passage of cold fronts are usually more rapid than the temperature rises that precede the passage of warm fronts.

(c) Temperature falls that follow the passage of cold fronts are usually most pronounced in situations where the frontal passage takes place during the late afternoon or evening.

(d) Pressure tendencies are not much help in locating stationary fronts.

(e) Fronts over the sea show up much more clearly on the 850-mb map than on the surface map.

(f) In mountainous areas the winds bear little relation to the isobars on the surface map.

(g) The combination of high sea level pressure and low 500-mb height is usually accompanied by below normal temperature.

(h) Large pressure and temperature changes occur prior to the passage of a warm front, but after the passage of a cold front.

(i) On the average, occluded fronts tend to be weaker than warm or cold fronts.

(j) Soundings taken in the vicinity of the jet stream do not exhibit a well-defined tropopause.

(k) The presence of a stable layer in a sounding may or may not be an indication that a front is present. What other information in the sounding can be used to determine whether a front is present?

(l) The northern boundary of the Gulf Stream is a favored region for the development of winter storms.

(m) In the northern hemisphere, the passage of either a warm or a cold front produces a windshift in the clockwise sense. In what sense are the windshifts in the southern hemisphere?

(n) Freezing rain is frequently observed in association with warm fronts, but rarely with cold fronts.

(o) Middle-latitude disturbances usually tilt westward with height.

3.2 (a) For the idealized frontal configuration shown in the accompanying figure, show that the slope of the front is given by $dz/dx = \beta/(\Gamma_w - \Gamma_F)$, where Γ_w is the lapse rate on the warm side of the front, Γ_F is the lapse rate within the frontal zone, and β is the absolute magnitude of the horizontal temperature gradient within the frontal zone. [Hint: Express the temperature difference between points A and D in terms of the gradients encountered along the path ABD and along the path ACD, and equate them.]

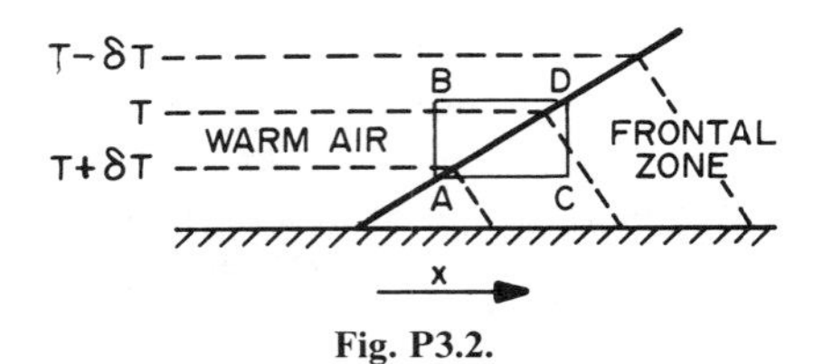

Fig. P3.2.

(b) Given a lapse rate of 7 deg km^{-1} in the warm air and zero in the frontal zone, and a horizontal temperature gradient of 10 deg per 100 km within the frontal zone, calculate the slope of the front.

Answer 14.3 m km^{-1}

3.3 (a) Show that for the idealized frontal zone shown in the figure the mean temperature of the layer BB′ is lower than that of the layer AA′ by the amount $\frac{1}{2}\beta x_B$, where β is the absolute magnitude of the horizontal temperature gradient within the frontal zone.

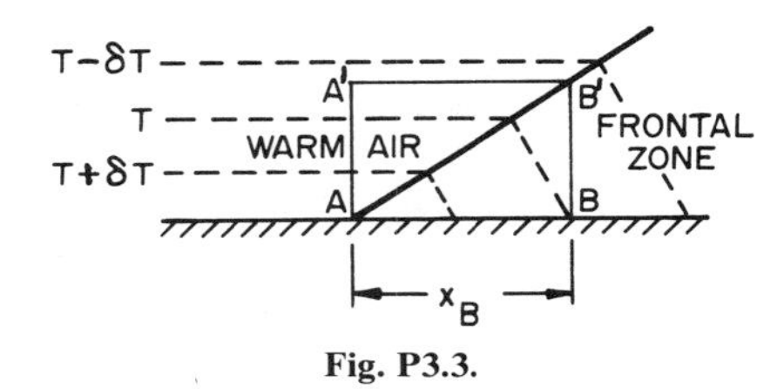

Fig. P3.3.

(b) If the pressures and altitudes at points A′ and B′ are equal, prove that the ratio of the pressure at point B to that at point A is given by

$$\frac{p_B}{p_A} = \left(\frac{p_A}{p_{A'}}\right)^{(\bar{T}_A - \bar{T}_B)/\bar{T}_B}$$

where p_A, p_B, and $p_{A'}$ are the pressures at points A, B, and A′, respectively, and $\bar{T}_A$ and $\bar{T}_B$ are the mean virtual temperatures of the layers AA′ and BB′, respectively.

3.4 On the basis of the expression derived in the previous problem, estimate the slope of the stationary front shown in Figs. 3.3a and 3.4a. (200 km to the north of this front the surface temperature is lower by $\simeq 10$ deg and the sea level pressure is higher by $\simeq 4$ mb than at the surface position of the front.) Assume that $\bar{T}_B \simeq 270°$K.

Answer $\simeq$ 100 mb per 100 km

Chapter

4

Atmospheric Aerosol and Cloud Microphysical Processes

The smallest meteorological entity observable without special equipment is probably a raindrop or snowflake. Yet from the perspective of a cloud microphysicist, the particles commonly encountered in precipitation are quite remarkable precisely because of their large sizes. To form raindrops, cloud particles have to increase in mass a million times or more and these same cloud particles are nucleated by aerosol† as small as 0.01 μm. In order to account for growth through such a wide range of sizes in such short time periods (as little as 10 min or so for convective clouds) it is necessary to consider a number of different physical processes. The scientific investigations of these processes has led to the development, over the past forty years, of the discipline of *cloud physics*.

Much of the early work in cloud physics was concerned with laboratory experiments, in which attempts were made to isolate processes of possible importance in clouds and investigate them through laboratory simulations. In more recent years, as increasingly sophisticated instrumentation for measuring the properties of clouds and precipitation from aircraft and by remote

† An aerosol is a suspension of solid or liquid matter (with small settling velocity) in a gaseous medium (air, in our case). We will often use the term synonymously for small solid and/or liquid particles in air.

sensing has become available, the emphasis in cloud physics has shifted to observations of real clouds and cloud systems. Also, since the advent of high-speed computers, considerable efforts have been spent in developing numerical models of clouds. Today, cloud physics is characterized by a healthy interplay between laboratory, field, and theoretical studies.

In this chapter we are concerned primarily with cloud microphysical processes. In the following chapter we will consider the dynamics of clouds and cloud systems.

We begin with a discussion of atmospheric aerosol, which is important not only in cloud microphysics but also in air pollution, since visibility is determined by the mass and size distribution of aerosol in the air. The next section (Section 4.2) is concerned with the formation of cloud droplets. Sections 4.3 and 4.4 are concerned with the structure of warm clouds and the ways in which cloud droplets and precipitation can grow in these clouds. In Section 4.5 we turn to cold clouds and describe the ways in which ice particles form and grow into solid precipitation (snowflakes, hail, and so on). Finally, in Section 4.6, we discuss the possible roles which cloud and precipitation particles play in the separation of electrical charges in the atmosphere and we consider the nature of lightning and thunder.

4.1 ATMOSPHERIC AEROSOL

4.1.1 Total concentrations

One of the oldest and most convenient techniques (which in various modified forms is still in widespread use) for measuring the concentrations of atmospheric aerosol is the Aitken† nucleus counter. In this instrument, saturated air is expanded rapidly so that it becomes supersaturated by several hundred percent with respect to water (see Problem 4.6). At these high supersaturations water condenses onto virtually all of the aerosol to form a cloud of small water droplets. The concentration of droplets in the cloud (which is close to the concentration of aerosol) can be determined by allowing the droplets to settle out onto a substrate, where they can be counted either under a microscope, or automatically by optical techniques. The concentration of aerosol measured with an Aitken nucleus counter is often referred to as the *Aitken nucleus count*.

Aitken nucleus counts near the earth's surface vary widely between different locations and they also fluctuate by more than an order of magnitude with time at any one site. Generally, they range from average values on the order of 10^3 cm^{-3} over the oceans, to 10^4 cm^{-3} over rural land areas, to 10^5 cm^{-3} or higher in polluted air over cities. These observations, together with the fact

† **John Aitken** (1839–1919) Scottish physicist, although originally an apprentice marine engineer. In addition to his pioneering work on atmospheric aerosol, he investigated cyclones, color, and color sensations.

that Aitken nucleus counts decline with increasing altitude (Fig. 4.1), indicate that the land is an important source of aerosol, with human and industrial activities being particularly prolific sources. It should be noted that the scale height for aerosol above about 10 km is approximately 7 km, the same as that for the atmosphere (see Section 1.3.1). However, below 5 km the scale height is only about 3 km.

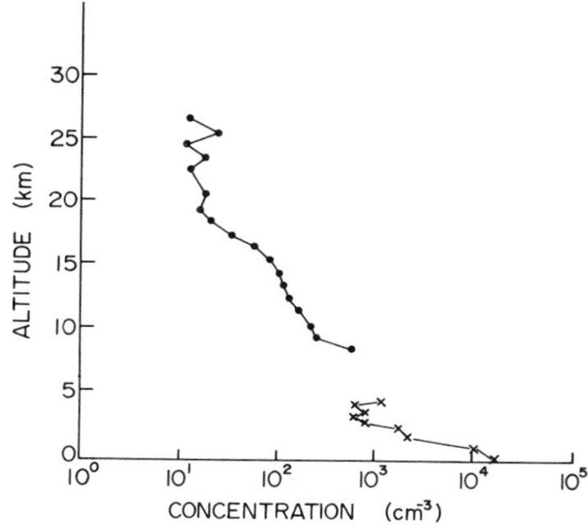

Fig. 4.1 Variations in the concentrations of Aitken nuclei with altitude over Western Germany on 5 October 1973. Upper curve was obtained from balloon measurements and lower curve from aircraft measurements. [From *Pure and Appl. Geophys.* **112**, 884 (1974).]

4.1.2 Size spectra

Atmospheric aerosol range in size from about 10^{-4} μm to tens of micrometers and, depending on the size range of particles considered and the location of the measurements, they can be found in concentrations ranging from about 10^7 to 10^{-6} cm^{-3}. In order to obtain measurements over these enormous ranges, several different techniques must be employed. Aerosol with diameters from 0.003 to 1 μm may be measured with a device (known as an electrical aerosol analyzer) in which the aerosol are first given a known electrical charge and are then collected in a controlled manner by the application of electric fields; variations in the magnitude of the collected charge are interpreted in terms of aerosol size distributions. Aerosol from about 0.3 to 30 μm in diameter can be sized by measuring the amount of light energy that they scatter. Aerosol may also be collected by impaction onto various surfaces and sized under optical or electron microscopes. The elemental compositions of the larger (>1 μm) aerosol may be determined by various analytical techniques (for example, energy dispersive analysis of X rays). In order to determine the atmospheric concentrations of aerosol collected by impaction, the collection efficiencies must be known. Larger aerosol are collected more efficiently than

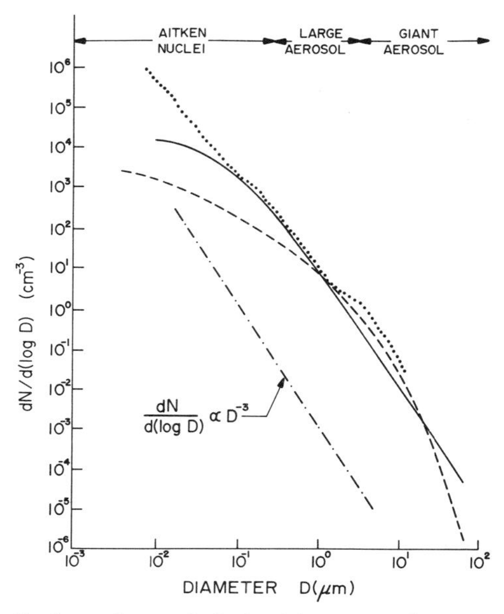

Fig. 4.2 Number distributions of aerosol obtained from averaging many sets of measurements made in continental air (——), marine air (– – –), and urban polluted air (· · ·). Also plotted is Eq. (4.1) with $\beta = 3$ but the line is displaced from the other curves for the sake of clarity.

smaller ones since the latter more readily follow the streamlines around the collecting obstacle. Consequently, the determination of aerosol concentrations by direct impaction is generally confined to aerosol with diameters greater than about 0.1 μm.

The averages of numerous sets of aerosol size measurements in continental, marine, and urban polluted air are shown in Fig. 4.2. The measurements are plotted in the form of a number distribution in which the ordinate is $dN/d(\log D)$, plotted on a logarithmic scale, and the abscissa is $\log D$, where N is the concentration of aerosol with diameters greater than D.†

Several interesting conclusions can be drawn from the results shown in Fig. 4.2:

- Concentrations of aerosol fall off very sharply with increasing size. Therefore, the total number concentration (namely, the Aitken nucleus count) is dominated by the smaller aerosol. (For this reason, aerosol with diameters less than 0.2 μm are termed *Aitken nuclei.*)

† For convenience we will assume the aerosol are spherical, although this is by no means always the case. Note that in this type of plot, if the ordinate were linear in N (which it is not in Fig. 4.2), the concentration of aerosol within a diameter interval $d(\log D)$ would be equal to the area under the curve in this interval.

- Those portions of the number distribution curves that are straight may be described by an expression of the form

$$\log\left[\frac{dN}{d(\log D)}\right] = \text{const} - \beta \log D$$

or, taking antilogs,

$$\frac{dN}{d(\log D)} = CD^{-\beta} \tag{4.1}$$

where C is a constant related to the concentration of the aerosol and $-\beta$ is the slope of the number distribution curve; the value of β generally lies between 2 and 4. Continental aerosol with diameters larger than about 0.2 μm follow (4.1) quite closely with $\beta \simeq 3$.
- The observed number distributions confirm observations from Aitken nucleus counts, which indicates that the total concentrations of aerosol are, on the average, greatest in urban polluted air and least in marine air.
- The concentrations of aerosol with diameters greater than about 2 μm (so-called *giant aerosol*) are, on the average, rather similar in continental, marine, and urban polluted air.

Further insights into aerosol size distributions can be obtained by plotting aerosol surface area or volume distributions rather than number distributions. In surface area distributions the ordinate is $dS/d(\log D)$ and the abscissa is $\log D$, where S is the total surface area of aerosol with diameters greater than D. In volume distributions the ordinate is $dV/d(\log D)$ and the abscissa $\log D$, where V is the total volume of aerosol with diameters greater than D.

If dN represents the number concentration of aerosol within the interval $d(\log D)$, the mass concentration dM of aerosol in this interval will be proportional to $D^3\, dN$, provided that the density of the aerosol is independent of diameter. If the number distribution follows (4.1) with $\beta = 3$, $D^3\, dN \propto d(\log D)$ and $dM/d(\log D) = \text{const}$. In this case, the aerosol in each logarithmic increment of diameter contribute equally to the total mass concentration of the aerosol. It follows from this result that, although aerosol with diameters from 0.2 to 2 μm (so-called *large aerosol*) are present in much higher number concentrations than giant aerosol, the giant and large aerosol in continental air make similar contributions to the total mass of aerosol. Also, despite their relatively large number concentration, Aitken nuclei contribute only about 10–20% to the total mass of aerosol since they do not increase in concentration with decreasing size as rapidly as indicated by (4.1).

Problem 4.1 If the aerosol number distribution is given by (4.1), derive expressions for (a) dN/dD, (b) $dS/d(\log D)$, and (c) $dV/d(\log D)$. Hence show that small fluctuations in the values of β about values of 2 and 3 will produce local maxima and minima in the surface and volume distributions, respectively.

Solution (a) From (4.1),

$$\frac{dN}{d(\log D)} = CD^{-\beta}$$

therefore

$$\frac{dN}{dD}\frac{dD}{d(\log D)} = CD^{-\beta}$$

or

$$\frac{dN}{dD} = CD^{-\beta}\frac{d(\log D)}{dD} = \frac{CD^{-\beta}}{\ln 10}\frac{d}{dD}(\ln D) = \frac{C}{\ln 10}\frac{1}{D}D^{-\beta}$$

therefore

$$\frac{dN}{dD} = \frac{C}{\ln 10}D^{-(\beta+1)}$$

(b)

$$dS = \pi D^2\, dN$$

therefore

$$\frac{dS}{d(\log D)} = \pi D^2 \frac{dN}{d(\log D)}$$

hence

$$\frac{dS}{d(\log D)} = \pi C D^{2-\beta}$$

(c)

$$dV = \frac{\pi}{6}D^3\, dN$$

therefore

$$\frac{dV}{d(\log D)} = \frac{\pi}{6}D^3 \frac{dN}{d(\log D)}$$

hence

$$\frac{dV}{d(\log D)} = \frac{\pi}{6}CD^{3-\beta}$$

From the expression derived in (b) we see that $dS/d(\log D)$ is an increasing function of D for $\beta < 2$ and a decreasing function of D for $\beta > 2$. Hence the surface distribution will reach a peak value when β passes through a value of 2. Similarly, from the expression derived in (c) above, we see that the volume distribution will attain a peak value when β passes through a value of 3.

The fact that small fluctuations in the slope of the aerosol number distribution about values of -2 and -3 (that is, $\beta = 2$ and 3, respectively) appear as local maxima and minima in the surface and volume distributions, respectively, are important features of the latter two plots which make them more useful in some respects than aerosol number distribution plots.

Shown in Fig. 4.3 are some surface and volume distributions based on aerosol measurements made in continental and urban polluted air in Denver, Colorado. It can be seen that these curves show far more structure, in the form of maxima

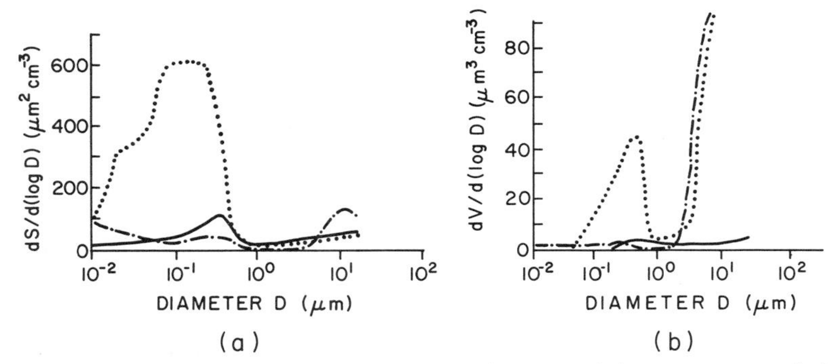

Fig. 4.3 Surface (a) and volume (b) distributions based on aerosol size measurements in Denver, Colorado. Urban polluted air (· · ·), continental air (——), and continental air with dust from mountains (–·–·–). [From *J. Air. Poll. Cont. Assoc.* **25**, 532 (1975).]

and minima, than do the number distribution curves shown in Fig. 4.2. We will see that maxima and minima in the surface and volume distribution curves can be associated with the sources and sinks of the aerosol.

4.1.3 Sources and sinks

The smallest aerosol (Aitken nuclei) originate primarily from combustion processes, of which the most important sources are probably those associated with human activities (although forest fires and volcanoes also contribute); hence, the high Aitken nucleus counts in cities and the high concentrations of aerosol with diameters below about 0.2 μm in urban polluted air (see Figs. 4.2 and 4.3). Moreover, near sources of fresh combustion, a peak is often found in the surface aerosol distribution curve around an aerosol diameter of about 0.01 μm (this peak does not appear in Fig. 4.3).

The fact that Aitken nuclei are present in appreciable concentrations in continental and marine air indicates the existence of sources other than combustion. One such source is the conversion of trace gases in the atmosphere into aerosol (so-called *gas-to-particle conversion*) which can occur through the nucleation of aerosol from supersaturated gases and by photochemical reactions associated with the absorption of solar radiation by molecules.† Gas-to-particle conversions may be enhanced by high relative humidity and the presence of liquid water. For example, the oxidation rate of sulfur dioxide to sulfate increases by a factor of about eight as the relative humidity increases from 70 to 80%. Also, sulfates can be produced by the reaction of sulfur dioxide and ammonia in cloud droplets; when the droplets evaporate they leave behind

† When high concentrations of certain man-made chemicals are present in the atmosphere, photochemical reactions can lead to the formation of smogs, such as those that are common in the Los Angeles Basin. Hydrocarbons, nitric oxides, and ozone play crucial roles in smog formation.

submicron sulfate particles. It has been suggested that this mechanism is the major source of sulfates in the atmosphere.

Aerosol may also originate at the surface of the earth as windblown dust (particularly in arid regions), by the emissions of pollens and spores from plants, and, over oceans, by the bursting of air bubbles. However, these mechanisms are thought to be more important as sources of large and giant aerosol rather than as sources of Aitken nuclei (see below).

In Fig. 4.3 there are prominent maxima in the surface and volume distributions in the size range of large aerosol (0.2–2 μm diam). These peaks are believed to be due primarily to the growth of Aitken nuclei by coagulation into the large aerosol size range where they have a long residence time. The peaks are therefore referred to as the *accumulation mode*. Another peak (the so-called *coarse particle mode*) occurs in the surface and volume distributions for aerosol with diameter greater than 1 or 2 μm. Due to the comparatively large settling speeds of these aerosol ($\simeq$0.01 and 1 cm s^{-1} for 2- and 20-μm-diam aerosol, respectively) their concentrations vary quite rapidly with time. The coarse particle mode is attributable to mechanical processes, such as wind erosion which produces dust, and industrial processes which produce fly ash and other large aerosol. Over the oceans, giant aerosol composed of sea-salt originate from drops ejected into the air when air bubbles in breaking waves burst at the ocean surface.† From 1 to 5 drops break away from each jet that forms when a bubble bursts (Fig. 4.4c) and these jet drops are thrown some 15 cm up into the air.

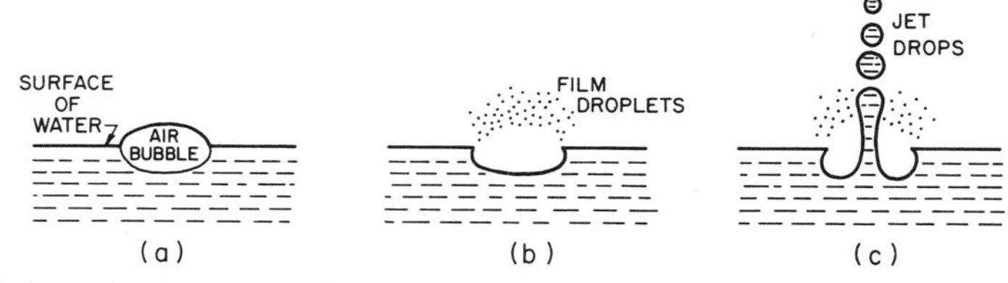

Fig. 4.4 Schematic diagrams to illustrate the manner in which film droplets and jet drops are produced when an air bubble bursts at the surface of the ocean. Some of the droplets and drops evaporate to leave sea-salt particles in the air.

Some of these drops subsequently evaporate and leave behind giant (>2 μm diam) sea-salt particles. Much smaller droplets are produced when the upper part of an air bubble film bursts at the ocean surface. Bubbles larger than about 2 mm in diameter each eject one or two hundred such droplets into the air when they burst (Fig. 4.4b). After evaporation these droplets leave behind sea-salt particles with diameters less than about 0.3 μm. The average rate of production of sea-salt particles over the oceans is estimated to be on the order of 100 cm^{-2} s^{-1}.

The differences in the nature of large and giant aerosol between ocean and land are illustrated in Figs. 4.5 and 4.6. Figure 4.5a shows aerosol collected by

† Some drops also enter the air by being torn from whitecaps, but due to the comparatively large size of these drops, their residence time in the atmosphere is very short.

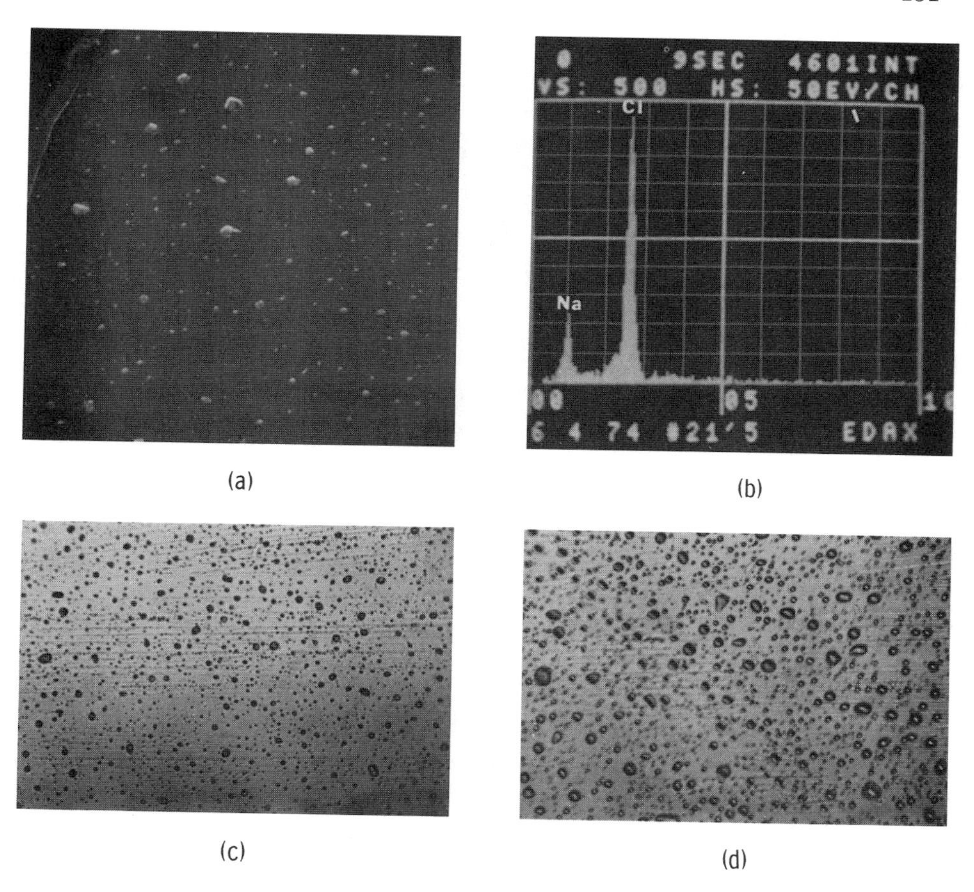

Fig. 4.5 (a) Scanning electron microscope photograph of aerosol collected at an altitude of 9 m above the surface of the Pacific Ocean. (b) Elemental analysis of a particle in (a), using energy dispersive analysis of X rays, shows that it consists primarily of sodium and chlorine. (c) Optical microscope photograph of aerosol collected over the ocean after being exposed to a relative humidity of 50%, and (d) a relative humidity of 95%. (Courtesy of Cloud Physics Group, University of Washington.)

impaction at a height of 9 m above the sea surface; it can be seen from Fig. 4.5b that the principal elements in the aerosol are sodium and chlorine, so it is reasonable to assume that they are sodium chloride particles which originated from the ocean surface. When these aerosol were exposed in the laboratory to a relative humidity of 95% most of them deliquesced (Fig. 4.5c and d), which is to be expected if they are sodium chloride. By comparison, aerosol collected in the eastern part of Washington State (a semiarid, rural area) are quite different in shape (Fig. 4.6a) and the elements they contain (Fig. 4.6b) indicate that they are probably silicates originating from soils. The majority of these aerosol did not deliquesce at 95% relative humidity (Fig. 4.6c and d).

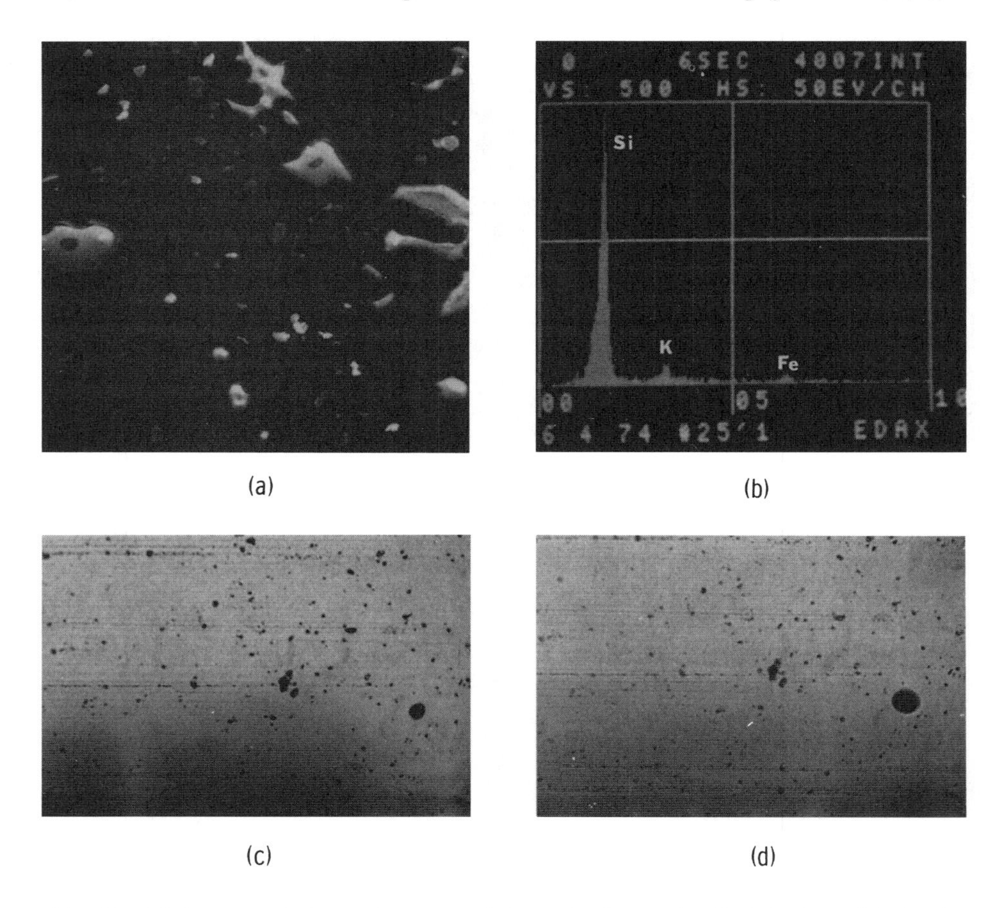

Fig. 4.6 (a) Scanning electron microscope photograph of aerosol collected at an altitude of 3 km over eastern Washington. (b) Elemental analysis of a particle in (a), using energy dispersive analysis of X rays, shows that it consists primarily of silicon, potassium, and iron. (c) Optical microscope photograph of aerosol collected over eastern Washington after being exposed to a relative humidity of 50%, and (d) a relative humidity of 95%. (Courtesy of Cloud Physics Group, University of Washington.)

Estimates of the world-wide production of aerosol from natural phenomena and human activities are listed in Table 4.1. The total mass of aerosol from natural phenomena was estimated to be about four times that from human activities in 1968.† However, by the year 2000 it is estimated that aerosol production from human activities will be about twice that in 1968. Even now, it is quite clear that urban aerosol are predominantly anthropogenic.

† It should be emphasized that these estimates are quite crude. Some workers have put the anthropogenic contribution to the atmospheric aerosol loading as low as 5% and others as high as 45%.

Table 4.1

Estimated world-wide aerosol production[a] due to natural phenomena and human activities in 1968[b]

Source	Aerosol diameter	
	$>5\ \mu m$	$<5\ \mu m$
(a) Natural phenomena		
Sea-salt	500	500
Gas-to-particle conversion	100	470
Windblown dust	250	250
Forest fires	30	5
Meteoric debris	10	0
Volcanoes (highly variable)	?	25
Total	890(+?)	1250
(b) Human activities (1968)		
Gas-to-particle conversion	25	250
Industrial processes	44	12
Fuel combustion (stationary sources)	34	10
Solid waste disposal	2	0.5
Transportation	0.5	2
Miscellaneous	23	5
Total	128.5	279.5

[a] In units of teragrams per year.
[b] From W. H. Matthews *et al.*, Eds., "Man's Impact on the Climate," MIT Press, Cambridge, Massachussets, 1971.

On the average, aerosol are removed from the atmosphere at about the same rate as they enter it. As mentioned previously, Aitken nuclei may be converted into larger aerosol by coagulation. Since the mobilities of aerosol decrease rapidly as their size increases, coagulation is essentially confined to aerosol less than about 0.2 μm in diameter. For example, model calculations show that the concentration of aerosol of average diameter 0.01 μm, initially present in a concentration of 10^5 cm^{-3} (for example, typical of urban air), should decrease by a factor of two within about 30 min due to coagulation, whereas it would take some 500 h for 0.2-μm-diam aerosol, initially present in concentrations of 10^3 cm^{-3}, to halve their concentration by coagulation. Although coagulation does not remove aerosol from the atmosphere, it modifies their size spectra and shifts Aitken nuclei into size ranges where they can be removed by other mechanisms.

Improvements in visibility that frequently follow periods of precipitation are due, in large part, to the removal (or scavenging) of aerosol by the precipitation particles. It is estimated that, on the global scale, precipitation processes account for about 80–90% of the mass of aerosol removed from the atmosphere. Prior to the formation of precipitation, aerosol serve as nuclei upon which cloud particles (water and ice) form (see Sections 4.2.2 and 4.5), and, as these particles grow, aerosol tend to be forced onto their surfaces by the diffusion field associated with the flux of water vapor to the growing cloud particles (the so-called diffusiophoretic force). Aerosol less than about 0.1 μm are collected most efficiently by diffusiophoresis. As precipitation particles fall through the air they collect aerosol by direct impaction. As mentioned in Section 4.1.2, the larger an aerosol particle the more efficiently it is collected by impaction. Raindrops collect aerosol greater than about 2 μm in diameter by impaction with reasonable efficiency. Aerosol are also removed by impaction onto obstacles on the earth's surface (for example, newly washed automobiles).

Fallspeeds of aerosol larger than about 1 μm in diameter are sufficiently large that gravitational settling (or dry fallout) is important as a removal process. For example, the fallspeeds of aerosol 1 and 10 μm in diameter are about 3×10^{-5} and 3×10^{-3} m s^{-1}, respectively (see Problem 4.7). It is estimated that some 10–20% of the mass of aerosol removed from the atmosphere is by gravitational settling.

Figure 4.7 summarizes much of the previous discussion and indicates the relationships between the maxima and minima of typical aerosol surface area

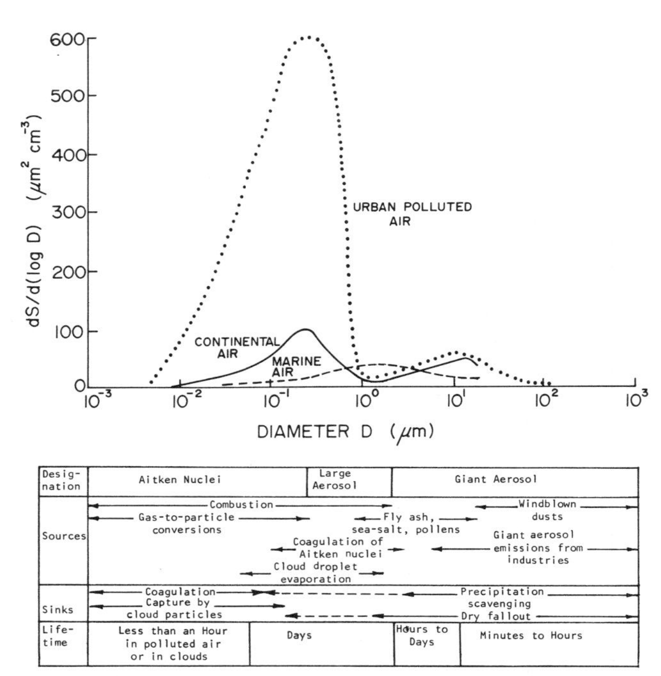

Fig. 4.7 Schematic curves of aerosol surface area distributions for urban polluted air, continental air, and marine air. Shown below the curves are the principal sources and sinks of atmospheric aerosol and estimates of their mean lifetimes in the troposphere. [Adapted from *Atmos. Environ.* **9**, 763 (1975).]

distributions for urban polluted, continental, and maritime air and the principal sources and sinks of atmospheric aerosol.

4.1.4 Effects on some atmospheric phenomena

Shown in Fig. 4.8 are the approximate size ranges of aerosol which play a role in atmospheric electricity, air chemistry, atmospheric radiation and optics, and cloud and precipitation processes. Aerosol involved in cloud and precipitation processes will be discussed in some detail in Section 4.2. In this section we describe briefly the role of aerosol in atmospheric electricity, radiative transfer, and chemistry, as summarized in Fig. 4.8.

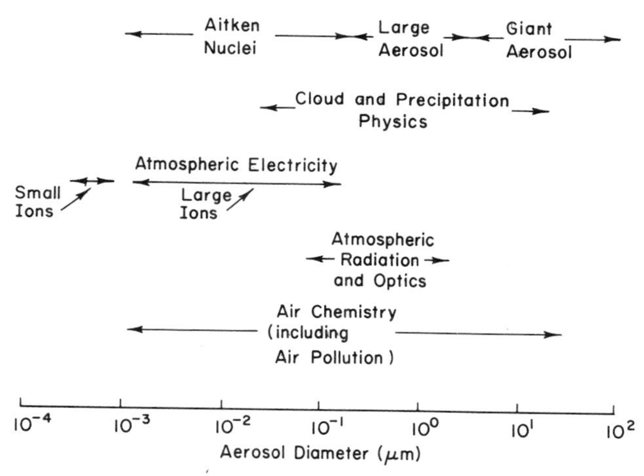

Fig. 4.8 Approximate size ranges of aerosol of importance in various atmospheric phenomena.

Molecular aggregates that carry an electric charge are called ions. The number density and type of ions in the air determines the electrical conductivity of the air which, in turn, affects the magnitude of the fair weather atmospheric electric field in the manner described in Section 1.4.2. Also, as we will discuss in Section 4.6, cloud and precipitation particles can become charged by collecting ions.

Ions in the lower atmosphere are produced primarily by cosmic rays, although very close to the earth's surface ionization due to radioactive materials in the earth and atmosphere also plays a role. Ions are removed by combining with ions of opposite sign. Atmospheric ions may be classified as small or large. Small ions, which are not much larger than molecular size, have electrical mobilities (defined as their velocity in a unit electric field) between 1×10^{-4} and 2×10^{-4} m s^{-1} for an electric field of 1 V m^{-1} at normal temperature and pressure (NTP).† Large ions, which approximately coincide with the size range

† Free electrons have much higher electrical mobilities than small ions but they do not remain unattached for any significant time in the atmosphere.

of Aitken nuclei, have electrical mobilities in the range from 3×10^{-8} to 8×10^{-7} m s^{-1} in a field of 1 V m^{-1} at NTP. Concentrations of small ions vary from about 40 to 1500 cm^{-3} at sea level, while concentrations of large ions vary from about 200 cm^{-3} in marine air to a maximum value of about 800,000 cm^{-3} in some large cities. It can be seen from these numbers that the electrical conductivity of the air (which is proportional to the product of the ion mobility and the ion concentration) is generally dominated by the small ions. However, when the concentrations of large ions and uncharged aerosol are large, as they are in cities, the concentration of small ions tends to be low due to the fact that they are captured by both large ions and uncharged aerosol. Consequently, the electrical conductivity of air is a minimum (and the atmospheric electric field a maximum) when the concentration of large ions and other Aitken nuclei is a maximum.

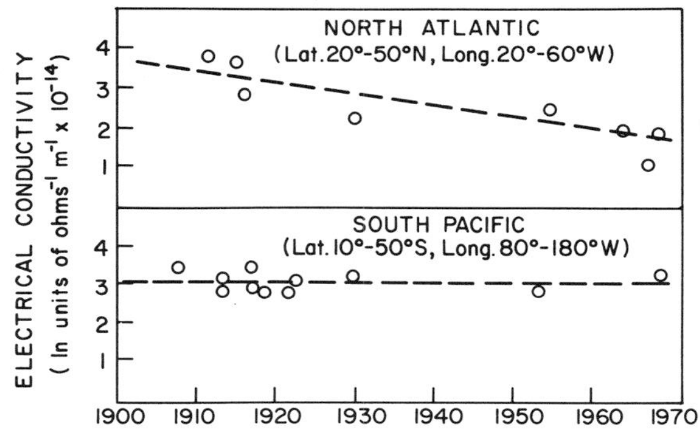

Fig. 4.9 Measurements of the electrical conductivity of the air over the North Atlantic and South Pacific Oceans. [From *J. Atmos. Sci.* **27**, 818 (1970).]

Since variations in the electrical conductivity of the air reflect changes in the concentrations of ions, measurements of the former may be related to fluctuations in atmospheric aerosol (primarily Aitken nuclei). Indeed, studies of atmospheric aerosol virtually began with studies of atmospheric electricity in the earlier part of this century. For example, large diurnal variations in the fair weather atmospheric electric field over cities may be related to the level of aerosol pollution. Thus, in cities the atmospheric electric field generally reaches peak values from 7 to 10 a.m. and from 7 to 9 p.m., when pollution is at its highest. Figure 4.9 shows a number of measurements of the electrical conductivity of air over the North Atlantic and South Pacific Oceans.† Over the North Atlantic there has been a continual decrease in the electrical conductivity

† The measurements prior to 1930 shown in Fig. 4.9 were made aboard the sailing vessel *Carnegie* and its predecessor *Galilee*. A tragic fire and explosion aboard the *Carnegie* on 18 November 1929, at the Island of Samoa, resulted in the sinking of the ship and the death of her captain. Much credit must go to the Carnegie Institute of Washington for having the foresight to initiate these valuable measurements over 60 yr ago.

through this century, whereas over the South Pacific there has been no change. The observed decrease of at least 20% in the electrical conductivity of the air over the North Atlantic is attributed to a doubling in the concentration of aerosol with diameters between 0.02 and 0.2 μm. This increase in aerosol concentration in the Northern Hemisphere may be due to anthropogenic sources. Unfortunately, there are not many other observations of sufficient duration, and in locations not dominated by local pollution, to check whether or not aerosol have increased in concentration on a world-wide basis. It is interesting to note that observations from many locations remote from cities in the Soviet Union show evidence of a systematic decrease, amounting to about 5% over the past 25 yr, in direct solar radiation. A similar decrease in direct solar radiation was apparent at "clean-air" stations over southeast Japan and in the western Pacific between 1948 and 1955 (when the measurements were terminated).

Aerosol play an important role in many chemical processes in the atmosphere. Solid aerosol provide surfaces upon which trace gases can be absorbed and then react, while liquid aerosol absorb gases which may then react together in solution. The role of aerosol in atmospheric chemistry is, perhaps, most dramatic in heavily polluted air. For example, when aerosol and sulfur dioxide from industrial and domestic sources accumulate in stable, moist air below a temperature inversion, the sulfur dioxide is converted into sulfates (including sulfuric acid). These chemical reactions produce the type of smog[†] that was common in London prior to the introduction of the Clear Air Act which forbade the burning of fossil fuels in open fires. The dense smog that persisted over London from the 5th to the 8th of December, 1952, caused some 4000 deaths due to aggravation of respiratory ailments.

Scattering and absorption of radiation by aerosol have important effects, ranging from the degradation of visibility to modifications in the transfer of solar radiation in the atmosphere. Since the radiative effects of aerosol will be considered in some detail in Section 6.8, we will mention them only very briefly here. The scattering of visible light by aerosol is dominated by the large aerosol ($0.2\ \mu\text{m} \leq D \leq 2\ \mu\text{m}$). Many of these aerosol increase in size due to deliquescence as the relative humidity increases (see, for example, Fig. 4.5c and d); therefore, when the relative humidity is high they form a haze which reduces visibility by scattering light. For example, the amount of light that sea-salt particles remove from a beam by scattering increases by about a factor of three as the relative humidity increases from 60 to 80%.

Although aerosol absorb and scatter only a small fraction of the total solar radiation passing downward through the earth's atmosphere (clouds and gases play a much more important role), a steady increase in global aerosol concentration might alter the radiation balance of the earth. To the extent that aerosol absorb solar radiation they will raise air temperatures; however,

[†] The term "smog" was introduced in 1905 by Des Voeux, a pioneer in air pollution control, to describe the aerosol formed when fog occurs together with smoke. More recently, the term has been extended to include the photochemical type of pollution which is common in Los Angeles.

aerosol also tend to decrease temperatures by scattering solar energy back into space. At the present time it is not possible to predict with any certainty whether the increase in atmospheric aerosol which appears to have taken place in the Northern Hemisphere in this century should result in a net increase or decrease in average global temperatures.

4.2 NUCLEATION OF WATER VAPOR CONDENSATION

Clouds form when air becomes *supersaturated*[†] with respect to liquid water (or in some cases with respect to ice); the most common means by which supersaturation occurs in the atmosphere is through the ascent of air parcels, which results in the expansion of the air and adiabatic cooling[‡] (see Section 2.6). Under these conditions, water vapor condenses onto some of the aerosol in the air to form a cloud of small water droplets.[§¶] In this section we are concerned with the ways in which the condensation of water vapor to liquid water is initiated (or nucleated) in the atmosphere.

4.2.1 Theory

We consider first the hypothetical problem (as far as the earth's atmosphere is concerned) of the formation of a pure water droplet by condensation from a supersaturated vapor without the aid of aerosol. In this process, which is referred to as *homogeneous* or *spontaneous nucleation* of condensation, the first stage in the growth process is a number of water molecules in the vapor phase coming together, through chance collisions, to form small embryonic water droplets large enough to remain intact.*

Let us suppose that a small embryonic water droplet of volume V and surface area A forms spontaneously from a supersaturated vapor at constant temperature and pressure. If μ_l and μ_v are the chemical potentials in the liquid and vapor phases, and n is the number of water molecules per unit volume of liquid,

† If the water vapor pressure in the air is e, the supersaturation (in percent) with respect to liquid water is $(e/e_s - 1)100$, where e_s is the saturation vapor pressure over a plane surface of liquid water. A supersaturation with respect to ice may be defined in an analogous way. When the term supersaturation is used without qualification, it will refer to supersaturation with respect to liquid water.

‡ The first person to explain the formation of clouds by the adiabatic cooling of moist air appears to have been the scientist and poet Erasmus Darwin (grandfather of Charles Darwin) in 1788.

§ It was widely accepted well into the second half of the 19th century that clouds are composed of numerous small bubbles of water! How else could it be explained that clouds float? Although John Dalton had suggested in 1793 that clouds may consist of water drops which are continually descending relative to the air, it was not until 1850 that James Espy clearly recognized the role of upward-moving air currents in suspending clouds.

¶ **James Pollard Espy** (1785–1860) Born in Pennsylvania. Studied law, but became a classics teacher at the Franklin Institute. Impressed by the meteorological writings of John Dalton, he gave up teaching to devote his time to meteorology. First to recognize the importance of latent heat release in sustaining cloud and storm circulations. Espy also made the first estimates of the dry and saturated adiabatic lapse rates based on experimental data.

* If Section 2.9 has been omitted move straight to Eq. (4.5) omitting the intervening proof.

the decrease in the Gibbs free energy of the system due to the condensation is $nV(\mu_v - \mu_l)$. Now, quite apart from any work associated with the change in volume of the system, work is done in creating the surface area of the droplet. This work may be written as $A\sigma$, where σ is the work required to create a unit area of vapor–liquid interface (called the *interfacial energy* between the vapor and the liquid or simply the *surface energy* of water). If this were an equilibrium transformation (which it is not) (2.111) would hold; that is, $A\sigma$ would be equal to $nV(\mu_v - \mu_l)$. Instead, the change in the Gibbs free energy will, in general, differ from the work term $A\sigma$. Let us write

$$\Delta E = A\sigma - nV(\mu_v - \mu_l) \tag{4.2}$$

then ΔE is the net increase in the energy of the system due to the formation of the drop. Combining (4.2) and (2.112) we obtain

$$\Delta E = A\sigma - nVkT \ln\left(\frac{e}{e_s}\right) \tag{4.3}$$

where e and T are the vapor pressure and temperature of the system and e_s the saturation vapor pressure over a plane surface of water at temperature T. For a droplet of radius R, (4.3) becomes

$$\Delta E = 4\pi R^2\sigma - \tfrac{4}{3}\pi R^3 nkT \ln\left(\frac{e}{e_s}\right) \tag{4.4}$$

In subsaturated air $e < e_s$; therefore, $\ln(e/e_s)$ is negative and ΔE is always positive and increases with increasing R (Fig. 4.10). In other words, the larger the embryonic droplet that forms in a subsaturated vapor the greater the increase in the energy of the system. Since a system approaches an equilibrium state by reducing its energy, the formation of droplets is clearly not favored under subsaturated conditions. Even so, due to random collisions of water molecules, very small embryonic droplets continually form (and evaporate) in

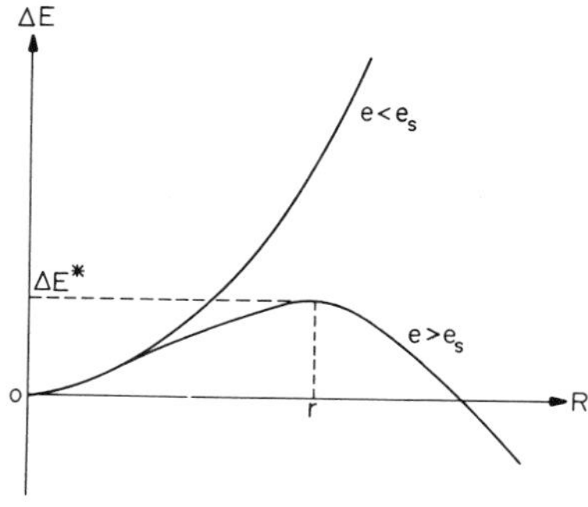

Fig. 4.10 The increase ΔE in the energy of a system due to the formation of a water droplet of radius R from water vapor with pressure e; e_s is the saturation vapor pressure with respect to a plane surface of water at the temperature of the system.

a subsaturated vapor but they do not grow large enough to become visible as a cloud of droplets.

When air is supersaturated, $e > e_s$ and $\ln(e/e_s)$ is positive so that ΔE in (4.4) can be either positive or negative depending upon the value of R. The variation of ΔE with R for this case is also shown in Fig. 4.10, where it can be seen that ΔE initially increases with increasing R, reaches a maximum value when $R = r$, and then decreases with increasing R. Hence, in supersaturated air, embryonic droplets with $R < r$ tend to evaporate, but droplets which manage to grow by chance to a radius which just exceeds r will continue to grow spontaneously by condensation from the vapor phase, since by so doing they will cause a decrease in the total energy of the system.

In the region just around $R = r$, a droplet can grow or evaporate infinitesimally without any change in the energy E of the system. We can obtain an expression for r in terms of e by setting $\partial(\Delta E)/\partial R = 0$ at $R = r$. Hence, from (4.4),

$$r = \frac{2\sigma}{nkT \ln(e/e_s)} \tag{4.5}$$

Equation (4.5) is referred to as Kelvin's formula, after Lord Kelvin who first derived it. We can use (4.5) in two ways. It can be used to calculate the radius r of a droplet which will be in (unstable) equilibrium† with air with a given water vapor pressure e. Alternatively, it can be used to determine the saturation vapor pressure e over a droplet of specified radius r. It should be noted that the relative humidity with which a droplet of radius r is in (unstable) equilibrium is $100e/e_s$, where e/e_s is given by (4.5). The variation of this relative humidity with droplet radius is shown in Fig. 4.11. It can be seen that a pure water droplet of radius 0.01 μm requires a relative humidity of 112.5% (that is, a supersaturation of 12.5%) in order to be in (unstable) equilibrium with its environment, while a droplet of radius 1 μm requires 100.12% relative humidity (that is, a supersaturation of 0.12%).

Since the supersaturations that develop in natural clouds due to the adiabatic ascent of air rarely exceed 1%, it follows from the preceding discussion that even embryonic droplets as large as 0.01 μm, which might form by the chance collision of water molecules, will be well below the critical radius required for survival at 1% supersaturation. Consequently, droplets do not form in natural clouds by the homogeneous nucleation of pure water. Instead they form by what is known as *heterogeneous nucleation* on atmospheric aerosol.‡§

† The equilibrium is unstable in the sense that if the embryonic droplet gains a molecule it will continue to grow by condensation and if it loses a molecule it will continue to evaporate.

‡ That aerosol play a role in the condensation of water was first clearly demonstrated by Coulier in 1875. His results were rediscovered independently by Aitken in 1881 (see p. 144).

§ **Paul Coulier** (1824–1890) French physician and chemist. Carried out research on hygiene, nutrition, and the ventilation of buildings.

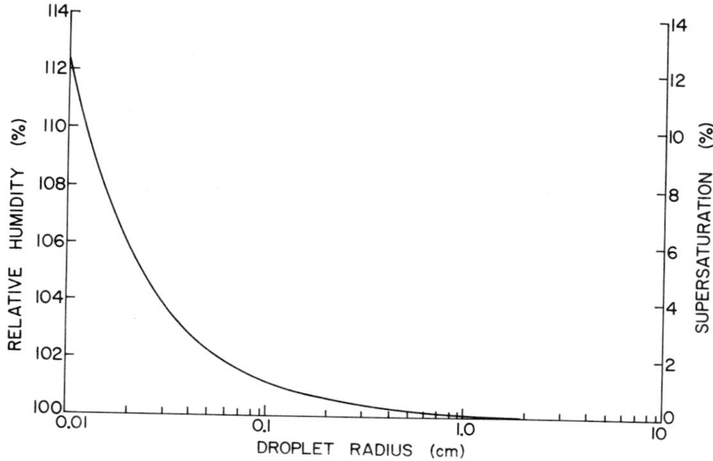

Fig. 4.11 The relative humidity and supersaturation (both with respect to a plane surface of liquid water) with which pure water droplets are in (unstable) equilibrium at 5°C.

As we have seen in Section 4.1, the atmosphere contains many aerosol ranging in size from submicron to several tens of micrometers. Those aerosol that are *wettable*† can serve as centers upon which water vapor can condense. Moreover, droplets can form and grow on these aerosol at much lower supersaturations than are required for homogeneous nucleation. For example, if sufficient water condenses onto a completely wettable aerosol 0.3 μm in radius to form a thin film of water over the surface of the aerosol, we see from Fig. 4.11 that the water film will be in (unstable) equilibrium with air which has a supersaturation of 0.4%. If the supersaturation were slightly greater than 0.4%, more water would condense onto the film and the droplet would increase in size.

Some atmospheric aerosol are soluble in water, so that they begin to dissolve when water condenses onto them. This so-called *solute effect* has an important effect on heterogeneous nucleation which we will now discuss.

The equilibrium saturation vapor pressure over a solution droplet (that is, a water droplet containing some dissolved salt; for example, sodium chloride or ammonium sulfate) is less than that over a pure water droplet of the same size for the following reason. The saturation vapor pressure is proportional to the concentration of water molecules on the surface of the droplet. Now, in a solution droplet some of the surface molecular sites are occupied by the molecules of salt (or ions, if the salt dissociates) and thus the vapor pressure is reduced by the presence of the solute. The fractional reduction in vapor pressure is given by the relation

$$\frac{e'}{e} = f \tag{4.6}$$

† A surface is said to be perfectly wettable if it allows water to spread out on it as a horizontal film. It is completely unwettable (or hydrophobic) if water forms spherical drops on its surface.

where e' is the saturation vapor pressure over a solution droplet containing a kilomole fraction f of pure water and e is the saturation vapor pressure over a pure water droplet of the same size and at the same temperature. The *kilomole fraction of pure water* is defined as the number of kilomoles of pure water in the solution divided by the total number of kilomoles (pure water plus salt) in the solution.

Let us consider now a solution droplet of radius r which contains a mass m of a dissolved salt of molecular weight M_s. If each molecule of the salt dissociates in water into i ions, the effective number of kilomoles of the salt in the droplet is im/M_s. If the density of the solution is ρ' and the molecular weight of water M_w, the number of kilomoles of pure water in the droplet is $(\frac{4}{3}\pi r^3 \rho' - m)/M_w$. Therefore, the kilomole fraction of water is

$$f = \frac{(\frac{4}{3}\pi r^3 \rho' - m)/M_w}{(\frac{4}{3}\pi r^3 \rho' - m)/M_w + im/M_s} = \left[1 + \frac{imM_w}{M_s(\frac{4}{3}\pi r^3 \rho' - m)}\right]^{-1} \tag{4.7}$$

Combining (4.5)–(4.7) (but replacing σ and n by σ' and n' to indicate the surface energy and number density of water molecules, respectively, in the solution) we obtain the following expression for the saturation vapor pressure e' over a solution droplet of radius r:

$$\frac{e'}{e_s} = \left[\exp \frac{2\sigma'}{n'kTr}\right]\left[1 + \frac{imM_w}{M_s(\frac{4}{3}\pi r^3 \rho' - m)}\right]^{-1} \tag{4.8}$$

Equation (4.8) may be used to calculate the vapor pressure e' [or relative humidity $100e'/e_s$ or supersaturation $(e'/e_s - 1)100$] of the air adjacent to a solution droplet of specified radius r. If we plot the variation of the relative humidity (or supersaturation) of the air adjacent to a solution droplet as a function of its radius, we obtain what is referred to as a *Köhler*[†] *curve*. Several such curves derived from (4.8) are shown in Fig. 4.12. Below a certain droplet size, the vapor pressure of the air adjacent to a solution droplet is less than that which is in equilibrium with a plane surface of pure water at the same temperature. As the droplets increase in size, the solutions become weaker, the Kelvin curvature effect becomes the dominant influence, and eventually the relative humidity of the air adjacent to the droplets becomes essentially the same as that over pure water droplets.

To illustrate further the interpretation of the Köhler curves let us consider curve 2 in Fig. 4.12, which corresponds to a solution droplet containing 10^{-19} kg of sodium chloride (dry radius 0.022 μm). If the solution droplet were 0.05 μm in radius, the relative humidity of the air adjacent to its surface would be 90%. Hence, if an initially dry sodium chloride particle of mass 10^{-19} kg were placed in air with 90% relative humidity, water vapor would condense onto the particle, the salt would dissolve, and a solution droplet with radius 0.05 μm

[†] **H. Köhler,** (1888–) Swedish meteorologist. Former Director of the Meteorological Observatory, University of Uppsala.

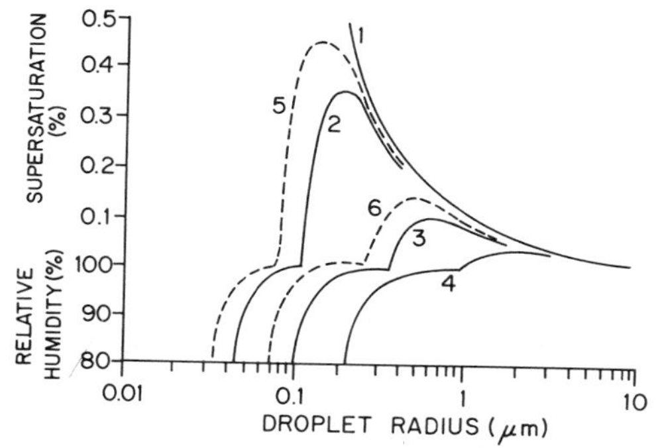

Fig. 4.12 Variations of the relative humidity and supersaturation of the air adjacent to droplets of (1) pure water and solution droplets containing the following fixed masses of salt: (2) 10^{-19} kg of NaCl, (3) 10^{-18} kg of NaCl, (4) 10^{-17} kg of NaCl, (5) 10^{-19} kg of $(NH_4)_2SO_4$, and (6) 10^{-18} kg of $(NH_4)_2SO_4$. [Adapted from S. I. Rasool, ed., "Chemistry of the Lower Atmosphere," Plenum Press, New York, 1973, p. 16.]

would form. This effect accounts for the deliquescence phenomenon illustrated in Fig. 4.5d. Similarly, if the air were supersaturated by 0.2%, a solution droplet with radius slightly in excess of 0.1 μm would form on the sodium chloride particle. In each of these two cases, the droplets that form are in stable equilibrium with the air since, if they grew a little more, the vapor pressures adjacent to their surfaces would rise above that of the ambient air and they would evaporate back to their equilibrium size. Conversely, if they evaporated a little, their vapor pressures would fall below that of the ambient air and they would grow back to the equilibrium size by condensation. Droplets small enough to be in stable equilibrium with the air are called *haze* droplets. All droplets in a state represented by points on the left hand side of the maxima in the curves shown in Fig. 4.12 are in the haze state. The formation of haze droplets in the atmosphere can result in considerable reduction of visibility (see Fig. 2.11).

Consider next a solution droplet in a situation represented by the peak of curve 2 in Fig. 4.12, where the supersaturation of the air is 0.36% and the droplet radius about 2×10^{-5} cm. If the droplet evaporated slightly, the supersaturation of the air adjacent to the droplet would fall below that of the ambient air so the droplet would grow by condensation back to its original size. If, on the other hand, a droplet at the peak in the curve should grow slightly, the supersaturation of the air adjacent to the droplet would fall below that of the ambient air, and therefore the droplet would grow by condensation. As a result of this growth, the supersaturation of the air adjacent to the droplet would further diminish and the droplet would continue to grow, passing through states represented by points to the right of the peak in curve 2. A droplet which has passed over the peak in its Köhler curve is said to have been *activated*. Once a droplet is activated it can grow rapidly to form cloud droplets by condensation as described in Section 4.4.1.

Finally, consider a situation in which the supersaturation of the air is 0.4%. In this environment a sodium chloride particle of mass 10^{-19} kg would first grow by condensation through states represented by points to the left of the peak in curve 2, during which time the supersaturation of the air adjacent to the droplet would rise. However, even when the particle reached the peak of curve 2, the supersaturation of the air adjacent to it would still be below that of the ambient air so the droplet would continue to grow by condensation out along the right hand side of curve 2. Clearly, any droplet growing along a curve which has a peak supersaturation lying below the supersaturation of the ambient air can form a cloud droplet. On the other hand, any droplet growing along a Köhler curve which intersects a horizontal line in Fig. 4.12, corresponding to the supersaturation of the air, can only form a haze droplet.

4.2.2 Cloud condensation nuclei

Aerosol which serve as the nuclei upon which water vapor condenses in the atmosphere are called *cloud condensation nuclei* (CCN). It follows from the previous discussion that the larger the size of an aerosol and the larger its water solubility (or, if it is insoluble, the more readily it is wetted by water), the lower will be the supersaturation at which it can serve as a CCN. Thus, in order to act as CCN at 1% supersaturation, completely wettable but water insoluble particles need to be at least about 0.1 μm in radius, whereas soluble particles can be as small as about 0.01 μm in radius. Because of these restrictions only a small fraction of atmospheric aerosol serve as CCN (about 1% in continental air and 10–20% in maritime air). Most CCN probably consist of a mixture of soluble and insoluble components (so-called *mixed nuclei*).

The concentrations of CCN active at supersaturations typical of those experienced in natural clouds can be measured with a *thermal diffusion chamber*. This device consists of a small chamber in which the upper and lower horizontal plates are kept wet and maintained at different temperatures, the lower plate being several degrees colder than the upper (Fig. 4.13a). Temperature within the chamber varies linearly with distance from the lower plate. Water vapor is distributed through the chamber by mixing and diffusion so that the density of water vapor molecules is uniform. Since the pressure of the water vapor at any point in the chamber is related to temperature by the ideal gas equation [see Eq. (2.12)], the vapor pressure also varies linearly with distance from the lower plate (Fig. 4.13b). The saturation water vapor pressure, on the other hand, does not vary linearly with temperature (see Fig. 2.7). Consequently, at any point in the chamber (other than at the top or bottom plates) the air is slightly supersaturated with respect to liquid water (Fig. 4.13b). For example, at the point in the chamber where the temperature is 26°C (Fig. 4.13b), the supersaturation is $(e/e_s - 1)100\%$. By changing the temperature difference between the two plates from about 2 to 7.5°C, with the lower plate held at 20°C, the maximum supersaturation in the chamber can be varied from about 0.2 to 2%.

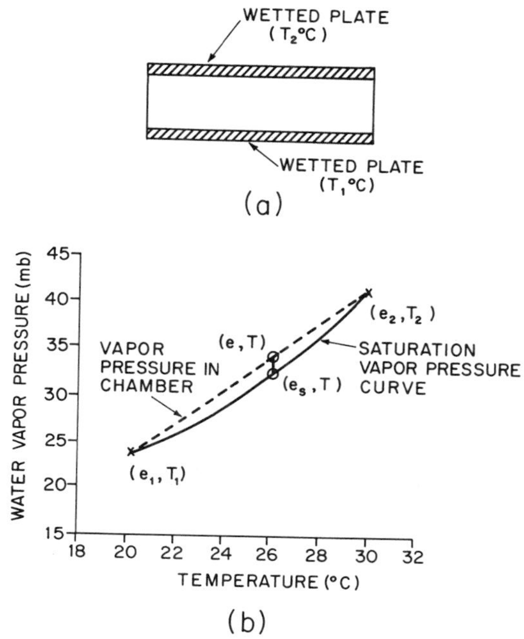

Fig. 4.13 (a) The thermal diffusion chamber. (b) Water vapor pressure in the chamber (---) and the saturation vapor pressure (——) as a function of the air temperature.

As a sample of air comes into equilibrium, the supersaturation near the center of the chamber approaches the maximum value determined by the temperature distribution. Small water droplets then form on those aerosol that act as CCN at the peak supersaturation in the chamber. (It is assumed that there is sufficient mixing in the chamber so that all the aerosol will pass through the region of maximum supersaturation.) The concentration of these droplets is determined either by photographing a known volume of the cloud and counting the number of droplets visible in the photograph or by measuring automatically the intensity of light scattered from the droplets. By repeating the above procedure with different temperature distributions in the chamber, the concentrations of CCN in the air at several different supersaturations (that is, the *CCN supersaturation spectrum*) can be determined.

World-wide measurements of CCN concentrations have not revealed any systematic latitudinal or seasonal variations. However, near the earth's surface continental air masses are generally significantly richer in CCN than are marine air masses (Fig. 4.14). For example, at 1% supersaturation the concentration of CCN in continental air is typically on the order of 500 cm^{-3} while in maritime air it is about 100 cm^{-3}. Concentrations of CCN over land decline by about a factor of five between the surface and 5 km, while over the same height interval concentrations of CCN over the ocean remain fairly constant.

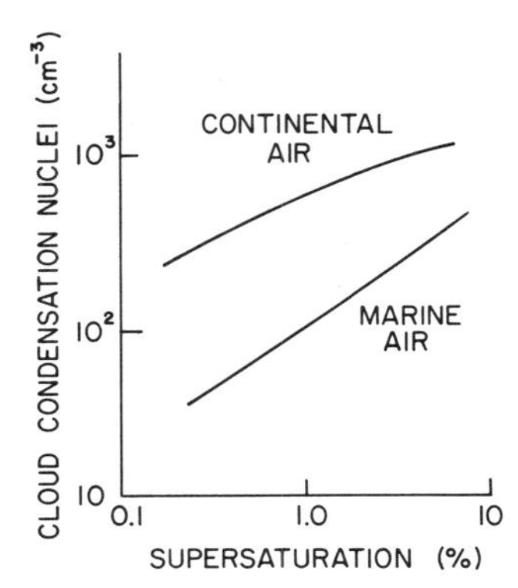

Fig. 4.14 Average cloud condensation nucleus spectra based on measurements made in continental and marine air masses. [From *J. Atmos. Sci.* **26**, 685 (1969).]

Ground-based measurements indicate that there is a diurnal variation in CCN concentrations, with a minimum at about 6 a.m. and a maximum at about 6 p.m.

The observations described provide some clues as to the origins of CCN. First of all it appears that the land acts as one source of CCN because of the higher concentrations of CCN over the land and their decrease in concentration with altitude. Some of the soil particles and dusts that enter the atmosphere probably serve as CCN but they do not appear to be a dominant source. Forest fires are sources of CCN; it has been estimated that the rate of production of CCN from burning vegetable matter is on the order of 10^{12}–10^{15} per kg of material consumed. Although sea-salt particles enter the air over the oceans by the mechanisms discussed in Section 4.1.3, they are not a primary source of CCN. Even at altitudes as low as 15 m above the surface of the ocean, and over surf, the concentrations of sea-salt particles are only a few percent of the concentrations of CCN. Certain industries are sources of CCN (for example, paper mills) but not all industrial pollutants act as CCN.

There appears to be a widespread and probably a fairly uniform source of CCN over both the oceans and the land. The nature of this source has not been definitely established. However, likely sources are the gas-to-particle conversion mechanisms mentioned in Section 4.1.3. As can be seen in Fig. 4.7, these mechanisms produce particles up to a few tenths of a micrometer in diameter which could act as CCN, particularly if they are soluble or wettable. Since some gas-to-particle conversion mechanisms require solar radiation, the rate of production of CCN by this mechanism might tend to reach a maximum in the afternoons, which could be the reason for the observed peak in CCN concentrations at 6 p.m.

Many CCN consist of sulfates; some of the sulfate may be produced in cloud droplets by the mechanism described in Section 4.1.3. Sulfate formed in this way is added to the mass of the CCN originally involved in the formation of the cloud so that, when the cloud evaporates, larger CCN, active at lower supersaturations, are left behind in the air.

4.3 THE MICROSTRUCTURE OF WARM CLOUDS

Clouds which lie completely below the 0°C isotherm are called *warm clouds*. Only liquid water droplets can form in warm clouds. In describing the microstructure of such clouds we are interested in the amount of liquid water per unit volume of air (called the *liquid water content* and usually expressed in grams per cubic meter†), the total number of water droplets per unit volume of air (called the *droplet concentration* and usually expressed as a number per cubic centimeter), and the size distribution of cloud droplets (called the *droplet spectrum* and usually displayed as a histogram of the number of droplets per cubic centimeter in various droplet size intervals). These three parameters are not independent; for example, if the droplet spectrum is known the droplet concentration and liquid water content can be deduced.

4.3.1 Measuring techniques

In principle, the simplest and most direct way of determining the microstructure of a warm cloud is to collect all the droplets in a measured volume of the cloud and then to size and count them under a microscope. For example, oil-coated slides may be exposed from an aircraft to cloudy air along a measured path length. Those droplets that collide with the slide and become completely immersed in the oil are preserved for subsequent analysis. However, the larger droplets may not be completely covered by the oil and they therefore may undergo evaporation. An alternative method is to obtain replicas of the droplets by coating a slide with magnesium oxide powder obtained by burning a magnesium ribbon near the slide; when water droplets collide with the slide they leave clear imprints, the sizes of which may be related to the actual sizes of the droplets.

It should be noted that direct impaction methods, of the type described above, bias against the smaller droplets since the latter tend to follow the streamlines around the collector and thereby avoid capture. This problem is aggravated at high aircraft speeds. Consequently, corrections have to be made, based on theoretical calculations. Recently, several automatic techniques have become available for sizing cloud droplets from an aircraft without collecting the droplets. These techniques are free from the collection problems described above, and they permit the cloud to be sampled continuously so that variations in cloud microstructure in space and time can be investigated more readily, and in real time.

Several techniques are available for measuring cloud liquid water contents from an aircraft. The most popular instrument is a device in which a wire, heated electrically to a high temperature, is exposed to the airstream. When cloud droplets impinge on the wire they are evaporated and thereby cool the wire; the magnitude of this cooling is measured and related to the liquid water

† Since the density of air is approximately 1 kg m^{-3}, a liquid water content of 1 g m^{-3} is approximately the same as 1 g kg^{-1}.

content. To ensure that this instrument responds only to changes in liquid water content, a differential method can be used in which one wire is exposed to the droplets but an exactly similar wire is shielded from them but otherwise exposed to the same airflow. The two wires form two arms of an electrical bridge which is balanced in clear air. Any imbalance in the current gives a measure of the liquid water content.

4.3.2 Effects of cloud condensation nuclei

In order to demonstrate the profound effects that cloud condensation nuclei (CCN) can have on the concentrations and size distributions of cloud droplets, we show in Fig. 4.15 the results of measurements in cumulus clouds which were growing in marine and continental air. Most of the marine cumulus had droplet concentrations less than 100 cm^{-3}, and none had droplet concentrations greater than 200 cm^{-3} (Fig. 4.15a). By contrast, some of the continental cumulus had droplet concentrations as high as 900 cm^{-3} and most had concentrations of a few hundred per cubic centimeter (Fig. 4.15c). These differences clearly reflect the higher concentrations of CCN present in continental air (see Section 4.2.2 and Fig. 4.14). Since the liquid water contents of maritime cumulus clouds do not differ significantly from those of continental cumulus, the higher droplet concentrations in the continental cumulus must result in a smaller average

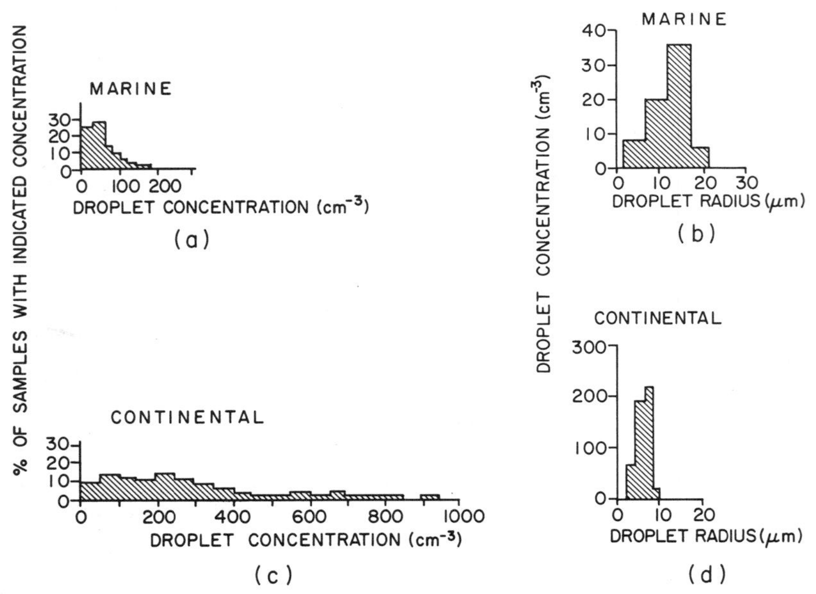

Fig. 4.15 (a) Percentage of marine cumulus clouds with indicated droplet concentrations. (b) Droplet size distributions in a marine cumulus cloud. (c) Percentage of continental cumulus clouds with indicated droplet concentrations. (d) Droplet size distributions in a continental cumulus cloud. Note change in ordinate from (b). [From *Tellus* **10**, 258–259 (1958).]

droplet size. By comparing the results shown in Fig. 4.15b and d, it can be seen that not only is the droplet size spectrum for the continental cumulus cloud much narrower than that for the marine cumulus but also the average droplet radius is significantly smaller. To describe this distinction in another way, we note that droplets with a radius of about 30 μm exist in concentrations of about 1 liter^{-1} in marine clouds whereas in continental clouds the radius has to be lowered to 20 μm before we can find droplets in concentrations of about 1 liter^{-1}. We will see shortly that these differences in microstructure have important effects on the development of precipitation in warm marine and continental cumulus clouds.

4.4 GROWTH OF CLOUD DROPLETS IN WARM CLOUDS

In warm clouds droplets can grow by condensation in a supersaturated environment and by colliding and coalescing with other cloud droplets. In this section we consider each of these two growth processes and assess the extent to which they can explain the formation of rain in warm clouds.

4.4.1 Growth by condensation

If a droplet has passed over the peak in its Köhler curve (Fig. 4.12), it can continue to grow by water condensing upon it provided that the vapor pressure in the ambient air exceeds the vapor pressure of the air adjacent to the droplet.

Let us consider first an isolated droplet, with radius r at time t, situated in a supersaturated environment in which the water vapor density at a large distance from the droplet is $\rho_v(\infty)$ and the water vapor density adjacent to the droplet is $\rho_v(r)$. If we assume that the system is in steady-state equilibrium, the rate of increase in the mass M of the droplet at time t is equal to the rate of flow of water vapor across any spherical surface of radius x centered on the droplet. Hence, if we define the diffusion coefficient D of water vapor in air as the rate of mass flow of water vapor across (and normal to) a unit area in the presence of a unit gradient in water vapor density, the rate of increase in the mass of the droplet is given by

$$\frac{dM}{dt} = 4\pi x^2 D \frac{d\rho_w}{dx}$$

where ρ_w is the water vapor density at distance $x(>r)$ from the droplet. Since dM/dt is independent of x, the above equation can be integrated as follows:

$$\frac{dM}{dt} \int_{x=r}^{x=\infty} \frac{dx}{x^2} = 4\pi D \int_{\rho_v(r)}^{\rho_v(\infty)} d\rho_w$$

or,

$$\frac{dM}{dt} = 4\pi r D[\rho_v(\infty) - \rho_v(r)] \tag{4.9}$$

Substituting $M = \frac{4}{3}\pi r^3 \rho_l$, where ρ_l is the density of liquid water, into this last expression we obtain

$$\frac{dr}{dt} = \frac{D}{r\rho_l}[\rho_v(\infty) - \rho_v(r)]$$

Finally, using the ideal gas equation for the water vapor, with some algebraic manipulation, we obtain

$$\frac{dr}{dt} = \frac{1}{r}\frac{D\rho_v(\infty)}{\rho_l e(\infty)}[e(\infty) - e(r)] \tag{4.10}$$

where $e(\infty)$ is the water vapor pressure in the ambient air well removed from the droplet and $e(r)$ the vapor pressure adjacent to the droplet.†

Strictly speaking, $e(r)$ in (4.10) should be replaced by e', where e' is given by (4.8). However, for droplets in excess of 1 μm or so in radius it can be seen from Fig. 4.12 that the solute effect and the Kelvin curvature effect are not very important, so that the vapor pressure $e(r)$ is approximately equal to the saturation vapor pressure e_s over a plane surface of pure water (which depends only on temperature). In this case, if $e(\infty)$ is not too different from e_s,

$$\frac{e(\infty) - e(r)}{e(\infty)} \simeq \frac{e(\infty) - e_s}{e_s} \equiv S$$

where S is the supersaturation of the ambient air (expressed as a fraction rather than a percentage). Hence (4.10) becomes

$$r\frac{dr}{dt} = G_l S \tag{4.11}$$

where G_l is given by

$$G_l = \frac{D\rho_v(\infty)}{\rho_l}$$

G_l may be considered a constant for a given environment at a fixed temperature. It can be seen from (4.11) that, for fixed values of the supersaturation S, dr/dt is inversely proportional to the radius r of the droplet. Consequently droplets growing by condensation initially increase in radius very rapidly but their rate of growth diminishes with time, as shown schematically by curve (a) in Fig. 4.16.

In a cloud we are concerned with the growth of a large number of droplets in a rising parcel of air. As the parcel rises it expands, cools adiabatically, and eventually reaches saturation with respect to liquid water. Further uplift

† Several assumptions have been made in the derivation of (4.10). For example, we have assumed that all of the molecules that land on the droplet remain there and that the vapor adjacent to the droplet is at the same temperature as the environment. Due to the release of latent heat of condensation, the temperature at the surface of the droplet will, in fact, be somewhat higher than the temperature of the air well away from the droplet. We have also assumed that the droplet is at rest; droplets which are falling with appreciable speeds will be ventilated and this will affect both the temperature of the droplet and the flow of vapor.

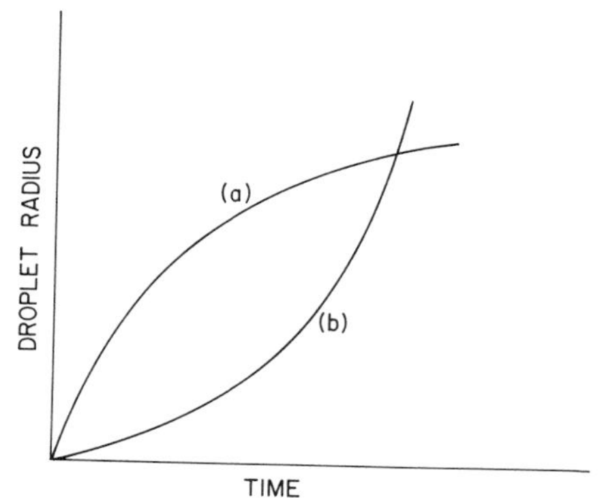

Fig. 4.16 Schematic curves of droplet growth by (a) condensation from the vapor phase and (b) coalescence of droplets.

produces supersaturation which initially increases at a rate proportional to the updraft velocity. As the supersaturation rises CCN are activated, starting with the most efficient ones. When the rate at which moisture is being made available by the adiabatic cooling equals the rate at which it is condensing onto the CCN and droplets, the supersaturation reaches a maximum value. The concentration of cloud droplets is determined at this stage (which occurs within 100 m or so of cloud base) and is equal to the concentration of CCN activated up to this point. Subsequently, the growing droplets consume water at a greater rate than it is made available by the cooling of the air so that the supersaturation begins to decrease. The haze droplets therefore evaporate slowly while the activated droplets continue to grow by condensation. Since the rate of growth of a droplet by condensation is inversely proportional to its radius, the smaller activated droplets grow faster than the larger droplets. Consequently, the sizes of the droplets in the cloud tend to become increasingly uniform with time (that is, the droplets approach a *monodispersed* distribution). This sequence of events is illustrated by the results of theoretical calculations shown in Fig. 4.17.

Comparisons between cloud droplet size distributions measured just a few hundred meters above the bases of nonprecipitating warm cumulus clouds and droplet size distributions computed assuming condensational growth for about 5 min show good agreement (Fig. 4.18). Note that the droplets produced by condensation during this time period only extend up to about 10 μm in radius. Moreover, as mentioned above, the rate of increase in the radius of a droplet growing by condensation decreases with time. It is clear, therefore, as first noted by Reynolds† in 1877, that growth by condensation alone in warm clouds

† **Osborne Reynolds** (1842–1912) Probably the outstanding English theoretical mechanical engineer of the 19th century. Carried out important work on hydrodynamics and the theory of lubrication. Studied atmospheric refraction of sound. The Reynolds number, which he introduced, is named after him.

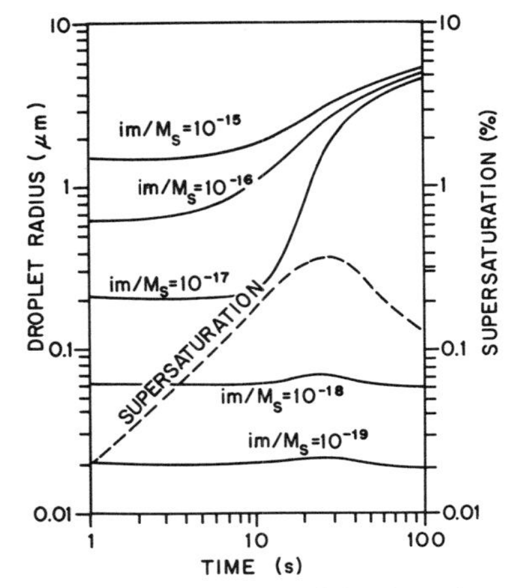

Fig. 4.17 Theoretical computations of the growth of cloud condensation nuclei by condensation in a parcel of air rising with a vertical speed of 60 cm s^{-1}. A total of 500 CCN cm^{-3} was assumed with im/M_s values [see Eq. (4.8)] as indicated. The variation with time of the supersaturation is also shown. [Based on data from *J. Met.* **6**, 143 (1949).]

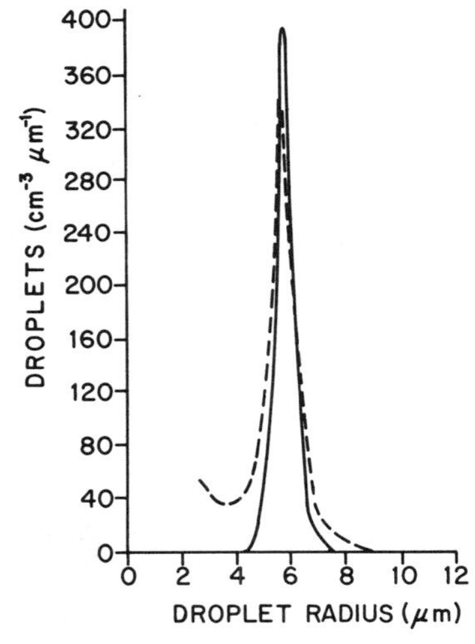

Fig. 4.18 Comparison of the cloud droplet size distribution observed 244 m above the base of a warm cumulus cloud (– – –) and the corresponding computed droplet size distribution assuming only condensational growth (——). [Adapted from Tech. Note No. 44, Cloud Physics Lab., Univ. of Chicago, p. 144.]

is much too slow to produce raindrops with radii of several millimeters. Yet rain does form in warm clouds, especially in the tropics. The enormous increases in size required to transform cloud droplets into raindrops is illustrated by the scaled diagram shown in Fig. 4.19. For a cloud droplet 10 μm in radius

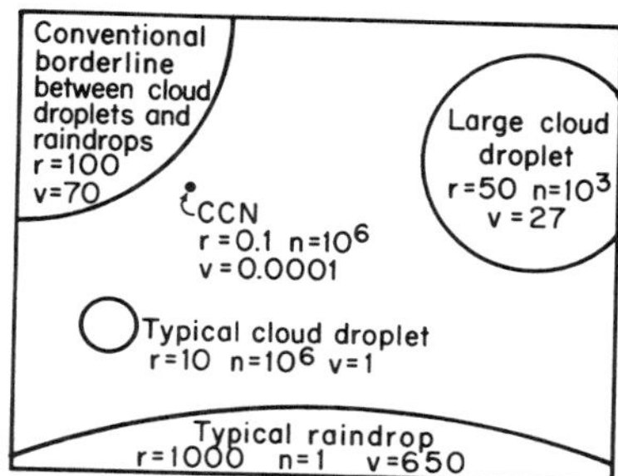

Fig. 4.19 Relative sizes of cloud droplets and raindrops; r is the radius in micrometers, n the number per liter of air, and v the terminal fall speed in centimeters per second. [From *Adv. in Geophys.* **5**, 244 (1958).]

to grow to a raindrop 1 mm in radius requires an increases in volume of one millionfold! However, only about one droplet in a million (say, about 1 liter^{-1}) in a cloud has to grow by this amount for the cloud to rain. The mechanism responsible for this selective growth in warm clouds is discussed in the next section.

4.4.2 Growth by collision and coalescence

In warm clouds the growth of droplets from the relatively small sizes achieved by condensation to the size of raindrops is achieved by the collision and coalescence of droplets.[†‡] Since the steady settling velocity of a droplet as it falls under the influence of gravity through still air (called the *terminal fall speed* of the droplet) increases with the size of the droplet (see Problem 4.7), those droplets in a cloud that are somewhat larger than average will have a higher than average terminal fall speed and they will collide with smaller droplets lying in their path.

Consider a single drop[§] of radius r_1 (called the *collector drop*) which is overtaking a smaller droplet of radius r_2 (Fig. 4.20). As the collector drop approaches the droplet the latter will tend to follow the streamlines around the collector drop and may thereby avoid capture. We define an effective collision cross section in terms of the parameter y shown in Fig. 4.20, which represents the critical distance between the centerline of the collector drop and the center of the droplet (measured at a large distance from the collector drop) such that

[†] As early as the 10th century a secret society of Basra ("The Brethren of Purity") suggested that rain is produced by the coalescence of cloud drops. In 1715 Barlow also suggested that raindrops form due to larger cloud drops overtaking and colliding with smaller droplets. These ideas, however, were not investigated seriously until the 20th century.

[‡] **Edward Barlow** (1639–1719) English priest. Author of "Meteorological Essays Concerning the Origin of Springs, Generation of Rain, and Production of Wind, with an Account of the Tide," John Hooke and Thomas Caldecott, London, 1715.

[§] In this section "drop" will refer to the larger and "droplet" to the smaller body.

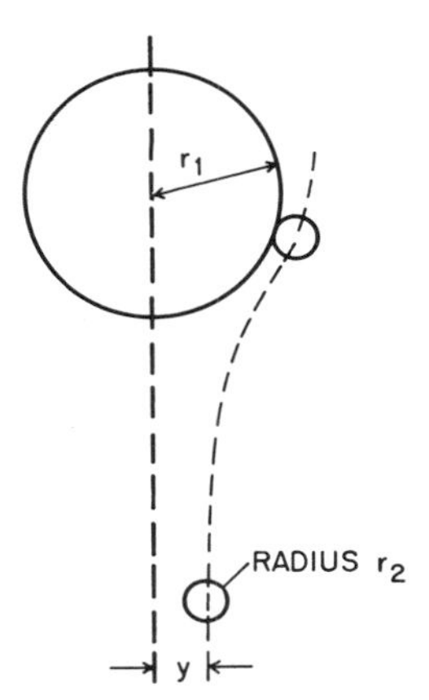

Fig. 4.20 Illustration of the maximum impact parameter y for a droplet of radius r_2 with a collector drop of radius r_1.

the droplet just makes a grazing collision with the collector drop. If the center of a droplet of radius r_2 is any closer than y to the centerline of a collector drop of radius r_1 it will collide with the collector drop; conversely, if it is at a greater distance than y from the centerline it will not collide with the collector drop. The effective collision cross section of the collector drop for droplets of radius r_2 is then πy^2, whereas the geometrical collision cross section is $\pi(r_1 + r_2)^2$. We therefore define the *collision efficiency* E of droplets of radius r_2 with a drop of radius r_1, as

$$E = \frac{y^2}{(r_1 + r_2)^2} \tag{4.12}$$

Determination of the values of the collision efficiency is a difficult mathematical problem, particularly when the drop and droplet are similar in size, in which case they strongly affect each other's motion. Some recently computed values for E are shown in Fig. 4.21 from which it can be seen that the collision efficiency increases markedly as the size of the collector drop increases, and that the collision efficiencies for collector drops less than about 20 μm in radius are quite small. When the collector drop is much larger than the droplet ($r_2/r_1 \ll 1$), collision efficiencies are small because the droplet tends to follow closely the streamlines around the collector drop. As the ratio r_2/r_1 increases, E initially increases because the larger droplets tend to move more nearly in straight lines rather than follow the streamlines around the collector drop. However, as r_2/r_1 increases from about 0.6 to 0.9, E falls off, particularly for the smaller collector drops, because the terminal fall speeds of collector drop and droplets approach one another so the relative velocity between them is very small. Finally, however, as r_2/r_1 approaches unity, E tends to increase again because two nearly equal sized drops strongly interact to produce a

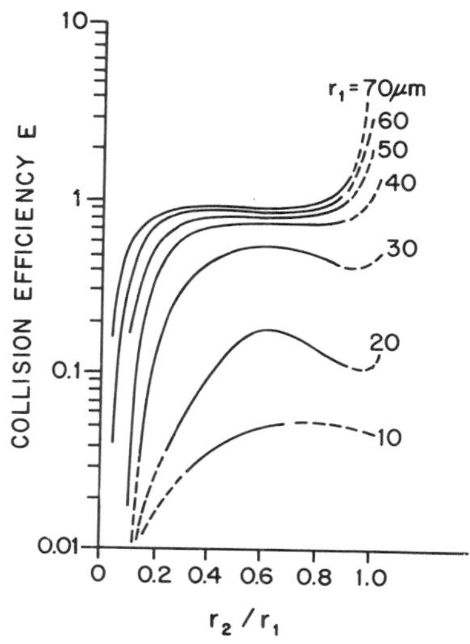

Fig. 4.21 Calculated values of the collision efficiency for collector drops of radius r_1 with droplets of radius r_2. The dashed portions of the curve represent regions of doubtful accuracy. [From *J. Atmos. Sci.* **30**, 112 (1973).]

closing velocity between them. Indeed, wake effects behind the collector drop can produce values of E greater than unity.

The next problem to be considered is whether or not a droplet is captured (that is, coalescence occurs) when it collides with a larger drop. It is known from laboratory experiments that, under certain conditions, small droplets can bounce off a water surface (Fig. 4.22). Bounce-off can occur when two droplets approach each other very closely if air becomes trapped between them so that

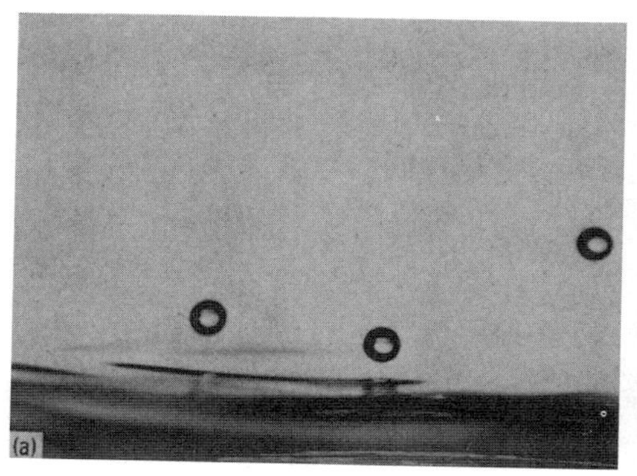

Fig. 4.22 (a) A stream of water droplets (entering from the right), about 100 μm in diameter, rebounding from a layer of water. (b) When the angle between the stream of droplets and the surface of the water is increased beyond a critical value, the droplets coalesce with the water. (Photo: P. V. Hobbs.)

the droplets interact and deform without actually touching.[†‡] In other words, the droplets can rebound on a cushion of air. If the cushion of air is squeezed out before rebound occurs, the droplets will make physical contact and coalesce.[§] We define the *coalescence efficiency* E' of a droplet of radius r_2 with a drop of radius r_1 as the fraction of collisions which result in a coalescence. The *collection efficiency* E_c is equal to EE'.

By comparing laboratory experimental measurements of collection efficiencies for droplets in free fall with theoretical collision efficiencies, estimates of coalescence efficiencies may be obtained. Such comparisons indicate that the coalescence efficiency E' falls off sharply as the sizes of the drops approach one another (Fig. 4.23).

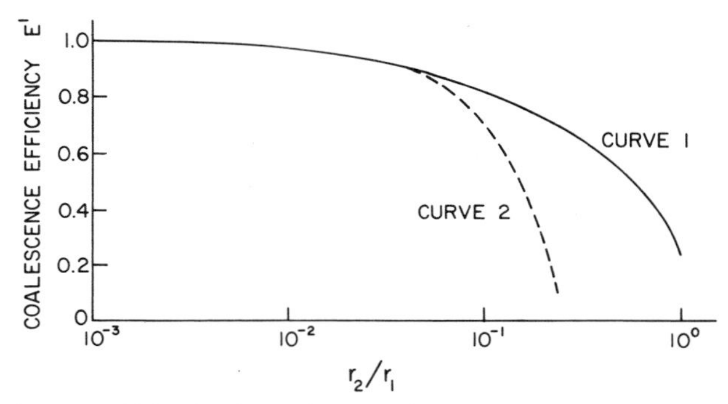

Fig. 4.23 Coalescence efficiencies for a droplet of radius r_2 with a drop of radius r_1. (——) 400 μm $< r_1 <$ 2000 μm, 20 μm $< r_2 <$ 100 μm. [From *J. Geophys. Res.* **76**, 2836 (1971).] (---) r_1 averages 50–100 μm. [From *J. Atmos. Sci.* **30**, 944 (1973).]

It is known from laboratory experiments that the presence of electric fields enhances coalescence. For example, in the experiment illustrated in Fig. 4.22, droplets which bounce on the water surface at a certain angle of incidence can be made to coalesce by applying an electric field of about 10^4 V m^{-1}. Similarly, coalescence is aided if the impacting droplets carry an electric charge in excess of about 10^{-14} C. Although the roles of electric fields and charges in enhancing the coalescence efficiencies of cloud droplets are not yet clear, present indications are that they are unlikely to be important except, perhaps, in thunderstorms.

† Lenard pointed out in 1904 that cloud droplets might not always coalesce when they collide, and he suggested that this could be due to a layer of air between the droplets or to electrical charges.

‡ **Phillip Lenard** (1862–1947) Austrian physicist. Studied under Helmholtz and Hertz. Professor of physics at Heidelberg and Kiel. Won the Nobel prize in physics (1905) for work on cathode rays. One of the first to study the charging produced by the disruption of water (for example, in waterfalls).

§ Even after two droplets have coalesced, the motions set up in their combined mass may cause it to subsequently break up into several droplets.

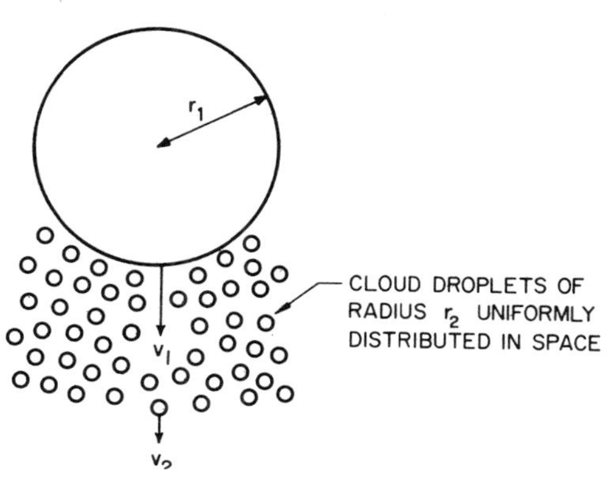

Fig. 4.24 Schematic diagram to illustrate the continuous collision model.

Let us now consider a collector drop of radius r_1 which has a terminal fall speed v_1 when it falls through still air. Let us suppose that this drop is falling in still air through a cloud of otherwise equal sized droplets of radius r_2, and terminal fall speed v_2 in still air. We will assume that the droplets are uniformly distributed in space and that they are collected uniformly at the same rate by all collector drops of a given size. This so-called *continuous collision model* is illustrated in Fig. 4.24. The rate of increase in the mass M of the collector drop due to collisions is then given by

$$\frac{dM}{dt} = \pi r_1^2 (v_1 - v_2) w_l E_c \tag{4.13}$$

where w_l is the liquid water content (in kg m^{-3}) of the cloud droplets of radius r_2. Substituting $M = \frac{4}{3}\pi r_1^3 \rho_l$ into (4.13), where ρ_l is the density of liquid water, we obtain

$$\frac{dr_1}{dt} = \frac{(v_1 - v_2) w_l E_c}{4\rho_l} \tag{4.14}$$

If $v_1 \gg v_2$ and we assume that the coalescence efficiency is unity, so that $E_c = E$, (4.14) becomes

$$\frac{dr_1}{dt} = \frac{v_1 w_l E}{4\rho_l} \tag{4.15}$$

Since v_1 increases as r_1 increases (see Problem 4.7) and E also increases with r_1 (see Fig. 4.21), it follows from (4.15) that dr_1/dt increases with increasing r_1; that is, the growth of a drop by collisions is an accelerating process, as illustrated by curve (b) in Fig. 4.16. It can be seen from Fig. 4.16 that for small cloud droplets growth by condensation is initially dominant but beyond a certain radius growth by collisions dominates.

If there is a steady updraft velocity w in the cloud, the velocity of the collector drop will be $w - v_1$ and the velocity of the cloud droplets $w - v_2$. Hence dr_1/dt

will still be given by (4.14) but the motion of the collector drops is given by

$$\frac{dh}{dt} = w - v_1 \tag{4.16}$$

where h is the height above a fixed level (say, cloud base) at time t. Eliminating dt between (4.15) and (4.16) and assuming $v_1 \gg v_2$ and $E_c = E$ we obtain

$$\frac{dr_1}{dh} = \frac{v_1 w_l E}{4\rho_l(w - v_1)}$$

or, if the radius of the collector drop at height H above cloud base is r_H and at cloud base is r_0,

$$\int_0^H w_l\, dh = 4\rho_l \int_{r_0}^{r_H} \frac{(w - v_1)}{v_1 E}\, dr_1$$

Hence, if we assume that w_l is independent of h,

$$H = \frac{4\rho_l}{w_l}\left[\int_{r_0}^{r_H} \frac{w}{v_1 E}\, dr_1 - \int_{r_0}^{r_H} \frac{dr_1}{E}\right] \tag{4.17}$$

If values of E and v_1 as a function of r_1 are known, (4.17) can be used to determine the value of H corresponding to any value of r_H and *vice versa*. We can also deduce from (4.17) the general behavior of cloud drops growing by collisions by the following qualitative argument. When the drop is still quite small $w > v_1$ and the first integral dominates over the second; H then increases as r_H increases; that is, a drop growing by collisions is carried upward in the cloud. Eventually, as the drop grows, v_1 becomes greater than w and the value of the second integral becomes larger than that of the first. H now decreases with increasing r_H; that is, the drop begins to fall through the updraft and, if it holds together, it will eventually pass through the cloud base and reach the ground as a raindrop. Some of the larger drops (with radius greater than 1 mm) may break up as they fall through the air, particularly those involved in collisions (see footnote on page 176). The resulting fragments may then grow and break up again; such a chain reaction may enhance the precipitation.

Problem 4.2 A drop enters the base of a cloud with a radius r_0 and, after growing with a constant collection efficiency by collisions while traveling up and down in the cloud, it reaches the cloud base again with a radius R. Show that R is a function only of r_0 and the updraft velocity w (assumed constant) in the cloud.

Solution Putting $H = 0$ and $r_H = R$ into the equation before (4.17) we obtain

$$0 = 4\rho_l \int_{r_0}^{R} \frac{w - v_1}{v_1 E}\, dr_1$$

or, since E and w are assumed to be constant,

$$w \int_{r_0}^{R} \frac{dr_1}{v_1} = \int_{r_0}^{R} dr_1 = R - r_0$$

Therefore,

$$R = r_0 + w \int_{r_0}^{R} \frac{dr_1}{v_1} \tag{4.18}$$

Since $\int_{r_0}^{R} dr_1/v_1$ is a function only of R and r_0, it follows from the last equation that R is a function only of r_0 and w.

In the early 1950s several investigators used (4.17) and similar equations to investigate the feasibility of rain formation by the collision–coalescence mechanism. A few drops, large enough to grow by collisions, were generally arbitrarily assumed to be present near the base of the cloud. Some of the earliest estimates of collision efficiencies were used, and the coalescence efficiency was assumed to be unity. These calculations indicated that rain should be able to form by the collision–coalescence mechanism within reasonable time periods in cumulus clouds. For example, in a cloud containing 1 g m^{-3} of water (consisting of 10-μm-radius droplets) and having an updraft velocity of 1 m s^{-1}, an initially small collector drop was predicted to grow to a radius of about 0.15 mm in about 45 min during its travel upward in the cloud to a height of 2.2 km. During the next 15 min the radius of the drop was predicted to increase to 0.75 mm (which corresponds to a fairly large raindrop) as it fell back to cloud base. With the same updraft velocity but a liquid water content of 0.5 g m^{-3}, the calculations indicated that the collector drop would be carried up to a height of about 3.2 km and finally emerge from cloud base with a radius of about 0.65 mm. With a liquid water content of 1 g m^{-3} and an updraft velocity of 0.1 m s^{-1} these simple model calculations indicated that a collector drop would only be carried 0.5 km above cloud base, from which height it would take nearly 2 h for it to fall back to cloud base where its radius would be only about 0.1 mm (which corresponds to a small drizzle drop). Hence, these model calculations indicated that warm clouds with strong updrafts should produce rain in a shorter time than clouds with weak updrafts, but clouds with strong updrafts must be quite deep in order for raindrops to be produced. Also, raindrops which form in deep, warm clouds with strong updrafts should be considerably larger than raindrops from shallower clouds with weak updrafts.

In the continuous collision model it is assumed that the collector drop collides in a continuous and uniform fashion with smaller cloud droplets which are uniformly distributed in space. Consequently, the continuous collision model predicts that all collector drops of the same size will grow at the same rate when they fall through the same cloud of droplets. An important advance in our understanding of the growth of drops in warm clouds has been made by replacing the continuous collision model by a *stochastic* (or *statistical*) *collision model.* The stochastic model allows for the fact that collisions are individual events, statistically distributed in time and space. Consider, for example, 100 droplets, initially the same size as shown on line 1 in Fig. 4.25. After a certain interval of time some of these droplets (let us say 10) will have collided with other droplets, so that the distribution will now be as depicted in line 2 of Fig. 4.25. Because of their larger size, these 10 larger droplets are now in a more

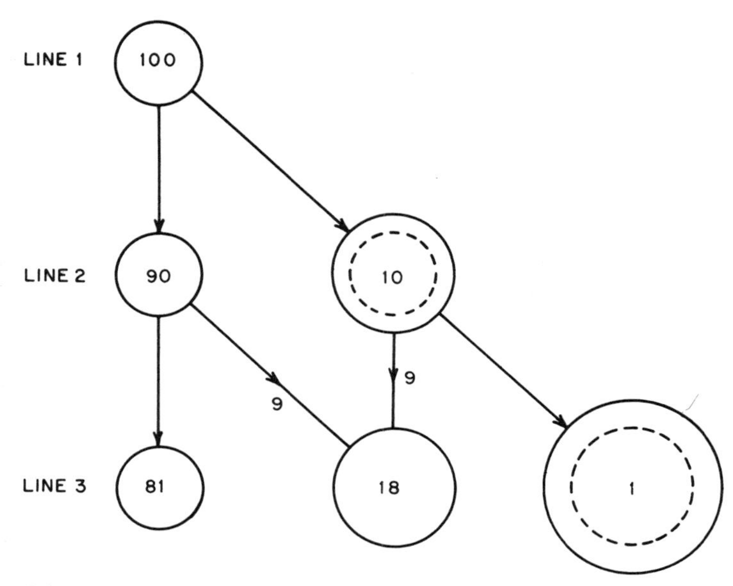

Fig. 4.25 Schematic diagram to illustrate broadening of droplet sizes by statistical collisions. [From *J. Atmos. Sci.* **24**, 689 (1967).]

favored position for making further collisions. The second collisions are similarly statistically distributed, giving a further broadening of the droplet size spectrum as shown in line 3 of Fig. 4.25 (where it has been assumed that in this time step, nine of the smaller droplets and one of the larger droplets on line 2 each had a collision). Hence, by allowing for a statistical distribution of collisions, we have developed three size categories of droplets after two time steps. This concept is extremely important, since it not only provides a mechanism for developing broad droplet size spectra from the fairly uniform droplet sizes produced by condensation, but it also reveals how a small fraction of the droplets in a cloud can grow much faster than average by statistically distributed collisions.

In recent years several workers have developed theoretical models which are designed to predict the life history of a cloud in a given environment. Much of what is presently known about both large- and small-scale cloud processes can be incorporated into these models and the calculations are carried out on large digital computers. Shown in Fig. 4.26 are some results from one such model illustrating the growth of drops, by condensation and stochastic collisions, in warm cumulus clouds forming in typical marine and continental air masses. We have already pointed out in Section 4.3 that the average droplet sizes are significantly larger and the droplet spectra much broader in marine cumulus than in nonraining continental cumulus (see Fig. 4.15). We have attributed these differences to the higher concentrations of CCN present in continental air (see Fig. 4.14). Figure 4.26 illustrates the effects of these differences in cloud

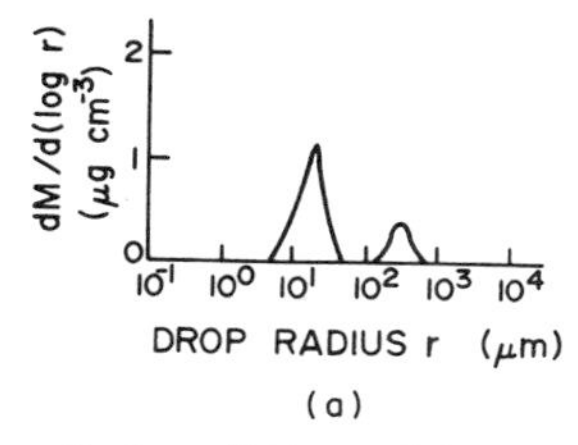

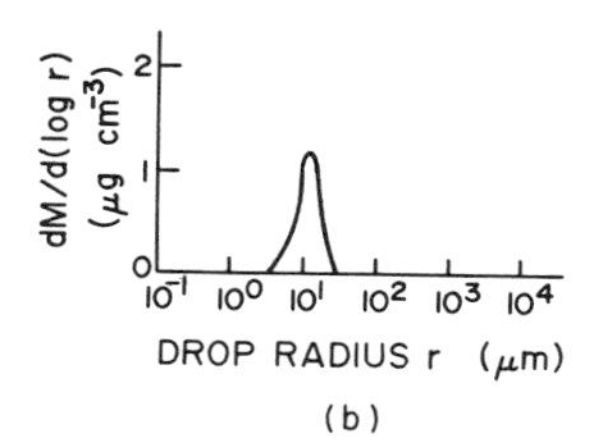

Fig. 4.26 Theoretical predictions of the mass spectrum of drops near the middle of (a) a warm marine cumulus cloud and (b) a warm continental cloud after 67 minutes of growth. (B. C. Scott and P. V. Hobbs, unpublished.)

microstructure on the development of larger-size drops. The CCN spectra used as input data to the two clouds were based on measurements, with the continental air having much higher concentrations than the marine air (about 200 versus 45 cm^{-3} at 0.2% supersaturation). It can be seen that after 67 min the cumulus cloud growing in the marine air has developed some drops between 100 and 1000 μm in radius (that is, raindrops), whereas the continental cloud does not contain any droplets greater than about 20 μm in radius. The results are different because a small fraction of the larger drops in the marine cloud (produced initially, perhaps, by condensation onto giant sea-salt particles or by the chance collisions of two smaller droplets) grow by collisions and coalescences, whereas in the continental cloud the number of drops large enough to grow effectively by collisions is insignificant. These model results support the observation that a marine cumulus cloud is more likely to rain than a continental cumulus cloud of the same depth.

4.5 THE MICROPHYSICS OF COLD CLOUDS

If a cloud extends above the 0°C level it is called a *cold cloud*. Even though the temperature may be below 0°C, water droplets can still exist in clouds, in which case they are referred to as *supercooled* droplets†‡§ (the term *undercooled* is probably preferable but the former term seems to be firmly entrenched). Cold clouds may also contain ice particles. If a cold cloud contains both ice particles and supercooled droplets it is said to be a *mixed cloud*; if it consists entirely of ice it is said to be *glaciated*.

† Saussure observed, around 1783, that water could remain in the liquid state below 0°C. A spectacular confirmation of the existence of supercooled clouds was provided by a balloon flight made by Barrel in 1850. He observed water droplets down to −10.5°C and ice crystals at lower temperatures.

‡ **Horace Bénédict de Saussure** (1740–1799) Swiss geologist, physicist, meteorologist, and naturalist. Traveled extensively, particularly in the Alps, made the second ascent of Mont Blanc (1787).

§ **Jean Augustine Barrel** (1819–1884) French chemist and agriculturalist. First to extract nicotine from tobacco-leaf.

In this section we are concerned with the origins and concentrations of ice particles in clouds, the various ways in which ice particles can grow, and the formation of precipitation in cold clouds.[†]

4.5.1 Nucleation of ice particles; ice nuclei

A supercooled droplet is in an unstable state. However, in order for freezing to occur, enough water molecules must collect together within the droplet to form an embryo of ice large enough to survive and grow. The situation is, in fact, analogous to the formation of a water droplet from the vapor phase discussed in Section 4.2.1. Thus, if an ice embryo exceeds a certain critical size, its further growth will produce a decrease in the total energy of the system. On the other hand, any increase in size of an ice embryo smaller than the critical size causes an increase in total energy; consequently, such embryos tend to break up.

If a water droplet contains no foreign particles it can only freeze by a process known as *homogeneous* (or *spontaneous*) *nucleation*, in which an ice embryo of critical size is formed by the chance aggregation of a sufficient number of water molecules in the droplet. Since the numbers and sizes of the ice embryos that form by chance aggregations increase with decreasing temperature, below a certain temperature (which depends on the volume of water considered) freezing by homogeneous nucleation becomes a virtual certainty. Homogeneous nucleation occurs at about $-36°C$ for droplets between about 20 and 60 μm in radius, and for droplets a few micrometers in radius it occurs at about $-39°C$. Hence, only in high clouds can freezing occur by homogeneous nucleation.

If a droplet contains a rather special type of foreign particle, called a *freezing nucleus*, it may freeze by a process known as *heterogeneous nucleation*[‡§] in which water molecules in the droplet collect onto the surface of the particle to form an icelike structure which may increase in size and cause the droplet to freeze. Since the formation of the ice structure is aided by the freezing nucleus, and the ice embryo also starts off with the dimensions of the freezing nucleus, heterogeneous nucleation can occur at much higher temperatures than homogeneous nucleation. Shown in Fig. 4.27 are the results of laboratory experiments on the heterogeneous freezing of water droplets. The droplets consisted of distilled water from which most, but not all, of the foreign particles were

[†] For detailed accounts of the nucleation and growth of ice and the roles of ice in the atmosphere the reader is referred to P. V. Hobbs, "Ice Physics," Oxford University Press, 1974, pp. 461–724.

[‡] Studies of the heterogeneous nucleation of ice date back to 1724 when Fahrenheit slipped on the stairs while carrying a flask of cold (supercooled) water and noticed that the water had become full of flakes of ice.

[§] **Gabriel Daniel Fahrenheit** (1686–1736) German instrument maker and experimental physicist. Lived in Holland from the age of 15 but traveled widely in Europe. Developed the thermometric scale which bears his name. Fahrenheit knew that the boiling point of water varies with atmospheric pressure [see Eq. (2.93)] and he constructed a thermometer from which the atmospheric pressure could be determined by noting the boiling point of water. Although Fahrenheit was an experimental physicist of the most practical bent he was made a *Fellow of the Royal Society*.

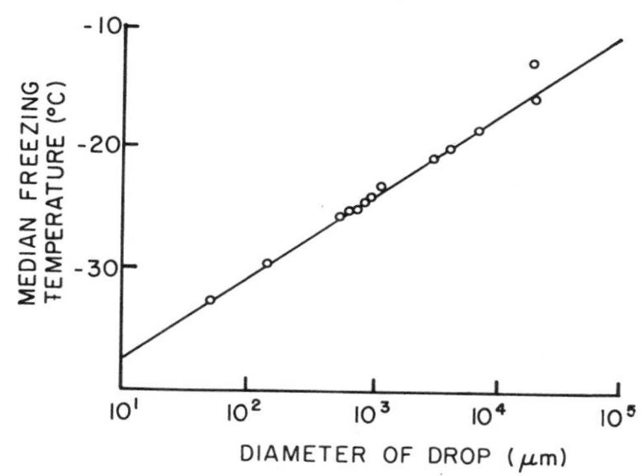

Fig. 4.27 Median heterogeneous freezing temperatures for water drops as a function of their size. [From *Proc. Phys. Soc., Lond.* **B66**, 690 (1953), copyright by the Institute of Physics.]

removed. A large number of droplets of each of the indicated sizes were cooled and the temperature at which half of the droplets had frozen was noted. It can be seen that this median freezing temperature increases as the size of the droplet increases. This is due to the fact that as the volume of the droplet increases the probability increases that the droplet will contain a freezing nucleus capable of causing heterogeneous nucleation at a given temperature.

We have assumed above that the freezing nucleus is contained within the droplet. However, cloud droplets may also be frozen if a freezing nucleus in the air comes into contact with the droplet, in which case freezing is said to occur by *contact nucleation* and the freezing nucleus is referred to as a *contact nucleus*. There is some evidence from laboratory experiments that a particle can cause a drop to freeze by contact nucleation at a temperature several degrees higher than it would if it were embedded in the drop.

Certain particles in the air also serve as centers upon which ice crystals form directly from the vapor phase. These particles are referred to as *deposition nuclei*. Ice can form by deposition† provided that the air is supersaturated with respect to ice and the temperature is low enough. If the air is supersaturated with respect to water, a suitable particle may serve either as a freezing nucleus (in which case liquid water first condenses onto the particle and subsequently freezes) or as a deposition nucleus (in which case there is no intermediate liquid phase, at least on the macroscopic scale).

Freezing nuclei, contact nuclei, and deposition nuclei are collectively referred to as *ice nuclei*. Also, if we wish to refer to an ice nucleating particle in general, without specifying its mode of action, we will call it an ice nucleus. It should be noted however that the *threshold temperature* at which a particle can

† The transfer of water vapor to ice is often referred to as *sublimation* in cloud physics. However, since chemists use this term to describe the evaporation of a solid, the term *deposition* is preferable.

cause ice to form depends, in general, upon the mechanism by which it nucleates the ice as well as upon the previous history of the particle.

Particles with molecular spacings and crystallographic arrangements similar to those of ice (which has a hexagonal structure) tend to have good ice nucleating ability. Most good ice nuclei are also virtually insoluble in water. Some inorganic soil particles (mainly clays) can nucleate ice at fairly high temperatures (that is, above $-15°C$) and they probably play an important role in nucleating ice in clouds. For example, in one study 87% of the snow crystals collected on the ground had clay mineral particles at their centers and more than half of these were kaolinite. Many organic materials are good ice nucleators. Recently it has been observed that decayed plant leaves contain copious ice nuclei, some active as high as $-4°C$. Ice nuclei active at $-4°C$ have also been found in sea water rich in plankton.

Several techniques exist for measuring the concentrations of particles in the air which are active as ice nuclei at a given temperature. A common method is to draw a known volume of air into a container and to cool it until a cloud is formed. The number of ice crystals forming at a particular temperature is then measured. In *expansion chambers* the cooling is produced by compressing the air and then suddenly expanding it, and in *mixing chambers* cooling is produced by refrigeration. In these chambers particles may serve as freezing, contact, or deposition nuclei. The number of ice crystals which appear in the chamber may be determined by illuminating a certain volume of the chamber and estimating visually the number of crystals in the light beam, by letting the ice crystals fall into a dish of supercooled soap or sugar solution where they can be detected and counted by the number of larger ice crystals that they produce, or by allowing the ice crystals to pass through a small capillary tube attached to the chamber where they produce audible clicks which can be counted electronically. In another technique, a measured volume of air is drawn through a Millipore filter which retains the particles in the air. The number of ice nuclei on the filter is then determined by placing it in a box held at a given supersaturation and temperature and counting the number of ice crystals that grow on the filter.

Some results of world-wide measurements of ice nuclei are shown in Fig. 4.28. Measurements made with mixing and expansion chambers indicate that concentrations of ice nuclei tend to be higher in the Northern than in the Southern Hemisphere. It should be noted however that ice nucleus concentrations can sometimes vary by several orders of magnitude over a period as short as several hours. On the average, the number N of ice nuclei per liter of air active at temperature T tends to follow the empirical relationship

$$\ln N = a(T_1 - T) \tag{4.19}$$

where T_1 is the temperature at which one ice nucleus per liter is active (typically about 253°K and a varies from about 0.3 to 0.8. For $a = 0.6$, (4.19) predicts that the concentration of ice nuclei increases by about a factor of 10 for every

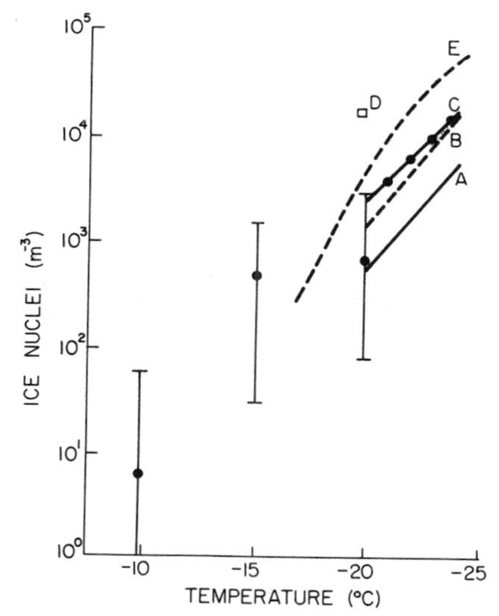

Fig. 4.28 Average ice nucleus concentrations in the Northern and Southern Hemispheres. A, Southern Hemisphere, expansion chamber; B, Southern Hemisphere, mixing chamber; C, Northern Hemisphere, expansion chamber; D, Northern Hemisphere, mixing chamber; [A–D from *Bull. Obs. Puy de Dome* **3**, 89 (1960)]; E, Antarctica, mixing chamber [from *J. Atmos. Sci.* **20**, 185 (1963)]. The vertical lines show the range and mean values (·) of ice nucleus concentrations based on Millipore filter measurements in many locations around the world [from *J. Rech. Atmos.* **2**, 41 (1970).]

4 deg fall in temperature. Since the total concentration of aerosol is on the order of 10^8 liter^{-1}, only about one particle in 10^8 acts as an ice nucleus at $-20°C$.

The concentration N_D of active deposition nuclei can be approximated by the empirical relationship

$$N_D = bS_i^{\alpha} \tag{4.20}$$

where b and α are constants and S_i is the supersaturation (in percent) with respect to ice. Values of α range from 3 to 8 depending on the source of the aerosol (for example, 3 for air over rural northeast Colorado and 8 for polluted air in St. Louis). Temperature is only an implicit parameter in (4.20). For example, in a cloud at water saturation, the supersaturation with respect to ice depends upon the temperature and the activity of the deposition nuclei as a function of temperature may be expressed by an equation of the form (4.19). However, (4.20) seems to be the more fundamental relationship for deposition nuclei. It is interesting to note that insofar as the experimental results for cloud condensation nuclei lie on straight lines in Fig. 4.14, their concentration is also related to supersaturation by expression such as (4.20), but in this case the supersaturation is with respect to water.

4.5.2 Occurrence and concentrations of ice particles in clouds; ice multiplication

The probability of ice particles being present in a cloud increases as the temperature decreases below 0°C, as illustrated in Fig. 4.29, which shows the combined results of a number of observations in different clouds. Clouds with tops at temperatures between 0 and −4°C generally consist entirely of supercooled droplets. It is in clouds such as these that aircraft are most likely to encounter severe icing conditions since supercooled droplets freeze when they collide with an aircraft. For cloud top temperatures of −10°C there is about a 50% probability of detecting ice, and below about −20°C there is better than 95% probability.

Measuring concentrations of ice particles in a cloud is difficult. An airborne method that has been used fairly frequently is to expose a moving strip of 16-mm movie film, covered with a solution of Formvar (a plastic) in ethylene dichloride, to the cloudy air so that ice particles impact on the film where they become embedded in the Formvar solution. After the ethylene dichloride evaporates a thin plastic skin remains. The ice evaporates through small holes in the skin but permanent plastic replicas of the ice particles are retained on the film (a similar technique can be used for cloud droplets). By counting the number of ice particles collected on a strip of film which has swept through a measured volume of cloud, the concentrations of ice particles in the cloud can be deduced.

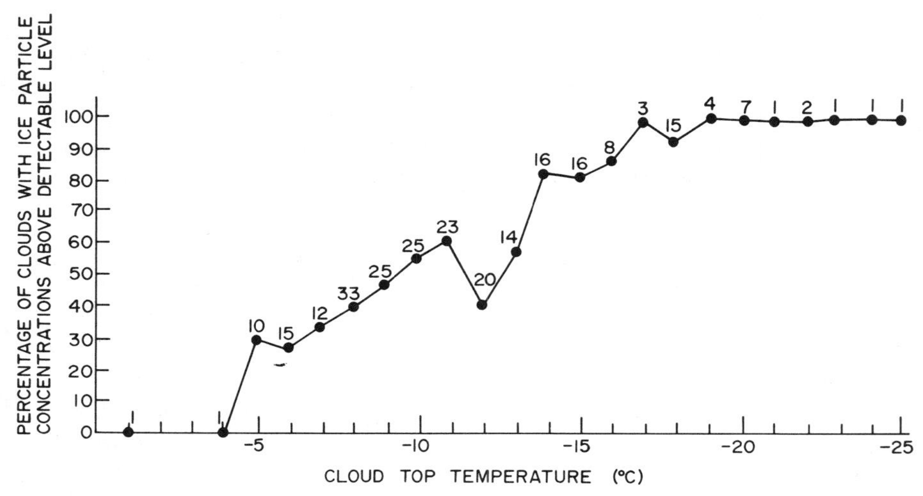

Fig. 4.29 Percentage chance of ice being detected in clouds as a function of the cloud top temperature. Results are based on field observations of 30 orographic cloud systems [from *J. Appl. Met.* **14**, 783 (1975)] and 228 cumulus clouds [from *Proc. Amer. Met. Soc. 1st National Conf. on Weather Modification*, Albany, New York, 1968, p. 306; *Quart. J. Roy. Met. Soc.* **96**, 487 (1970); *Proc. Intern. Conf. on Weather Modification*, Canberra, Australia, 1971, p. 5]. The number above each point is the total number of cloud samples for that temperature.

The main problem with this technique is that many of the ice particles, particularly the more delicate ones, fragment when they collide with the film. Recently an automatic method has been developed for determining the concentrations of ice particles from an aircraft. This method depends on the fact that two crossed Polaroid filters will normally not transmit any light directed onto them. When ice particles pass between the filters they cause a slight rotation in the plane of polarization of the light, thereby allowing light pulses to pass through the two filters. These pulses are detected by a photomultiplier tube.

Shown in Fig. 4.30 are a number of measurements of the concentrations of ice particles in clouds. Also shown are the concentrations of ice nuclei given by (4.19) with $a = 0.6$ and $T_1 = 253°K$. It can be seen from these results that on some occasions ice particles are not detected, even when ice nucleus measurements indicate that they should be. On other occasions they are present in concentrations several orders of magnitude greater than ice nucleus measurements would indicate.

The unexpectedly high concentrations of ice particles (100 liter^{-1} or more) observed in some clouds appear to be associated with older clouds, whereas young cumulus towers generally consist entirely of water droplets and generally require about 10 min before showing signs of glaciation. It also appears that high ice particle concentrations tend to be associated with the presence of droplet size distributions more typical of marine clouds (that is, a relatively broad droplet size distribution) and with the presence of ice particles on which liquid water is frozen (these are called *rimed* ice particles).

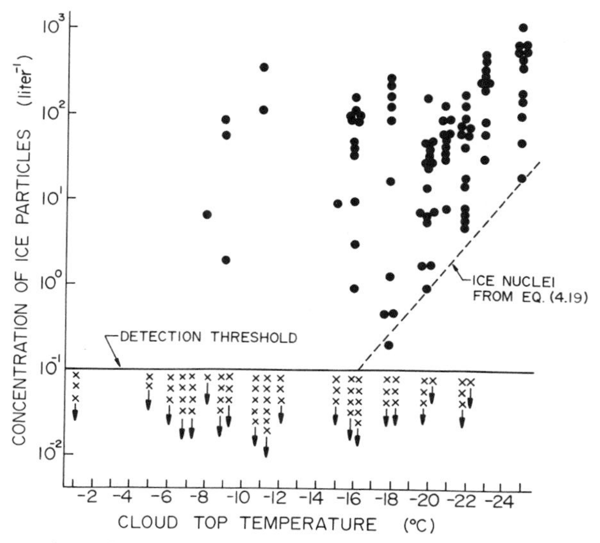

Fig. 4.30 Concentrations of ice particles in clouds against cloud top temperature: ●, measured value; ×, concentrations below detection threshold. Also shown (---) are the ice nucleus concentrations predicted by Eq. (4.19) with $a = 0.6$ and $T_1 = -20°C$. [From *J. Atmos. Sci.* **33**, 1362 (1976).]

Several explanations have been proposed to account for the high ice particle concentrations observed in some clouds. First, it is possible that the present techniques for measuring ice nuclei do not provide good estimates of the concentrations of ice nuclei active in natural clouds under certain conditions. It is also possible, however, that ice particles in clouds may increase in number without the action of ice nuclei, by what are termed *ice multiplication* mechanisms. For example, some ice crystals are quite fragile and may break up into many pieces when they are subjected to stress. However, at the present time the strongest contender for the primary ice multiplication mechanism in clouds is one which involves water droplets freezing by coming into contact with ice particles. When a supercooled droplet collides with an ice particle it freezes in two stages. In the first stage, which occurs almost instantaneously, a fine mesh of ice shoots through the droplet and freezes just enough water to raise the temperature of the droplet to 0°C. The second stage of freezing is much slower and involves the transfer of heat from the partially frozen droplet to the colder environment. During this stage an ice shell first forms over the surface of the droplet and then thickens progressively inward. As the droplet freezes inward, water is trapped in its interior; as this water freezes it expands and sets up large stresses in the ice shell. These stresses may cause the ice shell to crack and even explode, violently throwing off numerous small ice splinters, which may, in turn, grow by riming and eject further splinters. Laboratory experiments indicate that ice multiplication during riming should take place provided that the diameters of the droplets involved in the riming are greater than 23 μm, the temperatures are between -3 and -8°C, and the impact velocities are in excess of 0.7 m s^{-1}. Model calculations indicate that these conditions are most likely to be realized during the later stages of the life cycle of cumulus clouds that have a marine type of droplet size distribution.

Problem 4.3 Determine the fraction of the mass of a supercooled droplet which is frozen in the initial stage of freezing if the original temperature of the droplet is -20°C. What is the percentage increase in the volume of the droplet due to the initial freezing? (Latent heat of fusion = 3.3×10^5 J kg^{-1}; specific heat of liquid water = 4218 J deg^{-1} kg^{-1}; specific heat of ice = 2106 J deg^{-1} kg^{-1}; density of ice = 0.917×10^3 kg m^{-3}.)

Solution Let m be the mass of the droplet and dm the mass of ice which is frozen in the initial stage. Then the latent heat released due to freezing is $3.3 \times 10^5\, dm$. This heat raises the temperature of the unfrozen water and the frozen water from -20 to 0°C. Therefore,

$$3.3 \times 10^5 dm = [2106 \times 20 dm] + [4218 \times 20(m - dm)]$$

hence,

$$\frac{dm}{m} = \frac{4218}{(3.3 \times 10^5/20) - 2106 + 4218}$$

Therefore, 20% of the droplet is frozen during the initial stage of freezing.

Since the density of water is 10^3 kg m^{-3}, when mass dm of water freezes the increase in volume is $[(1/0.917) - 1]\, dm/10^3$. The fractional increase in volume of the mixture is therefore $[(1/0.917) - 1]\, dm/10^3 V$, where V is the volume of mass m of water. But $m/V = 10^3$;

therefore, the fractional increase in volume produced by the initial stage of freezing is $[(1/0.917) - 1]\, dm/m = [(1/0.917) - 1]0.2 = 0.018$ or 1.8%.

4.5.3 Growth of ice particles in clouds

(*a*) *Growth from the vapor phase.* In a mixed cloud dominated by supercooled droplets, the air is close to saturated with respect to liquid water and is therefore supersaturated with respect to ice. For example, air saturated with respect to liquid water at $-10°C$ is supersaturated with respect to ice by 10%, and at $-20°C$ it is supersaturated by 21%. These values are much higher than the supersaturations of cloudy air with respect to liquid water, which rarely exceed 1%. Consequently, in mixed clouds ice crystals grow from the vapor phase much more rapidly than do droplets.

The factors that control the mass growth rate of an ice crystal by deposition from the vapor phase are similar to those that control the growth of a droplet by condensation (see Section 4.4.1). However, the problem is more complicated because ice crystals are not spherical and therefore points of equal vapor density do not lie on a sphere centered on the crystal (as they do for a droplet).

For the special case of a spherical ice particle of radius r we can write, by analogy with (4.9),

$$\frac{dM}{dt} = 4\pi r D[\rho_v(\infty) - \rho_{vc}]$$

where ρ_{vc} is the density of the vapor just adjacent to the surface of the crystal, and the other symbols have been defined in Section 4.4.1. We can derive a more general expression for an ice crystal of arbitrary shape by exploiting the analogy between the vapor field around a crystal and the field of electrostatic potential around a charged conductor of the same size and shape. The leakage of charge from the conductor (the analog of the flux of vapor into or out of the crystal) is proportional to the electrostatic capacity C of the conductor, expressed in farads, which is entirely determined by the size and shape of the conductor. For a sphere, in SI units

$$\frac{C}{\varepsilon_0} = 4\pi r$$

where ε_0 is the permittivity of free space (8.85×10^{-12} $C^2\, N^{-1}\, m^{-2}$). Making use of this relationship we can rewrite the above expression for the mass growth rate of a spherical ice crystal as

$$\frac{dM}{dt} = \frac{DC}{\varepsilon_0}[\rho_v(\infty) - \rho_{vc}] \tag{4.21}$$

(4.21) is, in fact, quite general and can be applied to an arbitrarily shaped crystal of capacity C.

Provided that the vapor pressure corresponding to $\rho_v(\infty)$ is not too much greater than the saturation vapor pressure e_{si} over a plane surface of ice and the ice crystal is not too small, (4.21) can be written as

$$\frac{dM}{dt} = \frac{C}{\varepsilon_0} G_i S_i \tag{4.22}$$

where S_i is the supersaturation with respect to ice $\{[e(\infty) - e_{si}]/e_{si}\}$ and

$$G_i = D\rho_v(\infty) \tag{4.23}$$

The variation of $G_i S_i$ with temperature for the case of an ice crystal growing in air saturated with water is shown in Fig. 4.31. The product attains a maximum value at about $-14°C$, which is due mainly to the fact that the difference between the saturated vapor pressures over water and ice is a maximum near this temperature. Consequently, ice crystals growing by vapor deposition in mixed clouds increase in mass most rapidly at temperatures around $-14°C$.

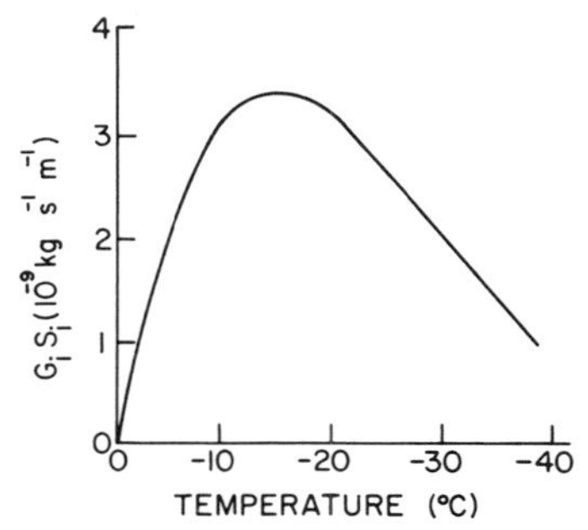

Fig. 4.31 Variation of $G_i S_i$ with temperature for an ice crystal growing in an environment at water saturation and a total pressure of 1000 mb.

Ice crystals growing from the vapor phase can assume a wide variety of shapes (or *habits*). However, the basic habits are either *platelike* or *prismlike*. The simplest platelike crystals are plane hexagonal plates (Fig. 4.32a) and the simplest prismlike crystals are solid columns which are hexagonal in cross section (Fig. 4.32b). Careful studies of the growth of ice crystals from the vapor phase under controlled conditions in the laboratory, and observations in natural clouds, have shown that the basic habit of an ice crystal is determined by the temperature at which it grows (Table 4.2). In the temperature range between 0 and $-50°C$ the basic habit changes three times. These changes occur near -4, -10, and $-22°C$. When the air is saturated or supersaturated with respect to water the basic habits become embellished. For example, between -4 and $-6°C$ prismlike crystals take the form of long thin needles, between -12 and $-16°C$ platelike crystals appear like ferns which are called dendrites (Fig. 4.32c), from -10 to $-12°C$ and -16 to $-22°C$ sector plates grow (Fig. 4.32d), and below $-22°C$ prismlike crystals are hollow columns. Ice crystals are exposed to continually changing temperatures and supersaturations as they fall through clouds. Therefore, even when they are growing solely by vapor deposition, they can assume quite complex shapes.

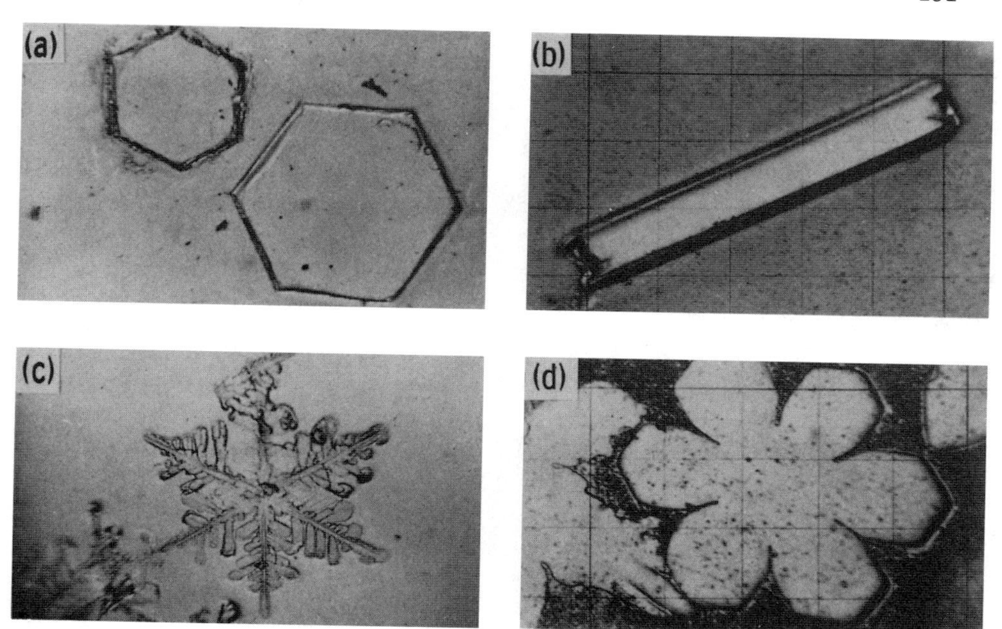

Fig. 4.32 Examples of ice crystals which have grown from the vapor phase: (a) hexagonal plates; (b) column; (c) dendrite (one of the six arms has broken off during collection); (d) sector plate. (Courtesy Cloud Physics Group, University of Washington.)

Table 4.2
Variations in the basic habits of ice crystals with temperature

Temperature (°C)	Basic habit	Types of crystal at slight water supersaturation
0 to −4	Platelike	Thin hexagonal plates
−4 to −10	Prismlike	Needles (−4 to −6°C) Hollow columns (−5 to −10°C)
−10 to −22	Platelike	Sector plates (−10 to −12°C) Dendrites (−12 to −16°C) Sector plates (−16 to −22°C).
−22 to −50	Prismlike	Hollow columns

(*b*) *Growth by riming; hailstones.* In a mixed cloud, ice particles increase in mass by colliding with supercooled droplets which then freeze onto them. This process is referred to as growth by *riming* and it leads to the formation of various rimed structures, some examples of which are shown in Fig. 4.33: (a) shows a needle which collected a few droplets on its leading edge as it fell through the air, (b) a uniformly, densely rimed column, (c) a densely rimed plate, and (d) a

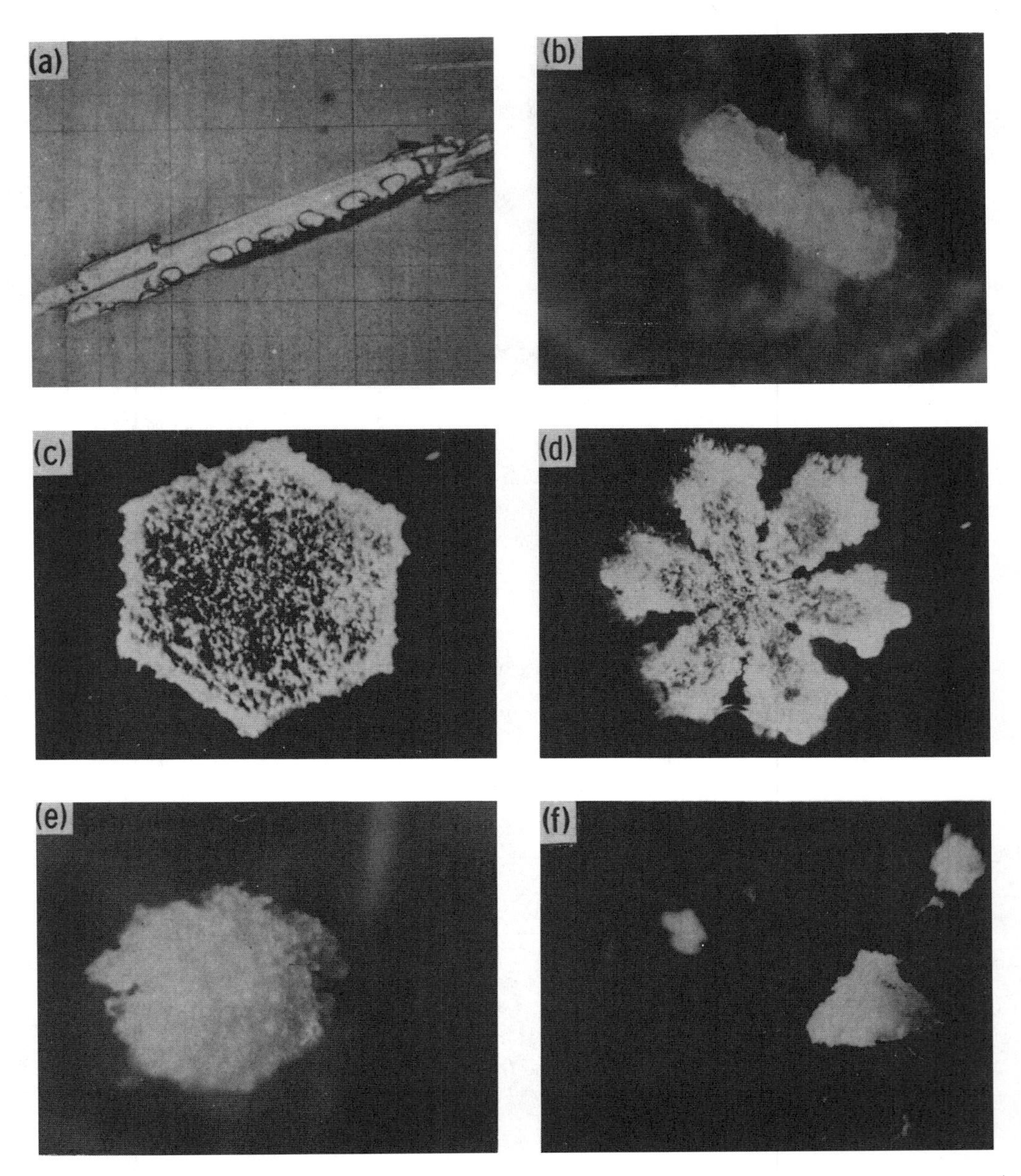

Fig. 4.33 (a) A lightly rimed needle: (b) densely rimed column; (c) densely rimed plate; (d) densely rimed stellar; (e) lump graupel; (f) cone graupel. (Courtesy Cloud Physics Group, University of Washington.)

densely rimed stellar crystal. When riming proceeds beyond a certain stage it becomes difficult to discern the original shape of the ice crystal. The rimed particle is then referred to as *graupel*. Examples of graupel are shown in Fig. 4.33e and f.

Hailstones represent an extreme case of the growth of ice particles by riming. They form in vigorous convective clouds which have high liquid water contents. Under extreme conditions hailstones as large as 13 cm in diameter and weighing over $\frac{1}{2}$ kg have been observed. However, hailstones about 1 cm in diameter are much more common. If a hailstone collects supercooled droplets at too great a rate, its surface temperature rises to 0°C and some of the water it collects will remain unfrozen. The surface of the hailstone then becomes covered with a layer of liquid water and the hailstone is said to grow *wet*. Under these conditions some of the liquid water may be shed in the wake of the hailstone but some may also be incorporated into a water–ice mesh to form what is known as *spongy hail*.

If a thin section is cut from a hailstone and viewed in transmitted light, it is often seen to consist of alternate dark and light layers (Fig. 4.34). The dark

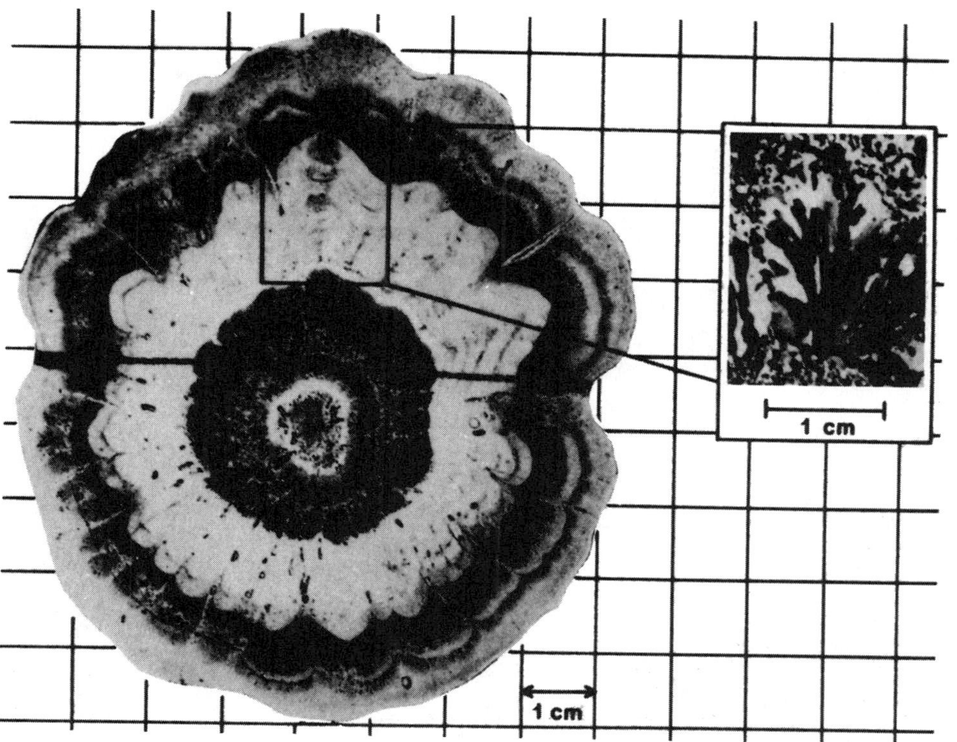

Fig. 4.34 Thin section through the growth center of a natural hailstone. [From *Quart. J. Roy. Met. Soc.* **92**, 10 (1966).]

layers are opaque ice containing numerous small air bubbles and the light layers are clear (bubble-free) ice. Clear ice is more likely to form when the hailstone is growing wet. Detailed examination of the orientation of the individual crystals within a hailstone (which can be seen when the hailstone is viewed between crossed polarizing filters—see inset to Fig. 4.34) can also reveal whether wet growth has occurred. It can be seen from Figs. 4.34 and 4.35 that the surface of a hailstone can contain fairly large lobes. Lobelike growth appears to be more pronounced when the accreted droplets are small and growth is near the wet limit. The development of lobes may be due to the fact that any small bumps on a hailstone will be areas of enhanced collection efficiencies for droplets.

(*c*) *Growth by aggregation.* The third mechanism by which ice particles grow in clouds is by colliding and aggregating with one another. Ice particles

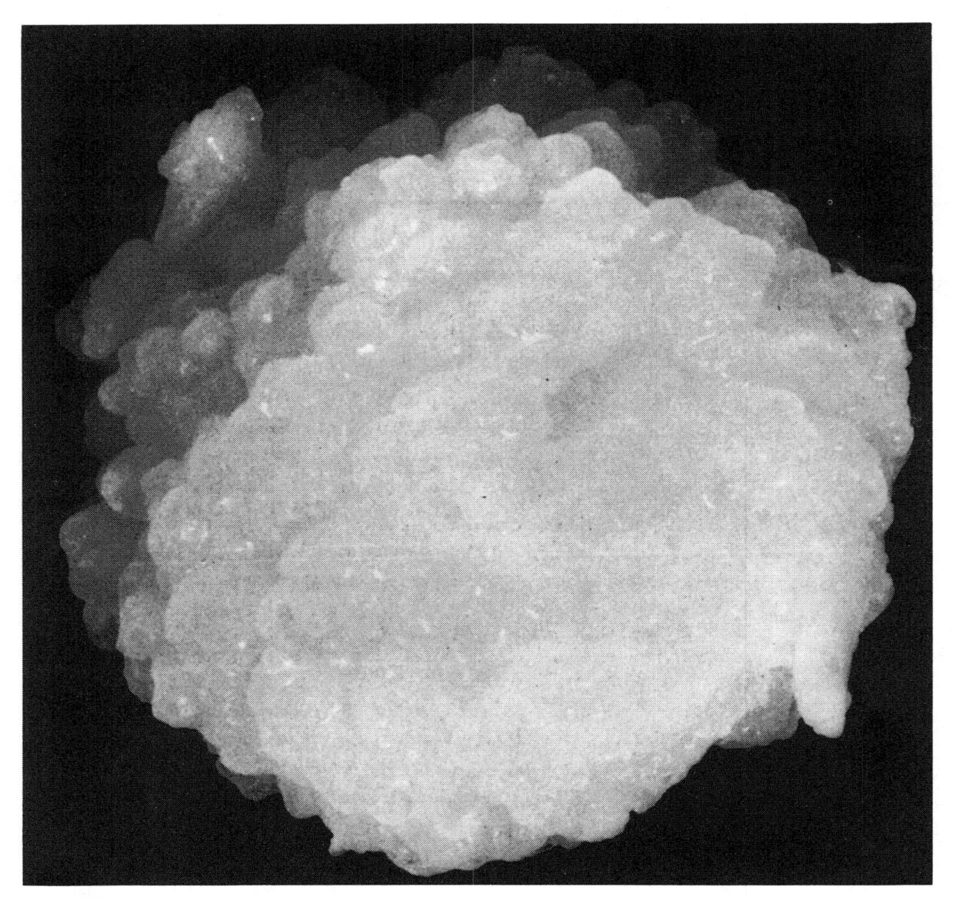

Fig. 4.35 Artificial hailstone, actual size, showing lobe structure. Growth was initially dry but tended toward wet growth as the stone grew. [From *Quart. J. Roy. Met. Soc.* **94**, 10 (1968).]

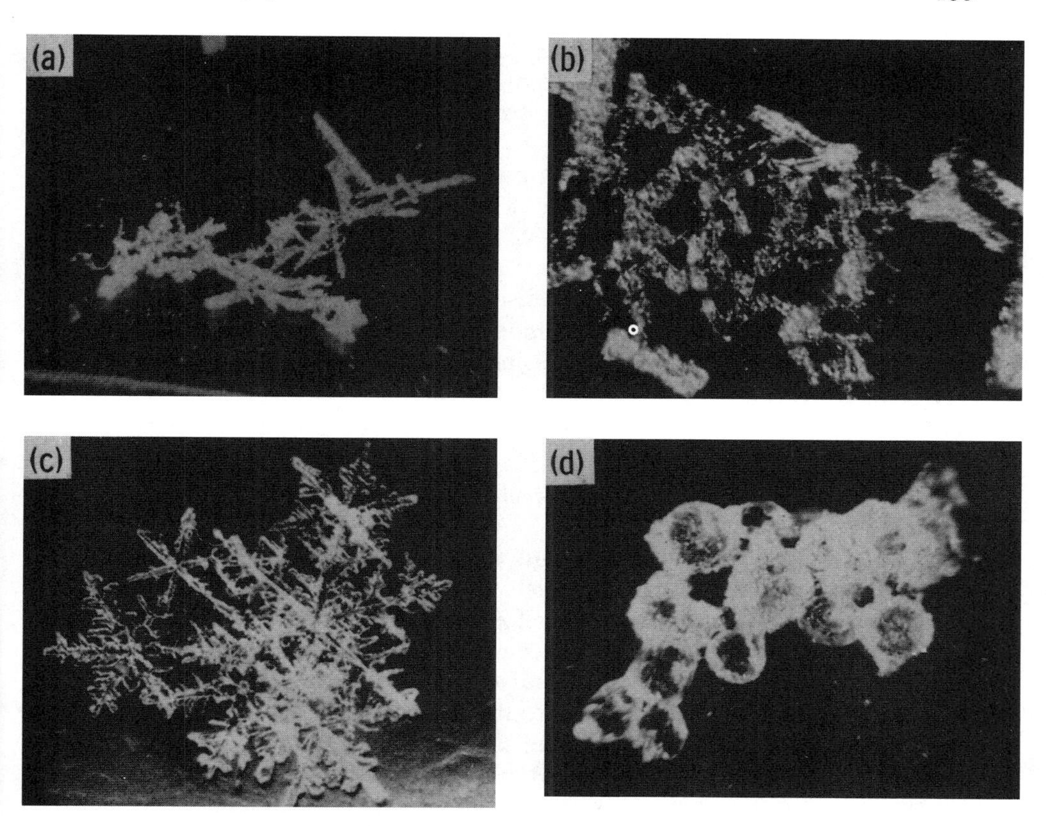

Fig. 4.36 Aggregates of (a) densely rimed needles; (b) unrimed to moderately rimed columns; (c) dendrites; (d) unrimed to densely rimed frozen drops. (Courtesy Cloud Physics Group, University of Washington.)

can collide with each other provided that their terminal fall speeds are different. The terminal fall speed of an unrimed prismlike ice crystal increases as the length of the crystal increases; for example, the fall speeds of needles 1 and 2 mm in length are about 0.5 and 0.7 m s^{-1}, respectively. In contrast, unrimed platelike ice crystals have terminal fall speeds virtually independent of their diameter. (This behavior can be explained as follows. The thickness of a platelike crystal is essentially independent of diameter; therefore, its mass varies linearly with cross-sectional area. Since the drag force acting on a platelike crystal also varies as the cross-sectional area of the crystal, the terminal fall speed, which is determined by a balance between the drag and the gravitational forces acting on a crystal, is independent of the diameter of a plate.) Consequently, unrimed platelike crystals are unlikely to collide with each other, unless they come close enough to be influenced by wake effects. The terminal fall speeds of rimed crystals and graupel are strongly dependent upon their degrees of riming and their dimensions. For example, graupel particles 1 and 4 mm in diameter have terminal fall speeds of about 1 and 2.5 m s^{-1}, respectively. It can be seen from this discussion that the collisions of ice particles in clouds are greatly enhanced if some riming has taken place.

The second factor that influences growth by aggregation is whether or not two ice particles will adhere together when they collide. The probability of adhesion is determined primarily by two factors: the types of ice particles and the temperature. Intricate crystals, such as dendrites, tend to adhere together because they become entwined on collision, whereas two solid plates tend to rebound. Apart from this dependence upon habit, the probability of two colliding crystals adhering increases with increasing temperature, adhesion being particularly likely at temperatures above about −5°C at which ice surfaces become quite "sticky." Some examples of ice particle aggregates are shown in Fig. 4.36.

4.5.4 Formation of precipitation in cold clouds

As early as 1784 Benjamin Franklin[†] suggested that "much of what is rain, when it arrives at the surface of the earth, might have been snow, when it began its descent . . . ." This idea was not developed until the early part of this century when Wegener,[‡] in 1911, stated that ice particles would grow preferentially by deposition from the vapor phase in a mixed cloud. Subsequently, Bergeron,[§] in 1933, and Findeisen,[¶] in 1938, developed this idea in a more quantitative manner and indicated the importance of ice nuclei in the formation of crystals. Since Findeisen carried out his field studies in northwestern Europe, he was led to believe that all rain originates as ice. However, as we have seen in Section 4.4, it is now known that rain can also form by the collision–coalescence mechanism.

We will now investigate the growth of ice particles to precipitation size in a little more detail. Application of (4.22) to the case of a hexagonal plate growing by deposition from the vapor phase in air saturated with respect to water at

[†] **Benjamin Franklin** (1706–1790) American scientist, inventor, statesman, and philosopher. Largely self-taught, and originally a printer and publisher by trade. First American to win international fame in science. Carried out fundamental work on the nature of electricity (introduced the terms "positive," "negative," "charges," "battery"). Showed lightning to be an electrical phenomenon (1752). Attempted to deduce path of storms over North America. Invented the lightning conductor, daylight savings time, bifocals, Franklin stove, and the rocking chair! First to study the Gulf Stream.

[‡] **Alfred Lothar Wegener** (1880–1930) German geophysicist and meteorologist. Author of "Thermodynamik der Atmosphäre," J. A. Barth, Leipzig, 1911. Studied the thickness of the polar ice cap and originated the theory of continental drift. Lost his life while attempting to cross Greenland.

[§] See p. 126.

[¶] **Theodor Robert Walter Findeisen** (1909–1945) German meteorologist. Director of Cloud Research, German Weather Bureau, Prague, Czechoslovakia, from 1940. Laid much of the foundation of modern cloud physics and foresaw the possibility of stimulating rain by introducing artificial ice nuclei. Disappeared in Czechoslovakia at the end of World War II.

$-5°C$ shows that the plate can obtain a mass of about 7 μg (that is, a diameter of about 1 mm) in half an hour (see Problem 4.23), thereafter its mass growth rate decreases significantly. A 7-μg ice crystal could form a small drizzle drop about 130 μm in radius if the updraft velocity of the air were less than the terminal fall speed of the crystal (about 0.3 m s^{-1}) and if the drop survived evaporation between cloud base and the ground. Clearly, however, the growth of ice crystals by deposition alone is not sufficiently rapid to produce large raindrops.

Unlike growth by deposition, the growth rates of an ice particle by riming and aggregation increase as the ice particle increases in size. A simple calculation shows that a platelike ice crystal, 1 mm in diameter, falling through a cloud with a liquid content of 0.5 g m^{-3}, could develop into a spherical graupel particle about 1 mm in diameter in about 10 min (see Problem 4.24). A graupel particle of this size, with a density of 100 kg m^{-3}, has a terminal fall speed of about 1 m s^{-1} and would melt into a drop about 230 μm in radius. The diameter of a snowflake can increase from 1 mm to 1 cm in about 30 min due to aggregation with ice crystals, provided that the ice content of the cloud is about 1 g m^{-3} (see Problem 4.25). An aggregated snow crystal with a diameter of 1 cm has a mass of about 3 mg and a terminal fall speed of about 1 m s^{-1}. Upon melting, a snow crystal of this mass would form a drop about 1 mm in radius.

We conclude from this discussion that the growth of ice crystals, first by deposition from the vapor phase in mixed clouds and then by riming and/or

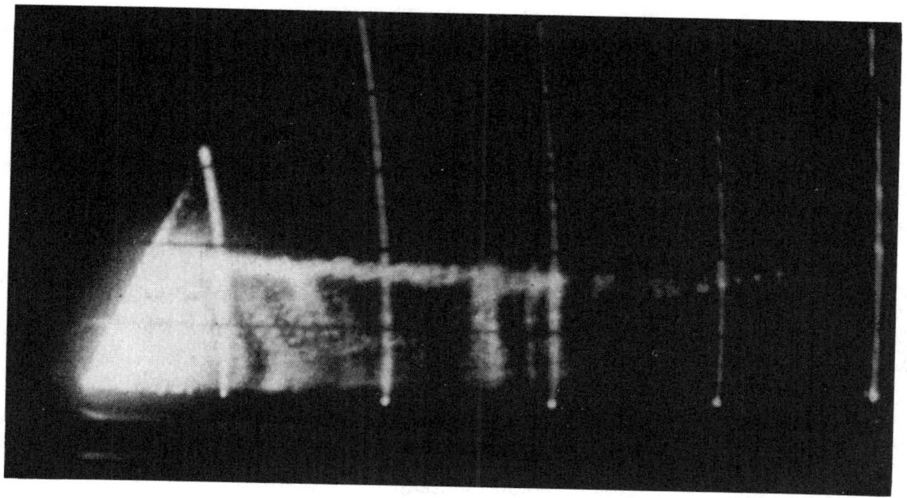

Fig. 4.37 Photograph of a conventional weather radar screen when the antenna was sweeping in the vertical plane. The melting band shows up as the bright horizontal line which corresponds to a region of high radar reflectivity. Radar returns (or echoes) from precipitation are also present below the melting layer. The bright arcs centered on the radar (which is located in the lower left-hand corner) are range markers. (Courtesy of the Weather Radar Group, MIT.)

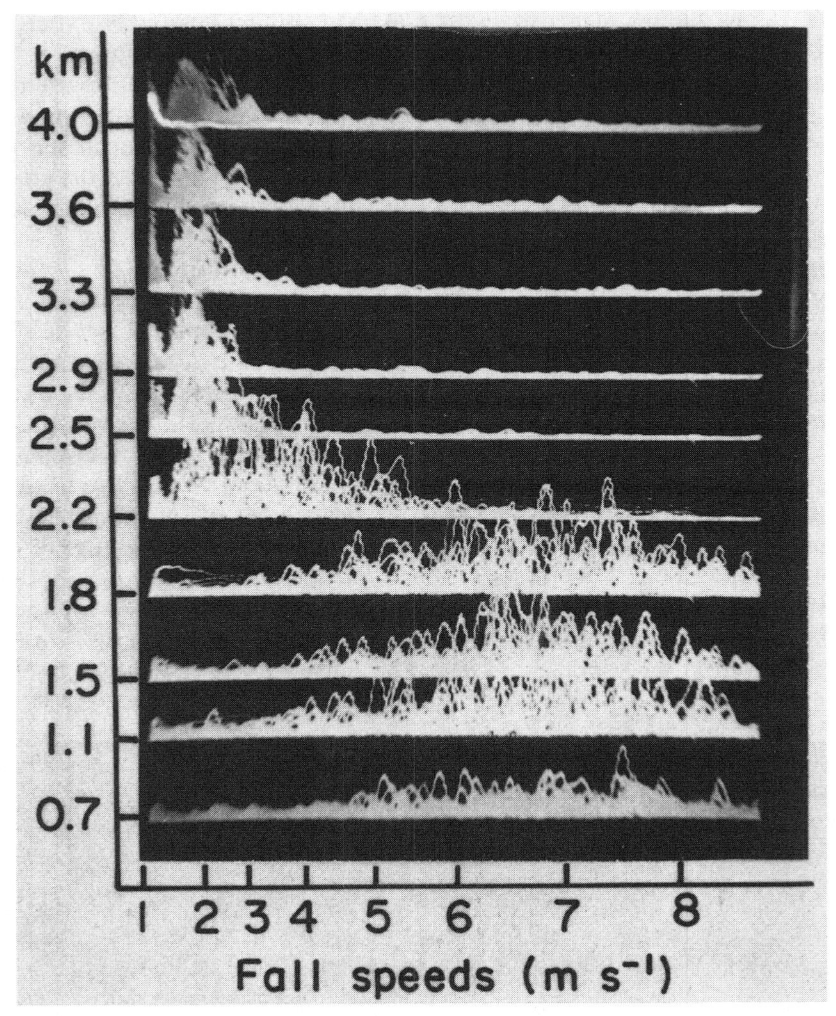

Fig. 4.38 The spectra of Doppler fall speeds for precipitation particles at ten heights in the atmosphere. The melting level is at about 2.2 km. (Courtesy of Cloud Physics Group, University of Washington.)

aggregation, can produce precipitation-sized particles in reasonable time periods (say about 40 min).

Problem 4.4 Derive an expression for the precipitation rate p (in meters of liquid water per second) due to snow crystals of mass m (in kilograms) which are present in concentration n (per cubic meter) and falling with a speed v (in meters per second).

Solution In 1 s the mass of water falling on the ground per unit area is $p\rho_l$ kg, where ρ_l is the density of water (in kilograms per cubic meter). Also, in 1 s the total mass of snow crystals falling on a unit area is given by mnv kg. Therefore,

$$p\rho_l = mnv$$

or, since $\rho_l = 10^3$ kg m^{-3},

$$p = 10^{-3}mnv$$

The role of the ice phase in producing precipitation in cold clouds is demonstrated very well by radar observations. For example, Fig. 4.37 shows a photograph of a conventional radar screen (on which the intensity of radar echoes reflected from atmospheric targets are displayed) while the radar antenna was sweeping in a vertical plane. The bright horizontal band corresponds to the level in the cloud where the ice particles were melting. The radar reflectivity is high around the melting level because, while melting, the ice particles become coated with a film of water which greatly increases their radar reflectively. When the crystals have melted completely they collapse into droplets, and their terminal fall speeds increase so that the concentration of particles is reduced. These changes result in a sharp decrease in radar reflectivity below the melting band.

The sharp increase in particle fall speeds produced by melting is illustrated in Fig. 4.38 which shows the spectrum of fall speeds of precipitation particles measured at a number of different heights with a vertically pointing Doppler radar.[†‡] At heights above 2.2 km the particles are ice with fall speeds centered around 2 m s^{-1}. At 2.2 km the particles are partially melted and below 2.2 km there are raindrops with fall speeds centered around 7 m s^{-1}

4.5.5 Classification of solid precipitation

The growth of ice particles by deposition from the vapor phase, riming, and aggregation can lead to a very wide variety of solid precipitation particles. A relatively simple classification of solid precipitation into ten main classes is shown in Table 4.3 together with recommended symbols. In this scheme, for example, a rimed stellar crystal 3–4 mm in diameter would be coded as F2rD3-4.

4.6 THUNDERSTORMS

The dynamical structure of thunderstorms will be described in the next chapter. Here we are concerned with the microphysical mechanisms which may be responsible for the electrification of thunderstorms and with the nature of lightning flashes and thunder.

[†] Doppler radars, unlike conventional meteorological radars, transmit coherent electromagnetic waves. From measurements of the difference in frequencies between the returned and transmitted waves, the velocity of the target (for example, precipitation particles) along the line of sight of the radar can be deduced. The radars used by the police for measuring the speeds of motor vehicles are based on the same principle.

[‡] **Christian Doppler** (1803–1853) Austrian physicist and mathematician. Noted for the discovery of the Doppler effect.

Table 4.3

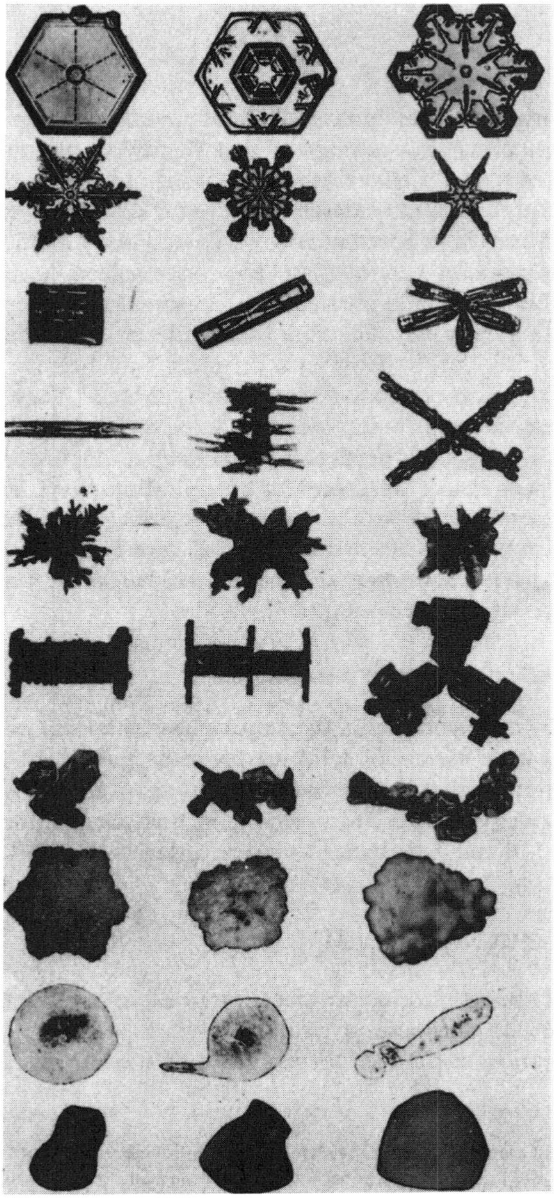

Typical forms	Symbol	Graphic symbol
	F1	
	F2	
	F3	
	F4	
	F5	
	F6	
	F7	
	F8	
	F9	
	F10	

[a] Suggested by the International Association of Hydrology's commission of snow and ice in 1951. (Photo: V. Schaefer.)

[b] Additional characteristics: p, broken crystals; r, rime-coated particles not sufficiently coated to be classed as graupel; f, clusters, such as compound snowflakes, composed of several individual snow crystals; w, wet or partly melted particles.

A Classification of solid precipitation[a,b,c]

Description
A plate is a thin, platelike snow crystal the form of which more or less resembles a hexagon or, in rare cases, a triangle. Generally all edges or alternative edges of the plate are similar in pattern and length.
A stellar crystal is a thin, flat snow crystal in the form of a conventional star. It generally has six arms but stellar crystals with three or twelve arms occur occasionally. The arms may lie in a single plane or in closely spaced parallel planes in which case the arms are interconnected by a very short column.
A column is a relatively short prismatic crystal, either solid or hollow, with plane, pyramidal, truncated, or hollow ends. Pyramids, which may be regarded as a particular case, and combinations of columns are included in this class.
A needle is a very slender, needlelike snow particle of approximately cylindrical form. This class includes hollow bundles of parallel needles, which are very common, and combinations of needles arranged in any of a wide variety of fashions.
A spatial dendrite is a complex snow crystal with fernlike arms which do not lie in a plane or in parallel planes but extend in many directions from a central nucleus. Its general form is roughly spherical.
A capped column is a column with plates of hexagonal or stellar form at its ends and, in many cases, with additional plates at intermediate positions. The plates are arranged normal to the principal axis of the column. Occasionally only one end of the column is capped in this manner.
An irregular crystal is a snow particle made up of a number of small crystals grown together in a random fashion. Generally the component crystals are so small that the crystalline form of the particle can only be seen with the aid of a magnifying glass or microscope.
Graupel, which includes soft hail, small hail, and snow pellets, is a snow crystal or particle coated with a heavy deposit of rime. It may retain some evidence of the outline of the original crystal although the most common type has a form which is approximately spherical.
Ice pellets (frequently called sleet in North America) are transparent spheroids of ice and are usually fairly small. Some ice pellets do not have a frozen center which indicates that, at least in some cases, freezing takes place from the surface inwards.
A hailstone[d] is a grain of ice, generally having a laminar structure and characterized by its smooth glazed surface and its translucent or milky-white center. Hail is usually associated with those atmospheric conditions which accompany thunderstorms. Hailstones are sometimes quite large.

[c] Size of particle is indicated by the general symbol D. The size of a crystal or particle is its greatest extension measured in millimeters. When many particles are involved (for example, a compound snowflake) it refers to the average size of the individual particles.

[d] In English, hail, like rain, refers to a number at one time, while hailstone, like raindrop, refers to an individual.

4.6.1 Charge generation

All clouds are electrified to some degree.[†‡] However, in vigorous convective clouds sufficient electrical charges can be separated to give rise to thunderstorms.

The distribution of charges in thunderstorms has been investigated with special radiosondes (called altielectrographs), with specially instrumented aircraft, and by measuring the changes in the electric field at the ground that accompany lightning flashes. These studies have shown that, on average, thunderstorms contain positive charges ($\sim +24$ C) in their upper regions, negative charges (~ -20 C) lower down but above the 0°C isotherm, and a much smaller pocket of positive charge ($\sim +4$ C) just below the melting level. This distribution is shown schematically in Fig. 4.39. The rate of generation of charge in a thunderstorm is believed to be about 1 C km^{-3} min^{-1}.

Although there have been a few observations of lightning from warm clouds, the vast majority of thunderstorms extend above the 0°C isotherm and contain both ice particles and supercooled water. In many cases, the onset of strong electrification follows the appearance of heavy precipitation within the cloud in the form of graupel or hailstones. Consequently, several of the more popular

[†] Benjamin Franklin, in July 1750, was the first to propose an experiment to determine whether thunderstorms are electrified. He suggested that a sentry box, large enough to contain a man and an insulated stand, be placed at a high elevation and that an iron rod 20–30 ft in length be placed vertically on the stand, passing out through the top of the box. He then proposed that if a man stood on the stand and held the rod he would "be electrified and afford sparks" when an electrified cloud passed overhead. Alternatively, he suggested that the man stand on the floor of the box and bring near to the rod one end of a piece of wire, held by an insulating handle, while the other end of the wire was connected to the ground. In this case, an electric spark jumping from the rod to the wire would be proof of cloud electrification. (Franklin did not realize the danger of these experiments; they can kill a man—and have done so—if there is a direct lightning discharge to the rod.)

The proposed experiment was set up in Marly-la-Ville in France by d'Alibard, and on 10 May 1752 an old soldier, called Coiffier, brought an earthed wire near to the iron rod while a thunderstorm was overhead and saw a stream of sparks. This was the first direct proof that thunderstorms are electrified. Joseph Priestley described it as "the greatest discovery that has been made in the whole compass of philosophy since the time of Sir Isaac Newton." (Since Franklin proposed the use of the lightning conductor in 1749, it is clear that by that date he had already decided in his own mind that thunderstorms were electrified.)

Later in the summer of 1752 (the exact date is uncertain), and before hearing of d'Alibard's success, Franklin carried out his famous kite experiment in Philadelphia and observed sparks to jump from a key attached to a kite string to the knuckles of his hand. By September 1752 Franklin had erected an iron rod on the chimney of his home and on 12 April 1753, by identifying the sign of the charge collected on the lower end of the rod when a storm passed over, he had concluded that "clouds of a thundergust are most commonly in a negative state of electricity, but sometimes in a positive state—the latter, I believe, is rare." No more definitive statement as to the electrical state of thunderstorms was made until the second decade of the 20th century when C. T. R. Wilson showed that the lower regions of thunderstorms are generally negatively charge while the upper regions are positively charged.

[‡] **Thomas Francois d'Alibard** (1703–1779) French naturalist. Translated into French Franklin's "Experiments and Observations on Electricity," Durand, Paris, 1756, and reenacted many of Franklin's experiments.

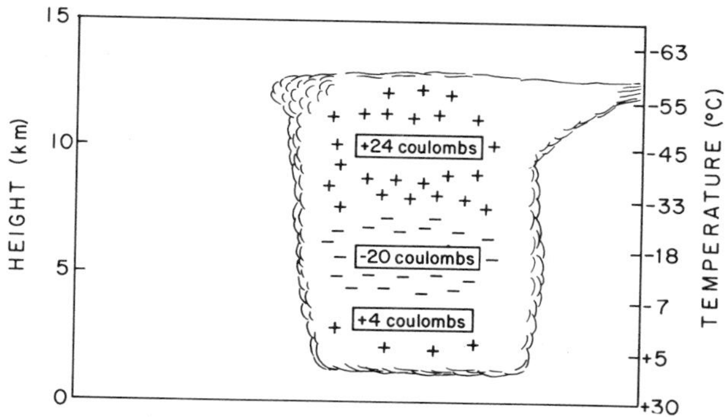

Fig. 4.39 Schematic diagram to show the distribution of charges in a thunderstorm.

theories for charge generation in thunderstorms assume that solid precipitation, in the form of graupel or hail, plays a key role. We will describe three such theories.†

The first two theories of thunderstorm electrifications we will describe depend upon the *thermoelectric effect* in ice; therefore, we will first give a brief explanation of this phenomenon. Consider a rod of ice warmed at one end and cooled at the other, so that a steady temperature difference ΔT is maintained between its two ends. Some of the water molecules in ice are always dissociated into positive and negative ions and the number of these ions is greater at higher temperatures. Therefore, the warmer end of the ice rod will contain more positive and negative ions than the colder end. Since ions migrate from regions of high concentration to regions of low concentration, both the positive and negative ions will tend to migrate from the warmer toward the colder end of the ice rod. However, the mobility of the negative ions in ice is essentially zero, whereas that of the positive ions is quite high. Therefore, the positive ions migrate toward the cold end where they build up a positive space charge which eventually prevents any further migration of positive ions into the region. Hence, under steady-state conditions, a potential difference ΔV is established along the rod, with the cold end positively charged and the warm end negatively charged. Laboratory experiments have shown that ΔV (expressed in millivolts) $\simeq 2\Delta T$. Therefore, a temperature difference of, say, 2°C across a piece of ice will produce a voltage difference of about 4 mV.

Consider now a hailstone (or graupel particle) falling through a mixed cloud of supercooled droplets and small ice crystals. The collision of ice crystals with the hailstone may charge it in the following way (Fig. 4.40a). The surface of the hailstone will be warmer than that of the ice crystals due to the latent heat of freezing released by the large number of supercooled droplets colliding with the hailstone. Therefore, during the time that an ice crystal is in contact with

† For a more detailed description of theories of thunderstorm electrification see B. J. Mason, "The Physics of Clouds," Oxford University Press, 1971, pp. 483–568.

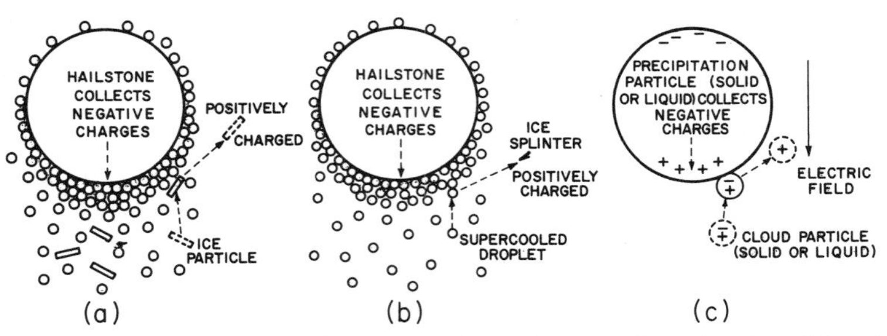

Fig. 4.40 Schematic diagrams (not drawn to scale) to illustrate three theories of thunderstorm electrification. (a) Ice particles collide with a hailstone whose surface is warmed by riming. Ice particles rebound with positive charge and hailstone receives negative charge. (b) Supercooled droplets collide with a hailstone. Ice splinters ejected during freezing of droplets carry positive charge. (c) Precipitation and cloud particles polarized by downward-directed electric field. Negative charge transferred to precipitation particles during contact, cloud particles rebound with positive charge.

the surface of the hailstone, the hailstone will become negatively charged and the ice crystal positively charged due to the thermoelectric effect. When the ice crystal rebounds from the hailstone it will retain this positive charge and, if its terminal fall speed is less than that of the updraft velocity of the air in the cloud, it will be carried to the upper regions of the cloud. In this way the upper regions of a convective cloud become filled with small positively charged ice crystals. Hailstones, on the other hand, become negatively charged by this mechanism and, due to their comparatively high fall speeds, carry these charges to the lower regions of the cloud. Simple calculations (see Problem 4.26) show that if each collision between ice crystals and small hailstones in a large convective cloud resulted in the latter receiving a charge of about -16 fC (approximately 10^5 elementary charges), the rate of charge generation would be equal to that in a typical thunderstorm (about 1 C km^{-3} min^{-1}). Unfortunately, laboratory experiments designed to simulate the above processes have yielded conflicting results. In one series of experiments the average charge per ice crystal collision that an ice surface received when it was bombarded with ice crystals and supercooled droplets was -170 fC, but in another series of experiments carried out by different workers the average charge was only -1.6×10^{-3} fC per ice crystal collision.

Another mechanism associated with hailstones falling through mixed clouds has been proposed as the principal charging mechanism in thunderstorms (Fig. 4.40b). We have already mentioned in Section 4.5.2 that, when a supercooled droplet freezes, numerous small ice splinters may be thrown into the air. Now consider a supercooled droplet which has collided with a hailstone and is in the second stage of freezing with the ice shell thickening inward. The inner surface of the shell is in contact with liquid water and is therefore at 0°C, while the outer surface of the ice shell is cooling toward the environmental temperature, which may be well below 0°C. Consequently, there is a temperature difference across the ice shell such that the thermoelectric effect will cause positive charges to accumulate on its outer surface. Since any ice splinters that break away from the drop come predominantly from the outer regions of the ice shell, they should carry a net positive charge and leave the hailstone with the corresponding negative charge. As before it is assumed that the small positively charged ice splinters are carried to the top of the cloud and the negatively charged hailstones to the lower regions of the cloud. Calculations show (see Problem 4.26) that, in order for this mechanism to generate charge in a thunderstorm at the required rate of 1 C km^{-3} min^{-1}, each droplet greater than about 15 μm in radius which collides with a hailstone would have to leave the latter with a charge of -1.6 fC. Some laboratory experiments indicate that charges of this magnitude are separated during riming but other experiments have failed to reveal any charging. Clearly, further studies are required to elucidate this charging mechanism.

Finally, we will describe briefly the *induction charging* theory of charge generation in thunderstorms (Fig. 4.40c). In the presence of the normal downward-directed fair weather electric field (see Section 1.4.2) both cloud and precipitation particles (solid or liquid) will be polarized so that their lower surfaces are positively charged and their upper surfaces negatively charged. Therefore, when cloud particles collide with downward-moving precipitation particles, negative charges will be transferred to the precipitation particles. Then, provided that the particles rebound, gravitational settling of the negatively charged precipitation particles and the upward movement of the small positively charged cloud particles could lead to the development of the two main charge centers in thunderstorms. It should be noted that, as the two main charge centers build up, they will reinforce the downward-directed electric field and, therefore, the magnitudes of the charges transferred during collisions will also increase. In other words, induction charging is a positive feedback mechanism which increases in importance with time. It is likely that initially noninduction charging mechanisms (such as the two mechanisms previously described) are dominant in the separation of electrical charges between colliding particles in clouds, but as the electric field grows induction charging becomes increasingly important and eventually dominates the exchange of charges between interacting particles in thunderstorms.

The three theories described above are concerned with the two main charge centers in thunderstorms. However, in some thunderstorms a small pocket of positive charge is observed just below the 0°C level (Fig. 4.39). This phenomenon is probably due to the charging of solid precipitation particles during melting. Laboratory experiments have shown that ice particles can receive large positive charges while melting due to the bursting of air bubbles. The splashing of water drops on melting ice particles can also leave the latter with large positive charges, particularly in the presence of electric fields.

4.6.2 Lightning and thunder

As electrical charges are separated in a cloud the potential gradient between the cloud and the ground, and between various regions of the cloud, increases and eventually exceeds that which the air can sustain. The resulting dielectric breakdown assumes the form of a lightning flash. Dielectric breakdown occurs for potential gradients of about 3 MV m^{-1} in dry air, and about 1 MV m^{-1} in the presence of water drops 1 mm in radius.

Cloud-to-ground lightning flashes originate near cloud bases in the form of an invisible (to the human eye) discharge, called the *stepped leader*, which moves downward toward the earth in discrete steps. Each step lasts for about 1 μs during which time the stepped leader advances about 50 m; the time interval between steps is about 50 μs. It is believed that the stepped leader is initiated by a local discharge between the small pocket of positive charge at the base of a thundercloud and the lower part of the negatively charged region (Fig. 4.41b). This discharge releases electrons which were previously attached to precipitation particles in the negatively charged region. These free electrons neutralize the small pocket of positive charge (Fig. 4.41c) and then move toward the ground

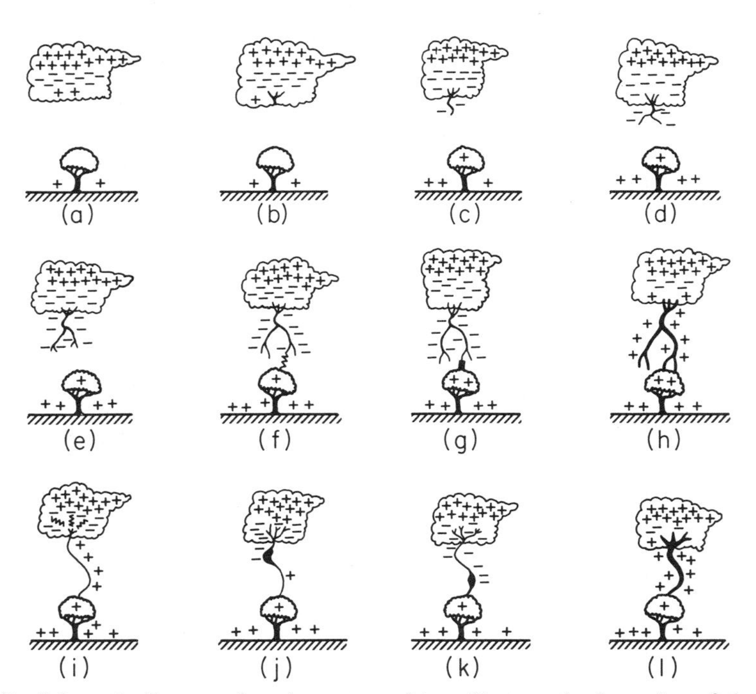

Fig. 4.41 Schematic diagrams (not drawn to scale) to illustrate the formation of the stepped leader (a)–(f), the first return stroke (g) and (h), the K and J streamers (i), the dart leader (j) and (k), and the second return stroke (l). [Adapted from M. Uman, "Understanding Lightning," Bek Tech. Pub. Inc., pp. 74–75, 78–79.]

(Fig. 4.41c–e). As the negatively charged stepped leader approaches the ground it induces positive charges on the ground, especially on protuding objects, and when it is 10–100 m from the ground a *traveling spark* moves up from the ground to meet it (Fig. 4.41f). After contact is made between the stepped leader and the traveling spark, large numbers of electrons flow to the ground and a highly luminous and visible *lightning stroke* propagates upward in a continuous fashion from the ground to the cloud along the path followed by the stepped leader (Fig. 4.41g and h). This flow of electrons (called the *return stroke*) is responsible for the bright channel of light which is observed as a lightning stroke. Since the stroke moves upward so quickly (in about 100 μs) the whole channel appears to the eye to brighten simultaneously. Despite the downward flow of electrons, both the return stroke and the ground, to which it is linked, remain positively charged in response to the remainder of the negative charge in the lower region of the cloud.

Following the first stroke, which carries the largest current (typically 1 or 2×10^4 A), subsequent strokes can occur along the same main channel, provided that additional electrons are supplied to the top of the previous stroke within 100 ms of the cessation of current. The additional electrons are supplied to the channel by so-called *K or J streamers* which move upward from the top of the previous stroke into progressively higher regions of the negatively charged area of the cloud (Fig. 4.41i). A negatively charged leader, called the *dart leader*, then moves continuously downward to the earth along the main path of the first-stroke channel and deposits further electrons on the ground (Fig. 4.41j and k). The dart leader is followed by another visible return stroke to the cloud (Fig. 4.41l). The first stroke of a flash generally has many downward-directed branches (Fig. 4.42a) because the stepped leader is, strongly branched; subsequent strokes usually show no branching since they follow only the main channel of the first stroke.

Most lightning flashes contain three or four strokes, separated in time by about 50 ms, which can eliminate the 20 C of charge in the lower region of a thundercloud. The charge-generating mechanisms within the cloud must then refurbish the charge before another stroke can occur.

In contrast to the lightning flashes described above, most flashes to very tall buildings are initiated by stepped leaders which start near the top of the building, move upward, and branch toward the base of a cloud (Fig. 4.42b). Lightning rods[†] protect tall structures from damage by routing the strokes to the ground through the rod rather than through the structure itself.

Lightning flashes also occur within thunderstorms (*intracloud discharges*) neutralizing the main positive and negative charge centers. Instead of consisting

[†] The use of lightning rods was first suggested by Benjamin Franklin in 1749, who refused to patent the idea or otherwise profit from their use. Lightning rods were first used in France and the United States in 1752. It is estimated that for rural buildings containing straw or having a thatched roof the danger of fire by lightning is reduced by a factor of 50 if the building is protected by a lightning rod. For houses roofed with tiles or slate the reduction factor is seven.

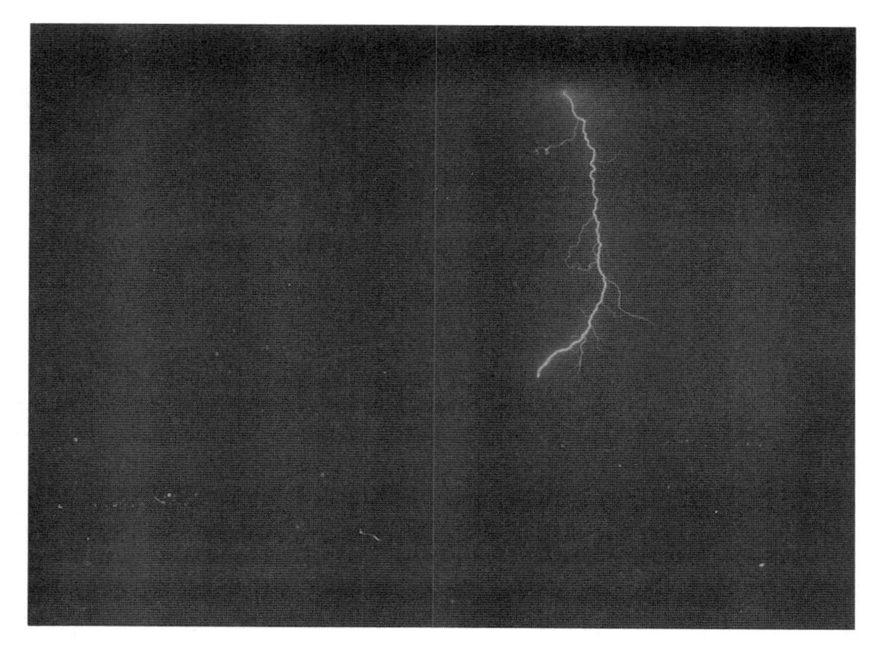

(a)

(b)

Fig. 4.42 (a) A time exposure of a cloud-to-ground lightning flash which was initiated by a stepped leader that propagated from the cloud to the ground. Note the downward-directed branches which were produced by the multibranched stepped leader. (b) A time exposure of a lightning flash from a tower on a mountain to a cloud above the tower. This flash was initiated by a stepped leader which started from the tower and propagated upward to the cloud. In contrast to (a), note the upward-directed branching. (Photos: Richard E. Orville.)

of several discrete strokes, an intracloud discharge generally consists of a single slowly moving spark or leader which travels between the positively and negatively charged regions in a few tenths of a second. This current produces a low but continuous luminosity in the cloud upon which may be superimposed several brighter pulses, each lasting about 1 ms. Tropical thunderstorms, which have relatively high cloud bases, produce about ten intracloud discharges for every cloud-to-ground discharge, but in temperate latitudes the frequencies of the two types of discharge are about the same.

The return stroke raises the temperature of the channel of air through which it passes to above 30,000°K in such a short time that the air has no time to expand. Therefore, the pressure in the channel increases almost instantaneously to 10, or perhaps 100, atm. The high-pressure channel then expands rapidly into the surrounding air and creates a very powerful shock wave (which travels faster than the speed of sound) and, farther out, a sound wave which is heard as *thunder*.[†‡] Thunder is also produced by stepped and dart leaders, but it is much weaker than that from return strokes. Thunder generally cannot be heard more than 25 km from a lightning discharge. At greater distances the thunder passes over an observer's head because it is generally refracted upwards (see Problems 4.27 and 4.28).

PROBLEMS

4.5 Explain or interpret the following:

(a) When the sun is low in the sky, the sun's rays can be seen when they pass through chinks in a cloud layer.

(b) At altitudes above about 3 km the inexperienced traveler often underestimates distances.

(c) Particles collected by impaction do not provide an unbiased sample of atmospheric aerosol.

(d) The air near a seashore is often very hazy and the visibility is poor.

(e) Hot objects in a dusty atmosphere are often surrounded by a thin, dust-free space.

(f) The concentration of aerosol in the plumes from some industrial sources do not decrease as rapidly as predicted by simple diffusion models.

(g) If the rate of removal of a particular toxic chemical from the atmosphere is high, it would be preferable to place a control limit on the amount of the chemical emitted into the atmosphere rather than on the concentration of the chemical in the air.

(h) Visibility is better in a shower of rain than in a cloud or mist.

(i) When the visibility is very good, the atmospheric electric field is very low.

(j) Fog is accompanied by high atmospheric electric fields.

(k) Small droplets of pure water evaporate in air, even when the relative humidity is 100%.

(l) A cupboard may be kept dry by placing a tray of salt in it.

(m) The air must be supersaturated in order for a cloud to form.

† This explanation for thunder was first given by Hirn in 1888.

‡ **Gustave Adolfe Hirn** (1815–1890) French physicist. One of the first to study the theory of heat engines. Established a small network of meteorological stations in Alsace which reported to him.

(n) Cloud condensation nucleus concentrations do not always vary in the same way as Aitken nucleus concentrations.

(o) Measurements of cloud microstructures are more difficult from fast-flying than from slow-flying aircraft.

(p) If the liquid water content of a cloud is to be determined from measurements of the droplet spectrum, particular attention should be paid to accurate measurements of the larger drops.

(q) Cloud droplets growing by condensation near the base of a cloud affect each other primarily by their combined influence on the environmental air rather than by direct interactions. [Hint: consider the average separation between small cloud droplets.]

(r) After landing on a puddle, raindrops sometime skid across the surface of the puddle for a short distance before disappearing.

(s) The presence of an electric field tends to raise the coalescence efficiency between colliding drops.

(t) Large raindrops falling through the air are not tear shaped (as often depicted) but are shaped more like a parachute.

(u) Raindrops reaching the ground cannot exceed a certain critical size.

(v) Ice crystals can be produced in a deep freeze container by shooting the cork out of a toy pop-gun.

(w) Large supersaturations with respect to water are rare in the atmosphere but large supercoolings of droplets are common.

(x) Large volumes of water rarely supercool by more than a few degrees.

(y) The Millipore filter technique may be used to distinguish deposition from freezing nuclei.

(z) Aircraft icing may sometimes be reduced by climbing.

(aa) Present techniques for measuring ice nucleus concentrations may not simulate atmospheric conditions very well.

(bb) The length of a needle crystal increases comparatively rapidly when it is growing by the deposition of water vapor.

(cc) Natural snow crystals are often comprised of more than one basic ice crystal habit.

(dd) Riming tends to be greatest at the edges of ice crystals (for example, Fig. 4.33 c and d).

(ee) Riming significantly increases the fall speeds of ice crystals.

(ff) Aggregated ice crystals have relatively low fall speeds for their masses.

(gg) The charging mechanism responsible for the small pocket of positive charge which exists below the melting level in some thunderstorms must be more powerful than the mechanism responsible for the generation of the main charge centers.

(hh) When the atmospheric electric field between cloud base and the ground is directed downward, precipitation particles reaching the earth generally carry a negative charge, and when the electric field is in the upward direction the precipitation is generally positively charged.

(ii) The presence of water drops in the air reduces the potential gradient required for dielectric breakdown.

4.6 Calculate the supersaturation reached in an Aitken nucleus counter if air in the counter, which is initially saturated with respect to water at 15°C, is suddenly expanded to 1.2 times its initial volume. You may assume that the expansion is adiabatic and use Fig. 2.7 to estimate saturation vapor pressures. Why does your calculation give the maximum possible supersaturation that can be attained for this expansion ratio?

Answer 157%

4.7 Aerosol with diameter D between about 2 and 40 μm experience a so-called *Stokes drag force* given by $3\pi\eta Dv$, where η is the viscosity of the air, and v is their velocity through air. Neglecting the density of air compared to the density ρ of an aerosol, derive an expression for the terminal fall speed v_s of the aerosol. Use this expression to calculate the terminal fall speeds of aerosol with diameters 1 and 10 μm. (Assume that $\rho = 10^3$ kg m^{-3}, $\eta = 1.7 \times 10^{-5}$ SI units.)

Answer $v_s = D^2\rho g/18\eta$; $v_s(1\ \mu\text{m}) = 3.2 \times 10^{-5}$ m s^{-1}; $v_s(10\ \mu\text{m}) = 3.2 \times 10^{-3}$ m s^{-1}

4.8 A particle of mass m passes horizontally through a small hole in a screen. If the velocity of the particle at the instant ($t = 0$) it passes through the hole is v_0, derive an expression for the horizontal velocity v of the particle at time t. You may assume that the drag force on the particle is the same as that given in Problem 4.7. Use this expression to deduce an expression for the horizontal distance (called the *stop distance*) the particle would travel beyond the hole.

Answer $v = v_0 \exp(-3\pi\eta Dt)/m$; stop distance $= v_0 m/3\pi\eta D$

4.9 A metal plate has a positive electric charge Q_0 at time $t = 0$. If the charge on the plate is gradually dissipated due to the collection of negative ions from the air, derive an expression for (a) the charge Q at time t in terms of the specific conductivity λ of the negative ions in the air and the permittivity ε_0 of free space, and (b) the time τ (called the negative ionic relaxation time) for Q to fall to $1/e$ of Q_0. If $\lambda = 4.3 \times 10^{-14}\ \Omega^{-1}\ \text{m}^{-1}$ near the earth's surface and $\varepsilon_0 = 8.85 \times 10^{-12}\ \text{C}^2\ \text{N}^{-1}\ \text{m}^{-2}$, calculate the value of λ near the earth's surface. [Hint: $\lambda = i/E$, where i is the current per unit area and E the electric field.]

Answer (a) $Q = Q_0 \exp(-\lambda t/\varepsilon_0)$; (b) $\tau = \varepsilon_0/\lambda$; (c) 3 min 26 s

4.10 Figure P4.10 indicates a film of liquid (for example, a soap film) on a wire frame. The area of the film can be changed by moving one of the wires. The surface tension of the liquid is defined as the force per unit length that the liquid exerts perpendicular to movable wire (as indicated by the arrow). If the surface energy of the liquid is defined as the work required to create a unit area of new liquid, show that the numerical values of the surface tension and surface energy of the liquid are the same.

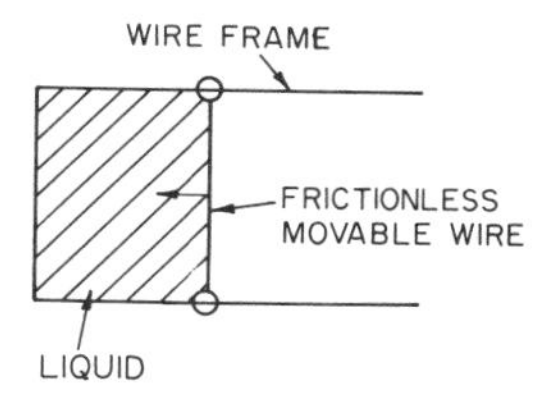

Fig. P4.10.

4.11 Show that the height of the critical free energy barrier ΔG^* in Fig. 4.10 is given by

$$\Delta G^* = \frac{16\pi\sigma^3}{3[nkT\ \ln(e/e_s)]^2}$$

4.12 Calculate the relative humidity of the air adjacent to a pure water droplet 0.2 μm in radius if the temperature is 0°C. (The interfacial energy of water at 0°C is 0.076 J m^{-2} and the number density of molecules in water at 0°C is 3.3×10^{28} m^{-3}.)

Answer 100.6%

4.13 Use the Köhler curves shown in Fig. 4.12 to estimate:

(a) the radius of the droplet that will form on a sodium chloride particle of mass 10^{-18} kg in air which is 0.1% supersaturated;

(b) the relative humidity of the air adjacent to a droplet of radius 0.04 μm which contains 10^{-19} kg of dissolved ammonium sulfate;

(c) the critical supersaturation required for an ammonium sulfate particle of mass 10^{-19} kg to grow beyond the haze state.

Answer (a) $\simeq 0.45\ \mu$m; (b) $\simeq 90\%$; (c) $\simeq 0.47\%$

4.14 Show that for a very weak solution droplet ($m \ll \frac{4}{3}\pi r^3 \rho'$), (4.8) can be written as

$$\frac{e'}{e_s} \simeq 1 + \frac{a}{r} - \frac{b}{r^3}$$

where $a = 2\sigma'/n'kT$ and $b = imM_w/\frac{4}{3}\pi\rho' M_s$. What is your interpretation of the second and third terms on the right-hand side of this expression? Show that in this case the peak in the Köhler curve occurs at

$$r \simeq \left(\frac{3b}{a}\right)^{1/2} \quad \text{and} \quad \frac{e'}{e_s} \simeq 1 + \left(\frac{4a^3}{27b}\right)^{1/2}$$

4.15 Assuming that cloud condensation nuclei are only removed from the atmosphere by first serving as the centers on which cloud droplets form, and subsequently grow to form precipitation particles, estimate the residence time of the cloud condensation nuclei in a column extending from the surface of the earth up to an altitude of 5 km. Assume that the annual rainfall is 100 cm and the cloud liquid water 0.3 g m^{-3}.

Answer On the order of a day

4.16 Derive an expression for the height h above cloud base of a droplet at time t, which is growing by condensation only in a cloud with a steady updraft velocity w and supersaturation S. [Hint: Use the expression for the terminal fall speed of a droplet derived in Problem 4.7 together with Eq. (4.11).]

Answer $h = wt - 2g\rho_l SGt^2/9\eta$

4.17 An isolated parcel of air is lifted from cloud base at 800 mb, where the temperature is 5°C, up to 700 mb. Use the pseudoadiabatic chart (back endpapers) to determine the amount of liquid water (in grams per kilogram of air) that is condensed out during this ascent. This is referred to as the *adiabatic liquid water content*. Why are actual liquid water contents in clouds usually less than the adiabatic values? Under what circumstances do you think the liquid water content in a cloud might be greater than the adiabatic value?

Answer 1.8 g kg^{-1}

4.18 A drop with an initial radius of 100 μm falls through a cloud containing 100 droplets per cm^3 which it collects in a continuous manner with a collection efficiency of 0.8. If all the cloud droplets have a radius of 10 μm, how long will it take for the drop to reach a radius of 1 mm? You may assume that for the drops of the size considered in this problem the terminal velocity v (in meters per second) of a drop of radius r (in meters) is given approximately by $v = 6 \times 10^3 r$. Assume that the cloud droplets are stationary and that the updraft velocity in the cloud is negligible.

Answer 76.3 min

4.19 If a raindrop has a radius of 1 mm at cloud base, which is located 5 km above the ground, what will be its radius at the ground and how long will it take to reach the ground if the relative humidity between cloud base and ground is constant at 60%. [Hint: Use Eq. (4.11) and the relationship between v and r given in Problem 4.18. If r is in micrometers the value of G_l in (4.11) is 10^2 for cloud droplets, but for the large drop sizes considered in this problem the value of G_l should be taken as 7×10^2 to allow for ventilation effects.

Answer 0.67 mm; 16.3 min

4.20 If the concentration of freezing nuclei which are active at temperature T is given by (4.19), show that the median freezing temperature of a number of drops should vary with their diameter as shown in Fig. 4.27. [Hint: The probability p of a drop freezing which contains n active ice nuclei is given by $p = 1 - \exp(-n)$].

4.21 A large number of drops, each of volume V, are cooled simultaneously at a steady rate $\beta(=dT/dt)$. If $p(V, t)$ is the probability of ice nucleation taking place in a volume V of water during a time interval t, derive a relationship between $p(V, t)$ and $\int_0^{T_t} J_{LS}\, dT$, where J_{LS} is the ice nucleation rate (per unit volume per unit time) and T_t the temperature of the drops at time t.

Answer $\ln[1 - p(V, t)] = -\dfrac{V}{\beta} \displaystyle\int_0^{T_t} J_{LS}\, dT$

4.22 A cloud which is cylindrical in shape has a cross-sectional area of 10 km^2 and a height of 3 km. The whole volume of the cloud is initially supercooled and the liquid water content is 2 g m^{-3}. If all of the water in the cloud is transferred onto ice nuclei present in a uniform concentration of 1 liter^{-1}, calculate the mass of each ice crystal produced and the total number of ice crystals in the cloud. If all the ice crystals precipitate and melt before they reach the ground, what will be the total rainfall produced?

Answer 2 mg; 3×10^{13}; 6 mm

4.23 Calculate the radius and the mass of an ice crystal after it has grown by deposition from the vapor phase for half an hour in a water-saturated environment at -5°C. Assume that the shape of the crystal may be approximated by a thin cylindrical disk of constant thickness 10 μm. [Hint: Use (4.22) and Fig. 4.31 to estimate the magnitude of $G_i S_i$. The electrostatic capacity C of a cylindrical disk of radius r is given by $C = 8r\varepsilon_0$, where ε_0 is the permittivity of free space.]

Answer 0.5 mm; 7.3 μg

4.24 Calculate the time required for an ice crystal, which starts off as a plane plate with an effective circular diameter of 1 mm and a mass of 0.01 mg, to grow by riming into a spherical graupel particle 1 mm in diameter if it falls through a cloud containing 0.5 g m^{-3} of small water droplets which it collects with an efficiency of 0.6. Assume that the density of the final graupel particle is 100 kg m^{-3} and that the terminal velocity v (in meters per second) of the crystal is given by $v = 2.4M^{0.24}$, where M is the mass of the crystal in milligrams.

Answer 9.6 min

4.25 Calculate the time required for the diameter of a spherical snowflake to increase from 1 mm to 1 cm if it grows by aggregation as it falls through a cloud of small ice crystals present in an amount 1 g m^{-3}. You may assume that the collection efficiency is unity, that the density of the snowflake is 100 kg m^{-3}, and that the difference in the fall speeds of the snowflake and the ice crystals is constant and equal to 1 m s^{-1}.

Answer 30 min

4.26 Determine the electric charge that would have to be separated for each collision of an ice crystal with a hailstone in a thunderstorm in order to generate charge at a rate of 1 C km^{-3} min^{-1}. Assume that the concentration of ice crystals is 10^5 m^{-3}, that their fall speed is negligible compared to that of the hailstones, and that they are collected with an efficiency of unity by the hailstones. The hailstones may be considered to be spherical, with a constant radius of 2 mm, and a density of 500 kg m^{-3}. The precipitation rate is 5 cm h^{-1}.

Carry out the same calculation for the case of supercooled droplets (larger than 15 μm in radius) colliding with hailstones if the former are present in concentrations of 10^6 m^{-3}. [Hint: Note the relationship derived in Problem 4.4.]

Answer 16 fC; 1.6 fC

4.27 If the velocities of sound in two adjacent thin layers of air are v_1 and v_2, a sound wave will be refracted at the interface between the layers (see adjacent figure) and

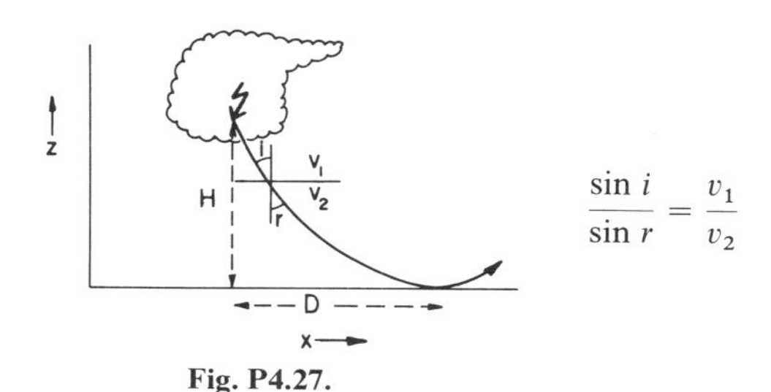

$$\frac{\sin i}{\sin r} = \frac{v_1}{v_2}$$

Fig. P4.27.

Use this relationship, and the fact that $v \propto \sqrt{T}$, where T is the air temperature in degrees Kelvin, to show that the equation for the path of thunder is given by

$$dx = -\left(\frac{T}{\Gamma z}\right)^{1/2} dz$$

where x and z are the coordinates indicated in the adjacent figure, Γ the temperature lapse rate in the vertical (assumed constant), and T the temperature at height z.

4.28 Use the expression derived in the previous problem to show that the maximum distance D at which a sound wave produced by thunder can be heard is given approximately by (see figure in Problem 4.27)

$$D = 2\left(\frac{T_0 H}{\Gamma}\right)^{1/2}$$

where T_0 is the temperature at the ground. Calculate the value of D given that $\Gamma = 7.5$ deg km^{-1}, $T_0 = 300$°K, and $H = 4$ km.

Answer 25.3 km

4.29 An observer at the ground hears thunder 10 s after he sees a lightning flash, and the thunder lasts for 8 s. How far is the observer from the closest point of the lightning flash and what is the minimum length of the flash? Under what conditions would the length of the flash you have calculated be equal to the true length? (Speed of sound is 0.34 km s^{-1}.)

Answer 3.4 km; 2.72 km

Chapter

5

Clouds and Storms

The most distinctive feature of the earth as seen from a satellite or a space vehicle is its cloud cover (see Fig. 1.1). At any one time, about one half of the earth's surface is covered by clouds which can occur at altitudes from the surface of the earth up to about 20 km.† In this chapter we first describe the principal types of clouds and the ways in which they are formed and modified. Then we consider the structure and dynamics of convective precipitation systems, ranging from the air-mass thunderstorm to the massive and highly organized "supercell" storms that produce hail and tornadoes. We then discuss the mesoscale structure and organization of precipitation in hurricanes and extratropical cyclonic storms. Finally, we review work on the artificial modification of clouds, precipitation, and storms. In the course of this chapter, we will need to consider a wide range of physical processes ranging from the microphysical phenomena discussed in the previous chapter up to those of synoptic scale.

† *Nacreous* (or *mother-of-pearl*) clouds occur at heights up to 30 km and *noctilucent* clouds occur at about 80 km. However, the exact compositions of these clouds are unknown.

5.1 CLOUD MORPHOLOGY

5.1.1 Mechanisms of formation

As explained in the previous chapter, clouds form in air which has become supersaturated with respect to liquid water or ice. By far the most common means by which air becomes supersaturated is through ascent accompanied by adiabatic expansion and cooling. The principal types of ascent, each of which produces distinctive cloud forms, are:

- Local ascent of warm, buoyant air parcels in a conditionally unstable environment which produces *convective*[†‡] *clouds*. These clouds have diameters ranging from about 0.1 to 10 km and air ascends in them with vertical velocities on the order of a few meters per second, although updraft speeds of several tens of meters per second can occur in large convective cloud systems. These lifting rates produce water (liquid or ice) contents on the order of 1 gram per cubic meter of air, although much higher values may occur. The lifetimes of convective clouds range from minutes to hours.
- Forced lifting of stable air which produces *layer clouds*. These clouds can occur at altitudes from ground level up to the tropopause and extend over areas of hundreds of thousands of square kilometers. Lifting rates range from a few centimeters per second to about 10 cm s^{-1}, and water contents are generally a few tenths of a gram per cubic meter or less. Layer clouds generally exist over periods of tens of hours.
- Forced lifting of air as it passes over hills or mountains produces *orographic clouds*. The resulting updraft velocities depend upon the speed and direction of the wind and the height of the barrier, but they can be several meters per second. Water contents are typically a few tenths of a gram per cubic meter. Orographic clouds may be quite transitory, but in steady winds they can exist for long periods of time.

In addition to formation in ascending air, clouds may also form by:

- The cooling of air below its dew point when it comes into contact with a cold surface. The most common examples are the formation of fog during clear, windless nights when the ground is cooled by radiation (*radiation fog*) and the formation of *advection fog* when warm air moves over colder sur-

[†] The term *convection* was first introduced by William Prout in 1834. The phenomenon of thermal convection in fluids was discovered by Count Rumford in 1797, as he sought to explain why "apple pie and apples and almonds mixed (a dish in great repute in England) remain hot a surprising length of time." The fact that locally heated parts of the atmosphere rise was noted by Benjamin Franklin in 1749.

[‡] **Count Rumford (Benjamin Thompson)** (1753–1814) American physicist and social scientist. At 13 apprenticed to a merchant in Salem, Massachusetts. Joined the British Army at 20, and emigrated to England as a loyalist in 1775. Pioneered soup kitchens for the poor while working for the Elector of Bavaria. Was made a Count of the Holy Roman Empire in 1791. Showed heat to be a mode of motion (rather than a fluid) and attempted to measure the mechanical equivalent of heat. Also developed the central steam and hot water heating systems. Married Lavoisier's widow. Founded the Royal Institution in England.

faces. The latter are particularly common at sea. As cooling is distributed upward from the ground by mixing, fog may lift to form a low-level layer of *stratus cloud* which is often not more than 500 m thick and featureless.

- The mixing of two parcels of air with different temperatures (see Problem 5.4). This process does not often form clouds because the temperature differences between adjacent air parcels are generally not large enough to produce saturation on mixing, unless both air masses are nearly saturated to begin with. This process is sometimes responsible for the formation of contrails and *arctic sea-smoke* (or *steam fogs*).
- Adiabatic expansion and cooling due to a rapid local reduction in pressure. This process is responsible for the formation of the *funnel clouds* associated with tornadoes and water spouts.

5.1.2 Types of clouds

The first published classification of clouds was that of Lamarck[†] in 1802, in which a limited number of interesting cloud forms were identified and named (in French). Although Lamarck's names were never adopted, his method of dividing the regions where clouds form into three layers is used in a modified form in the present international classification of clouds ("International Cloud Atlas," World Meteorological Organization, 1956). The basis of the international classification of clouds is the system proposed by Howard[‡] in 1803 who used the four Latin names *cumulus* (a heap or pile) for convective clouds, *stratus* (a layer) for layer clouds, *cirrus* (a filament of hair) for fibrous clouds, and *nimbus* for rain clouds, together with their compounds (for example, cirrocumulus, cirrostratus). In the international classification, nimbus, or *nimbo*, is used only in composite names to indicate precipitating clouds (for example, nimbostratus, cumulonimbus). Further composite names are used (for example, altostratus and altocumulus; where the prefix *alto* indicates middle-level clouds which occur from about 2 to 7 km in altitude).

The importance of orographic clouds was first clearly noted by Ley[§] in 1894, who also introduced the terms *lenticularis*, for lens-shaped clouds, and

[†] **Jean Babtiste Lamarck** (1744–1829) French naturalist. Founder of modern invertebrate zoology. Also devised one of the earliest evolutionary theories. Interested in the effects of climate on living organisms. Responsible for the establishment of a central meteorological data bank in France in 1800.

[‡] **Luke Howard** (1772–1864) English meteorologist, although a retail chemist by trade. One of the early pioneers of atmospheric physics. In addition to his work on the classification of clouds, he carried out some interesting studies on atmospheric electricity.

[§] **W. Clement Ley** (1840–1896) Early English meteorologist, although a clergyman by profession. Served as inspector of meteorological stations in England from 1879 to 1896. Ley stated, "My own earliest recollections are those of looking at the clouds, and forming infantine speculations as to the causes of their forms and movements, and of being reprehended for exposing myself to all states of weather for this purpose. The tendency was inveterate, and to this day I have spent nearly a twelfth part of my working existence in that occupation." Author of "Laws of the Winds Prevailing in Western Europe", E. Stanford, London, 1872 and "Cloudland", E. Stanford, London, 1894; the latter was prepared for press by his son during Ley's terminal (pulmonary) illness.

castellatus (now called *castellanus*), for middle clouds (usually altocumulus) which become convective and develop turrets.

It is not possible here to describe the many hundreds of different cloud types which can form; the reader is referred to the "International Cloud Atlas" for such a discussion. Instead, we wish to point out some of the insights into the processes responsible for the formation and modification of clouds which can be deduced from careful visual observations.†

5.1.3 Convective clouds

Shown in Fig. 5.1 are a series of photographs, taken over a period of 55 min, of a field of vigorous convective clouds. In the first photograph (Fig. 5.1a) the bases of the clouds are fairly well defined, which indicates that the air at lower levels is well mixed due to convective stirring, so that the height of the lifting condensation level is fairly uniform over a large area. While the clouds are young and growing they contain much fine detail and their boundaries are sharp, giving them a typical cauliflower appearance (Fig. 5.1a). At this stage the clouds consist mainly of liquid water, even though they may extend well above the 0°C isotherm. This is the case even for the large *cumulus congestus* in the background of Fig. 5.1a, which probably grew above a particularly intense "hot spot" on the ground.

As rising parcels of cloudy convective air lose their buoyancy their outlines become more ragged. In old cold clouds the ragged appearance may be enhanced by the presence of large numbers of ice particles for two reasons. First, in supersaturated cloudy air, ice particles grow faster than water droplets (see Section 4.5.3). Second, in the subsaturated air beyond the edges of a cloud, ice particles evaporate slower than water droplets because the equilibrium saturation vapor pressure over ice is less than that over water at the same temperature (see Fig. 2.7). On both scores, ice particles at the edges of a cloud survive longer than water droplets as they mix into the dry ambient air and they therefore produce more diffuse cloud boundaries. The first stages in this "glaciation" process can be seen in Fig. 5.1b where the upper edges of the largest cloud have become quite ragged. The upper regions of this cloud are also starting to be spread out horizontally by the wind at this level to form an *anvil.* Note also that the cloud bases have lowered, probably as a result of rising relative humidity produced by precipitation in the region below cloud base. At this stage the largest cloud has transformed into a *cumulonimbus*.

In Fig. 5.1c and d, regions on the right of the largest cloud have become increasingly diffuse as the concentrations of ice particles increase and the anvil becomes larger. A further stage in this development is seen in Fig. 5.1e where

† For a more extensive discussion along these lines the reader is referred to F. H. Ludlam and R. S. Scorer, "Cloud Study: a Pictorial Guide," John Murray Pub. Co., London, 1957, and R. S. Scorer, "Clouds of the World: a Complete Colour Encyclopedia," David and Charles Pub. Ltd., Newton Abbot, 1972.

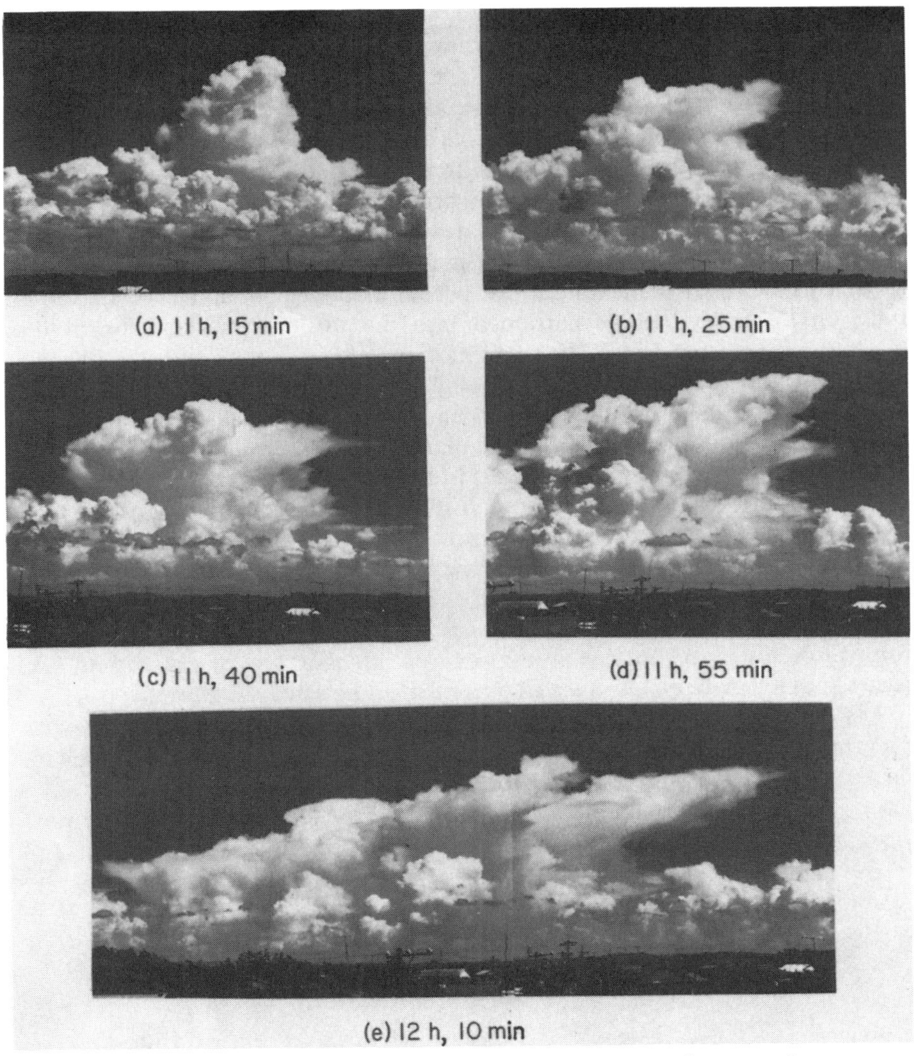

Fig. 5.1 This series of photographs shows the transformation process from cumulus congestus to cumulonimbus clouds during a period of 55 min. See text for discussion. (Photos from Y. Itoh and S. Ohta, "Cloud Atlas: An Artist's View of Living Cloud," Chijinshokan Co., Ltd., Tokyo, Japan, 1967.)

the upper regions of the cumulonimbus have *glaciated* (that is, they have become dominated by ice particles). The tops of vigorous convective clouds which reach to the tropopause often spread out horizontally over large areas at this level because of the strong positive static stability; the remnants of these anvils are called *anvil cirrus.* The heights of less vigorous convective clouds are often limited by temperature inversions at lower levels in the troposphere where the clouds may often spread out horizontally to form stratocumulus clouds (Fig.

2.11). Both anvil cirrus and stratocumulus clouds can restrict the heating of the ground and thereby inhibit the formation of new convective clouds.

Careful visual observations of convective clouds reveal that they are composed of a number of individual elements or *towers*, each of which goes through a life cycle of growth and decay in a matter of minutes. Numerous cloud towers can be seen in Fig. 5.1. These towers are produced by elements of rising buoyant air called *thermals*, which are well known to glider pilots (and birds). The growth of a thermal into a cloud tower is indicated in Fig. 5.2. As a thermal rises, it pushes environmental air away from its upper boundary. At the same time, environmental air is entrained into the turbulent wake beneath the thermal, as shown in Fig. 5.2b. These motions are such as to turn the thermal inside out and thereby produce thorough mixing. Also, some environmental air may be entrained through the sides and top of the thermal due to turbulent mixing. As a result of *entrainment*, the diameter of a thermal initially increases as it rises (as indicated by dashed lines in Fig. 5.2a). However, above the lifting condensation level the thermal, now visible as a rising cloud tower, generally ceases to widen because of the entrainment of cool, dry air. Evaporation of some of the cloud water results in cooling; the buoyancy of the thermal is therefore reduced and some of the air in it is left behind. By the time a thermal has been thoroughly diluted by being turned once inside out, its buoyancy is generally completely destroyed. By this time it has typically risen to a height above cloud base equal to about one and a half times the diameter of the thermal.

The evaporation that takes place at the boundary of a cloud causes cooling and therefore sinking motions. These downdrafts tend to inhibit thermals from rising above the lifting condensation level just outside the boundaries of a cloud; consequently, the boundaries remain fairly well defined. The net upward movement of air in convective clouds is compensated by slower subsidence of

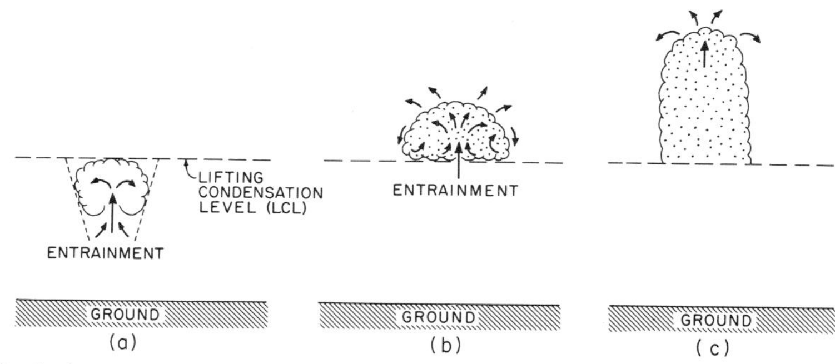

Fig. 5.2 A thermal at various stages in its development. The arrows indicate the air motions. (a) Below the lifting condensation level (LCL) the diameter of the (invisible) thermal increases as it rises. (b) Above the LCL the thermal becomes visible as a cloud tower. It ceases to widen because parts diluted by entrainment are left behind. (c) The cloud tower at its maximum height. Thereafter it becomes eroded, sinks, and evaporates. A cumulus cloud consists of a number of such thermals at various stages in their life history.

air over the much larger areas between the clouds. This subsidence produces warming and drying and hinders the growth of thermals in the regions between clouds. Therefore, there is a tendency for thermals to feed previously formed clouds. Moreover, since these same regions have been moistened by earlier thermals, evaporation of the newly rising thermals is reduced.

Problem 5.1 Assuming that the radius of a thermal, considered as a spherical bubble, is proportional to its height above the ground, show that the entrainment rate must be inversely proportional to the radius of the thermal and directly proportional to its upward velocity.

Solution Let the radius of the thermal at height z above the ground be b; then

$$b = \alpha z \tag{5.1}$$

where α is a constant. If m is the mass of a thermal at time t, the entrainment rate is given by

$$\frac{1}{m}\frac{dm}{dt} = \frac{1}{\frac{4}{3}\pi b^3}\frac{d}{dt}\left(\tfrac{4}{3}\pi b^3\right)$$

or

$$\frac{1}{m}\frac{dm}{dt} = \frac{3}{b}\frac{db}{dt} \tag{5.2}$$

From (5.1) and (5.2),

$$\frac{1}{m}\frac{dm}{dt} = \frac{3\alpha}{b} w \tag{5.3}$$

where $w = dz/dt$ is the vertical velocity of the plume. Hence, from (5.3), the entrainment rate of the plume is proportional to its vertical velocity and inversely proportional to its radius.

In recent years the microstructure of convective clouds has been investigated by means of *in situ* aircraft measurements. Shown in Fig. 5.3 are measurements of updraft velocity, liquid water content, and three droplet spectra made during

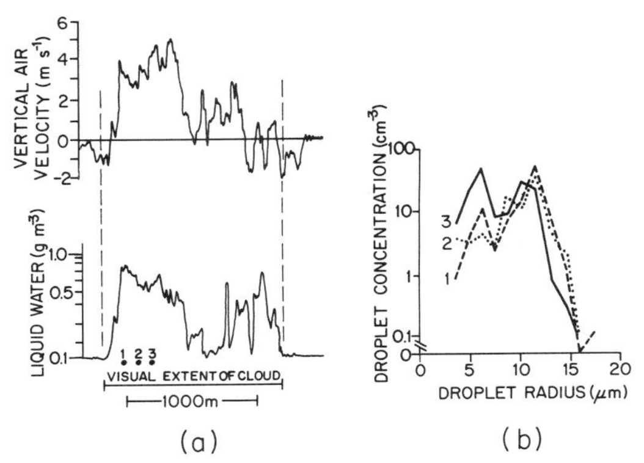

Fig. 5.3 Measurements of (a) vertical air velocities and liquid water contents and (b) droplet size distributions, through the middle of a cumulus cloud about 2 km in depth. The positions where the three droplet size distributions shown in (b) were measured are indicated by 1, 2, and 3 in (a). [From *J. Atmos. Sci.* **26**, 1053 (1969).]

a horizontal traverse through the middle of a small nonprecipitating cumulus cloud about 2 km deep. It can be seen that within the cloud the air is generally moving upward, whereas in the region surrounding the cloud the air motions are downward. The higher liquid water contents, which range up to about 1 g m^{-3}, are generally associated with the higher updraft velocities. The three droplet spectra shown in Fig. 5.3 were measured about 100 m apart in the region where the air velocity and liquid water content were relatively constant. The three spectra are remarkably similar and each shows a bimodal distribution with peak concentrations at droplet radii of about 6 and 11 μm. Since bimodal distributions are observed most commonly when the cloud is growing in a relatively unstable environment, it is thought that they may be produced by mixing of the cloudy air and the drier environmental air, particularly at the growing cloud top.

In larger convective clouds (for example, cumulus congestus and cumulonimbus), in which the updraft velocities can reach tens of meters per second, liquid water contents are generally several grams per cubic meter (values as high as 20 g m^{-3} have been reported in severe hailstorms) and the droplet concentrations several hundred per cubic centimeter.

Under certain conditions, the appearance of clouds can be influenced by convective overturning associated with downward motions produced by the weight of falling precipitation, and the cooling of the air as the precipitation

Fig. 5.4 Cumulonimbus mamma in evening light. (Photo: R. A. R. Tricker.)

evaporates into it. These downdrafts are visible as large protuberances called *mamma* (Latin for breast) which extend from the bases of clouds. Figure 5.4 shows a striking example of mamma, characteristic of the type that often forms on the under surface of the projecting portion of the anvil associated with severe convective storms. Less well developed mamma can often be seen on cirrus, cirrocumulus, altocumulus, altostratus, and stratocumulus.

5.1.4 Layer clouds

Although layer clouds, in the form of stratus, may be produced by the lifting of fogs, in temperate latitudes the most common mechanism for the formation of layer clouds over vast areas is the widespread ascent of air associated with the development of cyclones. The approach of a warm front provides a particularly good opportunity to observe layer clouds in various stages of development.

Ahead of the passage of a warm front on the ground, the gentle rise of the air in the warm sector over the denser air in the cold sector (see Fig. 3.5a) produces layer clouds which decrease in height as the front approaches. Thus, the first sign of an approaching warm front is the advance across the sky, usually from the west, of very high ($\simeq$9 km) *cirrus* clouds (Fig. 5.5). Cirrus clouds are

Fig. 5.5 Hooked-shaped cirrus clouds (cirrus uncinus) and nearly straight filaments of cirrus (cirrus fibratus) characteristic of an approaching warm front. (Photo: P. V. Hobbs.)

composed of ice particles which may be several millimeters in size, but the concentrations of the particles are rather low ($\simeq 0.1\ \text{cm}^{-3}$). As a result of the comparatively large sizes of the ice particles in cirrus, and the relatively low saturation vapor pressure of ice, ice particles often fall from them through distances of a kilometer or more before evaporating. The falling particles are visible as *fallstreaks* or *virga* (Fig. 5.6), which assume a characteristic hook, or comma, shape due to distortion by vertical wind shear. Cirrus fallstreaks may appear fairly substantial when viewed from the ground; however, because they contain only low concentrations of ice particles, they are often invisible to the aircraft pilot.

Fig. 5.6 Cirrus fallstreaks. Note the shearing of the fallstreaks which is caused by the wind. (Photo: L. F. Radke.)

As a front moves closer to the observer, the cirrus clouds give way to *cirrostratus*. Cirrostratus clouds may often be so thin as to be hardly discernible. However, under these conditions they often give rise to bright halos, of which the 22° halo† (Fig. 5.7) is the most common and is a well-known portent of stormy weather. The 22° halo is produced by the refraction of sunlight in

† 22° is the angular radius of the halo. This angular radius is roughly the same as that subtended at the eye by the distance between the top of the thumb and little finger when the fingers are spread wide apart and held at arm's length. (The reader is warned to not stare directly at the sun since this can cause eye damage.)

Fig. 5.7 22° and 46° halos formed in cirrostratus clouds. The exceedingly rare 8° halo (see Problem 5.6) may also be discerned very close to the sun. (Photo from Y. Itoh and S. Ohta.)

hexagonal prisms of ice as shown in Fig. 5.8. In principle, the halo should be a complete circle but it is rarely seen as such; it is usually brighter at the top or bottom or to the left or right. The sky immediately inside the halo is always darker than just outside. The inner edge of the 22° halo is fairly sharp and colored red; moving outward, the ring is colored yellow, green, white, and finally blue. Halos with an angular radius of 46° (Fig. 5.7) can also be produced by refraction in ice crystals (Fig. 5.8) but they are far less common and less bright than the 22° halo.†

† For an interesting, descriptive account of a wide range of atmospheric optical effects, the reader is referred to M. Minnaert's "The Nature of Light and Colour in the Open Air," Dover Pub. Inc., 1954. A more quantitative discussion is given by R. A. R. Tricker in "Introduction to Meteorological Optics," American Elsevier Pub. Co., New York, 1970.

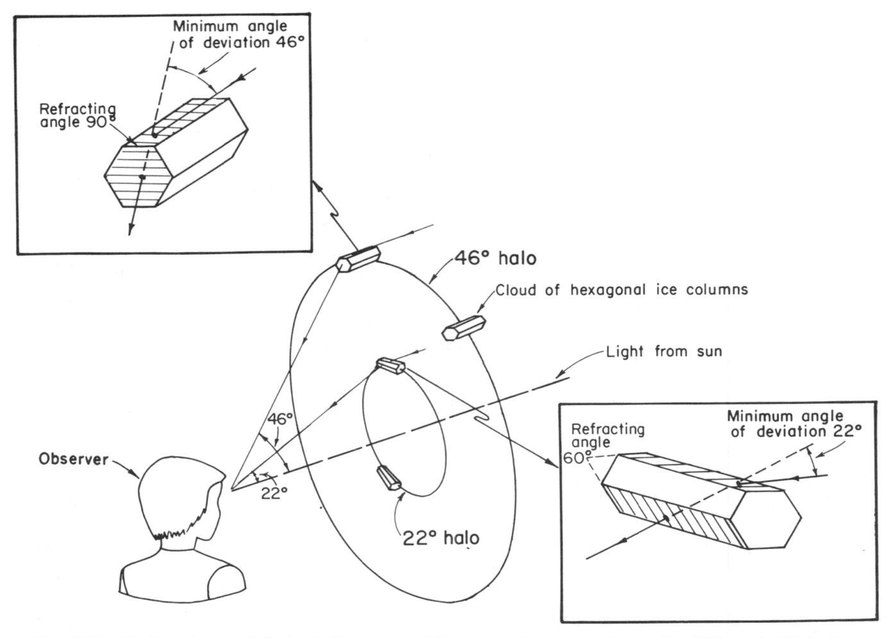

Fig. 5.8 Refractions of light in hexagonal ice crystals to produce the 22° and 46° halos.

Within a matter of hours cirrostratus may thicken and lower as a warm front approaches until the sun can barely be seen through the layer of ice crystals which may be 2–3 km thick and extend over an area of several hundred thousand square kilometers. At this stage the cloud is termed *altostratus* (Fig. 5.9). Halos are rarely seen in altostratus. However, one or more sequences of brilliant colored rings centered on and close to (generally less than an angular radius of about 15°) the sun or moon are sometimes produced by altostratus clouds. These rings are known as a *corona*; they are produced by the diffraction of light in small water droplets. If the cloud droplets are fairly uniform in size, several sequences of rings may be seen; the spacing of the rings depends on the droplet size. In each sequence, the inside ring is violet or blue and the outside ring is red, with other colors in between. A corona around the moon is quite common. However, coronas are just as common around the sun but one avoids looking in this direction (Newton first observed a solar corona by seeing its reflection in water).

As the altostratus thickens and lowers toward the ground it takes on a uniform, gray appearance (through which the sun can no longer be discerned) and begins to precipitate. The cloud is now termed *nimbostratus* (Fig. 5.10). At this stage, cloud layers are often present at several levels and ice particles from upper cloud layers may fall into lower cloud layers where they grow into

Fig. 5.9 Altostratus with pannus (or scud) cloud beneath. (Photo from Y. Itoh and S. Ohta.)

snow particles, melt as they pass through the 0°C level, and reach the ground as rain. The melting level can often be seen looking toward the horizon in the direction of the sun; since snow scatters more light than rain, the cloud is much darker just above the melting layer than below it (see Fig. 5.10). During the initial thickening of the altostratus and nimbostratus, lower-level clouds tend to be suppressed due to the reduction in the warming of the ground by the sun.

Fig. 5.10 Nimbostratus with virga, pannus, and precipitation. (Photo from Y. Itoh and S. Ohta.)

However, as the air is moistened by rain, fragments of low-level cloud, known as *pannus* or *scud*, often form (Figs. 5.9 and 5.10).

The distribution of liquid water in warm layer clouds, such as stratus, and in fogs is relatively uniform over large horizontal areas, and the liquid water contents are generally a few tenths of a gram per cubic meter. Some droplet spectra measured in fogs, which first formed over the water and then drifted on shore, are shown in Fig. 5.11. It can be seen that the radius of the droplets ranges from a few micrometers up to 30 to 40 μm and the liquid water contents range from about 0.05 to 0.1 g m^{-3}. The average droplet radius in fogs and nonraining stratus clouds is about 5 μm and the average size of the droplets generally increases with height into the cloud.

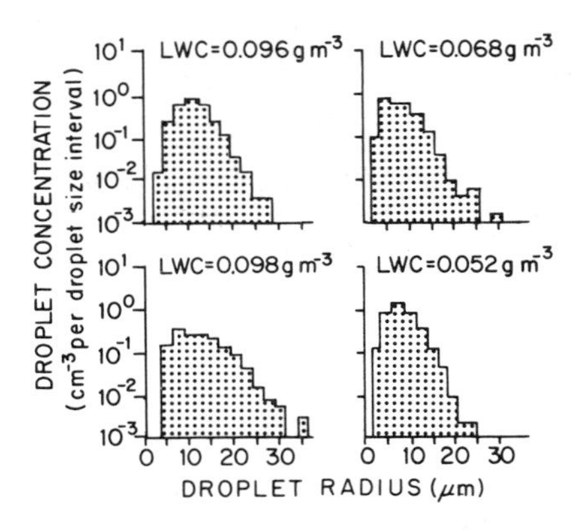

Fig. 5.11 Droplet size distributions measured in fogs on four occasions at Falmouth, Massachusetts. The liquid water contents (LWC) are shown for each case. [From *J. Appl. Met.* **10**, 485 (1971).]

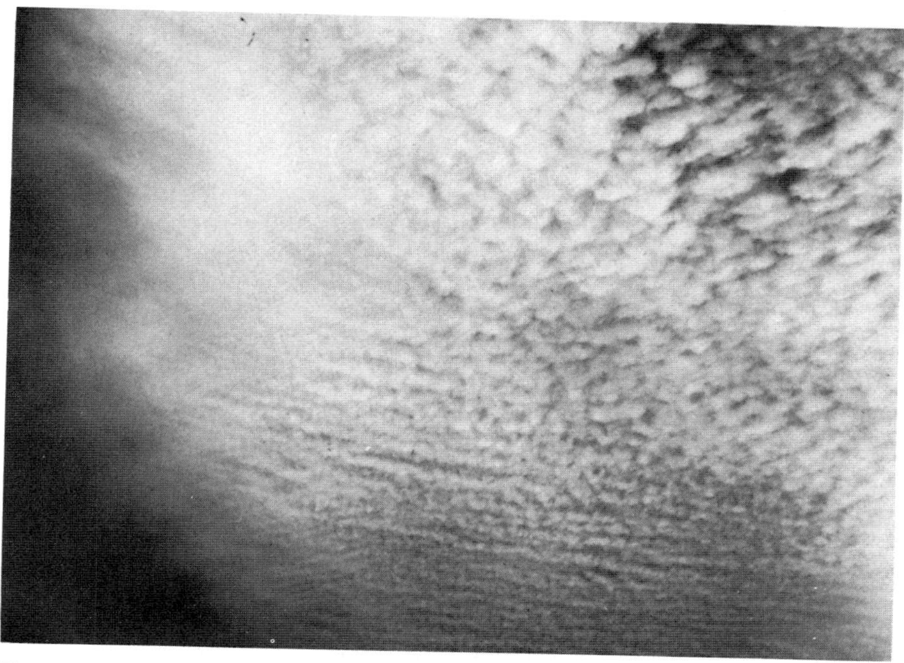

Fig. 5.12 Cirrocumulus. The clouds in the upper right-hand corner, which are broken up into small cumulus-type elements, are young clouds (probably only a few minutes old) and are composed of supercooled water droplets. Toward the horizon the clouds become more wavelike. On the left the clouds have glaciated. (Photo: P. V. Hobbs.)

Also falling under the general heading of layer clouds are *cirrocumulus*, *altocumulus*, and *stratocumulus*. These clouds are apparently affected by two basic types of motion which break them up into small cumulus-type elements and/or wavelike patterns. The first type of pattern is apparent in the cirrocumulus clouds shown in Fig. 5.12; similar patterns are common in altostratus and stratocumulus. The formation of cumulus-type elements in these clouds is thought to be due to small convective cells within the cloud layers which are produced when the bases of the clouds are warmed by radiation from the ground (aided, on occasions, by convective heat transfer) and the cloud tops are cooled as they radiate heat to space. The first contemporary observation of cells of this type being produced by convection was made by Thomson† in 1881 who observed them in a barrel of warm soapy water (the top of which was evaporating into the cooler air) which was being used for washing glasses in a pub!

† **James Thomson** (1822–1892) British engineer and physicist. His studies of whirling fluids led to improved understanding of atmospheric motions as well as to his invention of the vortex waterwheel, the centrifugal pump, a jet pump for draining low lands, and improvements in blowing fans. His younger brother (by two years) was Lord Kelvin.

Fig. 5.13 Lines of stratocumulus clouds associated with roll circulations in the mixed layer on a summer day over land. The prominent boundary between the black surface and the gray reflective surface marks the Carolina coastline with land toward the left. The low-level flow is onshore and nearly parallel to the lines of cumulus. (NASA photograph.)

These cells were subsequently investigated more carefully in the laboratory by Bénard[†] who pointed out their similarity to a "mackerel sky." The phenomenon was investigated theoretically by Rayleigh[‡] who showed that for a layer of given depth, containing a homogeneous fluid, there is a critical rate of differential heating above which cellular motion occurs. The cells are now called *Bénard cells* and they are said to be produced by *Rayleigh convection.*

If a fluid is heated from below, and at the same time subjected to a steadily increasing vertical shear of the horizontal flow, the classical Bénard cells eventually give way to long wavelike patterns, aligned along the direction of the shear vector, which resemble *cloud streets* or *rolls.* Stratocumulus clouds com-

[†] **Henri Bénard** (1874–1939) French physicist and meteorologist. Director of the Laboratory of Fluid Dynamics at the Sorbonne. One of the first to use gliders and movies of soaring birds to study atmospheric motions. He hated writing papers and published very little.

[‡] **Lord Rayleigh (John William Strutt, 3rd Baron)** (1842–1919) English mathematician and physicist. Best known for his work on the theory of sound and scattering of light. Discovered (with Ramsey) the presence of argon in the air, for which he won the Nobel prize in 1904.

Fig. 5.14 An example of billow clouds. (Photo: P. M. Saunders.)

monly form in cloud streets when winds are strong (greater than about 6 m s^{-1} at the surface) and the lapse rate is neutrally stable (Fig. 5.13). They tend to align along the mean direction of the wind, which generally does not differ greatly in direction from that of the wind shear, and have a typical horizontal spacing on the order of 10 km; the ratio of their horizontal wavelength to the depth of the convective layer is on the order of 10:1.

Wavelike motion can also develop within a stably stratified atmosphere, provided that the vertical wind shear exceeds some critical value which defines the onset of shear instability. The resulting motions take the form of *billows* oriented perpendicular to the wind shear. Billows can be detected in clear air by radars sensitive to changes in the refractive index of the air and they can also give rise to *billow clouds*,[†] particularly in altocumulus and cirrocumulus, an example of which is shown in Fig. 5.14. The mechanism of shear instability[‡] is discussed further in Section 9.6.1.

[†] Clement Ley was the first to suggest that billow clouds are due to shear instability.

[‡] In the laboratory, waves resulting from shear instability can be generated at the interface between two fluids of different density by tilting the vessel containing the fluids. These waves were first investigated by Kelvin and Helmholtz. Consequently, waves induced by shear instability in the atmosphere are often called *Kelvin–Helmholtz waves*.

5.1.5 Orographic clouds

Orographic clouds can form at various heights over very modest hills as well as large mountains.[†] They may be seen, on the average, about one day in three in hilly country in temperate latitudes. Their formation is due to the lifting of moist air above its lifting condensation level as the streamlines are perturbed by the mountain. The first wave in the streamlines forms over the mountain itself and is called the *mountain wave*; such a wave can produce *mountain-wave clouds* (Fig. 5.15). The liquid water content in a mountain-wave cloud generally

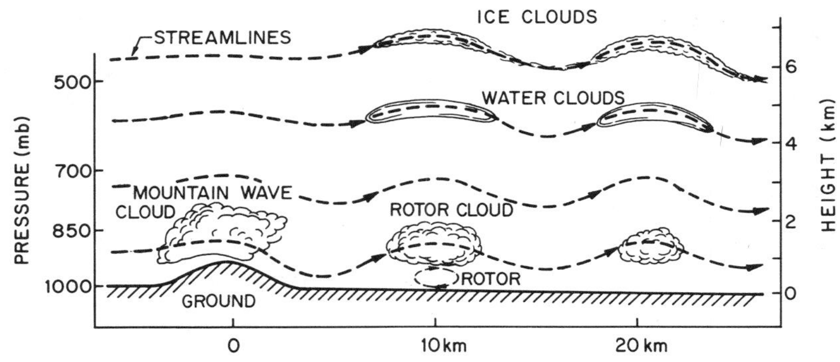

Fig. 5.15 A vertical cross section illustrating the formation of mountain and lee waves and an associated mountain-wave cloud (over the mountain) and lee-wave clouds (downwind of the mountain).

reaches a peak value over the windward slopes of the mountain (Fig. 5.16a), whereas, if the temperature of the cloud is sufficiently low, the ice content reaches a maximum value just downwind of the crest (Fig. 5.16b). Note that the cloud base on the leeward side is higher than on the windward side because precipitation on the windward side removes water from the air. If the air contains several moist layers, separated by drier layers of air, distinct mountain-wave clouds may form at several different heights over a mountain. Mountain-wave clouds can be particularly extensive and important over long ranges of mountains on the western sides of continents, such as the Sierras and Cascade Mountains in the western United States, where they augment clouds in cyclonic storm systems and produce heavy precipitation on the windward slopes. Over the leeward (eastern) slopes, on the other hand, clouds evaporate in the downward-moving air and precipitation is suppressed, giving rise to *rain shadow* areas.

Under suitable conditions the streamlines downwind of a mountain may go through a series of oscillations to produce a train of *lee waves* in which *lee-wave clouds* may form at various levels as illustrated in Figs. 5.15 and 5.17. In the troposphere the distance between lee waves is generally between 5 and 25 km,

[†]In the remainder of this section we will refer to the undulation in the terrain as a mountain.

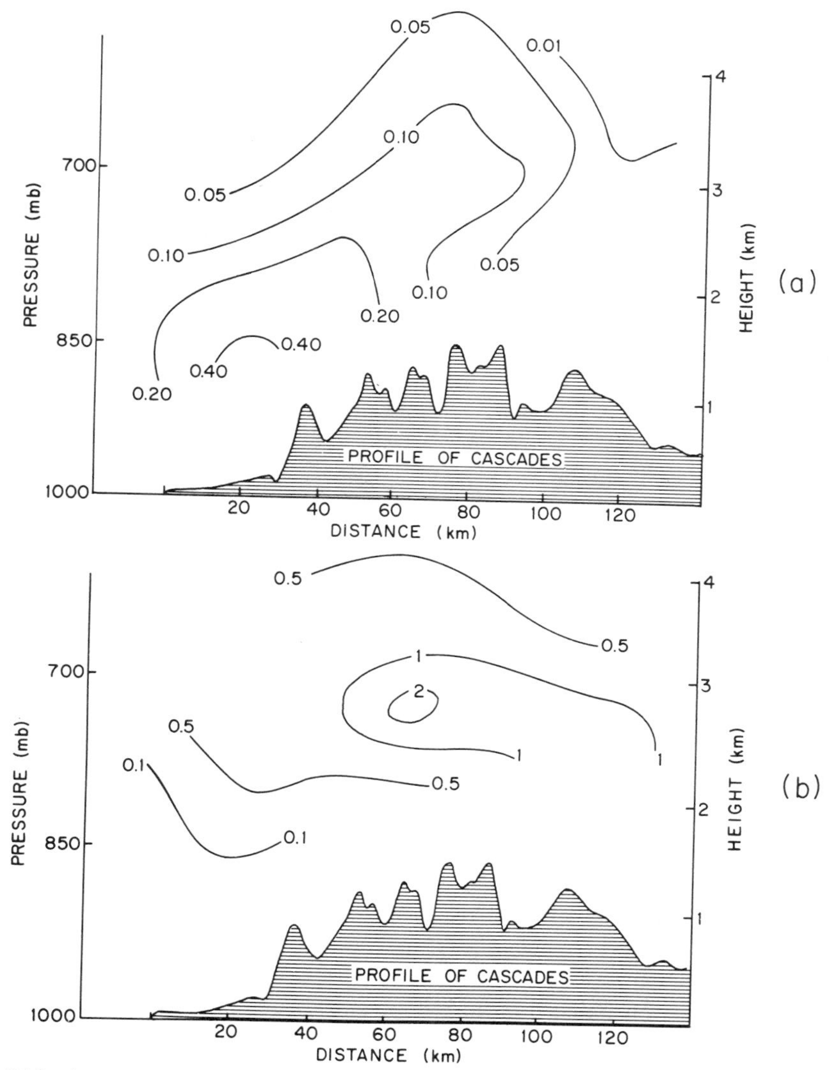

Fig. 5.16 Average distribution of (a) liquid water (in grams per cubic meter) in clouds over the Cascade Mountains and (b) ice particle concentrations (per cubic centimeter), when the wind is blowing from left to right. [From *J. Appl. Met.* **14**, 783 (1975).]

and in some cases more than six successive waves may be observed (see, for example, the waves in the satellite photograph shown in Fig. 7.11). To emphasize the large areas over which wave clouds can form, we show in Fig. 5.18 a satellite photograph of the eastern United States in which numerous lee-wave clouds can be seen over the Appalachian Mountains extending from West Virginia down to Georgia. Wave clouds remain relatively stationary with respect to the

Fig. 5.17 Lee-wave clouds forming downwind of the Rocky Mountains. Note the small portion of ragged ice clouds which can be seen above the smooth water clouds in the foreground. (Photo: National Center for Atmosphere Research.)

mountain, although the air is blowing through them with the speed of the wind; this fact can sometimes deceive an observer into believing that wave clouds are much higher than they actually are. Wave clouds, within an angular radius of about 25° from the sun, are often tinted with brilliant colors (green, purple-red, blue) known as *iridescent colors*. Iridescent colors are parts of a corona, and they are particularly common in newly formed wave clouds because the cloud droplets tend to be uniform in size, as explained in Section 4.4.1. Billow clouds (which may sometimes be confused with wave clouds) rarely exhibit iridescence because they are more turbulent and their droplet size distributions are broader.

It can be seen from Fig. 5.15 that the streamlines in lee-wave clouds have their steepest slope, and therefore the vertical air velocity reaches a maximum value, a few kilometers downwind of the lee slope of the mountain. It is in these

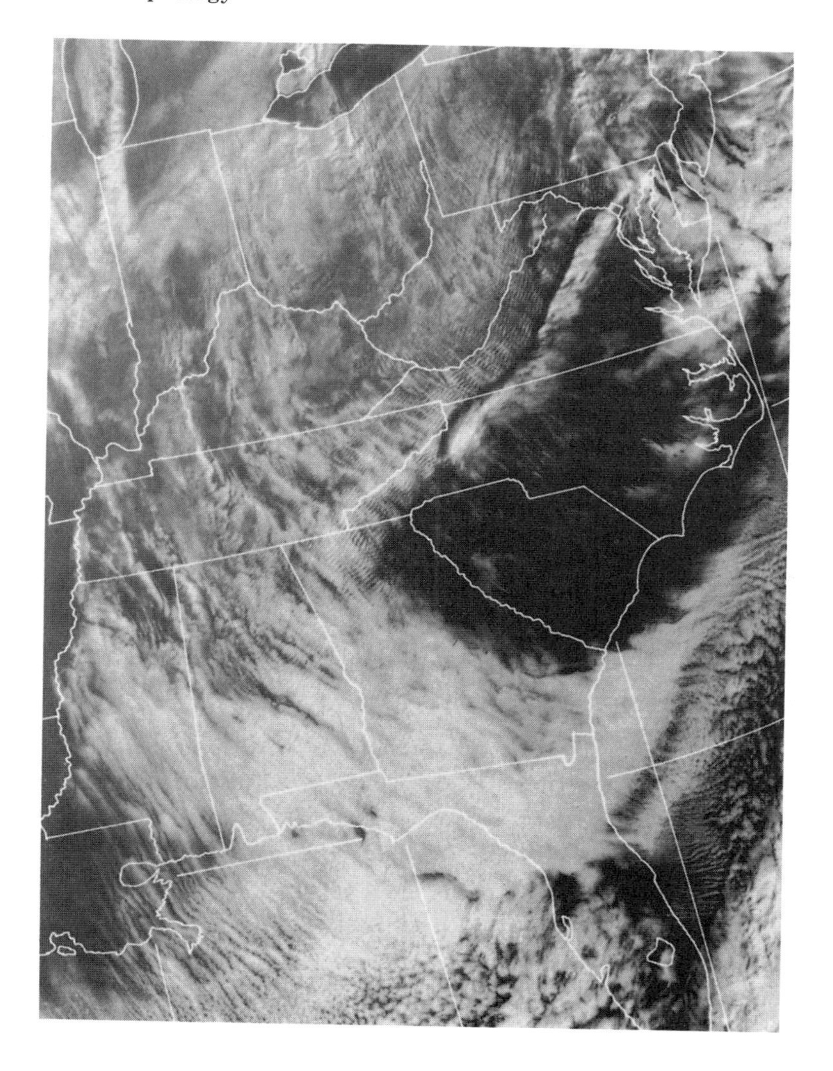

Fig. 5.18 NOAA-2 satellite photograph of the eastern United States showing extensive lee-wave clouds over the Appalachian Mountains. (Photo courtesy of Frances C. Parmenter and the Walter A. Bohan Co.)

regions that sailplanes have achieved record-breaking altitudes. Mountains produce disturbances in the streamlines to the greatest altitudes when the wind has a strong velocity component across the length of a long ridge. In this case, the waves can extend up to heights of 30 km where they occasionally produce nacreous clouds. Also, lee waves produced by long ridges may extend for large distances downwind without significant decreases in amplitude, provided that other mountains do not interfere. Isolated conical peaks generally produce

Fig. 5.19 Wave clouds over isolated peaks. Note the tendency of the clouds in the foreground to form a stack of disks. (Photo: P. V. Hobbs.)

disturbances only up to altitudes of 2–3 km and the amplitudes of the lee waves decrease fairly rapidly downwind of the mountain. However, isolated peaks can produce rather spectacular wave clouds within a wedge-shaped region fanning out from the mountain (Fig. 5.19).[†]

In the case of fairly high mountains, the winds that descend down the leeward slopes can be very strong and are known as *Föhn* winds in the Alps, *Chinooks* in North America, and by various other names in other localities. Also, farther downwind from the mountain, the winds may reverse their direction near the

[†] Note the resemblance of the disk-shaped wave clouds in Fig. 5.19 to the popular conception of a flying saucer. The first modern "sighting" of a flying saucer (1947) was made over Mt. Rainier, Washington, where disk-shaped wave clouds are very common.

ground due to the formation of a vortex, and produce what are known as a *rotor* and a *rotor cloud* (Fig. 5.15), which are very turbulent.

Early theories of mountain lee waves interpreted them as buoyancy oscillations in a stably stratified atmosphere, as illustrated in the following problem.

Problem 5.2 (a) Derive a formula describing the motion of an air parcel displaced from its equilibrium level in an unsaturated, stably stratified atmosphere, which is otherwise at rest.

(b) For what wind speed will resonance occur in air flowing over sinusoidal terrain with ridges 10 km apart, given a lapse rate of 5 deg km^{-1}.

Solution Let δz be the vertical displacement of an air parcel about its equilibrium level. Applying Newton's Law to a unit mass, we have

$$\frac{d^2}{dt^2}(\delta z) = F$$

where F is the net restoring force per unit mass associated with the static stability of the atmosphere. Substituting for F from the results of Problem 2.12, we obtain

$$\frac{d^2}{dt^2}\delta z + \frac{g}{T}(\Gamma_d - \Gamma)\,\delta z = 0$$

the solution of which describes a sinusoidal up and down motion called a *buoyancy oscillation* with period $2\pi/N$, where

$$N = \left[\frac{g}{T}(\Gamma_d - \Gamma)\right]^{1/2}$$

is called the *Brunt*†—*Väisälä*‡ *frequency*. In order for resonance to occur, the wind speed U must be such that the time τ required for the air to pass over one wavelength L of the mountain range is exactly equal to the period of a buoyancy oscillation; that is,

$$\tau = \frac{L}{U} = \frac{2\pi}{N}$$

or

$$U = \frac{NL}{2\pi} = \frac{L}{2\pi}\left[\frac{g}{T}(\Gamma_d - \Gamma)\right]^{1/2}$$

Substituting $L = 10$ km, $\Gamma = 5$ deg km^{-1}, $\Gamma_d = 9.8$ deg km^{-1}, and roughly estimating T as 250°K, we obtain $U \simeq 22$ m s^{-1}.

† **Sir David Brunt** (1886–1965) English meteorologist. First full time professor of meteorology at Imperial College, 1934–1952. His textbook "Physical and Dynamical Meteorology" (Cambridge University Press, London, 1934; 2nd ed., 1939) was one of the first modern unifying accounts of meteorology.

‡ **Vilho Väisälä** (1899–1969) Finnish meteorologist. Developed a number of meteorological instruments, including a version of the radiosonde in which readings of temperature, pressure, and moisture are telemetered in terms of radio frequencies. The modern counterpart of this instrument (manufactured by Väisälä Oy/Ltd.) recently won recognition as one of Finland's most successful exports.

It is now known that the vertical profiles of wind and temperature and the shape of the mountain also play important roles in the formation of orographic wave phenomena. The most favorable conditions for the formation of orographic waves are an increase in wind speed and a decrease in static stability from the lower to the upper troposphere. The stabilities associated with temperature inversions are particularly conducive to the formation of waves at the level of the inversion. If the width of the mountain range is comparable to the natural wavelength of the waves, the amplitude of the waves is increased by the resonance effect described above. Thus, lee waves are commonly observed when air flows over a mountain range with a width of 1–20 km, but flow over mountain ranges with greater widths does not, in general, produce lee waves, although such broad ranges may produce mountain waves to high levels in the atmosphere.

5.2 THE AIR-MASS THUNDERSTORM

It is only over the relatively warm and humid regions of the earth that cumulonimbus clouds commonly grow to thunderstorm proportions and assume major importance as rain producers. Thunderstorms occur widely over the more humid regions of the tropics and within warm, marine air-masses that drift over middle-latitude, continental regions during summer; for example, air from the Gulf of Mexico, which often drifts northward over the eastern United States. In mid-latitudes these isolated summer thunderstorms are frequently called *air-mass thunderstorms*, to distinguish them from the more highly organized thunderstorm complexes that occur in association with synoptic-scale disturbances (the subject of Section 5.3).

Air-mass thunderstorms were the subject of an intensive field program known as the Thunderstorm Project, which was carried out over Florida and Ohio during the late 1940s. Data on a large number of individual storms were composited together to construct an idealized model of the life cycle of a typical air-mass thunderstorm cell. In this model, which is depicted in Fig. 5.20, the life cycle of a cell is shown in terms of three stages. Most air-mass thunderstorms consist of several such cells, which grow and decay in succession, each having a lifetime of about half an hour.

In the "cumulus stage" (Fig. 5.20a) the cloud consists entirely of a warm, buoyant plume of rising air. The updraft velocity increases rapidly with height within the cloud and there is considerable entrainment through the lateral cloud boundaries. The top of the cloud moves upward with a velocity on the order of 10 m s^{-1}. Because of the large updraft velocities, supercooled raindrops may be present well above the freezing level (a situation potentially hazardous to aircraft because of the possibility of icing).

The mature stage of the life cycle of an air-mass thunderstorm (Fig. 5.20b) is characterized by a vigorous downdraft circulation which coincides with the region of heaviest rain. This downdraft circulation is initiated by the down-

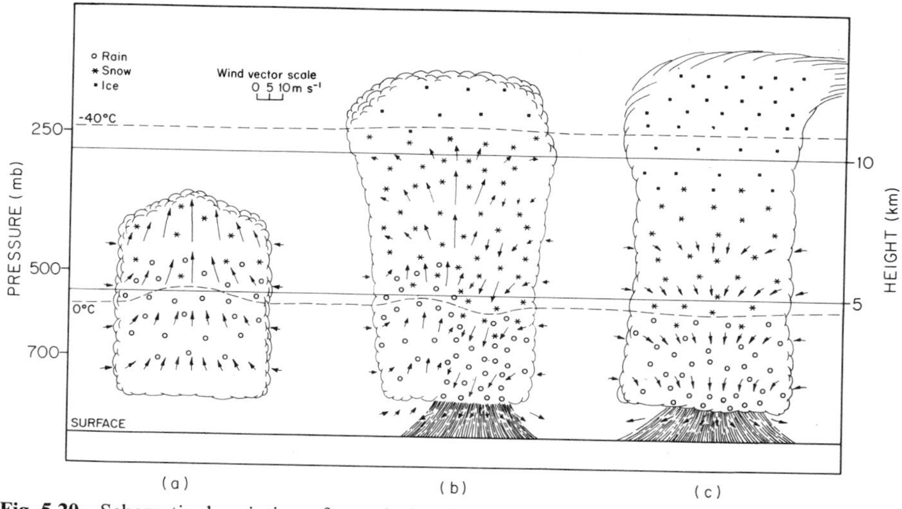

Fig. 5.20 Schematic description of a typical cell of an air-mass thunderstorm in three stages of its life cycle showing (a) cumulus stage, (b) mature stage, and (c) dissipating stage, based upon results of the Thunderstorm Project. The horizontal scale is compressed by about 30% relative to the vertical scale in the figure. (Adapted from "The Thunderstorm," U.S. Government Printing Office, 1949.)

ward frictional drag force induced by the drops.† Dry environmental air entrained into the downdraft circulation (on the right-hand side of Fig. 5.20b) and unsaturated air below cloud base are cooled by evaporation of falling precipitation. In some cases the resulting evaporative cooling is capable of greatly enhancing the negative buoyancy of the downdraft circulation. In the mature stage, supercooled raindrops still exist well above the freezing level in the updraft while snowflakes or soft hail pellets may be found below the freezing level in the downdraft. The maximum upward vertical velocities are in the middle of the cloud, with detrainment above that level. The top of the cloud approaches the tropopause and begins to spread out.

As precipitation develops throughout the cloud, the downdraft circulation gradually becomes more extensive until, in the dissipating stage (Fig. 5.20c), it occupies virtually the entire cloud. Deprived of a source of supersaturated updraft air, cloud droplets can no longer grow and, as a consequence, precipitation soon ceases and the cloud debris evaporates.

Data from the Thunderstorm Project indicate that only about 20% of the water vapor condensed in the updraft actually reaches the ground in the form of precipitation. The remainder evaporates in the downdraft or is left behind as cloud debris (including extensive patches of anvil cirrus) to evaporate into the ambient air.

† It is readily shown that the downward force due to the presence of the drops is independent of their terminal velocity and is simply equal to their weight. (See Problems 5.3k and 5.7.)

The air-mass thunderstorm cell is short lived and rarely produces destructive winds or hail because it contains within itself a fail-safe "self-destruct mechanism": namely, the downdraft circulation induced by the rainshaft. In the absence of vertical wind shear the thunderstorm has no way of ridding itself of the precipitation it produces without destroying the buoyant updrafts that feed it.

The distribution of electrical charges in thunderstorms, theories for their development, and the properties of lightning and thunder have been described in Section 4.6.

5.3 SEVERE STORMS

The vast majority of severe thunder storms that produce flash floods, large hail, high winds, and tornadoes form only in a convectively unstable environment in which there is considerable vertical wind shear between the low-level flow and the upper troposphere. In such an environment, convective precipitation systems can develop rapidly and persist for long periods of time in their mature stage without destructive interference between updrafts and downdrafts. Severe storms generally exhibit some type of mesoscale organization. Among the types frequently observed are the *squall line*, the *multicell storm*, and the *supercell storm*, each of which can be described in terms of an idealized model that provides some insight into the nature and causes of severe weather phenomena.

5.3.1 Squall lines

a. General description

Cumulonimbus towers are often arranged in long lines in which adjacent elements are so close together that to a first approximation the mesoscale system can be regarded as a "line thunderstorm." Such storms are often accompanied by gusty surface winds; hence the name *squall line*. They are frequently observed during summer over West Africa, south of the Sahara, and over certain middle-latitude land areas, including much of the central and eastern United States.

An idealized vertical profile of the environmental winds in the direction normal to a squall line is depicted in Fig. 5.21b, which applies to a squall line moving from left to right across the page. Note that the wind component in the direction of motion of the storm increases rapidly with height. The squall line moves with a speed characteristic of the winds in the middle troposphere; hence, it overtakes air in the prestorm environment at low levels, while it acts as a massive, slow-moving obstacle to the winds in the upper troposphere.

The idealized, two-dimensional air motion relative to the squall line is shown in Fig. 5.21c. The arrows directed into the storm from the right at low levels represent environmental air being overtaken by the squall line. This layer of warm, moist air is usually capped by a weak inversion which prevents widespread convection from breaking out in the prestorm environment. The ad-

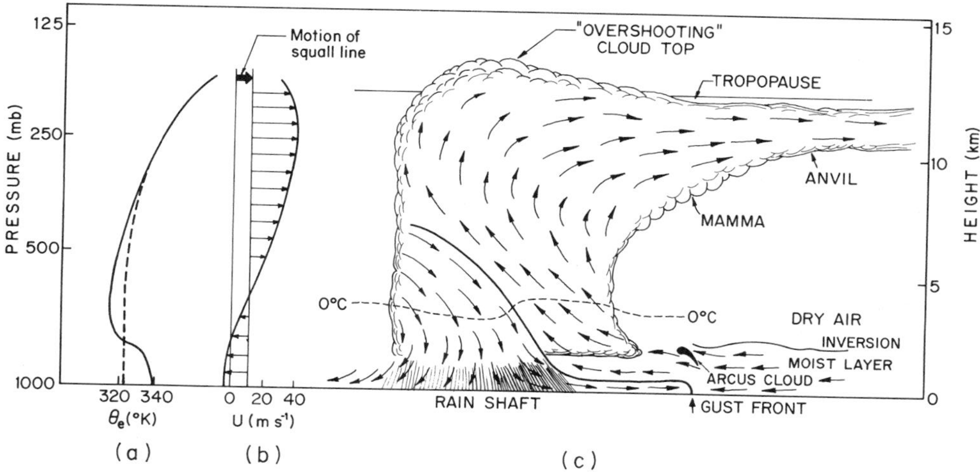

Fig. 5.21 Schematic description of a typical squall line moving from left to right, as indicated. (a) Vertical profile of equivalent potential temperature in prestorm environment (——) and behind squall line (– – –). (b) Vertical profile of wind component in the direction of motion of the squall line. The velocity scale is relative to the ground, while the arrows represent velocities relative to the squall line. (c) Cloud outline and air motions relative to the squall line.

vancing wedge of cold, downdraft air generated within the squall line lifts this warm, moist layer upward, beyond its lifting condensation level (which corresponds to the dark, sharply defined cloud base at the leading edge of the storm) up to its level of free convection, beyond which the air rises spontaneously under the force of its own buoyancy until it approaches the tropopause. As this air rises in the updraft, massive amounts of moisture are condensed out.

The arrows directed into the storm from the left at middle levels represent dry, environmental air with low equivalent potential temperature which is overtaking the squall line from the rear. As precipitation falls from the updraft into this dry air, it evaporates at a rapid rate and, in so doing, it lowers the temperature of the air toward its wet-bulb temperature. As the air cools it becomes negatively buoyant relative to its surroundings and begins to sink. The frictional drag of the falling precipitation produces an additional downward force, thus strengthening the downdraft.

Due to the presence of strong horizontal pressure gradient forces within the storm, neither the updraft nor the downdraft air maintains its original horizontal momentum relative to the squall line. The updraft air eventually acquires a high velocity toward the right, as shown in Fig. 5.21c, and spreads out ahead of the storm, creating an extensive anvil composed of evaporating ice crystals. To an observer on the ground the appearance of the anvil is usually the first sign of the approaching storm. Much of the cool, rather moist, downdraft air is left behind the storm in a shallow layer adjacent to the ground which may persist for many hours. Squall lines modify their large-scale environment by transporting vast quantities of air with high equivalent potential temperature upward from near the earth's surface and replacing it with downdraft air, which has a much lower equivalent potential temperature. The net effect of this vertical exchange is to reduce or eliminate the vertical gradient of equivalent potential temperature in the lower troposphere (see Fig. 5.21a), thus removing part or all of the convective instability that was present in the prestorm environment.

In contrast to their adversary roles in the air-mass thunderstorm, updrafts and downdrafts play complementary roles in the maintenance of squall lines over long periods of time. The downdraft is maintained by the precipitation falling into it from the updraft; the updraft, in turn, is maintained by the advancing wedge of downdraft air at the surface, which provides a mechanism for lifting the low-level air ahead of the storm up to its level of free convection. Such cooperation is possible only when the updraft and downdraft exhibit a strong slope as shown in Fig. 5.21c. Strong vertical wind shear in the squall line environment is a prerequisite for the development of such a sloping structure.

b. The gust front

A distinctive feature of most squall lines is the *gust front* (or *first gust*) which marks the leading edge of the advancing wedge of cold, downdraft air. In essence, the gust front is a form of "gravity current" or "density current" which can be simulated in the laboratory by allowing a layer of dense fluid to spread out laterally, under the action of gravity, along the bottom of a channel filled with a slightly lighter fluid. In many respects, the gust front resembles a sharp, mesoscale cold front. Its passage is marked by a pronounced windshift and the beginning of a brief period of sharply falling temperatures. The low-level outflow from the downdraft is usually strongest and deepest just behind the gust front, where surface wind gusts often exceed 25 m s^{-1} in the more intense storms.

The position of the gust front relative to the rest of the squall line is highly variable. In newly developing storms it tends to lie just ahead of the leading edge of the rain area, whereas in decaying storms it may outrun (or outlast) the cumulonimbus clouds that generated it and appear as an isolated phenomenon. Gust fronts are not unique to squall lines; they occur in conjunction with other types of large thunderstorm complexes. Weak gust fronts have even been observed in association with the passage of sharp, shallow cold fronts.

Under certain conditions the advancing layer of cold, dense air associated with the gust front is clearly visible to the ground-based observer. In regions of dry, exposed soil, strong surface winds immediately behind the gust front pick up large quantities of dust particles and distribute them throughout the layer of cold air. The resulting dust cloud clearly outlines the leading edge of the cold downdraft air, as shown in Fig. 5.22b. When sufficient moisture is present, condensation may take place along the leading edge of the gust front, where warm air is being lifted up and over the advancing dome of downdraft air, as shown in Fig. 5.22a. The resulting *arcus cloud* (also commonly called *roll cloud*) can be quite impressive, especially when it appears with a menacing cumulonimbus cloud in the background. The example shown in Fig. 5.22a was associated with thunderstorm outflow, while the one shown in Fig. 5.23 was

(a)

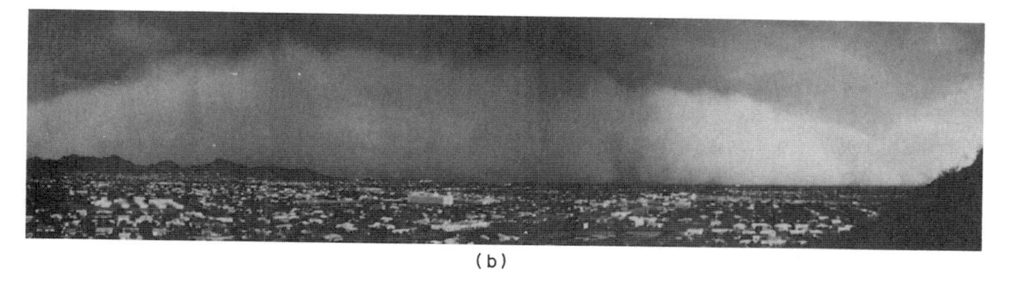

(b)

Fig. 5.22 Phenomena associated with gust fronts produced by downdrafts of thunderstorms: (a) arcus cloud over southwestern Minnesota; (b) dust cloud over Phoenix, Arizona. (Photos: S. B. Idso and C. Keyes.)

Fig. 5.23 Approaching arcus cloud associated with shallow cold front passing southward over Oklahoma. (Photo: R. L. Livingston.)

observed in conjunction with the passage of a shallow cold front which was not accompanied by thunderstorms. Arcus clouds can often be seen in satellite film loops, radiating outward from thunderstorm complexes. Intersections between arcus clouds, or between arcus clouds and fronts, appear to be preferred locations for the development of new thunderstorms.

5.3.2 Right-moving multicell storms

In the *multicell storm*, some degree of mesoscale organization is present, and yet most of the updrafts and downdrafts can still be identified with individual thunderstorm cells similar to those described in Section 5.2. Multicell storms come in a wide variety of shapes, sizes, and intensities. Included within this category are most air-mass thunderstorms as well as a large number of storms that might be classed as "severe." In this section we will describe a particular type of multicell storm structure which is characterized by thunderstorm movement directed systematically toward the right of the environmental air flow in the middle troposphere. Many of the severe storms that form over the central United States during summer exhibit this type of motion.[†] Through analyses of the movements of radar echoes it has been determined that the individual thunderstorm cells in such storms often move with the mid-tropospheric winds, while the whole multicell storm drifts toward the right of the winds. This peculiar type of behavior is related to the characteristic clockwise turning of the environmental wind with increasing height during severe storm situations.

We recall from Section 2.7.3 that, from a thermodynamic point of view, the synoptic conditions most favorable for the development of severe storms over the central United States are characterized by a moist, southerly flow from the Gulf of Mexico at low levels, with a drier westerly or southwesterly flow aloft, as depicted in Fig. 5.24a. Note the clockwise turning with increasing height. Now in many of these "severe right-moving storms" it is observed that the arrangement and movement of the individual thunderstorm cells is closely related to the mid-tropospheric wind vector $\mathbf{V}_M$, as shown in Fig. 5.24b. The cells are lined up perpendicular to $\mathbf{V}_M$ and their movement is in approximately the same direction as $\mathbf{V}_M$.

Because of the veering of the wind with height, the low-level inflow takes place preferentially along the right flank of the storm (that is, on the side facing to the right of the movement of the cells). The narrow band along which the inflow air is lifted up and over the cold downdraft air generated within the storm is a favored location for the development of new cells. As each new cell develops on the right flank of the storm, the older cells find themselves further and further removed from the source of the low-level inflow. The oldest cells on the left side of the storm are in a poor position to compete for the warm, moist inflow air

[†] Storms which propagate rapidly toward the left of the environmental flow in the middle troposphere have also been observed. On some occasions, single multicell storms have been observed to split into left- and right-moving storms.

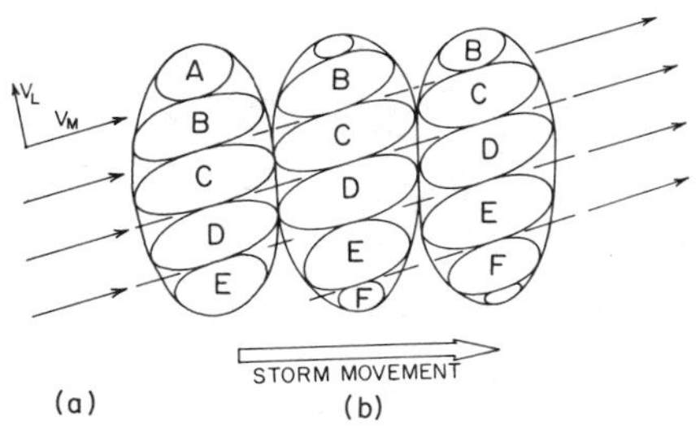

Fig. 5.24 Schematic illustration of right-moving multicell storm. (a) Vectors showing lower tropospheric (V_L) and midtropospheric (V_M) winds relative to the ground. (b) Positions of individual (lettered) cells at three successive times about 15 min apart. Note the development of new cell G and the dissipation of cells A and B.

needed to sustain their buoyant updrafts. As their updrafts weaken, these cells gradually dissipate.

The continuous generation of new cells on the right flank of the thunderstorm complex, together with the dissipation of old cells on the left flank, gives the storm an effective propagation toward the right of the mid-tropospheric winds.

5.3.3 Supercell storms

a. Structure

In a certain small fraction of severe storms, the mesoscale organization is so pervasive that the entire storm behaves as a single entity, rather than as a group of cells. It is these so-called *supercell storms* that account for most tornadoes and damaging hail. Most supercell storms move toward the right of the environmental winds. In this respect, the distinction between the multicell and supercell storm is that the former propagates toward the right in discrete jumps as individual cells form and dissipate, whereas the latter exhibits a continuous propagation.

Figure 5.25 shows several supercell storms in various stages of development as viewed by geostationary satellite. The topmost (and largest) cloud mass is associated with a dissipating storm. The smaller, but more solid cloud masses immediately below it are developing supercell storms, one of which produced a large tornado over Neosho, Missouri, several hours after the time of this photograph.

Figure 5.26 shows an intense supercell storm as viewed from a reconnaissance aircraft flying near the 500-mb level. The picture is taken looking toward the east. The updraft is clearly visible along the south (right) flank of the storm. Soon after this picture was taken the aircraft encountered large hail while attempting to fly below the anvil, to the left of the updraft.

A distinguishing feature of the supercell storm is the hook-shaped radar echo which typically forms along the right flank as shown in Fig. 5.27a. The echo-free region adjacent to the hook echo in Fig. 5.27a corresponds to the updraft.

Fig. 5.25 Supercell storms over Missouri 1725 CST 24 April 1975, as viewed by geostationary satellite at visible wavelengths, looking toward the north. The horizontal scale of the larger storms is on the order of 40 km and the horizontal resolution of the image is about 1 km. Some individual thunderstorm cells are visible as small, very bright cloud masses on the west side of the supercell storms. (NOAA Photo.)

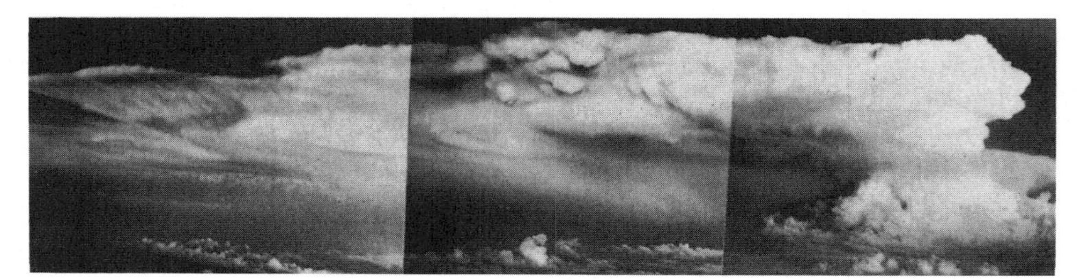

Fig. 5.26 A supercell storm near Salina, Kansas, 1915 CST 5 May 1965, as viewed from U.S. Weather Bureau reconnaissance aircraft, flying near the 500-mb level. The view is toward the east. The "tower" of dense clouds along the south (right-hand) side of the storm corresponds to the updraft. Mamma can be seen protruding downward from the anvil. (Photo: T. Fujita.)

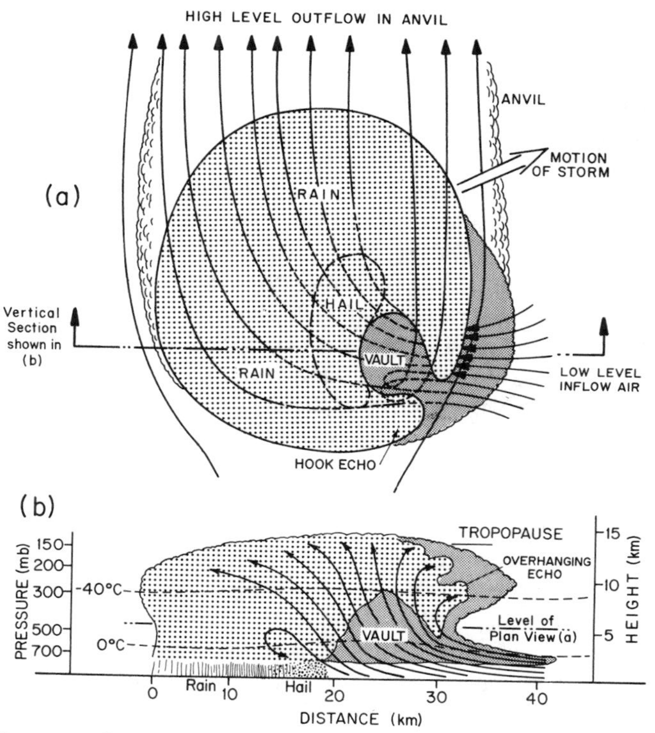

Fig. 5.27a,b Structure of a supercell storm (a) as seen from above and (b) as seen in a vertical cross section looking downstream in the direction of the midtropospheric winds. Areas of heavy stippling correspond to radar echoes; lighter stippling denotes clouds composed of small (non-reflective) cloud droplets. Arrows show projections of air motions in the updraft relative to the storm. Note that the arrows in (a) represent motion of air entering the storm at low levels and leaving at high levels, and that in (b) there is a component of air motion perpendicular to the page. [Adapted from *Quart. J. Roy. Met. Soc.* **102**, 499 (1976).]

Within this so-called *vault* region upward air velocities on the order of tens of meters per second sweep cloud droplets aloft to near the $-40°C$ level before they have time to grow large enough to reflect radar waves; hence the conspicuous absence of radar echoes. The dense, rather dark cloud of small water droplets in the vault region extends downward to the lifting condensation level where a well-defined cloud base is observed. The largest hail falls to the ground in a narrow band around the periphery of the vault.

Just as in the case of the squall line, the vertical shear of the environmental flow plays a crucial role in maintaining the sloping updraft, which is essential for the maintenance of the storm. The gust front marks the leading edge of the cold downdraft air which overtakes warm moist air adjacent to the earth's surface and lifts it to its level of free convection. In the more intense storms this ascending air acquires a pronounced counterclockwise rotation as it ascends through the vault region. The hook echo in Fig. 5.27a and the overhanging echo in Fig. 5.27b consist of large cloud droplets swept around the periphery of the vault by this counterclockwise circulation.

b. Tornado formation

Recent studies of severe storms over Oklahoma indicate that many tornadic storms are characterized by updrafts which show a pronounced counterclockwise (cyclonic) rotation through a deep layer extending from cloud base up to almost 10 km. These so-called *mesocyclones* are often identifiable in Doppler-radar returns and some of the stronger ones are even evident to ground-based observers watching the motions of clouds in the vicinity of the vault. Preliminary evidence, based on a small number of well-documented cases, suggests that the existence of a mesocyclone is often a precursor to tornado formation, which takes place beneath the updraft. In the cases that have been observed, the typical time lag between the initial detection of the mesocyclone and the touching down of the associated tornado funnel is on the order of a half hour. Thus it appears that the circulations associated with tornadoes may "spin up" rather gradually. Wind velocities are typically on the order of tens of meters per second in mesocyclones and approach values of 100 m s^{-1} at the outer periphery of tornado funnels. As far as can be determined, nearly all tornadoes rotate cyclonically[†] (in the same sense as the earth's rotation), while about 90% of observed waterspouts rotate cyclonically and dust devils seem to have no preferred sense of rotation. Apparently, the larger the system and the longer its lifetime the more influential is the earth's rotation in determining the sense of its rotation. The basis for this apparent relationship is discussed in Section 8.4.

c. Hail trajectories

It has long been hoped that detailed analysis of the structure of hailstones would provide definitive evidence concerning the environmental conditions under which they grew, and that from such evidence it might be possible to infer the trajectories of hailstones throughout their growth phase. However, as pointed out in Section 4.5.3, although the structure of a hailstone can provide some information on its modes of growth, the interpretation is by no means unambiguous. Therefore, the way in which the cloud dynamics controls the process of hail formation is still a matter of some controversy. Here we will examine one particular conceptual model that has been proposed in an attempt to explain the growth of large hailstones in supercell storms.

Figure 5.27c, d shows the trajectories of three hypothetical precipitation particles which might have grown in the supercell storm described above. Trajectory A is for a comparatively small particle which eventually exits from the storm through the anvil, B is the trajectory of a raindrop, and C is that of a large hailstone.

A particle following trajectory A, in the region of strong updraft velocity, is still considerably less than a millimeter in diameter by the time it reaches the

[†] There exists at least one well-documented case of an anticyclonically rotating tornado funnel.

$-40°C$ level, beyond which any further growth is unlikely. The terminal fall speed of this small particle is much less than the updraft velocity, and therefore its trajectory through the cloud is much the same as that of an air parcel. Most of the water vapor in the updraft condenses onto droplets that follow trajectories similar to A, which exit through the anvil. In this sense the supercell storm is rather inefficient as a mechanism for producing precipitation.

The precipitation particles located near the left-hand edge of the updraft in Figs. 5.27a and 5.27c are in a favored position for growth. In the first place, the updraft velocities in this region are considerably smaller than those near the center of the updraft, and therefore these particles have time to grow to about a millimeter in diameter before they reach the $-40°C$ level. As these particles execute the counterclockwise loop which takes them around the right-hand side of the vault, they encounter updraft velocities so small that they fall back into the main updraft as indicated in Fig. 5.27d. The overhanging echo on the right-hand side of the vault is largely due to the presence of these recycled particles which, by this time, are much larger than the particles making their first ascent through this region. After reentering the updraft, the particles are carried across the top of the vault, toward the left in Fig. 5.27d, growing rapidly in the presence of the extremely high liquid water contents in the updraft. The particles following trajectories B and C are assumed to have encountered similar environmental conditions up to this point in their life histories, the only difference being that

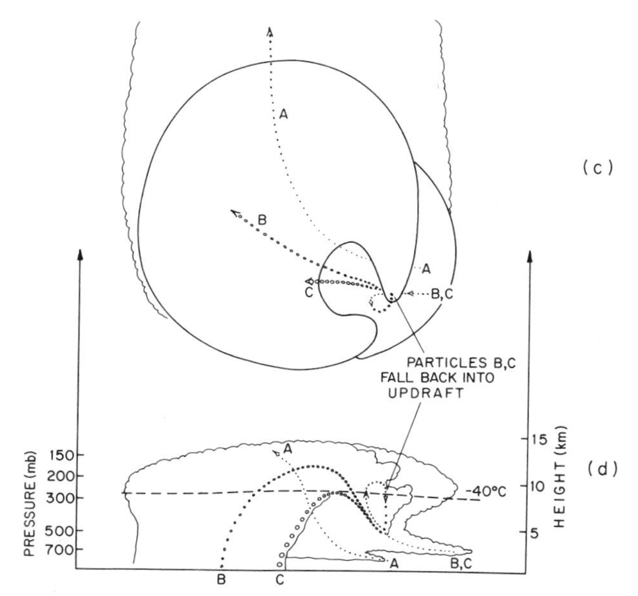

Fig. 5.27c,d Hypothetical trajectories of precipitation particles in the supercell storm described in Fig. 5.27a,b: (c) as seen from above and (d) as seen in a vertical cross section looking downstream in the direction of the midtropospheric winds. A represents a small ice crystal, B a raindrop, and C a large hailstone. [Adapted from *Quart. J. Roy. Met. Soc.* **102**, 499 (1976).]

the particle following trajectory C is assumed to be slightly larger. This small difference in size causes the two trajectories to diverge as they ascend toward the top of the vault. The heavier particle C has a higher terminal fall speed and therefore the updraft does not carry it as high. The minor difference in the heights of the trajectories has a profound influence upon the subsequent growth of the two particles. As the larger particle crosses the vault it falls through a region of extremely high liquid water content, while the smaller particle crosses over the vault at a level above the $-40°C$ isotherm, the upper limit of the liquid water. Thus, by the time the particles reach the ground, the one that followed trajectory B is still the size of an ordinary raindrop, small enough to melt on its way down, while the one following trajectory C, because it happened to be a little larger at a crucial point in its life history, grows into a large hailstone. (There may, perhaps, be some interesting analogies here with the life histories of certain successful and not-so-successful financial tycoons, or even scientists!)

The hail formation mechanism described above provides a plausible explanation for

- the well-defined "ceiling" of the vault, which presumably coincides with the trajectories of the lowest (and largest) hailstones;
- the position of the hail falling at the ground, relative to the vault;
- the fact that the large hailstones have distinct "embryos" several millimeters in diameter, which seem to have grown in a different part of the cloud from the rest of the hailstone.

5.4 HURRICANES

In hurricanes[†] convective precipitation is organized on two distinct scales: individual convective cells are grouped in mesoscale bands and the bands, in turn, are arranged in characteristic configurations on the scale of the hurricane itself. Because of its higher level of organization, the hurricane is far more efficient at producing precipitation than any of the mesoscale systems described in the previous section.

Most hurricanes display a high degree of circular symmetry in their pressure pattern and in the appearance of the cirroform cloud shield as viewed from satellites (Fig. 5.28). Centered on the axis of rotation is a distinct, circular, cloud-free *eye*, 20–50 km in diameter. Whirling around the eye is a ring of towering cumulonimbus clouds called the *eye-wall cloud*. The hurricane's strongest winds and highest rainfall rates are found within this ring. In intense hurricanes sea-level pressures as low as 880 mb have been observed within the eye, and sustained winds approaching 100 m s^{-1} have been recorded near the base of the eye-wall cloud, together with rainfall rates on the order of 10 cm h^{-1}

[†] Strictly speaking, the term hurricane applies only to storms in the Atlantic and Caribbean. Tropical cyclones in other parts of the world are known by various other regional names; for example, *typhoon* (western Pacific), and *cyclone* (Bay of Bengal).

Fig. 5.28 Hurricanes "Ione" (left) and "Kirsten" (right) in the eastern Pacific, 24 August 1974, as seen in a high-resolution visible image from a polar orbiting satellite. "Eyes" are partially visible near the center of both storms. The white, rather uniform cloud masses surrounding the eyes represent the tops of cirroform anvils generated by convective clouds in the eye-wall clouds and spiral rainbands. At larger distances from the storms convective clouds can be seen, arranged in spiral bands. Note also the cellular convection in the region of calm winds at the upper left. (NOAA Photo.)

Data from aircraft reconnaissance, weather radar, and "enhanced" satellite images, such as the one shown in Fig. 5.29, have revealed much concerning the detailed structure of hurricanes. In most of these storms the distributions of wind and rainfall exhibit large deviations from axial symmetry. Often the eye-wall cloud does not form a complete ring around the eye. In such situations the peak wind speeds observed just outside the eye may vary by as much as a factor of two between the eye-wall cloud and the open sector. At larger distances from the eye, most of the heavy rain and strong winds are concentrated within spiral *rainbands* such as those shown in Fig. 5.29. These bands are usually quasi-stationary with respect to the moving storm. The strongest band and the

Fig. 5.29 Visible satellite photo of Hurricane "Camille" (1969). The image has been processed to enhance the contrast between the brightest areas, which represent convective clouds, and areas not quite as bright, which represent high, layered overcast. (Courtesy of NOAA and T. Fujita.)

most strongly developed segment of the eye-wall cloud are usually found in the quadrant of the storm centered about 45° to the right (left) of the direction of storm movement in the Northern (Southern) Hemisphere.

An idealized radial cross section through a moderately intense hurricane is shown in Fig. 5.30. The distribution of cloud and precipitation is depicted on the left-hand side of the figure. The cumulonimbus towers represent the rain-bands and the eye-wall cloud. The strong updrafts within these rain areas are fed by the radial influx of mass and moisture within the subcloud layer, as indicated by the heavy arrows. The outflow from the updrafts exits from the hurricane by way of an immense cirroform anvil, which comprises the outline of the storm in the satellite photographs.

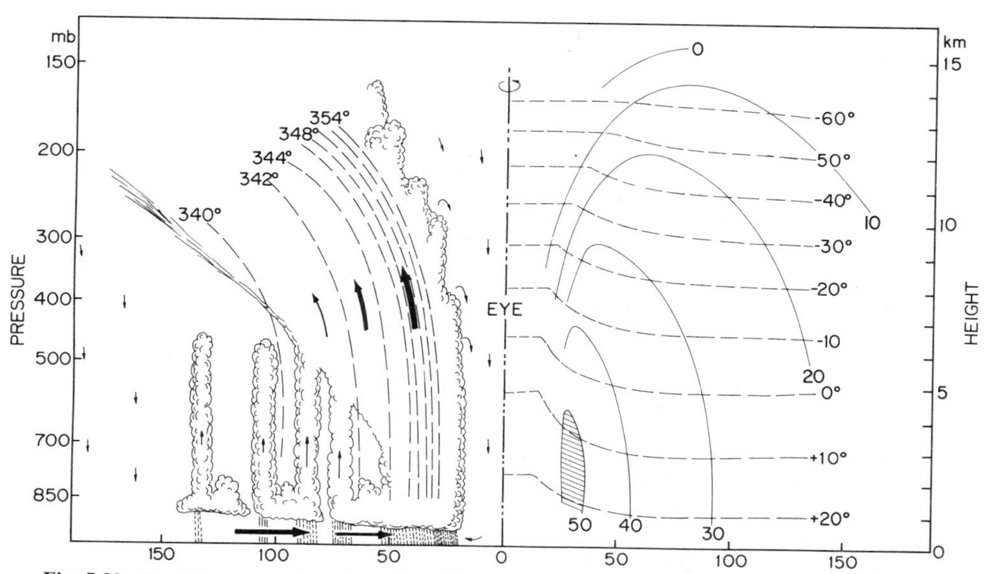

Fig. 5.30 Radial cross section through an idealized, axially symmetric hurricane. On left: radial and vertical mass fluxes are indicated by arrows, equivalent potential temperature in degrees Kelvin by dashed lines. On right: tangential velocity in meters per second is indicated by solid lines and temperature in degrees Celsius by the dashed lines. (Adapted from E. Palmén and C. W. Newton, "Atmospheric Circulation Systems," Academic Press, New York, 1969, p. 481.)

The air spiraling inward within the subcloud layer undergoes a steady increase in equivalent potential temperature (θ_e) due to the fluxes of water vapor and enthalpy from the sea surface. These fluxes are greatly enhanced by the strong winds and the highly disturbed sea states that prevail in the vicinity of the hurricane. In the example shown in Fig. 5.30, θ_e increases by 14° from the periphery of the hurricane to the eye-wall cloud. The air ascending moist-adiabatically in the updrafts and spiraling outward in the anvil maintains a nearly constant value of θ_e. Hence above the subcloud layer the isentropes of θ_e closely parallel the air trajectories (the arrows in Fig. 5.30).

The high equivalent potential temperatures in the interior of the storm are reflected in the temperature distribution shown on the right-hand side of Fig. 5.30. At any given level up to about 200 mb, the air within the eye is about 10° warmer than the air outside the hurricane at the same level.

Also shown on the right-hand side of Fig. 5.30 is the distribution of tangential wind speed. The strongest winds are found near the base of the eye-wall cloud. Wind speed decreases rapidly with height through the middle troposphere. In the upper troposphere the wind field is highly asymmetric with respect to the center of the hurricane, with the outflow air from the spiral rainbands often rotating in the opposite sense as the air between the bands. When averaged around the storm the tangential velocity at these levels is close to zero, as indicated in Fig. 5.30.

The hurricane circulation is generated and maintained by a cooperative interaction between the deep cumulus convection in the eye-wall cloud and spiral rainbands, and the large-scale circulation spiraling inward at low levels and outward at high levels. The low-level inflow provides a continuous source of warm moist air in the subcloud layer and forces it upward in the interior of the storm, while the latent heat released in the convection plays as essential role in maintaining the warm core structure which, in turn, is the energy source for the large-scale circulation. These interrelationships will be explained more fully in Section 9.4.3.

5.5 EXTRATROPICAL CYCLONIC STORMS

In Chapter 3 we described the synoptic-scale structure of extratropical cyclonic storms as manifested in the wind, pressure, and temperature fields. In this section we consider these same storms with emphasis upon their mesoscale structure, microphysical makeup, and associated cloud patterns.

5.5.1 Mesoscale structure and microphysical characteristics

Figure 5.31 shows the air motions and distribution of clouds in a mid-latitude cyclonic storm as envisaged in the classical Norwegian model which was described in Section 3.3.3. Ahead of the warm front on the ground, extensive layer clouds form as the air in the warm sector rides up over the denser, colder air. This overriding produces widespread and fairly uniform precipitation, which reaches a peak intensity just prior to the passage of the front. The microphysical structure of the clouds associated with a warm front, as depicted in the Norwegian model, is shown in Fig. 5.32. Ice crystals are assumed to nucleate at the lower temperatures in concentrations of about 1 liter^{-1} and to grow into precipitation particles by deposition from the vapor phase, as explained in Section 4.5.4. Following the passage of the warm front on the ground, the rain diminishes in intensity.

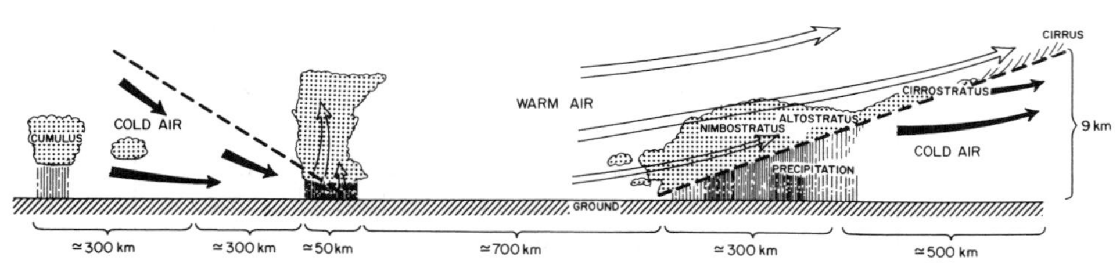

Fig. 5.31 Idealized vertical cross section (along AA′ in Fig. 3.11b) through a middle-latitude cyclone, according to the Norwegian model. (Note that the vertical scale is stretched by a factor of about thirty compared to the horizontal scale.)

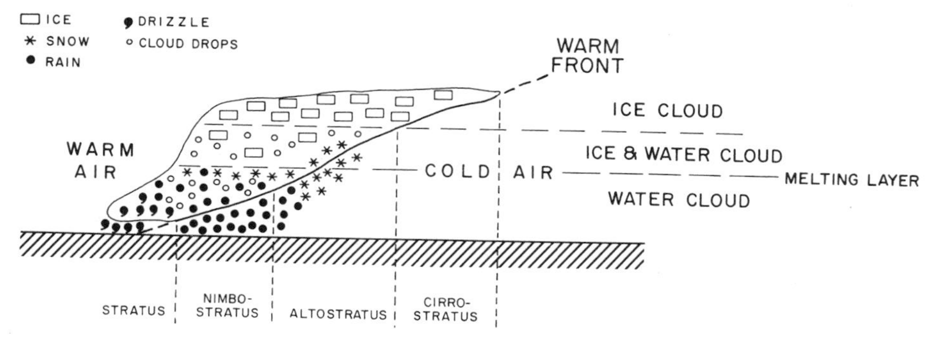

Fig. 5.32 The Norwegian model of the microphysical structure of a warm front.

The weather in the warm sector is highly dependent upon the moisture content of the warm air mass and the temperature of the underlying surface. If the air is humid and the underlying surface is cold, ground fog or stratus overcast is likely to be widespread. Such conditions are typical during winter, particularly at higher latitudes. On the other hand, if the warm air mass is relatively dry, or if the underlying surface is warm enough so that condensation does not take place at low levels, the warm sector may be mainly free of clouds. If the underlying surface becomes sufficiently warm (as, for example, on a summer afternoon) and if sufficient low-level moisture is present, isolated "air-mass" thunderstorms may break out in the warm sector.

The cold front is usually accompanied by another well-organized cloud system, which may take the form of layered clouds (altostratus and nimbostratus) or convective clouds (cumulonimbus), depending upon the static stability and the abruptness of the lifting that takes place when the dense cold polar air undercuts the warm air. Behind cold fronts the weather generally brightens up considerably, but in polar air of marine origin convective clouds can sometimes produce locally heavy precipitation. In the classical model of an occluded front, the two main cloud masses described above are merged into a single cloud system.

The structure and distribution of mesoscale rain areas within cyclonic storms has been documented, using radars, aircraft, and ground-based observations. Most of the precipitation in these storms is concentrated in mesoscale *rainbands* with typical horizontal areas on the order of 10^3–10^4 km^2 and lifetimes on the order of several hours. The more intense precipitation in rainbands is usually concentrated within a number of active regions, each of which contains several individual convective cells.

Much of the precipitation ahead of warm fronts tends to be concentrated in rainbands oriented parallel to the front. An example of such a band is indicated by the shaded area labeled A in Fig. 5.33. Figure 5.34 illustrates schematically how these bands might form. The precipitation is associated primarily with a stream of warm, moist air which originates in the warm sector, just ahead of the cold front. This stream of warm, moist air is confined below a layer of dry, subsiding air (indicated by the unshaded arrow) which is characterized by a lower equivalent potential temperature θ_e. Consequently, $d\theta_e/dz$ is negative and the air in the lower layers is convectively unstable. As the stream of warm, moist air rises over the warm front, the convective instability is released periodically in the mesoscale rainbands. Between the warm frontal rainbands, tongues of dry (low θ_e) air erode into the clouds.

If precipitation is present within the warm sector, it is likely to be associated with rainbands oriented parallel to the cold front; for example, note the band labeled B in Fig. 5.33. Cold frontal precipitation is usually concentrated in one or more rainbands oriented parallel to the front, as illustrated by bands C and D in Fig. 5.33. When the air behind a cold or occluded front is of marine origin,

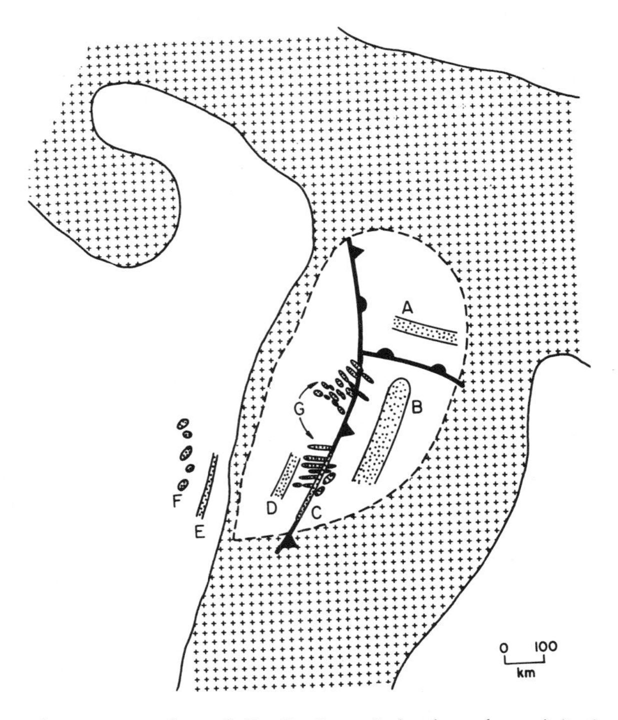

Fig. 5.33 Schematic representation of distribution of clouds and precipitation in a cyclonic storm which passed over Washington State on 27–28 November 1973. Cloud pattern observed by satellite is shown by (‡‡‡); not shown within dashed line for sake of clarity. The dashed line encloses the area where light rain fell: regions of heavy rain are shown by (:::); letters refer to features mentioned in text. Sea level fronts are shown. [Adapted from *Mon. Wea. Rev.* **104**, 868 (1976).]

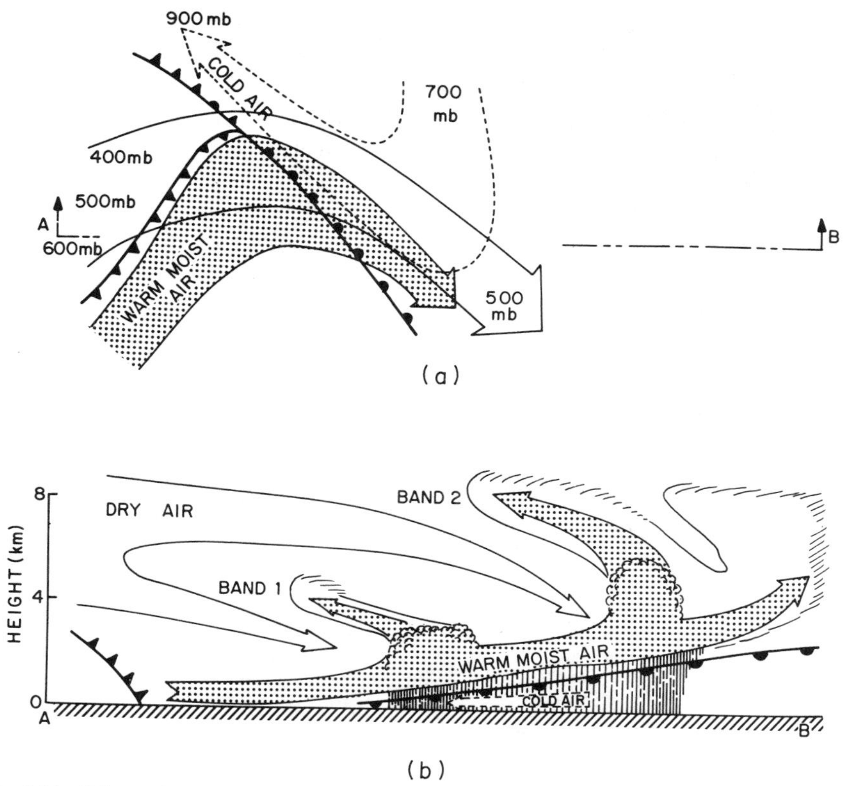

Fig. 5.34 Schematic diagram showing the airflow (relative to the moving frontal system) in a mature extratropical cyclonic storm. (a) Plan view, (b) vertical section along line AB in (a). [Adapted from *Quart. J. Roy. Met. Soc.* **99**, 228 and 240 (1973).]

widespread convective clouds may form. Sometimes this convective precipitation is organized into postfrontal rainbands (E and F in Fig. 5.33). Radar photographs revealing several postfrontal bands are shown in Fig. 5.35; the active regions within the bands are readily seen in Fig. 5.35a. Sometimes groups of smaller rainbands form in a wavelike pattern in frontal systems, such as those indicated by G in Fig. 5.33.

Ordinarily, the width of the major rainbands associated with cyclonic storms is on the order of tens of kilometers (e.g., bands A, B, and D in Fig. 5.33) and the rainfall rates within the bands are on the order of millimeters per hour. However, when conditions are favorable for organized, deep convection, the bands may be much narrower and the associated rainfall rates much higher. For example, band C in Fig. 5.33 was about 5 km in width and produced rainfall rates on the order of 30 mm h^{-1}. The squall line phenomenon discussed in Section 5.3.1 represents a very narrow, highly organized, and intense form of warm sector or cold frontal rainband.

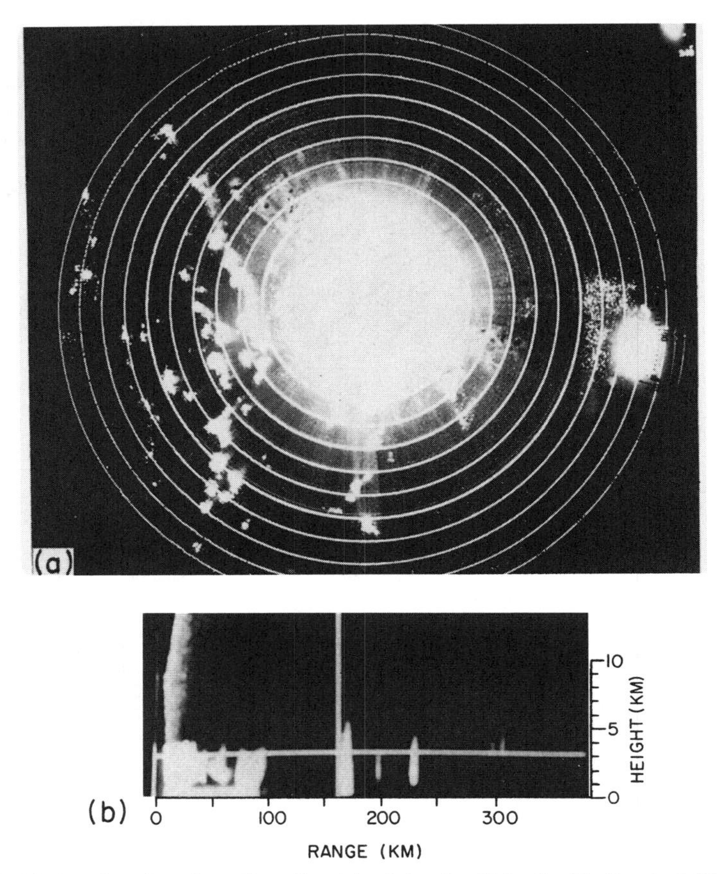

Fig. 5.35 (a) A radar plan view of postfrontal rainbands off the Pacific Coast of Washington. The radar, which is located in the center of the photograph, was situated on the Washington Coast and the bands can be seen to the west. The circles are range markers for every 10 nautical miles (18.5 km). The bright area in the center of the photograph is due to interference from the ground. (b) Corresponding range–height radar echoes along a vertical cross section looking west. Several rainbands can be seen: for example, one is at a range of about 175 km and extends from just above ground level to a height of about $5\frac{1}{2}$ km. The long uniform horizontal and vertical lines are calibration markers and the long fuzzy echo above the radar is due to interference. (Courtesy of Cloud Physics Group, University of Washington.)

Shown in Fig. 5.36 are streamlines of the airflow and the fluxes of water vapor in a vertical plane through a frontal system. These parameters were deduced from soundings of winds, temperature, and humidity taken every few hours as the frontal system passed over. The horizontal velocity of the frontal system itself has been subtracted from the airflow, so the streamlines shown in Fig. 5.36 are relative to the frontal system. Also shown in Fig. 5.36 is the precipitation measured at the ground; several rainbands marked by peaks in the precipitation

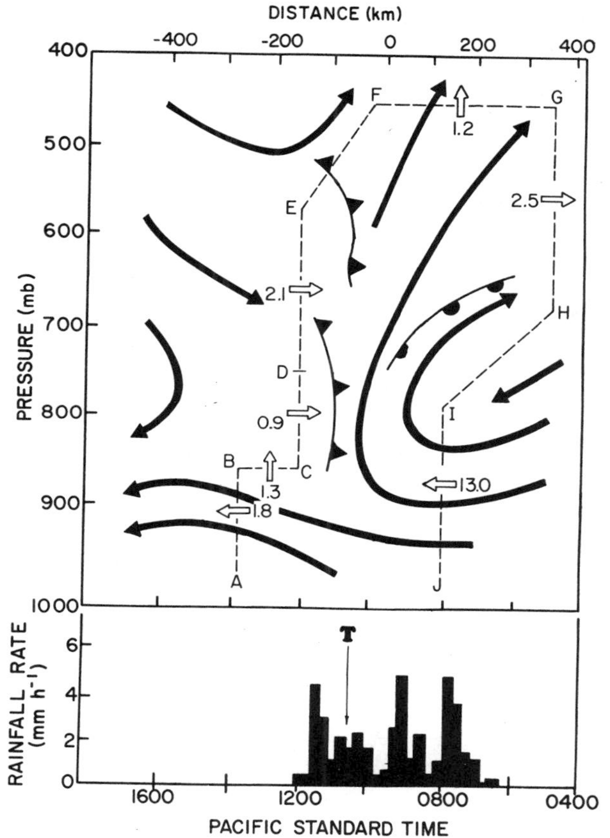

Fig. 5.36 Two-dimensional airflow in a frontal system. Streamlines relative to the front are shown by black arrows. Dashed line surrounds region of upward air motion associated with the frontal clouds. White arrows crossing dashed line show flow of water vapor (with numbers adjacent to the arrows showing the magnitudes of the water vapor fluxes, in units of 10 kg s^{-1}, flowing across areas bounded by the lines AB, BC, and so forth, and each 1 m in length normal to the cross section). The letter T marks the time of the surface pressure trough. [Adapted from *J. Amos. Sci.* **33**, 1921 (1976).]

rate are apparent. The region of upward air motions coincides with the period of precipitation on the ground and the maximum upward air motion is centered on the time of passage of the surface pressure trough. It can be seen from Fig. 5.36 that the primary source of moisture for the frontal clouds was the low-level air, below and ahead of the warm front, which was being overtaken by the frontal system. In the moving coordinate system of Fig. 5.36, this shows up as a flux of moisture through the line labeled IJ. The precipitating clouds were, in fact, fed by low-level (below 800 mb) marine air of tropical origin from the southwest. It is interesting to note the similarity in the airflow between Figs. 5.36 and 5.21.

Orography can play an important role in modifying the structure and distribution of precipitation in cyclonic storms. In unstable regions of a storm (for example, in the stream of warm, moist air in the warm sector), rainbands can be produced well upwind of a mountain as well as over the mountain itself. In stable air, orographic effects are confined closer to the mountain. The downward flow of air and evaporation of cloud water over the leeward slopes of a mountain produces the well-known *rain-shadow* effect. In addition, mountain ranges can interfere with the low-level flow of moist air in cyclones and thereby cut off the frontal clouds from their main source of moisture.

Shown in Fig. 5.37 are airborne measurements of the microstructure of a frontal band associated with an occluded frontal system. Significant amounts of liquid water are present only within the convective region near the forward edge of the main frontal cloud and at lower altitudes in the region of precipitation. The main body of the frontal cloud is composed primarily of ice particles. The measured concentrations of ice particles were high (indicating ice multiplication) and considerably in excess of the optimum concentrations required for the efficient growth of precipitable particles by vapor deposition; instead, growth by riming and aggregation played important roles. An interesting feature shown in Fig. 5.37 is the precipitation of ice particles, through the dry

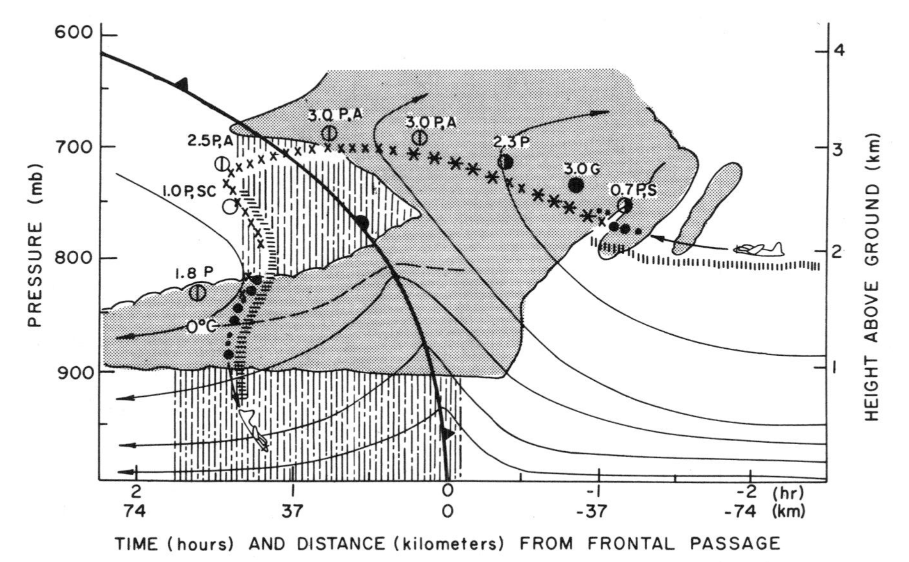

Fig. 5.37 Cloud microstructure of a frontal band associated with an occluded frontal system. Cloud liquid water content: • 0.1–0.2 g m^{-3}; ● >0.2 g m^{-3}. Ice particle concentrations: × $<$45 liter^{-1}; * 45–95 liter^{-1}; ○ >95 liter^{-1}. Maximum particle size in millimeters; ice particle types: N, needles and long columns; P, platelike; S, stellar; SC, short columns; AC, assemblages of columns; SP, assemblages of side planes; A, aggregates; G, graupel; H, hail. Degree of riming: ○, none; ⦶, light; ◑, moderate; ●, heavy. Turbulence zones indicated by (|||||). Note that the vertical scale is stretched with respect to the horizontal scale (as can be seen by the apparently near vertical descent of the aircraft!) [From *J. Atmos. Sci.* **32**, 1555 (1975).]

air of the prefrontal surge, from the upper cloud layer into the lower cloud deck. The upper layers of the cloud were therefore providing a source of ice particles which underwent further growth to augment precipitation in the lower cloud layers.

5.5.2 Cloud patterns as revealed by satellite imagery

Many of the distinctive cloud patterns that appear in photographs taken from satellites over extratropical latitudes are associated with various aspects of cyclonic storms. Narrow cloud bands, such as the one along the Texas coast and the much longer one that stretches across the southeastern Pacific Ocean in Fig. 1.1, are often related to cold fronts. Comma-shaped cloud masses are often associated with cyclonic storms during their earlier stages of development. An example ("upside down and backwards" because of its Southern Hemisphere location) is the cloud mass off the west coast of Chile in Fig. 1.1. As storms enter the occluded stage their cloud shields take on more distinctive spiral shapes. Note, for example, the cloud mass over eastern North America in Fig. 1.1. In decaying storms the remnants of the cloud masses tend to become wrapped up in tight spirals; several examples can be seen over the Southern Hemisphere oceans in Fig. 1.1.

Figure 5.38 shows an idealized model of the distribution of clouds observed in association with cyclonic storms that form along stationary fronts such as the Great Lakes storm described in Chapter 3. Figure 5.38a depicts the cloud pattern prior to the development of the wave along the front. In this early stage of development the idealized frontal configurations and upper-level flow patterns are reminiscent of Figs. 3.2a and 3.3a. Downwind of the trough of the upper-level wave the poleward edge of the cirrostratus cloud deck coincides with the axis of the jet stream.

In the satellite images the first indication of the development of a major wave along the front is the widening of the cirrostratus cloud deck together with the appearance of significant anticyclonic curvature in the jet stream, which lies along the poleward edge of the cirrostratus deck, as shown in Fig. 5.38b.

As the storm develops, the upper-level trough gradually overtakes the low-pressure center at the earth's surface. At about the time the jet stream passes over the surface low (Fig. 5.38c) a noticeable change takes place in the cloud pattern. A lower- or middle-level cloud deck (altostratus or stratocumulus) emerges from under the cirrostratus deck on the poleward side of the jet stream. It is usually at about this time that the storm enters the occluded stage. (The surface low over Buffalo, New York, in Fig. 3.3c corresponds roughly to this stage of development.) Precipitation from the lower deck is usually light and continuous but it can be moderate to heavy in more vigorous storms which have access to sufficient moisture. Most major winter snowstorms are associated with such cloud decks.

As the surface low develops farther back into the cold air, the upper and lower cloud decks become increasingly separate from one another and the

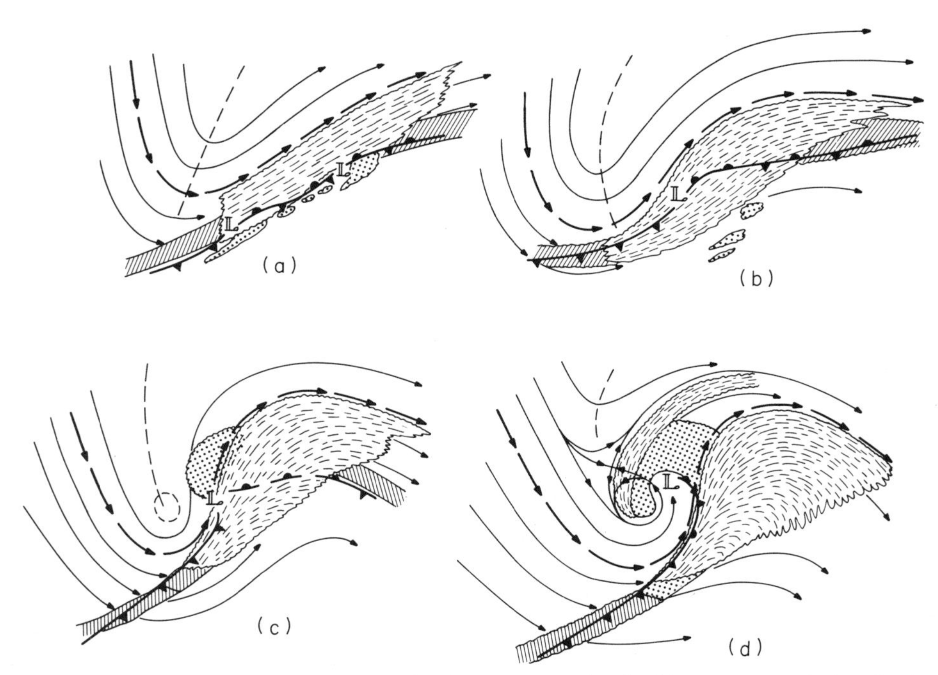

Fig. 5.38 Schematic description of the cloud pattern associated with an extratropical cyclonic storm which originates as a wave on a stationary front: (a) prior to the development of the storm, (b) the "open wave" stage, (c) the beginning of the occluded stage, and (d) the fully developed stage. The areas covered by ( ) represent cirrostratus cloud decks, (: : :) represent extensive low- or middle-level cloud decks, and (////) represent frontal cloud bands. The frontal symbols refer to the surface fronts and **L** refers to the position of the center of the low-pressure area at the surface. The arrows represent the flow in the vicinity of the 300-mb (9-km) level and the short heavy arrows denote the axis of the jet stream. (Adapted from lecture notes by Roger B. Weldon.)

combined cloud mass begins to assume a spiral shape. In the fully developed storm shown in Fig. 5.38d the trailing edge of the cloud mass lies close to the occluded front, while the poleward edge (which has begun to sprout its own cirrus deck) coincides with the line of confluence associated with the trough in the upper-level flow pattern. In the more vigorous storms, newly formed cirrus clouds may completely obscure the lower cloud deck.

In the decaying stage of the storm, the surface low and the upper-level vortex become coincident with one another. The occluded front gradually weakens and loses its identity as the low center becomes surrounded by cold air at the surface. The cyclonic circulation wraps the cloud mass around the storm center in a tightening spiral which disintegrates over a period of a day or two.

5.6 ARTIFICIAL MODIFICATION OF CLOUDS AND PRECIPITATION

We have seen in Sections 4.3–4.5 that the microstructures of clouds are strongly influenced by the concentrations of CCN and ice nuclei, and that the growth of precipitation particles is a result of certain instabilities which exist in the microstructures of clouds. These instabilities are of two main kinds. First, in warm clouds the larger drops increase in size at the expense of the smaller droplets due to growth by the collision–coalescence mechanism. Second, if ice particles exist in a certain optimum range of concentrations in a mixed cloud they grow by deposition at the expense of the droplets (and subsequently by riming and aggregation). In light of these ideas, the following techniques have been suggested whereby clouds and precipitation might be artificially modified:

- by the introduction of large hygroscopic particles or water drops into warm clouds (or *seeding*, as it is commonly called) in order to stimulate the growth of raindrops by the collision–coalescence mechanism;
- by the seeding of artificial ice nuclei into cold clouds (which may be deficient in natural ice nuclei) in concentrations of about 1 liter^{-1} in order to stimulate the production of precipitation by the ice crystal mechanism;
- by introducing comparatively high concentrations of artificial ice nuclei into cold clouds in order to reduce drastically the concentrations of supercooled droplets and thereby inhibit the growth of ice particles by deposition and riming. This should tend to dissipate the clouds and suppress the growth of precipitation particles.

Before turning to a brief account of the experiments that have been carried out on the artificial modification of clouds and precipitation, we must distinguish between two techniques which can be used for evaluating the results of such experiments: namely, *physical* and *statistical evaluation*. Physical evaluation involves determining, through careful field studies, the effects of a certain type of artificial seeding on the clouds and establishing through clearly documented cause-and-effect relationships, starting with nucleation in the clouds and ending with precipitation on the ground, the conditions under which precipitation can be artificially modified. The results obtained from physical evaluations can be incorporated into mathematical models which, in turn, might suggest further possibilities for modification.

Statistical evaluation involves carrying out artificial seeding under certain specified conditions (for example, cloud top temperatures within a certain range, winds from a certain direction, and so on) and comparing statistically the precipitation in a certain target area on the artificially seeded occasions with the precipitation on other occasions when the same conditions existed but artificial seeding was not carried out. This procedure is best carried out through what is known as a *randomized experiment* in which, if the specified conditions are met, a "go-period" is defined. However, artificial seeding is only carried out on a mathematically randomized number of the go-periods (determined, for example, by tossing an "unbiased" coin). The precipitation which falls on the remainder of the (unseeded) go-periods is used as a control against which to compare the precipitation that fell during the seeded go-periods. Any differences between the precipitation during the seeded and unseeded go-periods may then be assigned a statistical significance. For example, it might be found that the precipitation on the seeded days was 10% greater than on the unseeded days and that this difference was significant at the 1% level (namely, that there was a probability of 1% that the observed difference of 10% between the precipitation on the seeded days and the unseeded days could occur by chance).

Ideally, an artificial modification experiment should commence with physical evaluation to determine whether a potential for modification exists and, if so, under what conditions. The results of the physical evaluation should then be used as the basis for determining the go-periods in a statistical experiment. Unfortunately, very few cloud seeding experiments have been conducted in this way; consequently, the majority of experiments have been inconclusive.

5.6.1 Modification of warm clouds

The introduction of large water drops (radius $\simeq 0.25$ mm) into the tops of clouds is not a very efficient method for producing rain (see Problem 5.14) and, in any case, large quantities of water are required. A preferable technique is to introduce small water droplets (radius $\simeq 30\ \mu$m) or hygroscopic particles (for example, common salt) into the base of a cloud; these may then grow by condensation and then by collisions and coalescence, as they are first carried up and then fall down through the cloud (see Problem 5.15).

In the 1950s water drop and hygroscopic particle seeding experiments on warm clouds were carried out in several countries. In some cases rain appeared to be initiated by the seeding, but since neither extensive physical nor statistical evaluations were carried out the results were inconclusive.

A hygroscopic seeding experiment has been carried out in three areas in India where sodium chloride particles were released from the ground at a rate of about $10^{10}\ s^{-1}$. Control and target areas were defined upwind and downwind of the central seeding locations, and comparisons were made between rainfall in these two areas for seeded and unseeded days. In 16 out of the total of 18 seasons for which the experiment was conducted, the rainfall in the target area was 42% greater than in the control area (significant at better than the 0.5% level). Unfortunately, very little physical evaluation was carried out in this experiment and it was not known whether the salt particles reached cloud bases. In fact, to account for the 42% increase in rainfall, every salt particle released from the ground generators would have had to reach the clouds and to grow to a raindrop 2.1 mm in diameter. It seems unlikely that this could have occurred.

Seeding with hygroscopic particles has also been used in attempts to improve visibility in warm fogs. Since the visibility in a fog is inversely proportional to the number concentration of droplets and to their total surface area, visibility can be improved by decreasing either the concentration or the size of the droplets. When hygroscopic particles are dispersed into a warm fog they grow by condensation (causing partial evaporation of some of the fog droplets) and they then settle out under the influence of gravity. Calculations indicate that hygroscopic particles about 15 μm in radius, dispersed at a rate of 5 g m^{-2}, should improve the visibility in a warm fog from about 100 m to 1 km within 10 min of seeding, but thereafter the visibility will deteriorate again due to the fog reforming. Fog clearing by this method has not been widely used due to its expense and lack of dependability.

At the present time, the most effective methods for dissipating warm fogs are "brute force" approaches involving evaporating the fog droplets by ground-based heating (for example, burning hydrocarbon fuels). During World War II, fog dissipating systems of this type (called FIDO) were installed at 15 airfields in England and were found to be highly successful although expensive.

5.6.2 Modification of cold clouds

By far the greatest effort to modify clouds and precipitation has been directed at triggering the ice crystal mechanism in cold clouds. We have seen in Section 4.5 that this mechanism depends on the coexistence in clouds of supercooled droplets and ice particles, for under these conditions the ice particles may increase to precipitation size rather rapidly. We have also seen in Section 4.5 that the concentrations of ice nuclei in the air are rather small at high temperatures, and on some occasions there may be less than the number required for the efficient initiation of the ice crystal mechanism. Under these conditions it should be possible to induce a cloud to rain by seeding it with artificial ice nuclei or some other material that will cause ice to appear. This is, in fact, the scientific basis for most of the rain-making experiments that have been carried out in the past.

A material suitable for seeding cold clouds was first discovered in July 1946 in Project Cirrus, which was carried out by the General Electric Company under the direction of Langmuir.† One of Langmuir's colleagues (V. Schaefer) observed that when a small piece of Dry Ice (solid carbon dioxide) is dropped into a cloud of supercooled droplets in a deep-freeze box, numerous small ice crystals are produced and the cloud is quickly transformed into ice. The reason for this transformation is not that Dry Ice acts as an ice nucleus in the usual sense of this term, but because it has such a low temperature ($-78°C$) numerous ice crystals form in its wake by homogeneous nucleation. For example, a pellet of Dry Ice 1 cm in diameter falling through air at $-10°C$ produces about 10^{11} ice crystals before it evaporates.

The first field trials using Dry Ice were made in Project Cirrus on 13 November 1946 when about $1\frac{1}{2}$ kg of crushed Dry Ice were dropped along a line about

† **Irving Langmuir** (1881–1957) American physicist and chemist. Spent most of his working career as an industrial chemist in the General Electric Research Laboratories in Schenectady, New York. Made major contributions to several areas of physics and chemistry and won the Nobel prize in Chemistry in 1932 for work on surface chemistry. His major preoccupation in later years was meteorology and particularly weather modification. His outspoken advocacy of large-scale effects of cloud seeding involved him in much controversy toward the end of his life.

5 km long into a layer of supercooled altocumulus cloud. Snow was observed to fall from the base of the seeded cloud for a distance of about $\frac{1}{2}$ km before it evaporated in the dry air.

Because of the large numbers of ice crystals that it can produce, Dry Ice is most suitable for *overseeding* cold clouds rather than producing ice crystals in concentrations of about 1 liter^{-1} for the purpose of enhancing precipitation. When a cloud is overseeded it is converted completely into ice crystals and the cloud is then said to be *glaciated*. The ice crystals in a glaciated cloud are generally quite small and, since there are no supercooled droplets present, the air cannot be supersaturated with respect to ice. Therefore, rather than the crystals growing (as they would in a mixed cloud at water saturation) they tend to evaporate. Consequently, by seeding with Dry Ice, large areas of supercooled cloud or fog can be dissipated (Fig. 5.39). This technique is now used routinely for clearing supercooled fogs at a number of airports around the world.

Following the demonstration that supercooled clouds can be modified by Dry Ice, Bernard Vonnegut, who was also working with Langmuir, began searching for artificial ice nuclei. In this search he was guided by the expectation that an effective ice nucleus should have a crystallographic structure similar to that of ice. Examination of crystallographic tables revealed that silver iodide fulfilled this requirement. Subsequent laboratory tests showed that silver iodide could act as an ice nucleus at temperatures as high as $-4°C$.

Fig. 5.39 A γ-shaped path cut in a layer of supercooled cloud by seeding with Dry Ice. (Photo by courtesy of the General Electric Company, Schenectady, New York.)

The seeding of natural clouds with silver iodide was first tried as part of Project Cirrus on 21 December 1948. Pieces of burning charcoal impregnated with silver iodide were dropped from an aircraft into about 16 km^2 of supercooled stratus cloud 0.3 km thick, at a temperature of $-10°C$. The cloud was converted into ice crystals by less than 30 g of silver iodide!

Many artificial ice nucleating materials are now known (for example, lead iodide, cupric sulfide) and some organic materials (for example, phloroglucinol, metaldehyde) are more effective as ice nuclei than silver iodide. However, silver iodide has been used in most artificial seeding experiments to date.

In the thirty years that have passed since the first cloud seeding experiments, many more experiments have been carried out all over the world. Most of them have been concerned with increasing precipitation. It is now well established that the concentrations of ice crystals in clouds can be increased by seeding with artificial ice nuclei and that, under certain conditions, precipitation can be artificially initiated in some clouds. However, the important question that has yet to be unequivocally answered is: under what conditions (if any) can seeding with artificial ice nuclei be employed to produce significant increases in precipitation over large areas?

Although it is probably still premature to attempt to answer this question, some general indications are beginning to emerge. Let us consider first the seeding of convective clouds with artificial ice nuclei in concentrations of about 1 $liter^{-1}$ in an attempt to increase the efficiency of the ice crystal mechanism for producing precipitation. It appears that this type of seeding might be more effective in modifying precipitation from continental-type cumulus clouds than from marine-type cumulus. A likely reason is that the narrowness of the droplet size distribution in continental cumulus clouds renders the collision–coalescence mechanism for the production of rain inefficient (see Section 4.2.2); therefore, the formation of precipitation in these clouds may often be dependent upon the ice crystal mechanism. Moreover, again because of the lack of appreciable numbers of large (>25 μm radius) drops in continental clouds, ice multiplication mechanisms may not operate (see Section 4.5.2). Consequently the production of precipitation in continental cumulus clouds may sometimes be hindered by the lack of sufficient concentrations of natural ice nuclei, particularly those that are effective at temperatures above about $-16°C$.

One of the few long-term randomized cloud seeding experiments that appears to have produced significant increases in rainfall has been carried out in Israel, where winter continental-type cumulus clouds were seeded with silver iodide from aircraft. During the period 1961–1967 the precipitation over the entire experimental area of the Israeli Project was 15% greater on seeded days than on unseeded days, and this difference was significant at the 5.4% level. By contrast, an experiment in Missouri, in which summer cumulus clouds were seeded with silver iodide from an aircraft on a randomized basis, failed to show any significant increases in rainfall on the seeded days. Physical studies revealed that the clouds over Missouri were marine in their characteristics.

So far we have only discussed the role of artificial ice nuclei in modifying the microstructures of cold clouds. When large volumes of a convective cloud are glaciated by overseeding, the resulting release of latent heat provides added buoyancy to the cloudy air. Now suppose that prior to seeding the height of a cloud was restricted by a stable layer produced by a temperature inversion. Then the sudden release of the latent heat of fusion caused by artificial seeding might provide enough buoyancy to push the top of the cloud through the inversion into a region where cloudy air is unstable. The cloud top might then rise to much greater heights than it would have done naturally (Fig. 5.40). Careful randomized statistical studies of seeded and unseeded cumulus clouds have shown that this type of "explosive" growth can be produced by artificial seeding and, more importantly, that it can be predicted with skill. The deeper a cloud the more water it processes and the greater will be the chance that it will produce precipitation. Hence, the production of "explosive" growth provides another possible mechanism by which precipitation from cumulus clouds might be increased. In a series of randomized experiments carried out in Florida in 1968 and 1970, it was found that the precipitation (as measured by radar) from isolated cumulus clouds about 5 km in diameter which were artificially seeded to induce explosive growth, was about twice that from the unseeded clouds (results significant at better than the 5% level). The seeded clouds rained more than the control clouds because they were bigger, longer lasting clouds, not because their rainfall rates were significantly higher. However, rainfall from isolated cumulus clouds in Florida accounts for less than 10% of the total rainfall; most of the rain is produced by convective cloud systems which are organized on the mesoscale. Experiments are now underway in Florida to determine whether "explosive" growth induced by artificial seeding can be used to enhance the organization and merging of individual cumulus clouds into larger, interacting, mesoscale cloud systems to produce significant increases in precipitation over areas of about 10^4 km^2.

There is slowly mounting evidence that under certain conditions it may be possible to increase the precipitation (generally snow) from winter orographic cloud systems by seeding with artificial ice nuclei. Let us consider a relatively simple situation in which air is lifted over a mountain barrier and the only cloud particles that form are ice particles which grow by deposition. The optimum conditions for the production of precipitation over the windward slopes of the mountain will then occur when the rate at which condensate is made available by the lifting and cooling of the air is just equal to the mass growth rate of the ice particles by deposition. Above a certain temperature the concentrations of ice particles in the clouds may be less than that required for the optimum release of precipitation. In this case the addition of artificial ice nuclei might be expected to increase snowfall. However, below some critical temperature the natural concentrations of ice particles may already be greater than that required for the most effective release of precipitation. In this case, artificial seeding might be expected to decrease precipitation. In a randomized winter cloud seeding experiment (in which silver iodide was dispersed from ground-based generators

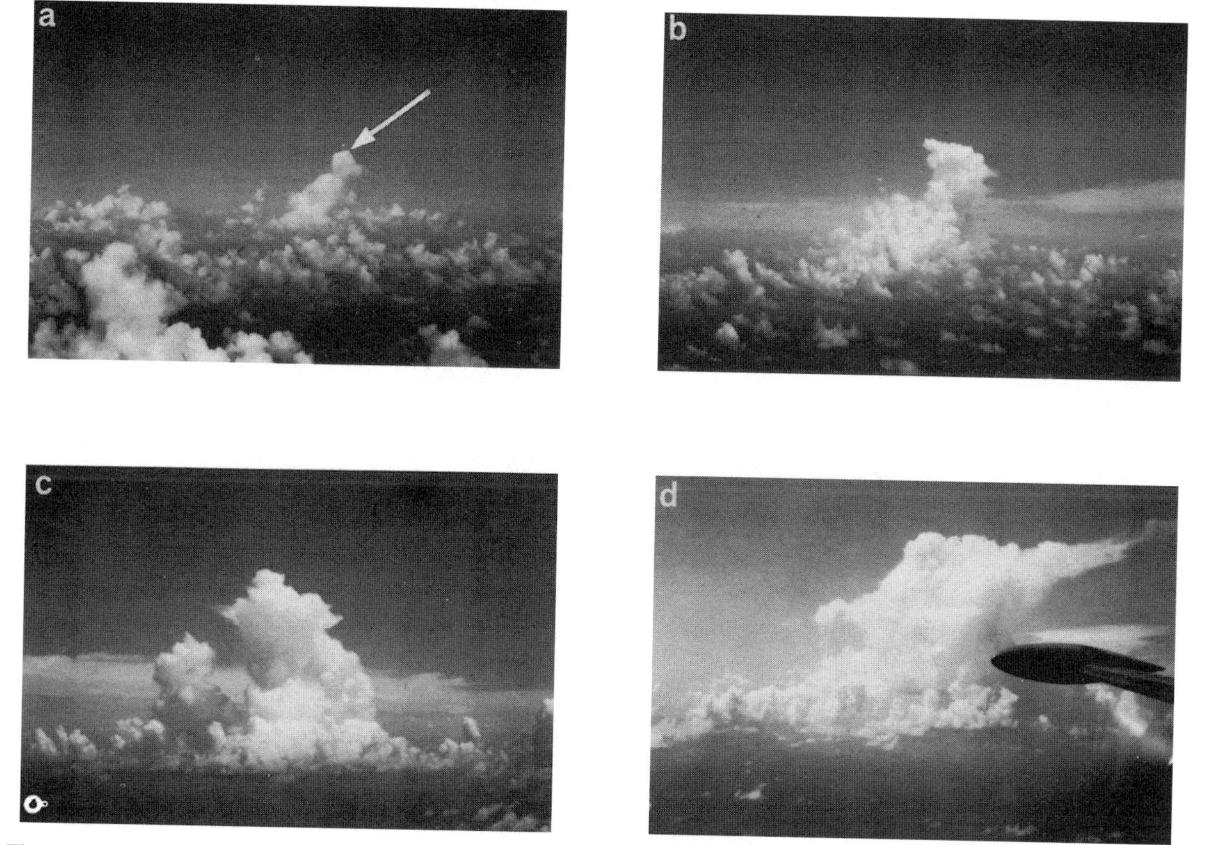

Fig. 5.40 The "explosive" growth of a cumulus cloud following seeding with silver iodide. (a) 1401 h, about ten min after the highest cloud in the center of the photograph was seeded; (b) 1410 h; (c) 1420 h; (d) 1439 h. (Photo: J. Simpson.)

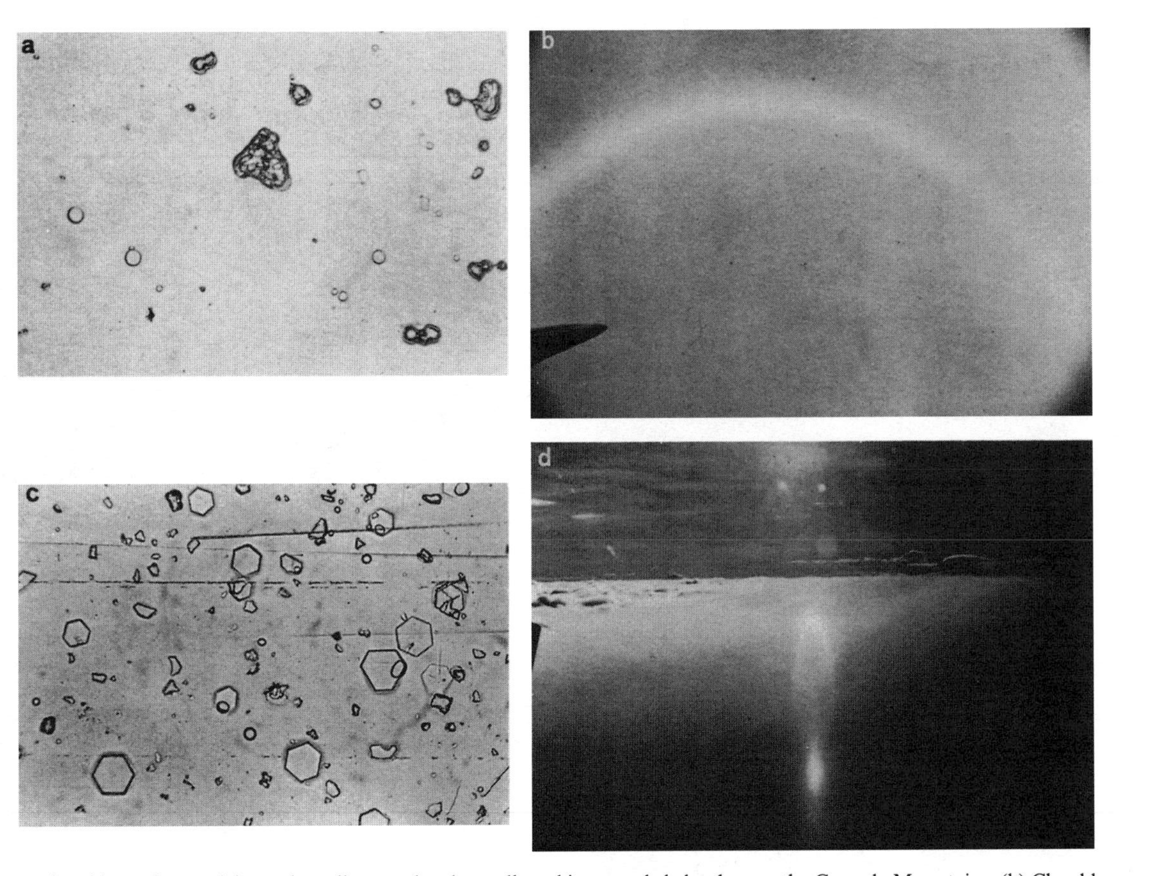

Fig. 5.41 (a) Large rimed irregular particles and small water droplets collected in unseeded clouds over the Cascade Mountains. (b) Cloud bow produced by the refraction of light in small water droplets. Following heavy seeding with artificial ice nuclei, the particles in the cloud were converted into small unrimed plates (c) which markedly changed the appearance of the clouds. In (d) the uniform cloud in the foreground is the seeded cloud and the more undulating cloud in the background is the unseeded cloud. In the seeded cloud, optical effects due to ice particles (portion of the 22° halo, lower tangent are to 22° halo and subsun) can be seen. (Courtesy of Cloud Physics Group, University of Washington.)

upwind of the mountains) carried out in the Colorado Rockies from 1960–1970, precipitation in the target area was 75% greater (significant at the 4.5% level) on the seeded days when the 500-mb temperature was between −11 and −20°C, and 11% less (significant at the 15% level) when the 500-mb temperature was between −27 and −39°C.

Another possibility exists for the beneficial modification of snowfall from orographic cloud systems in which the growth of ice particles by riming plays an important role. Rimed ice particles have comparatively large terminal fall speeds (~ 1 m s^{-1}) and therefore they follow fairly steep trajectories as they fall to the ground. If clouds on the windward side of a mountain are overseeded with artificial ice nuclei or Dry Ice, supercooled droplets can be virtually eliminated and growth by riming significantly reduced. In the absence of riming, the ice particles grow by deposition from the vapor phase and their fall speeds are reduced to about 0.5 m s^{-1}, so that the wind tends to carry them farther across the mountain. In this way it may be possible to divert snowfall from the windward slopes of certain mountain ranges (where precipitation is often heavy) to the drier leeward slopes. A series of case studies carried out in the Cascade Mountains of Washington State have yielded encouraging results (Fig. 5.41).

5.6.3 Modification of severe storms

(*a*) *Hail suppression.* The physical principle behind the majority of experiments designed to reduce the damage produced by hailstones is as follows. Seeding with artificial nuclei should tend to increase the number of small ice particles competing for the available supercooled droplets. Therefore, seeding should result in a reduction in the average size of the hailstones. It is also possible that, if a hailstorm is overseeded with extremely large numbers of ice nuclei, the majority of the supercooled droplets in the cloud will be nucleated, and the growth of hailstones by the accretion of droplets will be significantly reduced.

The results of experiments in the United States, Switzerland, and France on hail suppression have not been particularly encouraging. However, Soviet scientists have reported much more promising results. In experiments carried out in the Caucasus and Transcaucasus, with silver iodide and lead iodide shot from antiaircraft guns, more than a 50% decrease in hail damage has been reported. Unfortunately, the Soviet experiments have not been carried out in a manner that permits statistical evaluation.

A five-year hail-suppression program, using ground-based silver-iodide generators, has been carried out in Argentina on a randomized basis. When all the results were considered together, no marked effect of the seeding was discernible. However, when seeding was carried out in the vicinity of cold fronts, there was an apparent decrease in hail damage of 70%. On the other hand, in non-cold-front weather the results showed an increase of 100% in hail damage on seeded days. An attempt to repeat the Soviet experiments in Colorado has so far failed to reveal any significant decrease in hail due to artificial seeding. Physical studies carried out in conjunction with the Colorado experiments indicate that most of the hail in this region is produced by supercell storms of the type described in Section 5.3.3. If large hailstones in those storms are indeed produced by the natural selection mechanism described in Section 5.3.3, one would not expect the Soviet seeding technique to be effective, since the number of particles attaining the size of large hailstones would be limited, not by the availability of liquid water but, rather, by the short time that precipitation particles remain below the $-40°$ isotherm.

(*b*) *Lightning suppression.* Two main methods have been tried in attempts to reduce the intensity and frequency of lightning flashes. One method consists of seeding clouds with silver iodide, with hopes that the extra ice crystals produced by the seeding will serve as additional points for small electrical discharges (known as corona discharges) which will increase the leakage current between the two main charge centers in the thunderstorm. In a field program based on this idea, carried out from 1960 to 1961, the number of cloud-to-ground discharges occurring on seeded days was 38% less than on unseeded days. Intracloud and total lightning were less by 8 and 21%, respectively, on seeded days. However these differences were only significant at the 25% level. In further trials carried out from 1967 to 1968, there were 66% fewer cloud-to-ground discharges, 50% fewer intracloud discharges, and 54% less total lightning in the seeded than in the unseeded storms. Of possibly greater importance from the point of view of reducing forest fires caused by lightning was the fact that the seeded storms had a larger percentage of short-duration lightning strokes.

In the other method under investigation for lightning suppression, millions of tiny metallic needles are released into a thunderstorm in an effort to increase corona discharges. Laboratory experiments show that the onset of corona discharge on a 10-cm-long chaff fiber occurs at 30 kV m^{-1}, which is about thirty times lower than the field required to initiate lightning discharges in clouds. In statistically randomized field experiments, in which about 10^7 chaff fibers, 10 cm long, were released into the bases of 10 thunderstorms and 18 storms were control (unseeded) cases, the observed lightning discharges in the seeded clouds were about one third or less of those in the control clouds.

(*c*) *Hurricanes.* For several years now a project has been underway to investigate the feasibility of reducing the intensity of winds in hurricanes. It is postulated that if a hurricane is seeded with sufficient quantities of artificial ice nuclei just outside the eye-wall cloud the resulting release of latent heat of fusion will broaden the "warm core" of the storm and spread out the region of low pressure in the eye. This, in turn, will lead to an outward movement of the region of maximum rotating winds and a corresponding reduction in their intensity.

Three hurricanes have been artificially seeded with silver iodide as part of this project ("Esther" in 1961, "Beulah" in 1963, and "Debbie" in 1969). All these storms showed some reductions in wind speeds following artificial seeding (30% in the case of "Debbie"). However, since the reductions in the winds were not outside the range of natural variability, the effectiveness of this approach cannot yet be properly evaluated.

5.6.4 Inadvertent modification

Certain industries release large quantities of heat, water vapor, and cloud-active aerosol (CCN and ice nuclei) into the atmosphere. Consequently, these effluents might modify the formation and structure of clouds and affect precipitation. For example, the effluents from a single paper mill in Pennsylvania profoundly affects the valley in which it is situated, and adjoining areas out to about 30 km (Fig. 5.42). A study has been made of the effects on clouds and precipitation of the effluents from a large paper mill in Washington State. The mean annual precipitation in a region downwind of this paper mill was more than 30% greater, relative to the surrounding regions, during the period 1947–1966 (when the amount of paper produced was rising rapidly) than it was from 1924 to 1946 (when paper production was very low). Further studies revealed that the mill emits about 10^{17} s^{-1} of large CCN and that clouds situated in the plume from this paper mill contain higher concentrations of cloud drops with diameters greater than 30 μm than clouds unaffected by the plume. The higher concentrations of larger drops should increase the efficient of the collision–coalescence mechanism for the production of precipitation. In addition, the heat and water vapor from the mill might play a triggering role in increasing precipitation downwind. By contrast, the smoke from the burning of

Fig. 5.42 The cloud in the valley in the background formed due to effluents from a paper mill. In the foreground, the cloud is spilling through a gap in the ridge into an adjacent valley. (Photo: C. L. Hosler.)

agricultural waste products in sugar cane fields in Australia contains large concentrations of small CCN; these tend to produce a continental-type droplet distribution in clouds and thereby decrease rainfall.

High concentrations of ice nuclei have been observed in the plumes from steel mills. It has also been suggested that airborne lead particles from automobiles using gasolines containing tetraethyl lead antiknock agents can combine with trace quantities of atmospheric iodine to form significant concentrations of lead iodide particles which act as ice nuclei. As yet, there has been no definitive studies of the effects of anthropogenic ice nuclei on clouds and precipitation.

Recently the effects of large cities on the weather have attracted considerable interest. Here the possible interactions are extremely complex and unlikely to be unraveled in the near future since, in addition to being areal sources of aerosol, gases, heat, and water vapor, large cities also modify the radiative properties of the earth's surface, the moisture content of the soil, and the surface roughness. The existence of urban "heat islands," several degrees higher in temperature than adjacent less densely populated regions, has long been known. More recent studies have revealed anomalies in precipitation around some large cities. For example, within 5–10 city diameters downwind of St. Louis, Missouri, rainfall is higher by 30%, thunderstorms more frequent by 25%, heavy rainstorms are 50% more common, and there are 200% more hailstorms than in areas upwind of the city.

PROBLEMS

5.3 Explain or interpret the following:

(a) When the sun heats the ground wetted by rain, wisps of cloudy air sometimes form just above the puddles.

(b) Clouds often form in a bathroom when the taps are running.

(c) Wisps of clouds are often seen in the vortices which trail from the tips of aircraft propellers and wings.

(d) Ahead of cyclonic storms, widespread layer clouds often form at several different levels with clear air between them.

(e) Birds tend to soar in circles on hot windless days but along straight lines on windy days.

(f) Altitude records for gliders are often set downwind of mountains.

(g) Cirrus trails from wave clouds do not exhibit mamma.

(h) When cloud streets occur there is usually no rain.

(i) For cloud "streets" to occur the clouds must be evaporating rather rapidly.

(j) Lee-wave clouds are common ahead of an approaching warm front.

(k) The downward frictional drag force induced by raindrops falling at their terminal velocity is equal to the weight of the drops. (Remember that for a drop falling at its terminal velocity, the sum of the vertical forces is equal to zero. The volume occupied by such a drop is a negligible fraction of the volume of the air.)

(l) Only rather severe thunderstorms have tops reaching 15 km in middle latitudes whereas ordinary thunderstorm cells in the tropics reach that level.

(m) The tops of cumulonimbus clouds are capable of rising higher than their level of neutral buoyancy.

(n) Towering cumulus clouds containing large amounts of supercooled water can sometimes be induced to grow to higher levels by seeding them with artificial ice nuclei.

(o) Squall lines frequently form ahead of cold fronts but rarely behind them.

(p) Severe storms over the central United States occur most frequently during springtime.

(q) Over most land areas, thunderstorms are most frequent during the afternoon hours.

(r) The fair weather electric field usually increases during the afternoon.

(s) Most African squall lines move westward whereas most of those over the United States move eastward.

(t) Most severe storms in the central United States move toward the right relative to the winds in the middle troposphere.

(u) Research aircraft can penetrate more safely into the vault of a supercell storm at cloud base than they can at higher levels.

(v) Hurricanes never form over land and they dissipate quickly when they move over land or cold water.

(w) Hurricanes are warm core despite the fact that air parcels ascending in the eye-wall cloud experience rapid cooling due to adiabatic expansion.

(x) In New York a day or two of fair weather almost always follows the passage of a fast-moving cold front whereas in London or Seattle the weather under these conditions may be showery.

(y) When the ground is snow covered, weather in the warm sector of cyclonic storms is usually characterized by widespread fog.

(z) At any given time, most of the global precipitation is falling in an area that covers only a few percent of the globe.

(aa) In occluded storms that resemble the one shown in Fig. 5.38d the clearing following the frontal passage is quite abrupt.

(bb) Satellite pictures are not much help for locating warm fronts.

(cc) Warm fogs are more difficult to dissipate than cold fogs.

(dd) Despite its effectiveness in nucleating ice, lead iodide is not a very desirable material to use in cloud seeding experiments.

(ee) The targeting of snowfall by artificial seeding in specified areas in mountainous terrain is more difficult in cyclonic storms than in simple orographic flow situations.

(ff) The seeding of thunderstorms with artificial ice nuclei might increase the occurrence of lightning rather than decrease it.

(gg) Precipitation and severe storms may be more frequent downwind than upwind of a large city.

5.4 Two unsaturated parcels of air of equal mass, with mixing ratios of 7.5 and 4.0 g kg^{-1}, mix together at a pressure of 1000 mb. If the parcel with the higher mixing ratio has a temperature of 10°C, determine from a pseudoadiabatic chart (see back endpapers) the maximum temperature that the other parcel can have if a cloud is to form in the mixed air.

Answer 2°C

5.5 Show that if a light beam passes through a hexagonal prism of ice in the manner shown for the 22° halo in Fig. 5.8, the angle of minimum deviation of the beam is about 22°. If the light beam passes through the base and out of one of the hexagonal sides of an ice prism, as shown for the 46° halo in Fig. 5.8, prove that the angle of minimum deviation of the beam is approximately 46°. (Take the refractive index of ice as 1.3.) Why is the angle of minimum deviation appropriate in deducing the angular radius of halos? Why is the 22° halo far more common than the 46°? [Hint: The refractive index n of a prism is given by

$$n = \frac{\sin \frac{1}{2}(D + A)}{\sin \frac{1}{2}A}$$

where A is the refracting angle of the prism and D the angle of minimum deviation of a light beam which is refracted by the prism.]

5.6 By considering the angle of minimum deviation of a light beam which passes through a bullet-shaped ice crystal as indicated in the accompanying diagram, show that a halo with an angular radius of about 8° may be formed. (Take the refractive index of ice as 1.3.)

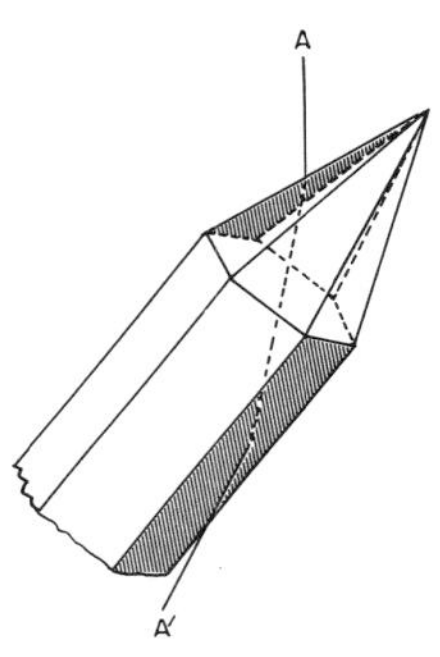

Fig. P5.6 A bullet ice crystal and a typical ray path AA′ necessary to produce the 8° halo.

5.7 The air at the 500-mb level in a cumulonimbus cloud has a temperature of 0°C and a liquid water content of 3 g m^{-3}. (a) Assuming that the water drops are falling at their terminal velocities, calculate the downward frictional drag force per unit mass of air. (b) Express this downward force in terms of a (negative) virtual temperature correction. (In other words, compute the temperature that air without liquid water would have to be in order to be as heavy as the air in question.) [Hint: At temperatures around 0°C the virtual temperature correction due to the presence of water vapor in the air can be neglected.]

Answer 0.046 N kg^{-1}; 1.28 deg

5.8 Surface air with a dew point of 25°C (about the highest ever observed except over very limited regions of the tropics) is lifted in a strong updraft. If the terminal velocities of the cloud droplets are negligible in comparison to the updraft velocity, what is the highest possible liquid water content that can be generated in the updraft?

Answer About 7 g m^{-3}

5.9 Air in a strong updraft of a cumulonimbus cloud is ascending at a rate of 30 m s^{-1} at the time it reaches the level (just below the tropopause) where its temperature drops below that of the ambient air. (a) Calculate the maximum possible distance that the cloud top can penetrate beyond its level of neutral buoyancy if the environmental lapse rate is isothermal and the temperature is 210°K. (At these temperatures the dry and moist adiabatic lapse rates are virtually the same.) (b) Estimate the temperature of this "overshooting" cloud top. (c) What factors or processes might prevent the cloud top from reaching the level calculated in (a)?

Answer 1410 m; 196.2°K

5.10 Air with a relative humidity of 20% at 500 mb and $-20°C$ is entrained into a cloud. It remains at 500 mb while cloud droplets are evaporated into it. To what temperature can it be cooled by this process. Suppose the same parcel of air is carried down to 1000 mb in a downdraft. What will be its temperature if it remains saturated? What will be its temperature if its relative humidity is 50% when it reaches the ground?

Answer $-22.5°C$; 12°C; 25°C

5.11 A hurricane with a central pressure of 940 mb is surrounded by a region with a pressure of 1010 mb. The storm is located over an ocean region. At 200 mb the pressure anomaly vanishes (that is, the 200-mb surface is perfectly flat). Estimate the average temperature difference between the hurricane and its surroundings in the layer between the surface and 200 mb. Assume that the mean temperature of this layer outside the hurricane is $-3°C$ and ignore the virtual temperature correction.

Answer 12 deg

5.12 The rainfall rate in the eye-wall cloud of a hurricane is 10 cm h^{-1}. The air at cloud base has a pressure of 900 mb, a temperature of 28°C, and vapor mixing ratio of 20 g kg^{-1}. Estimate the updraft velocity at the base of the eye-wall cloud, assuming that entrainment above cloud base is negligible and that all the water vapor is condensed out and falls as rain.

Answer 1.34 m s^{-1}

5.13 Determine the minimum quantity of heat that would have to be supplied to each cubic meter of air in order to evaporate a fog containing 0.3 g m^{-3} of liquid water at 10°C. [Hint: You will need to inspect the saturation vapor pressure of air as a function of temperature, shown on the back endpapers of the book.]

Answer 1270 J

5.14 If 40 liters of water in the form of drops 0.5 mm in diameter were poured into the top of a cumulus cloud and all of the drops grew to a diameter of 5 mm before they emerged from the base of the cloud which had an area of 10 km^2, what would be the amount of rainfall induced? Can you suggest any physical mechanisms by which the amount of rain produced in this way might be greater than this problem suggests?

Answer 4×10^{-3} mm

5.15 Compare the increase in the mass of the drops in the previous problem with those of drops 20 μm in radius which are introduced at cloud base, travel upward, then downward, and finally emerge from cloud base with a diameter of 5 mm.

Answer Factors of 10^3 and 2×10^6, respectively

5.16 If hailstones of radius 5 mm were present in concentrations of 10 m^{-3} in a certain region of a hailstorm where the hailstones were competing for the available cloud water, determine the size of the hailstones which would be produced if the concentrations of hailstone embryos were increased to 10^4 m^{-3}.

Answer 0.5 mm

5.17 A supercooled cloud is completely glaciated at a particular level by artificial seeding. Derive an approximate expression for the increase in the temperature ΔT at this level in terms of the original liquid water content w_l (in grams per kilogram), the specific heat of the air c, the latent heats of fusion L_f, and deposition L_d of the water substance and the original saturation mixing ratio with respect to water w_s and the final saturation mixing ratio with respect to ice w_i (in grams per kilogram).

Answer $c\,\Delta T = w_l(10^{-3}L_f) + (w_s - w_i)(10^{-3}L_d)$

5.18 Ignoring the second term on the right-hand side of the answer given in the above problem, calculate the increase in temperature produced by glaciating a cloud containing 2 g kg^{-1} of water.

Answer 0.6 deg

5.19 On a certain day, towering cumulus clouds are unable to penetrate above 500 mb because of the presence of a weak stable layer in the vicinity of that level, in which the environmental lapse rate is 5 deg km^{-1}. If these clouds are successfully seeded with silver iodide so that all the liquid water in them (assumed to be 1 g m^{-3}) is frozen, how much higher will their tops rise? Under what conditions might such seeding be expected to result in a significant increase in precipitation?

Answer Roughly $\frac{1}{4}$ km

Chapter

6

Radiative Transfer

Virtually all the exchange of energy between the earth and the rest of the universe takes place by way of radiative transfer. The earth and its atmosphere are constantly absorbing solar radiation and emitting their own radiation to space. Over a long period of time the rates of absorption and emission are very nearly equal; that is to say, the earth–atmosphere system is very nearly in radiative equilibrium with the sun.

Radiative transfer also serves as a mechanism for exchanging energy between the atmosphere and the underlying surface and between different layers of the atmosphere. It plays an important role in a number of chemical reactions in the upper atmosphere and in the formation of photochemical smogs. The transfer properties of visible radiation determine the visibility, the color of the sky, and the appearance of clouds. Infrared radiation emitted by the atmosphere and intercepted by satellites is the basis for remote sensing of the atmospheric temperature structure.

In this chapter we will introduce the fundamental principles of radiative transfer. The chapter begins with a brief qualitative discussion of the electromagnetic spectrum and the processes by which individual molecules absorb and emit radiation. The next section defines the basic terminology, symbols, and units required for the quantitative description of radiation as a function of wavelength and orientation in space. There follows a brief treatment of Planck's law and subsidiary relations describing blackbody radiation. Kirchhoff's law is then introduced, together with examples showing how the absorption spectrum of an atmosphere influences the radiative equilibrium temperature of the surface of a planet. The remainder of the chapter deals with radiative transfer within the atmosphere. Particular emphasis is placed on the concept of optical depth, applied both to the vertical profile of absorption of solar radiation and to remote sensing of the spectrum of radiation emitted by the atmosphere. The final section deals with scattering by particles in the atmosphere.

6.1 THE SPECTRUM OF RADIATION

Electromagnetic radiation may be viewed as an ensemble of waves which travel through a vacuum at the speed of light $c^* = 3 \times 10^8$ m s^{-1}. The waves may exhibit a continuous range of wavelengths, and the totality of all possible wavelengths is called the *electromagnetic spectrum.* Radiation can often be identified with a particular part of the spectrum by noting the effects it produces when it impinges upon certain materials. For example, the cells on the retina of the human eye are sensitive to a radiation with a rather narrow range of wavelengths which we call *light.* They are capable of distinguishing among a number of subranges of wavelength within the visible part of the spectrum. Radiation or light associated with wavelengths within a particular subrange is spoken of as having a particular color. Much of the terminology that is used to describe the electromagnetic spectrum is based on this relationship between wavelength and color.

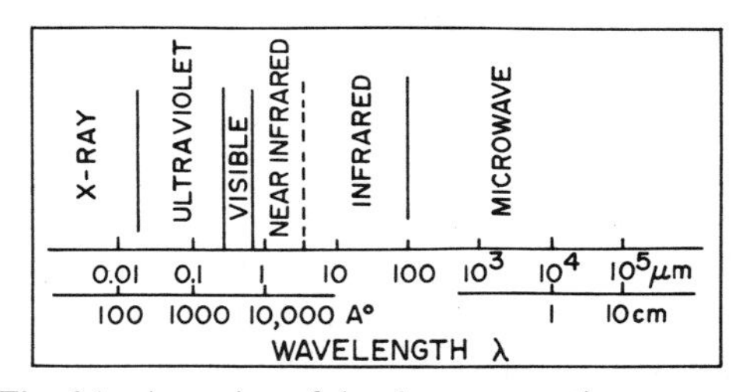

Fig. 6.1 A portion of the electromagnetic spectrum.

Figure 6.1 shows the portion of the electromagnetic spectrum that is of interest in the discussion of the atmospheric energy budget. Various ranges are labeled in accordance with the nomenclature we will be using in this and the subsequent chapter. Table 6.1 shows the relationship between wavelength and color in the visible part of the spectrum.

Table 6.1
Wavelengths and the corresponding colors[a]

Color	Wavelength interval (μm)	Typical wavelength (μm)
Violet	0.390–0.455	0.430
Dark blue	0.455–0.485	0.470
Light blue	0.485–0.505	0.495
Green	0.505–0.550	0.530
Yellow-green	0.550–0.575	0.560
Yellow	0.575–0.585	0.580
Orange	0.585–0.620	0.600
Red	0.620–0.760	0.640

[a] From K. Ya. Kondratiev, "Radiation in the Atmosphere," Academic Press, New York, 1969.

6.2 ABSORPTION AND EMISSION OF RADIATION BY MOLECULES

Any isolated molecule possesses a certain amount of energy, exclusive of that associated with its motion through space. Most of this is in the form of kinetic energy and electrostatic potential energy of electrons revolving in orbits about the nuclei of the individual atoms. There are additional, lesser energy amounts associated with vibration of the individual atoms about their mean positions in the molecule and rotation of the molecule about its center of mass.

Quantum mechanics predicts that only certain configurations of electron orbits are permitted within each atom, and only certain vibrational frequencies and amplitudes, and certain rotation rates are permitted for a particular molecule. Each possible combination of electron orbits, vibration, and rotation may be identified with a particular energy level, which represents the sum of the three types of energy. A molecule may undergo a transition to a higher energy level by absorbing electromagnetic radiation. Likewise, it may drop to a lower energy level by emitting radiant energy. Only certain discrete changes in energy level are allowable, which are predicted by quantum theory, and these changes are the same whether energy is being absorbed or emitted.

Quantum theory also predicts that energy transmitted by electromagnetic radiation exists in discrete units called *photons*. The amount of energy W associated with a photon of radiation is given by

$$W = h\nu \tag{6.1}$$

where ν is the frequency of the radiation (that is, the number of waves passing a given point per unit time, expressed in units of s^{-1}), and h is Planck's[†] constant,

[†] **Max Planck** (1858–1947) German physicist. Studied under Helmholtz and Kirchhoff. Made professor of physics at the University of Kiel and University of Berlin at ages 27 and 31, respectively. Played an important role in the development of quantum theory. Awarded the Nobel prize in 1918.

which is equal to 6.626×10^{-34} J s. Since radiation travels at the speed of light c^*, it follows that frequency and wavelength are related by the expression

$$\nu = c^*/\lambda \tag{6.2}$$

Thus the amount of energy contained in a photon of radiation is inversely proportional to the wavelength of the radiation.†

Since an isolated molecule can only absorb and emit energy in discrete amounts corresponding to the allowable changes in its energy level, it follows that it can interact only with radiation having certain discrete wavelengths. Thus the absorption and emission properties of an isolated molecule can be described in terms of a *line spectrum* consisting of a finite number of extremely narrow absorption or emission lines, separated by gaps in which the absorption and emission of radiation are not possible.

Most of the absorption lines associated with orbital changes involve X-ray, ultraviolet, and visible radiation. The characteristic yellow light emitted by sodium compounds when they are heated in a flame is produced by one of these lines. Vibrational changes are usually associated with infrared wavelengths, while the rotational transitions, which involve the smallest amounts of energy, tend to be associated with radiation in the microwave region. Some molecular species such as CO_2, H_2O, and O_3 have structures which permit them to absorb and emit a photon of radiant energy when they undergo a simultaneous rotation–vibration transition. These molecules exhibit *line clusters* consisting of myriads of individual, closely spaced absorption lines in the infrared region of the electromagnetic spectrum. Other molecules such as O_2 and N_2 cannot interact with radiation in this way, and consequently their absorption spectra do not exhibit many lines in the infrared region.

In addition to the processes described above, there are two other possible ways in which an atom or molecule can absorb or emit electromagnetic radiation:

- A molecule may absorb radiation, the energy of which is sufficient to cause it to break down into its atomic components. Unstable atoms may also combine to form a more stable molecule, disposing of their excess energy in the form of radiation in the process. In these so-called *photochemical reactions*, the absorption or emission of electromagnetic radiation plays a crucial role in providing or removing energy. An example of such a reaction is

  $$O_2 + h\nu = O + O \quad (\lambda < 0.2424\ \mu m)$$

  where $h\nu$ indicates the energy associated with the photon and λ is the wavelength of the radiation. Unlike the changes discussed previously, photochemical reactions may involve a continuum of wavelengths of radiation

† The reciprocal of the wavelength is the *wave number*, which gives the number of wavelengths in a unit distance. Thus the frequency of radiation and the amount of energy contained in a photon are directly proportional to the wave number.

provided that the wavelength is short enough that a photon of energy will raise the chemical energy of the molecule to the threshold level where photodissociation can occur. Any excess energy is imparted to the kinetic energy of the atoms, which serves to increase the temperature of the gas. Most photochemical reactions in the earth's atmosphere involve ultraviolet and visible radiation.

- All atoms are capable of being ionized by radiation with sufficiently short wavelengths. This process, called *photoionization*, requires photons with enough energy to strip one or more of the outer electrons from their orbits around the atomic nucleus. Like photochemical reactions, photoionization may involve radiation with a continuum of wavelengths up to the value corresponding to the threshold energy level. Ionizing radiation is usually associated with wavelengths shorter than about 0.1 μm.

As a consequence of Heisenberg's† uncertainty principle, which states the impossibility of accurately determining simultaneously both the position and velocity of a particle, the individual lines in the absorption spectrum of an isolated molecule have finite widths. Within a gas the width of absorption lines is greatly enhanced by *Doppler broadening* associated with the random molecular motions and by *collision broadening* which results from interactions between the electrostatic force fields of individual molecules during the process of collision.‡ At levels below 30 km in the earth's atmosphere the width of absorption lines is largely determined by the amount of collision broadening. Within liquids and solids the interaction between the force fields of individual molecules is so strong that absorption and emission take place throughout a continuous spectrum of wavelengths, as opposed to a line spectrum.

6.3 QUANTITATIVE DESCRIPTION OF RADIATION

The rate of energy transfer by electromagnetic radiation is called the *radiant flux*, which has units of energy per unit time: joules per second or watts. For example, the radiant flux from the sun is about 3.90×10^{26} W. By dividing the radiant flux by the area through which it passes, we obtain the *irradiance* which is expressed in units of watts per square meter and denoted by the symbol E.§ The irradiance of electromagnetic radiation passing through the outermost limits of the visible disk of the sun (which has an approximate radius of

† **Werner Heisenberg** (1901–) German physicist. Founded quantum mechanics in 1925. Awarded the Nobel prize for physics in 1932.

‡ The amount of Doppler broadening depends upon the root mean square velocity of the gas molecules, which is directly proportional to the square root of the absolute temperature. The amount of collision broadening depends upon the frequency of molecular collisions, which is directly proportional to the pressure of the gas.

§ Prior to the widespread adoption of the SI system, this quantity was widely referred to as *flux density* and denoted by the symbol F. In the earlier meteorology literature it is commonly expressed in units of calories per square meter per minute.

7×10^8 m) is given by

$$\frac{3.90 \times 10^{26}}{4\pi(7 \times 10^8)^2} = 6.34 \times 10^7 \quad \mathrm{W\,m^{-2}}$$

In discussing the theory of radiative transfer, we will have occasion to speak of the radiation having wavelengths within a particular infinitesimal wavelength interval of the electromagnetic spectrum. The irradiance per unit wavelength interval at wavelength λ is called the *monochromatic irradiance* E_λ, which has the units of watts per square meter per micrometer. With this definition, the irradiance is readily seen to be

$$E = \int_0^\infty E_\lambda \, d\lambda \tag{6.3}$$

Thus, if E_λ is plotted as a function of λ as in Fig. 6.2, the area under the curve between any two wavelengths is proportional to the contribution of the radiation in that wavelength band to the total irradiance.

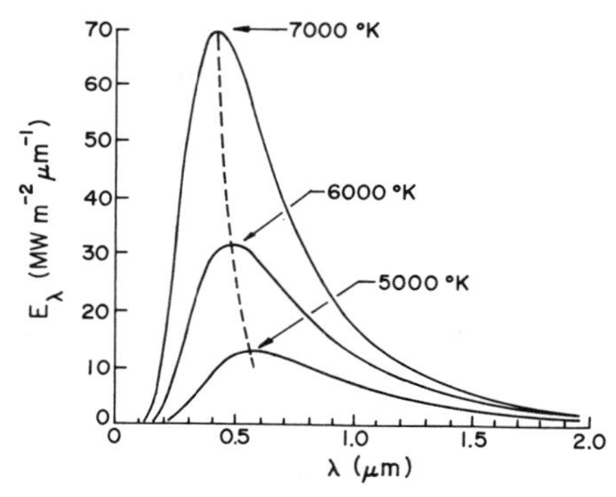

Fig. 6.2 Emission spectra for blackbodies with temperatures as indicated. [Adapted from R. G. Fleagle and J. A. Businger, "An Introduction to Atmospheric Physics," Academic Press, New York (1963), p. 137.]

6.3.1 Diffuse and parallel beam radiation†

In general, the irradiance upon an element of surface area may consist of contributions which come from an infinity of different directions. It is sometimes necessary to identify the part of the irradiance that is coming from directions within some specified infinitesimal arc of solid angle $d\omega$. For this purpose it is useful to define the *radiance* L, which is the irradiance per unit solid angle, expressed in watts per square meter per steradian.‡

† In introductory survey courses at the undergraduate level, this subsection can be omitted without serious loss of continuity provided that Section 6.7 is omitted also.

‡ In literature which predates the adoption of the SI system, this quantity is commonly referred to as *intensity* and denoted by the symbol I.

In order to express quantitatively the relationship between irradiance and radiance, it is necessary to define the *zenith angle* ϕ, which is the angle between the direction of the radiation and the normal to the surface in question. The component of the radiance normal to the surface is then given by $L \cos \phi$. The irradiance represents the combined effects of the normal component of the radiation coming from the whole hemisphere; that is,

$$E = \int_0^{2\pi} L \cos \phi \, d\omega \tag{6.4}$$

In order to indicate more clearly the nature of this integration over solid angle, it may be helpful to review some solid geometry. Let us picture an infinitesimal area, located at the center of a sphere of unit radius, which is emitting or receiving radiation, as pictured in Fig. 6.3. Let the orientation of the area define the equatorial plane so that the normal to the surface passes through the poles of the sphere. The zenith angle ϕ is then identical to co-latitude. An element of solid angle $d\omega$ is, by definition, identical to a unit of area on the surface of the unit sphere. Such an area can be expressed in spherical coordinates in the form

$$d\omega = \sin \phi \, d\phi \, d\theta$$

where θ is the longitude or azimuth angle. The integral of $d\omega$ over the upper hemisphere is given by

$$\int_0^{\pi/2} \sin \phi \, d\phi \int_0^{2\pi} d\theta = 2\pi$$

which is readily seen to be equal to half the area of the sphere of unit radius. Using the definition of solid angle, (6.4) can be written in the form

$$E = \int_0^{2\pi} \int_0^{\pi/2} L \cos \phi \sin \phi \, d\phi \, d\theta$$

The following two problems illustrate the relationship between irradiance and radiance.

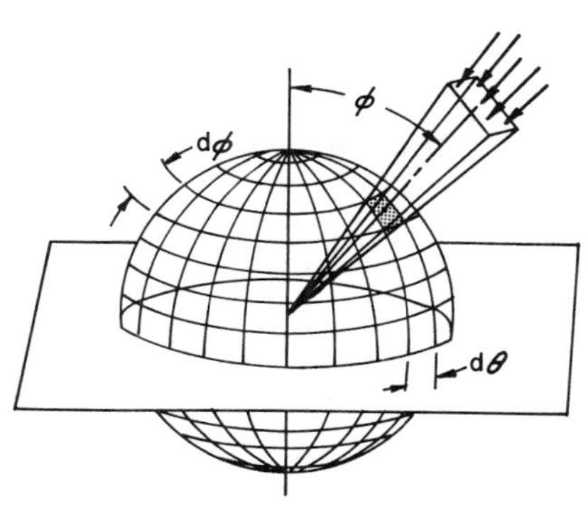

Fig. 6.3 Relationship between radiance and irradiance. The radiation is impinging on a unit area in the "equatorial plane" of a sphere of unit radius.

Problem 6.1 Radiation is emitted from an infinite plane surface with uniform radiance in all directions. What is the irradiance on a plane parallel to the surface?

Solution

$$E = L \int_0^{2\pi} \cos\phi \, d\omega = L \int_0^{2\pi} \int_0^{\pi/2} \cos\phi \sin\phi \, d\phi \, d\theta$$

$$= 2\pi L \int_0^{\pi/2} \cos\phi \sin\phi \, d\phi = \pi L$$

Problem 6.2 Prove that the radiance of solar radiation is independent of distance from the sun, provided that the distance is large and that radiation emitted from each elemental area on the sun is independent of zenith angle.

Solution The flux of outgoing radiation passing through a sphere of radius R, concentric with the sun (given by $4\pi R^2 E$), must be independent of radius, since virtually no energy is absorbed in space and the planets absorb only an infinitesimal fraction of the emitted solar radiation. Therefore, $E \propto R^{-2}$.

At large distances from the sun, the radiance L is approximately equal to $E/\delta\omega$, where $\delta\omega$ is the arc of solid angle subtended by the sun when viewed from a point in space. By definition

$$\delta\omega = D/R^2$$

where D is the cross-sectional area of the solar disk.

Thus $L = E \times (R^2/D)$ and, since $E \propto R^{-2}$, it follows that L is independent of radius. From the data given previously in this section, it is easily shown that the radiance of solar radiation is approximately equal to 2.02×10^7 W m^{-2} sr^{-1}.

Radiation emanating from a source that subtends a finite arc of solid angle is called *diffuse radiation.* In the limiting case of radiation coming from a concentrated source, the radiance, as defined previously, approaches infinity and the angle subtended by the source approaches zero, (that is, all the radiation is coming from practically the same direction). Emission from such a concentrated source is sometimes referred to as *parallel beam radiation.* For most purposes it is sufficiently accurate to treat the solar radiation reaching the earth as a parallel beam, which effectively eliminates the integration over solid angle.

6.3.2 Measurement of radiation

Several methods may be used for measuring the intensity of radiation. For example, if the radiation is allowed to fall simultaneously upon a black and upon a white surface, the black surface absorbs most of the radiation, while the white surface reflects most of the radiation. The difference in the rates of increase in the temperatures of the two surfaces gives a measure of the incident radiation. In another class of radiation instrument a thin blackened surface, supported inside a large polished case, is exposed to the radiation. The temperature of the blackened surface then changes until its rate of heat loss is equal to the rate at which it gains heat from the incident radiation. In order that the change in temperature be determined only by the radiation intensity, convection currents inside the case must be eliminated and the loss of heat from the blackened surface must be independent of the air temperature and wind speed. If the case is fairly massive and its outer surface highly polished, the latter requirements can be approached. The temperature of the blackened surface may be monitored with a thermocouple or thermopile which has a reference point located inside the case. In order to determine the intensity of radiation within specified wavelength intervals, the radiation can first be passed through high-quality optical filters which only allow transmission of those wavelengths.†

6.4 BLACKBODY RADIATION

A *blackbody* is a hypothetical body‡ comprising a sufficient number of molecules absorbing and emitting electromagnetic radiation in all parts of the electromagnetic spectrum so that

- all incident radiation is completely absorbed (hence the term black), and
- in all wavelength bands and in all directions the maximum possible emission is realized.

The amount of radiation emitted by a blackbody is uniquely determined by its temperature, as described by *Planck's law*, which states that the monochromatic irradiance of radiation emitted by a blackbody at (absolute) temperature T is given by

$$E_\lambda{}^* = \frac{c_1}{\lambda^5[\exp(c_2/\lambda T) - 1]} \tag{6.5}$$

where $c_1 = 3.74 \times 10^{-16}$ W m^2 and $c_2 = 1.44 \times 10^{-2}$ m °K. Blackbody radiation is *isotropic*; that is to say, the radiance is independent of direction.

When $E_\lambda{}^*$ is plotted as a function of wavelength for any given temperature, the resulting spectrum of monochromatic irradiance displays the characteristic shape illustrated in Fig. 6.2, with a sharp short wavelength cutoff, a steep rise to the maximum, and a more gentle dropoff toward longer wavelengths. Throughout most of the wavelength range, the exponential term in (6.5) is much larger than unity; therefore,

$$E_\lambda{}^* \simeq c_1 \lambda^{-5} \exp(-c_2/\lambda T)$$

Only at the long wavelength end of the curves (well to the right of the peaks) does it become necessary to use the full expression (6.5).

† For further details on radiation measuring instruments the reader is referred to the "Handbook of Meteorological Instruments," Her Majesty's Stationery Office, London, 1956.

‡ The term *body* refers to a coherent mass of material which can be regarded as having uniform temperature and composition (for example, a layer of gas of a specified thickness, or the surface layer of a mass of solid material).

6.4.1 The Wien displacement law

Using the above approximation of (6.5) it can be shown that the wavelength of peak emission for a blackbody at temperature T is given by

$$\lambda_m = 2897/T \tag{6.6}$$

where λ_m is expressed in micrometers and T in degrees Kelvin. (See Problem 6.14.) Through (6.6), which is known as the *Wien*† *displacement law*, it is possible to estimate the temperature of a radiation source from a knowledge of its emission spectrum. Let us consider a specific example.

Problem 6.3 The wavelength of maximum solar emission is about 0.475 μm, which corresponds to blue light. Use the Wien displacement law to compute the "color temperature" of the sun.

Solution

$$T = 2897/\lambda_m = 2897/0.475 = 6100°\text{K}$$

The sun appears more yellow than blue because of the asymmetry in the shape of the blackbody curve (that is, most of the radiation is emitted at wavelengths longer than that of the peak monochromatic irradiance). Stars cooler than the sun emit maximum radiation at longer wavelengths (that is, they appear more red) and vice versa.

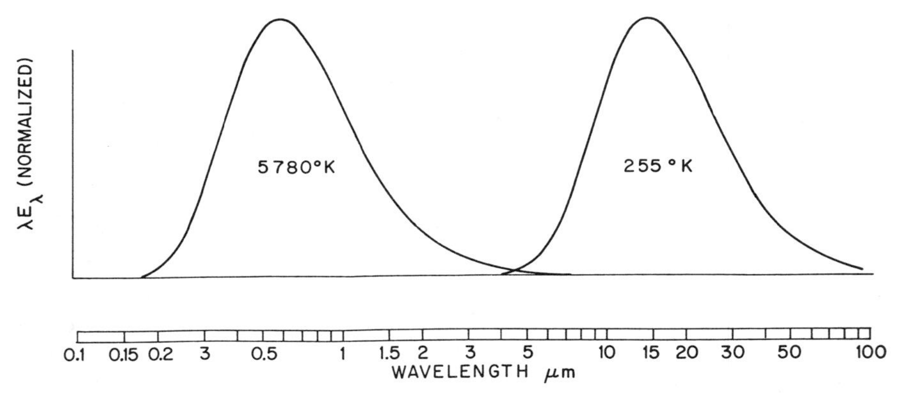

Fig. 6.4 Normalized blackbody spectra representative of the sun (left) and earth (right), plotted on a logarithmic wavelength scale. The ordinate is multiplied by wavelength in order to make area under the curves proportional to irradiance. [Adapted from R. M. Goody, "Atmospheric Radiation," Oxford Univ. Press (1964), p. 4.]

An important consequence of the Wien displacement law is the fact that solar radiation is concentrated in the visible and near-infrared parts of the spectrum, while radiation emitted by the planets and their atmospheres is largely confined to the infrared. This separation of wavelengths is evident in Fig. 6.4, which

† **Wilheim Wien** (1864–1928) German physicist. Received the Nobel prize in 1911 for the discovery in (1893) of the displacement law named after him. Also made the first rough determination of the wavelength of X rays (1905).

shows the normalized blackbody curves for temperatures of 5780°K and 255°K, plotted on a common wavelength scale. The nearly complete absence of overlap between the curves justifies dealing with solar and planetary radiation separately in many problems of radiative transfer. (We make implicit use of this separation in formulating Problems 6.7 and 6.8.)

6.4.2 The Stefan–Boltzmann law

The *blackbody irradiance*, obtained by integrating (6.5) over all wavelengths, is given by

$$E^* = \sigma T^4 \tag{6.7}$$

where σ is a constant (called the Stefan†–Boltzmann constant) which has a value of 5.67×10^{-8} W m^{-2} deg^{-4}. Equation (6.7) is called the *Stefan–Boltzmann law*.

The dependence of blackbody emission upon temperature is apparent in Fig. 7.11b, which shows an infrared satellite photograph based upon radiation in the 10.5–12.5-μm wavelength band, which is near the peak of the blackbody curve for terrestrial radiation. The photograph is a negative, so that dark areas represent regions of strong emission (high temperatures). The contrast between the warm waters of the Gulf Stream and the colder water adjacent to the coastline is quite evident in the figure. Middle and high cloud decks show up clearly because their tops are colder than the earth's surface.

Problem 6.4 The average irradiance of solar radiation reaching the earth's orbit is 1.38×10^3 W m^{-2}. Nearly all the radiation is emitted from the outer-most visible layer of the sun, which has a mean radius of 7×10^8 m. Calculate the *equivalent blackbody temperature* or *effective temperature* of this layer (that is, the temperature a blackbody would have to have in order to emit the same amount of radiation). The mean distance between the earth and sun is 1.5×10^{11} m.

Solution We first calculate the irradiance at the top of the layer, making use of the inverse square dependence of irradiance upon distance from the sun as deduced in Problem 6.2:

$$E = 1.38 \times 10^3 \times \left[\frac{1.5 \times 10^{11}}{7 \times 10^8}\right]^2 \ \text{W m}^{-2} = 6.34 \times 10^7 \ \text{W m}^{-2}$$

Note that this result is consistent with the figure given previously in Section 6.3. The equivalent blackbody temperature T_E can then be calculated directly from (6.7), assuming that this radiation is being emitted by a blackbody. Therefore,

$$T_E = \left(\frac{E}{\sigma}\right)^{1/4} = \left(\frac{6.34 \times 10^7}{5.67 \times 10^{-8}}\right)^{1/4} = (1118)^{1/4} \times (10^{12})^{1/4} = 5780°\text{K}$$

† **Joseph Stefan** (1835–1893) Austrian physicist. Became professor of physics at the University of Vienna at age 28. Originated the theory of diffusion of gases as well as carrying out fundamental work on radiation.

Problem 6.5 Calculate the equivalent blackbody temperature of the earth, assuming a *planetary albedo* of 0.30. (The planetary albedo is the fraction of the total incident solar radiation that is reflected back into space without absorption.) Assume that the earth is in *radiative equilibrium*, so that there is no net energy gain or loss due to radiation.

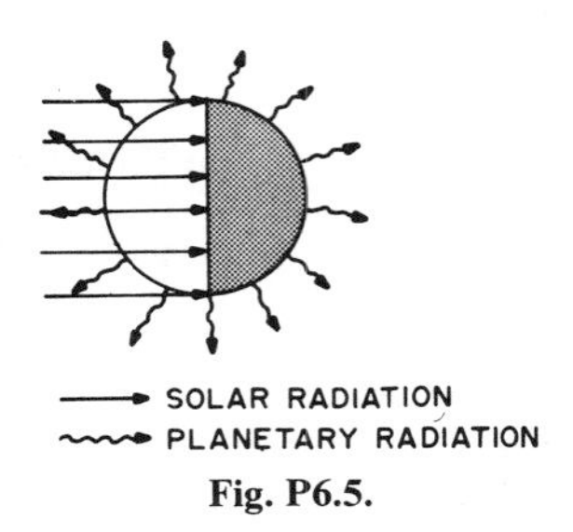

Fig. P6.5.

Solution Let S be the irradiance of solar radiation incident upon the earth (1380 W m^{-2}); E, irradiance of planetary radiation emitted to space; R_E, radius of the earth; and A, planetary albedo. For radiative equilibrium, incoming flux = outgoing flux (see Fig. P6.5)

$$(1 - A)S\pi R_E^2 = E4\pi R_E^2$$

Therefore,

$$E = (1 - A)S/4 = 241 \quad \text{W m}^{-2}$$

Then we proceed as in the previous problem to obtain the equivalent blackbody temperature $T_E = 255°\text{K}$. This result, together with corresponding data for some of the other planets, is listed in Table 6.2.

Table 6.2

Equivalent blackbody temperature T_E of some of the planets[a]

Planet	Distance from sun (in units of 10^6 km)	Albedo	T_E(°K)
Mercury	58	0.06	442
Venus	108	0.78	227
Earth	150	0.30	255
Mars	228	0.17	216
Jupiter	778	0.45	105

[a] Based upon radiative equilibrium considerations. From R. M. Goody and J. C. G. Walker, "Atmospheres," Prentice-Hall, Inc., Englewood Cliffs, New Jersey, 1972, with modifications.

6.5 ABSORPTIVITY AND EMISSIVITY

It was mentioned in the previous section, and it should be reemphasized, that blackbody radiation represents the upper limit to the amount of radiation that a real substance may emit at a given temperature. At any given wavelength

λ, one can speak of the *emissivity*†

$$\varepsilon_\lambda \equiv E_\lambda / E_\lambda^* \tag{6.8}$$

which is a measure of how strongly the body radiates at that wavelength. By definition, the emissivity of a blackbody is unity at all wavelengths, and the emissivity of any real substance is between zero and one, and may vary with wavelength.

It is also useful to define a "*gray body*" *emissivity*:

$$\varepsilon \equiv E/E^* = E/\sigma T^4 \tag{6.9}$$

Here the term "gray" arises from the neglect of the wavelength (color) dependence of the emissivity. (Strictly speaking, a *gray body* is a substance whose emissivity is independent of wavelength.)

It is possible to define corresponding quantities called the *absorptivity* a_λ and the *gray body absorptivity* a which are measures of the ratio of the irradiance absorbed by a particular body to that which is incident upon it. By definition, the absorptivity of a blackbody is equal to one at all wavelengths.

6.5.1 Kirchhoff's law

Let us consider the steady-state situation pictured in Fig. 6.5 with two parallel plates, infinite in extent, separated by empty space. The plate on the left radiates as a blackbody, while the plate on the right is a gray body with absorptivity a and emissivity ε. The surfaces which face outward are perfectly insulated and the plates are in radiative equilibrium.

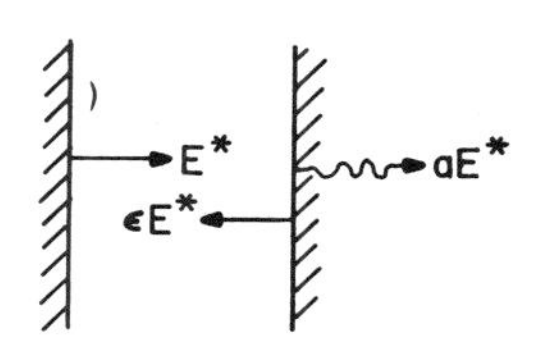

Fig. 6.5 Hypothetical situation pictured in proof of Kirchhoff's law. Arrows indicate irradiance emitted by the "black" plate (left) and "gray" plate (right). Radiation reflected by the "gray" plate is not shown.

Let us begin by assuming that the radiative equilibrium temperatures of the two plates are different. Now let us suppose that some substance which conducts heat is temporarily inserted between the plates, so that heat is allowed to flow from the hotter one to the colder one. Now if radiation acts to restore radiative equilibrium, it must produce a net transfer of heat from the colder

† Strictly speaking, the emissivity E_λ defined by (6.8) should be called the *irradiance emissivity*. An analogous definition can be made in terms of the radiance. Only in the special case of parallel beam radiation are the radiance and irradiance emissivities the same. In this text, except in Section 6.7, the term *emissivity* applies to irradiance emissivity as defined in (6.8). The same remarks apply to absorptivity, reflectivity, and transmissivity, which are defined later in this section.

plate back to the hotter one, in violation of the Second Law of Thermodynamics (see Section 2.8). Therefore, the radiative equilibrium temperatures of the two plates must be the same.

Now in the radiative equilibrium situation pictured in Fig. 6.5, the irradiance $E^* = \sigma T^4$ from the "black" plate is impinging upon the "gray" plate on the right and the portion aE^* is being absorbed. Since both plates are at the same temperature, the emission from the plate on the right is εE^*. For thermal equilibrium, each plate must emit as much radiation as it absorbs; hence, $\varepsilon E^* = aE^*$, from which it follows that ε and a are equal.

By means of a somewhat more involved argument it is possible to establish the more stringent relationship

$$a_\lambda = \varepsilon_\lambda \tag{6.10}$$

which is known as *Kirchhoff's*† *law*. In qualitative terms, Kirchhoff's law states that materials which are strong absorbers at a particular wavelength are also strong emitters at that wavelength; similarly, weak absorbers are weak emitters.

The validity of Kirchhoff's law is not dependent upon whether a body is in thermal and radiative equilibrium. (These conditions were assumed in the above proof only as a matter of convenience.) It can be applied not only to opaque surfaces, but also to gases, provided that the frequency of molecular collisions is large in comparison to the frequency of individual absorption and emission events. In the earth's atmosphere this condition is fulfilled up to altitudes of about 60 km.

Problem 6.6 A completely gray flat surface on the moon with an absorptivity of 0.9 is exposed to direct, overhead solar radiation.

(a) What is the radiative equilibrium temperature of the surface?

(b) If the actual temperature of the surface is 300°K, what is the net irradiance immediately above the surface?

Solution Since the moon has no atmosphere, the solar radiation constitutes the total incident upon the surface.

(a) For radiative equilibrium

$$E(\text{absorbed}) = E(\text{emitted})$$

$$a \times 1380 \quad \text{W m}^{-2} = \varepsilon\sigma T_E^4$$

where

$$a = \varepsilon, \quad \text{by Kirchhoff's law}$$

Therefore,

$$T_E = \left(\frac{1380}{\sigma}\right)^{1/4} = 395°\text{K}$$

(b) The temperature of the surface is lower than the radiative equilibrium value, so the net radiation will be downward:

$$E_{net}\downarrow = a(1380)\downarrow - a\sigma T^4\uparrow = 1242 - 412 = 830 \quad W\ \text{m}^{-2}$$

† **Gustav Kirchhoff** (1824–1887) German physicist. Was made professor of physics at the University of Breslau at age 26. In addition to his work in radiation, he made fundamental discoveries in electricity and spectroscopy. Also developed (with Bunsen) spectrum analysis. Discovered cesium and rubidium.

6.5.2 Selective absorbers and emitters

Note that the radiative equilibrium temperature of the "gray" surface in the above problem is independent of the absorptivity. When the absorptivity is a function of wavelength, the situation may be considerably different, as illustrated in the following problem.

Problem 6.7 A flat surface is subject to overhead solar radiation, as in the previous problem. The absorptivity is 0.1 for solar radiation and 0.8 in the infrared part of the spectrum, where most of the emission takes place. (Here we make use of the spectral separation between solar and planetary radiation that was shown in Fig. 6.4.) Compute the radiative equilibrium temperature.

Solution

$$E(\text{absorbed}) = E(\text{emitted})$$
$$0.1 \times 1380 \quad \text{W m}^{-2} = 0.8\sigma T_E^{\,4}$$
$$T_E = 235°\text{K}$$

(It should be noted that this value is not typical of the radiative equilibrium temperature of such a surface on earth, since we have not yet taken into account the presence of the atmosphere.)

This result applies to substances like snow which are relatively weak absorbers at visible and near-infrared wavelengths but strong absorbers in the infrared. In general, the radiative equilibrium temperature of such substances is lower than that of a gray body exposed to the same incident solar radiation. However, when a gas that displays the same type of absorption spectrum is a constituent of a planetary atmosphere, the result is quite different. Let us consider the following example:

Problem 6.8 Calculate the radiative equilibrium temperature of the earth's surface and atmosphere assuming that the atmosphere can be regarded as a thin layer with an absorptivity of 0.1 for solar radiation and 0.8 for terrestrial radiation. Assume that the earth's surface radiates as a blackbody at all wavelengths.

Solution (a) Let x be the irradiance emitted from the earth's surface; y, the irradiance emitted by the atmosphere (both upwards and downwards); and E, the net solar irradiance absorbed by the earth–atmosphere system as calculated in Problem 6.5 (241 W m^{-2}).

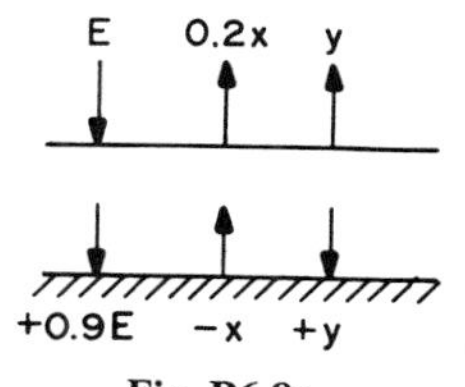

Fig. P6.8a.

(See Fig. P6.8a.) At the earth's surface, radiative equilibrium requires that

$$0.9E + y = x$$

while the radiation balance for the earth–atmosphere system leads to the relation

$$0.2x + y = E$$

Solving these simultaneous equations we obtain $x = 1.58E$ and $y = 0.69E$. Thus, for the earth's surface,

$$\sigma T^4 = 1.58 \times 241 \quad \mathrm{W\ m^{-2}}$$

Therefore, $T = 277°\mathrm{K}$. For the atmosphere,

$$0.8\sigma T^4 = 0.69 \times 241 \quad \mathrm{W\ m^{-2}}$$

Therefore, $T = 245°\mathrm{K}$

(b) This problem can also be solved by the following iterative method, which proves to be somewhat more tedious, but possibly more enlightening. For each unit of the incoming radiation E, 0.10 units are absorbed in the atmosphere and the remaining 0.90 are transmitted to the surface of the planet, where they are absorbed. For radiative equilibrium, these same 0.90 units must be emitted from the surface of the planet, in the form of longwave (infrared) radiation. Of these, the fraction 80% (0.72 units) is absorbed in its passage through the atmosphere, and 20% (0.18 units) passes through directly into space without absorption. In these processes the atmosphere has absorbed 0.10 units of shortwave radiation and 0.72 units of longwave radiation which it must dispose of; half of these (0.41 units) are emitted to space and the other half (0.41 units) back to the surface of the planet. The 0.41 units emitted to the surface are absorbed and reemitted in the form of longwave radiation. Of this, 80% (0.33 units) is absorbed by the atmosphere and the remaining 20% (0.08 units) passes directly into space. Succeeding iterations are shown in Fig. P6.8b, with the calculations rounded to the nearest hundredth unit. On the next iteration, only 16 units are returned to the surface, and this amount continues to diminish with each succeeding iteration until its size is negligible for all practical purposes. Note that the total flux of longwave radiation to space must equal the incoming solar flux, since neither the surface nor the atmosphere has been allowed to store any heat.

Next we compute the total longwave radiation emitted by the ground, which is $(0.90 + 0.41 + 0.16 + \cdots) = 1.58$ units, and the corresponding total for the atmosphere which is $(0.41 + 0.17 + 0.06 + \cdots) = 0.69$ units. Now we can proceed as in (a).

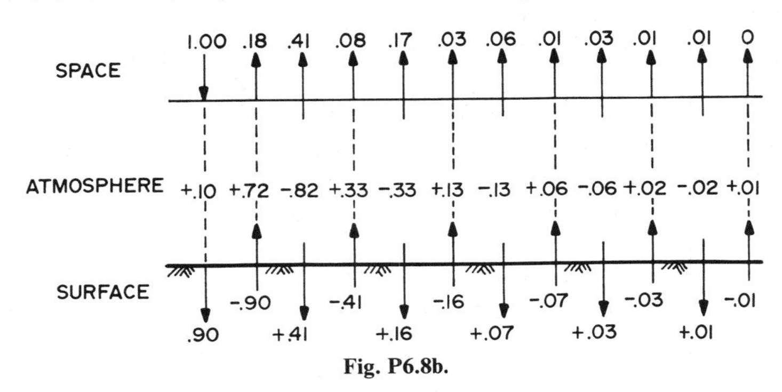

Fig. P6.8b.

This problem is a common one in radiative transfer. Both methods of solution can readily be extended to a multilayer atmosphere for which one can compute the vertical profile of radiative equilibrium temperature. (For further examples, see Problems 6.27 and 6.28.)

Note that the temperature at the surface of the earth computed in Problem 6.8 is considerably higher than the effective temperature obtained in Problem 6.5, which would have been the surface temperature in the absence of an atmosphere.

Whenever a gas that is a weak absorber in the visible and a strong absorber in the infrared is a constituent of a planetary atmosphere, it contributes toward raising the surface temperature of the planet.† The warming results from the fact that incoming radiation can penetrate to the ground with relatively little absorption, while much of the outgoing longwave radiation is "trapped" by the atmosphere and emitted back to the ground. In order to satisfy the radiation balance the surface must compensate by emitting more radiation than it would in the absence of such an atmosphere. To emit more it must radiate at a higher temperature. It is easily verified that a truly gray atmosphere would not produce this effect (see Problem 6.26).

6.5.3 Disposition of monochromatic radiation

The monochromatic radiation incident upon any *opaque* surface is either absorbed or reflected; that is to say, for any wavelength

$$E_\lambda(\text{absorbed}) + E_\lambda(\text{reflected}) = E_\lambda(\text{incident})$$

Dividing each term in this expression by the incident monochromatic irradiance, we obtain

$$a_\lambda + r_\lambda = 1 \tag{6.11}$$

for all wavelengths, where a_λ is the absorptivity, as defined previously, and r_λ, the fraction of the flux that is reflected, is called the *reflectivity* of the surface. At any wavelength, strong reflectors are weak absorbers (for example, snow at visible wavelengths), and weak reflectors are strong absorbers (for example, asphalt at visible wavelengths). The reflectivities for selected surfaces for the wavelengths of solar radiation are listed in Table 6.3.

Table 6.3

Reflectivity (in percent) of various surfaces in the spectral range of solar radiation[a]

Surface	Reflectivity
Bare soil	10–25
Sand, desert	25–40
Grass	15–25
Forest	10–20
Snow (clean, dry)	75–95
Snow (wet and/or dirty)	25–75
Sea surface (sun >25° above horizon)	<10
Sea surface (low sun angle)	10–70

[a] Based upon data presented in K. Ya. Kondratiev, "Radiation in the Atmosphere," Academic Press, New York, 1969.

† This warming is commonly, but misleadingly, referred to as the "*greenhouse effect.*" Greenhouses attain higher temperatures than the outside air primarily because the glass cover restricts the vertical movement of the air that is heated by solar radiation. Fleagle and Businger ("An Introduction to Atmospheric Physics," Academic Press, New York, 1963, pp. 153–154) suggest that trapping of the radiation by the earth's atmosphere be referred to as the "*atmosphere effect.*"

The monochromatic irradiance incident upon a nonopaque layer may be reflected, absorbed, or transmitted. Thus, by analogy with (6.11), we may write

$$a_\lambda + r_\lambda + \tau_\lambda = 1 \tag{6.12}$$

where τ_λ, the fraction of the irradiance that is transmitted through without absorption, is called the *transmissivity* of the layer.

6.6 ATMOSPHERIC ABSORPTION OF SOLAR RADIATION

In the absence of scattering, the absorption of parallel beam radiation as it passes downward through a horizontal layer of gas of infinitesimal thickness dz is proportional to the number of molecules per unit area that are absorbing radiation along the path. This relationship can be expressed in the form

$$da_\lambda \equiv -\frac{dE_\lambda}{E_\lambda} = -k_\lambda \rho \sec \phi \, dz \tag{6.13}$$

where ρ is the density of the gas and ϕ is the zenith angle. Here absorbed monochromatic irradiance is expressed as an incremental amount of depletion of the incident beam. In (6.13), dE_λ and dz are both negative quantities, so da_λ is positive. The product $\rho \sec \phi \, dz$ is the mass within the volume swept out by a unit cross-sectional area of the incident beam as it passes through the layer, as pictured in Fig. 6.6. The *absorption coefficient* k_λ is a measure of the fraction of the gas molecules per unit wavelength interval that are absorbing radiation at the wavelength in question. We recall from Section 6.2 that k_λ is a function of composition, temperature, and pressure of the gas within the layer. It has units of square meters per kilogram, which makes the product $k_\lambda \rho \, dz$ dimensionless.

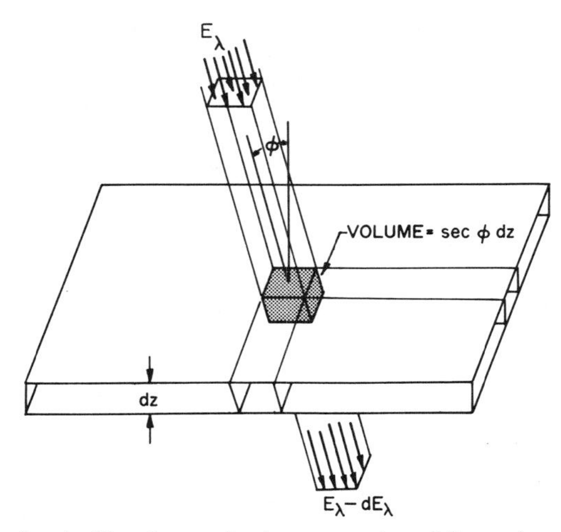

Fig. 6.6 Depletion of an incident beam of unit cross section while passing through an absorbing layer of infinitesimal thickness.

Now we integrate (6.13) from the level z up to the "top" of the atmosphere (denoted by the infinity symbol). The result is

$$\ln E_{\lambda\infty} - \ln E_{\lambda} = \sec\phi \int_z^{\infty} k_{\lambda}\rho \, dz \tag{6.14}$$

Taking the antilog of both sides and rearranging, we obtain

$$E_{\lambda} = E_{\lambda\infty} \exp(-\sigma_{\lambda}) \tag{6.15}$$

where

$$\sigma_{\lambda} = \sec\phi \int_z^{\infty} k_{\lambda}\rho \, dz$$

This relation, often referred to as *Beer's*[†] *Law*, states that irradiance decreases monotonically with increasing path length through the layer. The dimensionless quantity σ_{λ} is called the *optical depth*, or *optical thickness*, depending upon the context in which it is used.[‡] It is a measure of the cumulative depletion that the beam of radiation has experienced as a result of its passage through the layer. In penetrating downward through a planetary atmosphere to an optical depth of one, or in passing through a layer of unit optical thickness, the monochromatic irradiance of the incident beam is diminished by a factor e.

The transmissivity of the layer of gas lying above the level z is given by

$$\tau_{\lambda} \equiv E_{\lambda}/E_{\lambda\infty} = e^{-\sigma_{\lambda}} \tag{6.17}$$

and it follows that, in the absence of scattering, the absorptivity

$$a_{\lambda} = 1 - \tau_{\lambda} = 1 - e^{-\sigma_{\lambda}} \tag{6.18}$$

approaches unity exponentially with increasing optical depth. At wavelengths close to the center of absorption lines, k_{λ} is large so that a very short (density weighted) path length is sufficient to absorb virtually all the incident radiation. At wavelengths away from absorption lines, a path length many orders of magnitude longer may be required to produce any noticeable absorption.

Problem 6.9 Parallel beam radiation is passing through a layer 100 m in thickness, containing a gas with an average density of 0.1 kg m^{-3}. The beam is directed at 60° from the normal to the layer. Calculate the optical thickness, transmissivity, and absorptivity of the layer at wavelengths λ_1, λ_2, and λ_3 for which the absorption coefficients are 10^{-3}, 10^{-1}, and 1 m^2 kg^{-1}, respectively.

Solution The density weighted path length is given by

$$u = \sec\phi \int \rho \, dz$$

$$= 2 \times 0.1 \text{ kg m}^{-3} \times 100 \text{ m} = 20 \text{ kg m}^{-2}$$

[†] **August Beer** (1825–1863) German physicist, noted for this work in optics.

[‡] In some references optical depth is defined as

$$u \equiv \sec\phi \int_z^{\infty} \rho \, dz \tag{6.16}$$

We will refer to u as the *density weighted path length*. If k_{λ} is independent of path length, it is clear from (6.15) and (6.16) that

$$\sigma_{\lambda} = k_{\lambda} u$$

(where the integration extends from the bottom to the top of the layer). Since k_λ is not a function of path length through the layer, the optical thickness is given by

$$\sigma_\lambda = k_\lambda \sec\phi \int^{\infty} \rho \, dz = k_\lambda u$$

and

$$\tau_\lambda = e^{-\sigma_\lambda} = e^{-k_\lambda u}, \qquad a_\lambda = 1 - \tau_\lambda = 1 - e^{-k_\lambda u}$$

Substituting for k_λ and u in the above equations, we obtain the results in the accompanying tabulation.

	$\lambda = \lambda_1$	$\lambda = \lambda_2$	$\lambda = \lambda_3$
σ_λ	0.02	2	20
τ_λ	0.98	0.135	2×10^{-9}
a_λ	0.02	0.865	1.00

6.6.1 The absorption spectrum of a layer of gas

The nonlinear relationship between absorptivity and opitcal depth causes the individual lines in the absorption spectrum (that is, a plot of a_λ versus λ) to broaden and merge into *absorption bands* as the (density weighted) path length increases. This transformation is illustrated in Fig. 6.7, which shows the shape of an idealized spectral line for three different path lengths, where k_λ is assumed to be independent of z within the layer. For sufficiently short path lengths, (6.15), (6.16), and (6.18) reduce to

$$a_\lambda \simeq \sigma_\lambda = k_\lambda u$$

so that there is a linear relationship between absorptivity and density weighted path length at all wavelengths. As the path length increases the relationship gradually becomes nonlinear, with a_λ asymptotically approaching unity as virtually complete absorption or "saturation" is realized over an ever-widening band of wavelength.

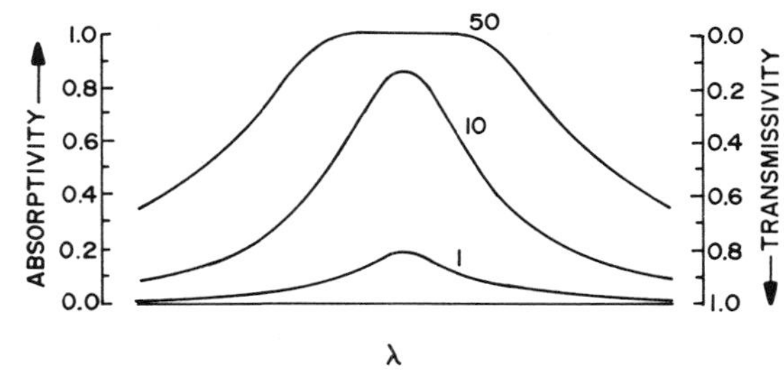

Fig. 6.7 Absorption spectrum of an idealized spectral line for three different density weighted path lengths with ratios as indicated in the figure. [Adapted from *Meteorol. Monographs* **4**, 23, Amer. Meteorol. Soc., Boston (1960), p. 5.]

As the widening of individual absorption lines progresses, adjacent lines begin to overlap, first partially and then completely. Overlapping occurs first within the line clusters associated with rotation–vibration transitions, which were discussed in Section 6.2. Thus, for a certain range of values of u, the line clusters are manifested as absorption bands, in which substantial absorption takes place over a continuous range of wavelengths. The quasi-transparent regions of the spectrum that lie between the major line clusters are sometimes called "*windows.*" Ultimately, as $u \rightarrow \infty$, even the absorption bands merge so that $a_\lambda \rightarrow 1$ at all wavelengths and the gas absorbs and emits as a blackbody.

6.6.2 Indirect determination of the solar spectrum

Indirect calculation of the spectrum of solar radiation incident on the top of the atmosphere, on the basis of ground-based measurements, provides an interesting example of the application of Beer's law. Such calculations were made quite successfully many years before direct measurements of undepleted solar radiation were available from satellites.

Equation (6.14) can be rewritten in the form†

$$\ln E_\lambda = \ln E_{\lambda\infty} - \sec \phi \int_z^\infty k_\lambda \rho \, dz$$

Over the course of a single day E_λ is measured at frequent intervals at a ground station. During this period the numerical value of the integral in the above

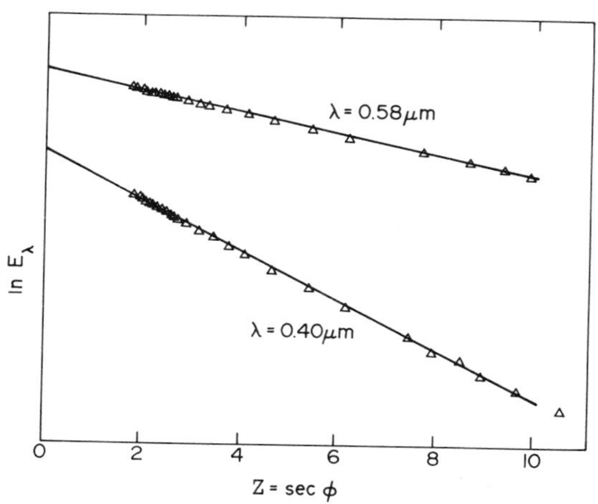

Fig. 6.8 Monochromatic irradiance of solar radiation measured at the ground as a function of secant of solar zenith angle under clear, stable atmospheric conditions, at Tucson, Arizona, 12 December 1970. [From *J. Appl. Meteor.* **12**, 376 (1973).]

† Depletion of solar radiation is a result of both scattering and absorption by atmospheric constituents. Therefore the coefficient k_λ in this expression includes implicitly both effects.

expression changes very little in comparison to the large changes in solar zenith angle. Thus, to a good approximation, the above expression assumes the form

$$\ln E_\lambda = A - BZ$$

where $Z = \sec\phi$ and A and B are constants; that is to say, when the individual data points for E_λ are plotted on a logarithmic scale as a function of $\sec\phi$, they tend to fall along a straight line as illustrated in Fig. 6.8. Since the (density weighted) path length is directly proportional to Z, it is possible to deduce the monochromatic irradiance upon the top of the atmosphere simply by extending the straight line that makes the "best fit" to the data points back to $Z = 0$ (that is, zero path length through the atmosphere).

6.6.3 The vertical profile of absorption

The following simplified mathematical model was first proposed by Chapman† to explain the vertical distribution of absorption of monochromatic radiation in planetary atmospheres.‡ Let us consider the case of parallel beam radiation from an overhead sun, penetrating into a well-mixed isothermal atmosphere in which k_λ is independent of height. In an isothermal atmosphere (2.2) and (2.26) can be combined to give

$$\rho = \rho_0 \exp(-z/H) \tag{6.19}$$

where ρ_0 is the density at sea level. (We will ignore the fine distinction between geopotential height and geometric height.) Substituting for density in the expression (6.15) for optical depth, we obtain

$$\sigma_\lambda = k_\lambda \rho_0 \int_z^\infty \exp(-z/H)\, dz$$

which can be integrated to give a more explicit expression for optical depth as a function of height

$$\sigma_\lambda = Hk_\lambda \rho_0 \exp(-z/H) \tag{6.20}$$

The incident radiation absorbed within any differential layer of the atmosphere is given by

$$dE_\lambda = E_{\lambda\infty} \tau_\lambda\, da_\lambda \tag{6.21}$$

where τ_λ is the transmissivity of that portion of the atmosphere that lies above the layer in question. After substituting from (6.17) for τ_λ, (6.13) for da_λ, and (6.19) for ρ, the differential absorption assumes the form

$$dE_\lambda = (E_{\lambda\infty} k_\lambda \rho_0) e^{-z/H} e^{-\sigma_\lambda}\, dz$$

† **Sydney Chapman** (1888–1970) English geophysicist. Made important contributions to a wide range of geophysical problems, including geomagnetism, space physics, photochemistry, and diffusion and convection in the atmosphere.

‡ In an elementary course the mathematical development can be omitted without serious loss of continuity.

Substituting for $\exp(-z/H)$ from (6.20), we obtain an expression for the absorption per unit thickness of the layer as a function of optical depth,

$$\frac{dE_\lambda}{dz} = \frac{E_{\lambda\infty}}{H}\sigma_\lambda e^{-\sigma_\lambda}$$

Now at the level where the absorption is strongest,

$$\frac{d}{dz}\left(\frac{dE_\lambda}{dz}\right) = \frac{E_{\lambda\infty}}{H}\frac{d}{dz}(\sigma_\lambda e^{-\sigma_\lambda}) = 0$$

Performing the indicated differentiation, it is readily shown that $\sigma_\lambda = 1$; that is to say, the strongest absorption takes place at the *level of unit optical depth.*

The shape of the vertical profile of absorption rate $\partial E_\lambda/\partial z$ is shown in Fig. 6.9, together with profiles of E_λ and ρ. We recall from (6.13) that

$$\frac{\partial E_\lambda}{\partial z} \propto E_\lambda \rho$$

The scale for optical depth is shown at the right-hand side of the figure. At levels for which $\sigma_\lambda \ll 1$ the incoming beam is virtually undepleted, but the density is so low that there are too few molecules to produce very much absorption. At levels for which $\sigma_\lambda \gg 1$, there is no shortage of molecules, but there is very little radiation left to absorb. The larger the value of the absorption coefficient k_λ the smaller the density required to produce significant amounts of absorption and the higher the level of unit optical depth. For small values of k_λ the radiation may reach the bottom of the atmosphere long before it reaches the level of unit optical depth.

The assumption of an isothermal atmosphere with a constant absorption coefficient was helpful in simplifying the mathematics in the above derivation. However, it turns out that for realistic vertical profiles of T and k_λ the above result is still at least qualitatively valid; that is, most of the absorption takes place along that portion of the ray path for which the optical depth is of order unity.

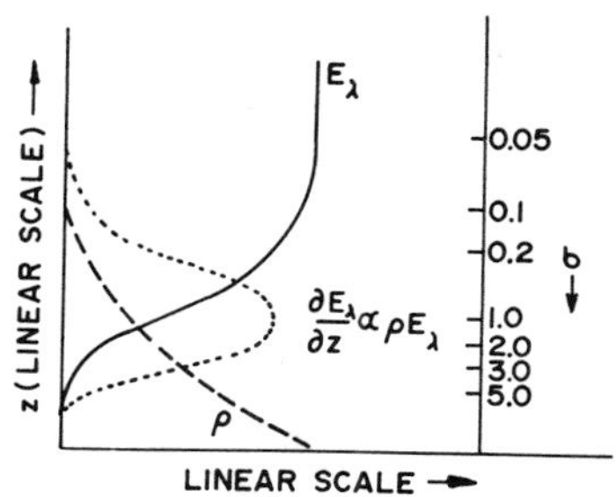

Fig. 6.9 Incident radiation E_λ, rate of absorption $\partial E_\lambda/\partial z$, and air density ρ as a function of height and optical depth in an isothermal atmosphere with k_λ independent of height.

6.7 ATMOSPHERIC ABSORPTION AND EMISSION OF INFRARED RADIATION[†]

The description of the transfer of terrestrial radiation through the atmosphere is complicated by the fact that such radiation is diffuse and, therefore, calculations of irradiance generally involve an integration over solid angle. In this section we will evade the geometrical problems associated with the treatment of diffuse radiation[‡] in order to concentrate on the physical processes that govern absorption and emission of infrared radiation in planetary atmospheres.

All the relationships derived for parallel beam radiation in the previous section can be applied to diffuse radiation, with the following modifications:

- irradiance E_λ should be replaced by radiance L_λ wherever it appears, and
- a_λ and τ_λ should be understood as fractions of the *radiance* along a particular path length that are absorbed by or transmitted through the layer in question.

6.7.1 Schwarzchild's equation

Aside from the geometrical differences discussed above, there is a very important distinction between the transfer of solar and terrestrial radiation through the atmosphere. At the wavelengths of solar radiation, atmospheric emission is negligible, and only absorption needs to be considered. However, at the wavelengths of terrestrial radiation, absorption and emission are equally important and must be considered simultaneously.

The absorption of terrestrial radiation along an upward path through the atmosphere is described by (6.13) with the sign reversed and with radiance substituted for irradiance:

$$-dL_\lambda = L_\lambda k_\lambda \rho \sec \phi \, dz \tag{6.22}$$

The emission of radiation from a gas can be treated in much the same manner as the absorption. Making use of Kirchhoff's law and (6.13) it is possible to write an expression analogous to (6.22) for the emission,

$$dL_\lambda = L_\lambda{}^* \, d\varepsilon_\lambda = L_\lambda{}^* \, da_\lambda = L_\lambda{}^* \, k_\lambda \rho \sec \phi \, dz \tag{6.23}$$

where $L_\lambda{}^*$ is the blackbody monochromatic radiance specified by Planck's law. Now we subtract the absorption from the emission to obtain the net contribution of the layer to the monochromatic radiance of the radiation passing upward through it:

$$dL_\lambda = -k_\lambda (L_\lambda - L_\lambda{}^*) \rho \sec \phi \, dz \tag{6.24}$$

This expression, known as *Schwarzchild's equation*, is the basis for computations of the transfer of infrared radiation.

[†] In an elementary survey course this section can be omitted without serious loss of continuity.

[‡] A more advanced treatment of this subject is given by Fleagle and Businger, "An Introduction to Atmospheric Physics," Academic Press, New York, 1963, pp. 155–161.

For an isothermal gas, with constant k_λ, (6.24) may be integrated to obtain

$$(L_\lambda - L_\lambda^*) = (L_{\lambda 0} - L_\lambda^*) \exp(-\sigma_\lambda) \tag{6.25}$$

where $L_{\lambda 0}$ is the radiance incident on the layer from below. This expression shows that L_λ should exponentially approach L_λ^* as the optical thickness of the layer increases. For a layer of infinite optical thickness the emission from the top is L_λ^* regardless of the value of $L_{\lambda 0}$; that is to say, such a layer behaves as a blackbody, as defined at the beginning of Section 6.4.

6.7.2 Remote temperature sensing

By means of satellites, the spectra of infrared radiation emitted by the earth and the neighboring planets can be monitored. From a careful analysis of these radiation data it is possible to infer the gross features of the vertical profile of atmospheric temperature as a function of latitude and longitude. To illustrate the relationship between temperature and radiance, let us first consider a qualitative example.

Figure 6.10 shows the spectra of infrared radiation emitted from three locations on the planet Mars (a–c) and one location on Earth (d). These spectra are

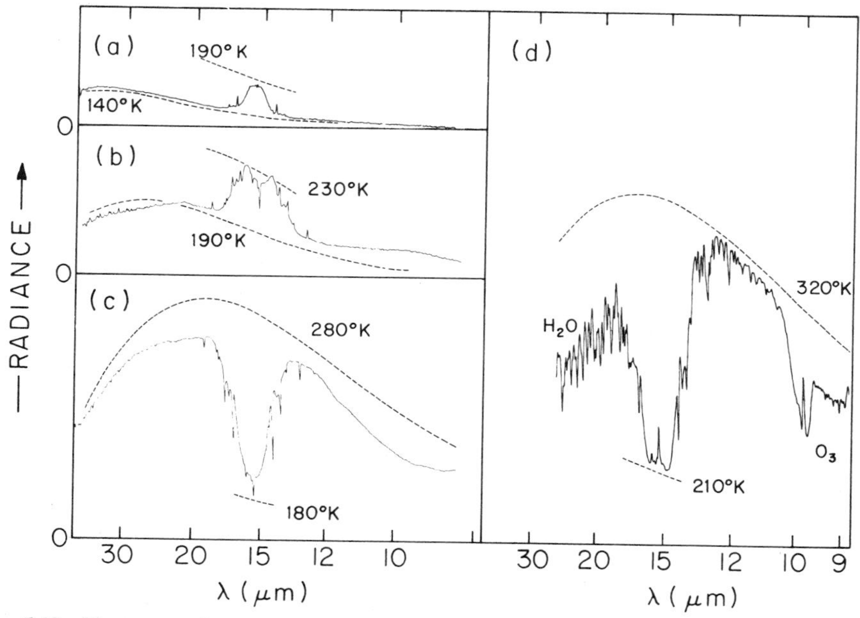

Fig. 6.10 Planetary radiance spectra observed from space. (a) Martian north polar region during winter, (b) Martian south polar region during summer under dust storm conditions, (c) Martian middle-latitude location under clear conditions; all from Mariner space probe. (d) Sahara Desert region on earth from Nimbus 4 satellite. All spectra are plotted as linear functions of wave number and the radiances have been multiplied by the appropriate factor to retain a linear relationship between area under the curve and radiance. Dashed curves represent selected blackbody spectra with temperatures as indicated. [Adapted from *Icarus* **17**, 425 (1975).]

plotted as a linear function of wavenumber (the reciprocal of wavelength), and therefore the shape of the blackbody curves is somewhat different from those shown in previous figures. The features of primary interest in the following discussion are

- the prominent absorption band extending from about 12 to 18 μm, and
- the "spike" near 15 μm, which corresponds to a cluster of strong absorption lines.

Both features are associated with carbon dioxide.

The spectrum shown in Fig. 6.10a was taken over one of the Martian polar cap regions where the surface temperature is close to 140°K. In this spectrum, the absorption band stands out above a smooth continuum of background radiation emitted from the surface of the planet. Note the similarity of the background continuum to the 140°K blackbody curve. Within the absorption band the radiation reaching the space probe is emitted by carbon dioxide in the lower atmosphere (below $\sim$20 km). The fact that the radiance within this band is large in comparison to the background continuum indicates that the lower atmosphere is considerably warmer than the surface of the planet in this region; that is to say, there is a low-level temperature inversion.

The spectrum shown in Fig. 6.10b was taken in middle latitudes where the surface temperature is close to 190°K, as indicated by the background continuum. Here again the elevated absorption band is indicative of a low-level temperature inversion. The spike near 15 μm appears as a local minimum in the emission spectrum. At wavelengths of the absorption lines that contribute to this spike, carbon dioxide is such a strong absorber and emitter that even the rarified Martian atmosphere above 20 km is capable of radiating as a blackbody. The fact that the monochromatic radiance at the wavelength of the spike is less than that at neighboring wavelengths indicates that the Martian upper atmosphere is colder than the lower atmosphere. The absence of the spike in (a) indicates that in that sounding the upper and lower regions of the atmosphere must have been at about the same temperature.

The spectrum shown in Fig. 6.10c was taken in the Martian tropics. The reader is invited to make his/her own inferences concerning the temperature structure in this region of the Martian atmosphere.

Because of the presence of a number of different absorbers in the earth's atmosphere, the emission spectrum shown in Fig. 6.10d is somewhat more complicated than the Martian spectra. The band between 10.5 and 12 μm corresponds to a spectral "window" through which radiation from the earth's surface can penetrate to space with relatively little absorption. Hence, the radiances in this wavelength band are indicative of the effective temperature of the underlying surface, in this case the Sahara Desert. Emission in the carbon dioxide band is indicative of the tropospheric temperature, which is considerably lower than the surface temperature. Radiation at the wavelength of the 15-μm spike is emitted by carbon dioxide in the lower stratosphere where temperature increases with height. Hence this spike appears as a peak in the absorption spectrum.

The purpose of this qualitative example is to demonstrate that, by sampling the radiation emitted at a number of wavelengths which correspond to different atmospheric absorptivities, one can infer some of the characteristics of the vertical profile of temperature in a planetary atmosphere.

The relationship between the vertical temperature profile and the infrared emission spectrum can be expressed in a more quantitative form. Consider a satellite which measures the monochromatic intensity emitted from directly below. The amount of radiation contributed by each incremental layer dz is given by

$$dL_\lambda = L_\lambda{}^* \tau_\lambda \, da_\lambda \tag{6.26}$$

which is simply the total emission from the layer, as given in (6.23), multiplied by the transmissivity of that part of the atmosphere which lies above the layer in question. By exploiting the similarity between (6.26) and (6.21) it is readily shown that in an isothermal atmosphere (in which $L_\lambda{}^*$ is independent of height), dL_λ/dz is largest at the level where $\sigma_\lambda = 1$; that is to say, most of the monochromatic radiance impinging upon the satellite is emitted by layers near the level of unit optical depth. Most of the radiation emanating from deeper layers is absorbed on its way up through the atmosphere; there is not enough mass far above the level of unit optical depth to emit very much radiation.

Equation (6.26) can be integrated over height to obtain the total radiance impinging upon the satellite from below:

$$L_\lambda = \tau_{\lambda 0} a_{\lambda 0} L_{\lambda 0}^* + \int_0^\infty \tau_\lambda L_\lambda{}^* k_\lambda \rho \, dz \tag{6.27}$$

where the 0 subscript refers to the earth's surface. The integral can be approximated as the sum of the contributions of N layers of finite thickness, each of them isothermal, so that $L_\lambda{}^*$ can be taken outside the integral. With this approximation, (6.27) can be written in the form

$$L_\lambda \simeq \alpha_0 L_{\lambda 0}^* + \alpha_1 L_{\lambda 1}^* + \alpha_2 L_{\lambda 2}^* + \cdots + \alpha_N L_{\lambda N}^*$$

where

$$\alpha_0 = \tau_{\lambda 0} a_{\lambda 0}$$

and

$$\alpha_i = \tau_{\lambda i} \int k_{\lambda i} \rho_i \, dz_i \qquad (i = 1, N)$$

The integration is carried out from the bottom to the top of the layer in question. The term $\tau_{\lambda i}$ is the transmissivity of the atmosphere lying above the ith layer. Making use of (6.15) it is also possible to write the coefficients in the form

$$\alpha_i = \tau_{\lambda i} \sigma_{\lambda i}$$

where $\sigma_{\lambda i}$ is the optical thickness of the ith layer. The α_i can be determined quite accurately from data on average atmospheric composition as a function of height.

Now for each of the layers in (6.27) the $L_\lambda{}^*$ in various wavelength bands are related by Planck's law. Therefore, if measurements of L_λ are available at N different wavelengths, it should be possible, in principle, to solve the resulting set of simultaneous equations to obtain $L_\lambda{}^*$ in each layer. The temperatures corresponding to these blackbody radiances can then be determined from Planck's law. In practice, accurate solutions can be obtained only if each of the N equations contains a sufficient amount of information that is not redundant with the other equations. Thus there is an upper limit to the amount of vertical resolution that can be obtained with remote sensing. Even with this limitation it is usually possible from radiance measurements in cloud-free areas to infer average temperatures over layers a few kilometers thick to within about 1 deg.

6.8 SCATTERING OF SOLAR RADIATION

It is possible to formulate an expression analogous to (6.13) for ds_λ, the fraction of parallel beam radiation that is scattered when passing downward through a layer of infinitessimal thickness: namely,

$$ds_\lambda \equiv dE_\lambda/E_\lambda = KA \sec\phi \, dz \tag{6.28}$$

where K is a dimensionless coefficient, and A is the cross-sectional area that the particles in a unit volume present to the beam of incident radiation. If all the particles which the beam encounters in its passage through the differential layer were projected onto a plane perpendicular to the incident beam, the product $A \sec\phi \, dz$ would represent the fractional area occupied by the particles. Thus, K plays the role of a *scattering area coefficient* which measures the ratio of the effective scattering cross section of the particles to their geometric cross section. In the absence of absorption, (6.28) can be integrated to obtain expressions analogous to (6.14), (6.15), (6.17), and (6.18) for τ_λ and s_λ. On any given occasion a variety of particle shapes and a whole spectrum of particle sizes are likely to be present simultaneously.† Nevertheless, it is instructive to consider the idealized case of scattering by spherical particles of uniform radius r, for which the scattering area coefficient K can be prescribed on the basis of theory. It is convenient to express K as a function of a dimensionless *size parameter* $\alpha = 2\pi r/\lambda$, which is a measure of the size of the particles in comparison to the wavelength of the incident radiation. Figure 6.11 shows a plot of α as a function of r and λ.

The scattering area coefficient K depends not only upon the size parameter but also upon the index of refraction of the particles responsible for the scattering. Figure 6.12 shows K as a function of α for two widely differing refractive indices.

For the special case of $\alpha \ll 1$ (the extreme left-hand side of Fig. 6.12), Rayleigh showed that, for a given value of refractive index, $K \propto \alpha^4$ and the scattered radiation is evenly divided between the forward and backward hemispheres. It can be seen from Fig. 6.11 that the scattering of solar radiation by air molecules falls within this so-called *Rayleigh scattering* regime. It is interesting to compare

† See discussion of atmospheric aerosol in Section 4.1.

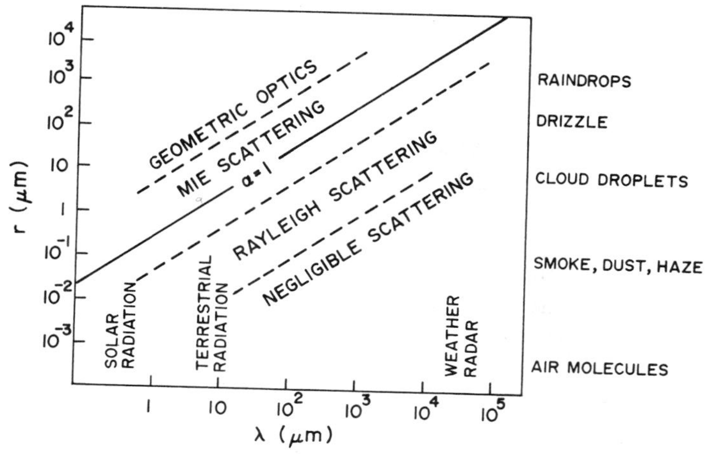

Fig. 6.11 Size parameter α as a function of wavelength of the incident radiation and particle radius.

the relative amounts of scattering of dark blue ($\lambda \simeq 0.47$ μm) and red ($\lambda \simeq 0.64$ μm) light by air molecules:

$$\frac{K(\text{blue})}{K(\text{red})} = \left(\frac{0.64}{0.47}\right)^4 = 3.45$$

The predominance of short wavelengths in the radiation scattered by air molecules is responsible for the blueness of the sky, shadows, and distant objects. Similarly, the predominance of longer wavelengths in solar radiation transmitted through the atmosphere without scattering imparts a reddish or orange cast to objects viewed in direct sunlight, particularly around sunrise and sunset when the path length through the atmosphere is long. The scattering of microwave radiation by raindrops also falls within the Rayleigh scattering regime. The sharp increase in K with increasing drop size makes it possible to discriminate between precipitation size drops and smaller cloud droplets. This principle is exploited in weather radar.

When α is greater than about 50 (the value of the abcissa at the extreme right-hand side of Fig. 6.12), $K \simeq 2$ and the angular distribution of scattered radiation

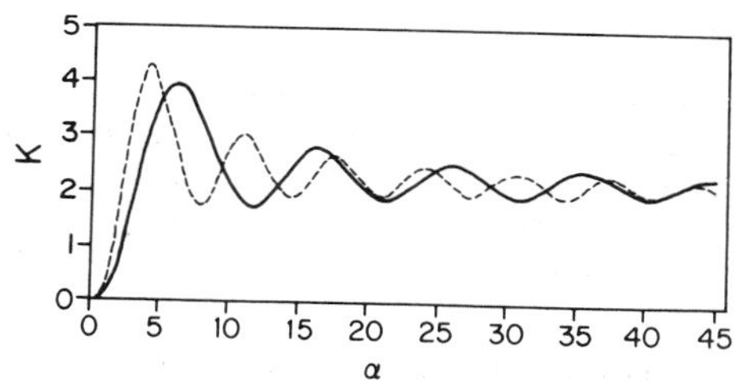

Fig. 6.12 Scattering area coefficient K as a function of size parameter α for refractive indices of 1.330 (——) and 1.486 (– – –). [Adapted from *J. Opt. Soc. Amer.* **47**, 1010 (1957).]

can be described by the principles of geometric optics. The scattering of visible radiation by cloud droplets, raindrops, and ice particles falls within this regime and produces a number of distinctive optical phenomena such as rainbows, halos, and so forth.

For intermediate values of the size parameter between about 0.1 and 50, the scattering phenomenon must be described in terms of the more general theory developed by Mie.† Within this so-called *Mie regime*, K exhibits the oscillatory behavior shown in Fig. 6.12. The angular distribution of scattered radiation is very complicated and varies rapidly with α, with forward scattering predominating over back scattering. The scattering of sunlight by particles of haze, smoke, smog, and dust usually falls within the Mie regime. If the particles are rather uniform in size, the scattered sunlight may be either bluish or reddish in hue, depending upon whether $\partial K/\partial \alpha$ is positive or negative at the wavelengths of visible light. Usually such particles exhibit a spectrum of sizes wide enough to span several maxima and minima in the plot of $K(\alpha)$, thus rendering the scattered light neutral or whitish in color.

6.9 THE ROLE OF RADIATIVE TRANSFER IN THE GLOBAL ENERGY BALANCE

The various physical processes considered in Sections 6.6–6.8 can be categorized in terms of the matrix in Table 6.4. Most of the elements in the matrix can be identified with a distinct class of problems within the field of radiative transfer. For most purposes, solar and terrestrial radiation can be treated independently because they fall within different ranges of wavelength. Furthermore,

Table 6.4

Relative importance of various radiative transfer processes in the global energy balance[a]

	Solar radiation		Terrestrial radiation	
Atmospheric constituent	Absorption	Scattering	Absorption and emission	Scattering
Air molecules	1	2	1[c]	3
Aerosols	2[b]	2	2	3
Clouds	2	1[c]	1[c]	3

[a] One (1) denotes processes of primary importance, 2 denotes processes of secondary importance, and 3 denotes processes of little or no importance.

[b] Plays an essential role in the production of photochemical smog.

[c] Has application to the remote sensing of the atmosphere by satellites.

† **Gustav Mie** (1868–1957) German physicist. Carried out fundamental studies on the theory of electromagnetic scattering and kinetic theory.

in many problems it turns out that absorption and scattering can be treated independently. For example,

- in some cases one of the processes may be negligible (for example, the scattering of terrestrial radiation by air molecules—see Fig. 6.11),
- one of the processes may be clearly dominant over the other one (for example, cloud droplets are so efficient at absorbing infrared radiation that clouds are effectively "black" to terrestrial radiation), or
- both effects may involve such a small fraction of the incident radiation that the interaction between them can be ignored (for example, the absorption and scattering of solar radiation, at most wavelengths, by air molecules and thin layers of aerosol).

The ability to separate the general problem of radiative transfer in planetary atmospheres into a number of smaller, more tractable problems is often exploited in the design of radiation experiments, the interpretation of radiation measurements, and the formulation of theoretical models.

In Table 6.4 the various elements in the matrix are classified according to their importance in the global energy balance. The leading terms involve

- scattering of solar radiation by clouds, through its effect upon the planetary albedo;
- absorption of solar radiation by gases (primarily O_3 in the upper atmosphere and H_2O in the lower atmosphere);
- absorption of terrestrial radiation by gaseous constituents of the atmosphere mainly CO_2, H_2O, and O_3), which produces the "greenhouse effect" described in Section 6.5.2;
- absorption of terrestrial radiation by clouds, which enhances the greenhouse effect.

These mechanisms will be considered in more detail in the next chapter. In any careful treatment of the global energy balance the five additional elements identified by "2" in Table 6.4 should also be taken into account.

Although radiative transfer plays a crucial and very complex role in the global energy balance, other physical processes must also be important, as evidenced by the fact that the observed temperatures at the earth's surface and throughout much of the atmosphere are far from radiative equilibrium. In the next chapter we will discuss the global energy balance from a broader perspective in an attempt to explain the observed temperature distribution.

PROBLEMS

6.10 Explain or interpret the following in terms of the principles discussed in this chapter.

(a) The absorptivity of a gas is a function of temperature and pressure.

(b) The gases in the sun's atmosphere emit isotropic radiation, yet solar radiation entering the earth's atmosphere may be regarded as parallel beam.

(c) The colors of stars are related to their temperatures whereas the colors of the planets are not.

(d) The equivalent blackbody temperature of Venus is lower than that of Earth, even though Venus is nearer to the sun.

(e) The "color temperature" of the sun is slightly different from its "equivalent blackbody temperature."

(f) Low clouds emit more infrared radiation than high clouds of comparable thickness.

(g) Satellite pictures of cloud patterns, based on infrared radiation near 10 μm, often resemble the negatives of pictures taken with visible radiation.

(h) The equivalent blackbody temperature of the earth is lower than the mean temperature at the earth's surface by about 30 deg.

(i) The presence of a cloud cover tends to favor lower daytime temperatures and higher nighttime temperatures.

(j) There has been some concern that a long-term increase in the amount of CO_2 in the atmosphere, caused by the burning of fossil fuels, might tend to raise the mean temperature of the earth's atmosphere.

(k) On a clear, still night (other factors being the same) the surface temperature drops more rapidly when the air above is dry than when it is moist, even before dew begins to form.

(l) Temperature inversions tend to form at night immediately above the tops of cloud layers.

(m) Frost may form on the ground when the temperature a few meters above the ground is well above freezing.

(n) Light colored clothing is frequently worn in hot climates.

(o) Individual absorption lines are not always apparent in the absorption spectrum.

(p) The technique described in Section 6.6.2 cannot be used to estimate the monochromatic intensity of solar radiation at ultraviolet wavelengths.

(q) Film manufacturers warn that color pictures taken near sunrise or sunset may appear reddish.

(r) The smoke of a cigarette appears blue when blown immediately into the air but appears white if it has been kept in the mouth for some time.

(s) When the atmosphere is dominated by a well-aged, dense aerosol cloud (for example, from a volcanic emission or a large forest fire), bright objects such as the sun or moon may appear blue or blue-green.

(t) Smoke from an automobile appears blue against a dark background but yellow against a light background (for example, the sky).

(u) Clouds of dust (or insects) are seen much better looking in the direction of the sun than looking away from it.

(v) Under certain conditions the solar irradiance upon the earth's surface can be substantially greater than the solar irradiance upon the top of the atmosphere. (State under what conditions.)

(w) Cells of convective motion tend to form in stratocumulus clouds.

6.11 What fraction of the radiative flux emitted by the sun does the earth intercept?

Answer 4.5×10^{-10}

6.12 The distance R between the earth and the sun varies by about 3.3%, between a maximum in early July and a minimum in early January. Show that the corresponding seasonal change in effective temperature T_E is 1.65% or about 4 deg. [Hint: Show that

$$\frac{dT_E}{T_E} = -\frac{1}{2}\frac{dR}{R}\Big]$$

6.13 A body is emitting radiation with the following idealized spectrum of monochromatic irradiance:

$$\begin{aligned} \lambda &< 0.35\ \mu\text{m}, & E_\lambda &= 0 \\ 0.35\ \mu\text{m} < \lambda &< 0.50\ \mu\text{m}, & E_\lambda &= 1.0\ \text{W m}^{-2}\ \mu\text{m}^{-1} \\ 0.50\ \mu\text{m} < \lambda &< 0.70\ \mu\text{m}, & E_\lambda &= 0.5\ \text{W m}^{-2}\ \mu\text{m}^{-1} \\ 0.70\ \mu\text{m} < \lambda &< 1.00\ \mu\text{m}, & E_\lambda &= 0.2\ \text{W m}^{-2}\ \mu\text{m}^{-1} \\ \lambda &< 1.00\ \mu\text{m}, & E_\lambda &= 0 \end{aligned}$$

Compute the irradiance.

Answer 0.31 W m^{-2}

6.14 Derive the Wien displacement law [Eq. (6.6)] from the simplified approximation of Planck's law [Eq. (6.5)]. Note that the constant is slightly different.

6.15 Suppose that the sun were somewhat bluer in color, so that its wavelength of maximum monochromatic irradiance were 0.400 μm instead of 0.475 μm. Estimate the radiative equilibrium temperature of the earth under these conditions, assuming that the albedo remains the same.

Problems 6.16–6.22 all deal with the same physical situation, and are interrelated to varying degrees.

6.16 A small, perfectly black, spherical satellite is in orbit around the earth at a height of 2000km. What angle does the earth subtend when viewed from the satellite? [Hint: Consider the diagrams in Fig. P6.16.]

Answer 2.21 sr

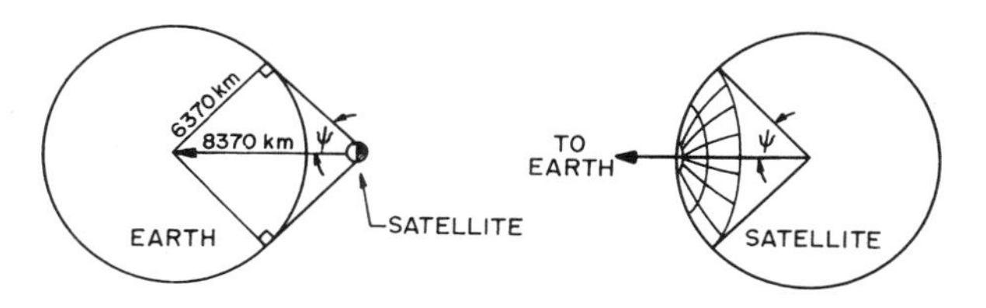

Fig. P6.16.

6.17 If the earth radiates as a blackbody with an effective temperature $T_E = 255°\text{K}$, calculate the radiative equilibrium temperature of the satellite in Problem 6.16 when it is in the earth's shadow. [Hint: Let dQ be the amount of heat imparted to the satellite by the irradiance dE received through an infinitesimal element of solid angle $d\omega$. Then

$$dQ = \pi r^2\, dE = \pi r^2 L\, d\omega$$

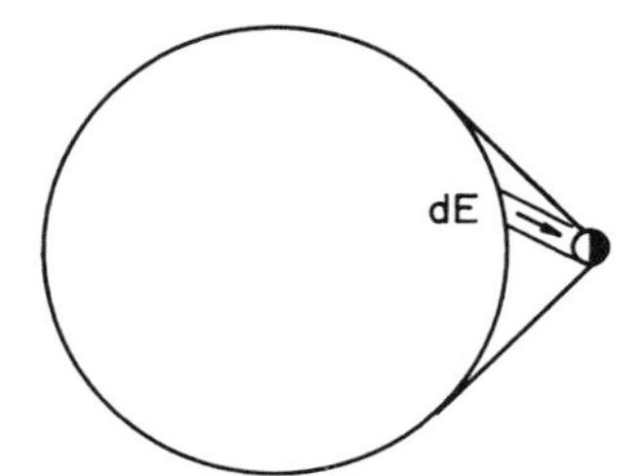

Fig. P6.17.

where r is the radius of the satellite and L is the radiance of the radiation emitted by the earth. Use the results of Problem 6.1 to evaluate L. Integrate the above expression over the entire arc of solid angle which the earth subtends, using the results of Problem 6.16 and noting that L is independent of direction. This gives the total energy absorbed by the satellite per unit time,

$$Q = 2.21 r^2 \sigma T_E{}^4$$

Finally, show that the temperature of the satellite is given by

$$T_s = T_E\left(\frac{2.21}{4\pi}\right)^{1/4}\Bigg]$$

Answer 166°K

6.18 Solve Problem 6.17, using the method developed in Problem 6.5 in the text. Show that this leads to a satellite temperature

$$T_s = T_E\left[\frac{1}{4}\left(\frac{6370}{8370}\right)^2\right]^{1/4} = 158°\text{K}$$

Can you explain why this method underestimates the temperature of the satellite?

6.19 Show that the approximate method in Problem 6.18 converges to the correct answer when the distance d between the satellite and center of the earth is large in comparison to the radius of the earth R_E. [Hint: Show that for the limiting case, in which d is much larger than the radius of the earth, the earth subtends a solid angle $\pi R_E{}^2/d^2$. Use this limiting value in Problem 6.17 and show that the symbolic form of the answer becomes identical to that of Problem 6.18.]

6.20 What is the radiative equilibrium temperature of the satellite in Problem 6.16 at the moment after it passes completely out of the earth's shadow. (Note that at this time the earth is still completely dark when viewed from the satellite.)

Answer $T_s = 289$°K

6.21 The satellite in Problem 6.16 has a mass of 100 kg, a radius of 1 m, and a specific heat of 10^3 J kg^{-1} deg^{-1}. Calculate the rate at which the satellite heats up immediately after it passes out of the earth's shadow. (Assume that the satellite passes out of the earth's shadow instantaneously.)

Answer 0.043 deg s^{-1}

6.22 (a) If the satellite in Problem 6.16 had an absorptivity of 0.5 at all wavelengths, how would the answers in Problems 6.17, 6.20, and 6.21 be affected?

(b) Suppose that it were desirable to keep the temperature of the satellite as uniform as possible throughout its orbit. How might the surface coating of the satellite be designed to accomplish this? Would it be easier to reduce the maximum temperature, or to raise the minimum temperature? Explain.

6.23 An opaque surface with the following absorption spectrum is subjected to the radiation described in Problem 6.13:

$$\lambda < 0.70: \quad a_\lambda = 0$$
$$\lambda > 0.70: \quad a_\lambda = 1$$

How much of the irradiance is absorbed? How much is reflected?

Answer 0.06 W m^{-2} absorbed; 0.25 W m^{-2} reflected

6.24 Describe how one might measure the absorptivity of a flat surface for (a) solar and (b) infrared radiation given a knowledge of the temperature of the surface and an instrument which is capable of measuring irradiance. (Assume that measurements can be taken sufficiently close to the ground so that atmospheric effects are negligible.)

6.25 (a) For the situation depicted in Fig. 6.5, prove that

$$\frac{E_1'}{a_1} = \frac{E_2'}{a_2}$$

where E_1' and E_2' are the irradiances of the emission from the two plates and a_1 and a_2 are the corresponding absorptivities. Make use of the fact that the two plates are in radiative equilibrium at the same temperature, but do not use Kirchhoff's law, and do not assume that one of the plates is "black." [Hint: Consider the total irradiances E_1 from plate 1 to plate 2, and E_2 from plate 2 to plate 1. The problem can be worked without explicitly dealing with the multiple reflections of radiation between the plates.]

(b) Make use of the results of (a) and the definition of *absorptivity* to show that a plate that is a perfect absorber emits the maximum possible amount of radiation. How does this result justify the definition of a *blackbody*, given in Section 6.4?

6.26 Consider a planet with a perfectly black surface and an atmosphere with an absorptivity a_S for incoming solar radiation and a_L for outgoing long-wave radiation. Prove that the radiative equilibrium surface temperature of the planet is increased by the presence of the atmosphere if $a_L > a_S$ and decreased if $a_L < a_S$. [Hint: by equating the upward and downward irradiances at various levels, show that the irradiance from the surface is proportional to the factor $(2 - a_S)/(2 - a_L)$.]

6.27 Consider a planet with an albedo of 0.30 and an atmosphere consisting of two isothermal layers, each with an absorptivity of zero at the wavelength of solar radiation, and 0.5 in the wavelength range for planetary radiation. Assume that the surface of the planet emits as a blackbody.

(a) Show that the upward irradiances of planetary radiation are $E/3$ for the top layer, $E/2$ for the bottom layer, and $5E/3$ for the surface of the planet, where E is the irradiance of outgoing planetary radiation, as defined in Problem 6.5.

(b) Calculate the radiative equilibrium temperatures of the surface of the planet and the two atmospheric layers. (Note that the "effective temperature" of the planet remains the same as in Problem 6.5.)

Answer 290°K; 255°K; 231°K

6.28 (a) Extend the results of Problem 6.27 to an atmosphere with N layers, each having an absorptivity of zero for solar radiation and 0.5 for planetary radiation. Show by induction that the upward irradiance of radiation emitted by successive layers, starting with the topmost one, are $E/3, E/2, 2E/3, \ldots, E(1 + N)/6$. Show that the surface of the planet emits radiation with an irradiance $E(1 + N/3)$.

(b) How many of these semiopaque layers would be required to account for the surface temperature of about 700°K observed on the planet Venus, assuming that the surface and the atmosphere are in radiative equilibrium and the surface radiates as a blackbody?

Answer $N = 276$

6.29 (a) Show that the irradiance emitted by the topmost layer of a multilayered atmosphere, transparent in the visible, is given by the expression $aE/(2 - a)$, where a is the absorptivity of the layer in the infrared, and E is the irradiance of planetary radiation emitted to space.

(b) By making use of the Stefan–Boltzmann law for an infinitesimally thin topmost layer, show that in the absence of short-wave absorption, the radiative equilibrium temperature of the top of a planetary atmosphere reaches a limiting value of

$$T^* = (0.5)^{1/4} T_E$$

where T_E is the effective temperature of the planet.

6.30 Show that for an optically thin layer (that is, $\sigma \ll 1$), $a_\lambda \simeq \sigma_\lambda$.

6.31 A certain gas has an absorption coefficient of 0.01 $m^2\ kg^{-1}$ for all wavelengths. What fraction of a beam of incident radiation is absorbed in passing vertically through a layer containing 1 kg m^{-2} of the gas? How much gas would the layer have to contain in order to absorb half the incident radiation?

Answer 1%; 69.4 kg m^{-2}

6.32 (a) Show that if k_λ and g are assumed to be independent of height, the optical depth at pressure level p is given by

$$\sigma_\lambda(p) = k_\lambda p \sec \phi / g$$

(b) Prove that in an isothermal atmosphere with k_λ independent of height, optical depth increases exponentially with geometric depth.

(c) A hypothetical planetary atmosphere is composed entirely of the gas described in Problem 6.31. Estimate the pressure level and height at which a beam of radiation oriented at 30° from the zenith encounters an optical depth of unity. The atmosphere has a surface pressure of 1000 mb, a gravitational acceleration of 10 m s^{-2}, and an isothermal temperature profile with a scale height of 10 km.

Answer 8.66 mb; 47.5 km

(d) Compute the heights above which 10, 50, and 90% of the incoming radiation at wavelength λ is absorbed. [Hint: First show that these levels correspond to optical depths of 0.105, 0.693, and 2.30, respectively.]

Answer 70 km; 51.1 km; 38.3 km

6.33 (a) What fraction of the incident radiation with wavelength λ is absorbed in passing through the layer of the atmosphere extending from optical depth $\sigma_\lambda = 0.2$ to $\sigma_\lambda = 4.0$?

Answer 80.0%

(b) Of the monochromatic radiance emitted to space by an isothermal atmosphere, what fraction is emitted by the layer extending from optical depth $\sigma_\lambda = 0.2$ to $\sigma_\lambda = 4.0$?

Answer 80.0%

(c) For an isothermal atmosphere, through how many scale heights does the layer described in (a) and (b) extend for zero zenith angle?

Answer 3.00

6.34 Prove that for isotropic radiation, the irradiance absorptivity of an optically thin layer (that is, $\sigma_\lambda \ll 1$) is approximately equal to twice the radiance absorptivity at zero zenith angle.

6.35 Monochromatic irradiance can be defined in terms either of wavenumber k or wavelength λ such that area under the spectrum, plotted as a linear function of k or λ, respectively, is proportional to irradiance. Show that $E_k = \lambda^2 E_\lambda$.

6.36 What is the smallest size of uniform spherical particles with a refractive index of 1.33 that could impart a bluish cast to transmitted white light?

Answer $\simeq 0.6\ \mu m$

6.37 Consider a model cloud consisting of monodispersed droplets 40 μm in diameter in concentrations of 1 cm^{-3}. How long a path length through such a ud is required to deplete a beam of visible radiation by a factor e by scattering?

Answer ~ 400 m

Chapter

7

The Global Energy Balance

The atmosphere plays a complex and extremely important role in the global energy balance. In the previous chapter it was shown that the average temperature of the surface of the earth is considerably higher than it would be in the absence of the atmosphere. This so-called "greenhouse effect" is a consequence of the radiative transfer properties of water vapor, carbon dioxide, and clouds.

In this chapter we begin by considering the various sources and sinks in the globally averaged energy balance for the atmosphere as a whole. Then, in the next two sections, we discuss separately, and in more detail, the energetics of the upper atmosphere and the troposphere, still from a globally averaged perspective. Section 7.4 is devoted to a discussion of the interaction between the atmosphere and the underlying surface, with emphasis on air–sea interactions. The final section provides some specific examples of how the atmosphere responds to the constantly changing distribution of energy sources and sinks, on various time scales. Discussion of the role of atmospheric motions in the horizontal redistribution of energy on the earth's surface is reserved for Section 9.9.

7.1 THE GLOBALLY AVERAGED ATMOSPHERIC ENERGY BALANCE

Despite the vast amounts of energy continually being added to and taken away from the atmosphere by radiation and other energy transfer processes, the storage of energy in the atmosphere is not systematically increasing or decreasing. Thus, to a very close approximation, in the long-term average, there is a balance or cancellation between the energy sources and energy sinks for the atmosphere as a whole; that is,

$$\{\overline{S^+{}_1 + S^+{}_2 + \cdots}\} = \{\overline{S^-{}_1 + S^-{}_2 + \cdots}\} \tag{7.1}$$

where the $S^+{}_i$ denote the various energy sources, the $S^-{}_i$ the sinks, the brackets $\{\ \}$ an integration over the entire mass of the atmosphere, and the overbar an average over a long time period like a year. We would like to express this energy balance requirement in terms of specific physical processes.

7.1.1 The sources and sinks

Since the atmosphere is very nearly in hydrostatic equilibrium, the rate of diabatic heating or cooling of an air parcel can be related to the rate of change of the enthalpy and geopotential of the parcel through (2.54) which can be written in the form

$$\dot{H} = \frac{d}{dt}(c_p T + \Phi) \tag{7.2}$$

where $\dot{H}$ is the diabatic heating rate in watts per kilogram, and the other symbols are as defined in Chapter 2. The quantity $c_p T + \Phi$ appears frequently in expressions relating to atmospheric energetics. It represents the sum of the internal and gravitational potential energy per unit mass, plus the term $p\alpha$, which may be viewed as the work done against the environment in expanding a unit mass of air from zero volume to its present volume at constant pressure (see Section 2.3.3). It is called the *static energy* even though, strictly speaking, it is a hybrid expression involving energy plus work.

The heating rate $\dot{H}$ in (7.2) represents the combined effect of the following physical processes:

- absorption of solar radiation;
- absorption and emission of terrestrial radiation;
- release of the latent heat of condensation of water vapor;
- exchange of heat with the surroundings due to mixing by random molecular motions (*conduction*);
- exchange of heat with the surroundings due to mixing by organized fluid motions (*convection*).

We can represent these processes symbolically in the form

$$\dot{H} = \dot{H}_{\mathrm{R}} + \mathrm{LH} + S_{\mathrm{h}} \tag{7.3}$$

where $\dot{H}_{\mathrm{R}}$ represents the net radiative heating rate, LH the rate of latent heat release, and S_{h} the rate of heating due to conduction and convection. The latent heat term can be related to the time rate of change of the water vapor mixing ratio of the parcel in the following manner: The time rate of change of mixing ratio is divided into two components:

$$\frac{dw}{dt} = \left(\frac{dw}{dt}\right)_{\mathrm{P}} + \left(\frac{dw}{dt}\right)_{\mathrm{E}}$$

The first term on the right-hand side represents the effect of phase changes between vapor and liquid (or solid) and the second term represents the effect of exchanges of water vapor molecules with the surroundings due to conduction and convection. The heating due to latent heat release can then be written in the form

$$\mathrm{LH} = -L\left(\frac{dw}{dt}\right)_{\mathrm{P}} = -L\frac{dw}{dt} + L\left(\frac{dw}{dt}\right)_{\mathrm{E}}$$

It is convenient to denote the exchange term as follows:

$$S_{\mathrm{m}} \equiv L\left(\frac{dw}{dt}\right)_{\mathrm{E}}$$

Incorporating the last two expressions into (7.3), we obtain

$$\dot{H} = \dot{H}_{\mathrm{R}} - L\frac{dw}{dt} + S_{\mathrm{h}} + S_{\mathrm{m}}$$

Then, combining this expression with (7.2) and rearranging, we obtain the useful result

$$\frac{d}{dt}(c_p T + \Phi + Lw) = \dot{H}_{\mathrm{R}} + S_{\mathrm{h}} + S_{\mathrm{m}} \tag{7.4}$$

where the quantity in parentheses is called the *moist static energy*. As defined here, the moist static energy of an air parcel is the sum of its enthalpy, gravitational potential energy, and *latent heat content* (Lw). It has the convenient property of being unaffected by condensation processes, which merely redistribute energy between the Lw and $c_p T$ terms.

For the atmosphere as a whole, in the long-term mean, the moist static energy is not systematically changing; therefore,

$$\frac{d}{dt}\overline{\{c_p T + \Phi + Lw\}} = 0 \tag{7.5}$$

From (7.4) it follows that

$$\overline{\{\dot{H}_{\mathrm{R}} + S_{\mathrm{h}} + S_{\mathrm{m}}\}} = 0 \tag{7.6}$$

which is a more specific statement of the energy balance requirement (7.1) for the earth's atmosphere as a whole.

7.1.2 The balance at the earth's surface

The globally averaged and time-averaged net upward flux of moist static energy per unit area from the earth's surface is given by

$$[\overline{F}] = [\overline{E_{\mathrm{n}}}] + [\overline{F_{\mathrm{h}}}] + [\overline{F_{\mathrm{m}}}] \tag{7.7}$$

where E_n represents the net irradiance at the earth's surface, F_h and F_m are the fluxes of *sensible* and *latent heat*, respectively, and the bracket operator [ ] denotes an average over the surface of the globe. The net irradiance can be divided into short-wave and long-wave components as follows:

$$[\overline{E_n}] = [\overline{E_L\uparrow}] - [\overline{E_L\downarrow}] - [\overline{E_s}] \tag{7.8}$$

where E_s is the solar radiation actually absorbed at the earth's surface (the incident minus the reflected), $E_L\uparrow$ the emission of infrared radiation from the earth's surface, and $E_L\downarrow$ the downward emission from the atmosphere.

From the principle of the conservation of energy it is evident that

$$4\pi R_E{}^2[\overline{F}] = -\frac{\partial \overline{W}}{\partial t} \tag{7.9}$$

where R_E is the average radius of the earth and W the combined internal, potential, kinetic, chemical, and nuclear energies of the solid earth, the oceans, the polar icecaps, and the biosphere. Let us consider the relative importance of the various physical processes that contribute to changes in the total energy stored in this system.

- The burning of fossil fuels and the fission in nuclear reactors is currently releasing energy at a rate equivalent to a globally averaged flux on the order of 0.02 W m^{-2}. Projected rates of nuclear energy production a century from now range up to about 1 W m^{-2}.
- The rate of leakage of geothermal energy through the earth's crust is on the order of 0.06 W m^{-2} in the global average. (On Jupiter this process is apparently a major term in the energy balance—see Problem 7.8.)
- Between late winter and late summer, the upper layers of the ocean warm by an average of about 5 deg. The oceans cover about 70% of the globe and the temperature rise takes place within a layer on the order of 100 m in thickness. The energy absorbed in heating the oceans is equivalent to a net downward flux of

$$[\overline{F}] = \left(\frac{5 \text{ deg}}{180 \text{ day}}\right) \times \frac{4.2 \times 10^3 \text{ J kg}^{-1} \text{ deg}^{-1} \times 10^3 \text{ kg m}^{-3} \times 100 \text{ m} \times 0.70}{8.65 \times 10^4 \text{ s day}^{-1}}$$

or roughly 100 W m^{-2} for an average over an individual season in one hemisphere. Over the course of an individual year, the net change in globally averaged sea surface temperature is certainly not more than a few tenths of a degree and probably much less. Within the last century the globally averaged sea surface temperature has never risen or fallen faster than a few tenths of a degree per decade. On the basis of these rates of change, it is easily verified that the globally averaged fluxes associated with temperature changes in the upper layers of the ocean are not more than a few watts per square meter for an average over an individual year, and not more than a few tenths of a watt per square meter for the average over a decade.

- Phase changes associated with the buildup or decay of continental ice sheets absorb and release large amounts of energy over long time periods. At present the mass of ice in the Antarctic and Greenland icecaps is equivalent to a layer about 60 m thick if it were spread evenly over the surface of the globe. Let us assume that over the course of a thousand years ($\sim 3 \times 10^{10}$ s) these icecaps either disappear or double in size (an enormous rate of change, as we will show later). The corresponding globally averaged energy flux is

$$[\bar{F}] = \left(\frac{60 \text{ m}}{3 \times 10^{10} \text{ s}}\right) \times 3.34 \times 10^5 \text{ J kg}^{-1} \times 10^3 \text{ kg m}^{-3} \simeq 0.6 \text{ W m}^{-2}$$

On the basis of these arguments, it is reasonable to conclude that, on the average over a decade, the net flux of energy through the earth's surface is more than two orders of magnitude smaller than the other terms in (7.7), which are on the order of 30–100 W m^{-2}. Thus, to a close approximation the globally averaged and time-averaged energy balance at the earth's surface requires that

$$[\overline{E_s}] + [\overline{E_L\downarrow}] = [\overline{E_L\uparrow}] + [\overline{F_h\uparrow}] + [\overline{F_m\uparrow}] \tag{7.10}$$

where the overbar denotes an average over a decade or longer.

If the net flux of energy through the earth's surface is negligible and the energy stored in the earth's atmosphere is not systematically changing, it follows that in the global average the net flux of energy through the top of the atmosphere must be very small. Furthermore, since radiative transfer accounts for virtually all the exchange of energy between the earth and the rest of the universe,† it follows that the earth–atmosphere system must be very close to radiative equilibrium as assumed in Problem 6.5. Thus, to a close approximation the outgoing infrared irradiance is given by

$$[\bar{E}] = [\overline{(1 - A)S/4}] \tag{7.11}$$

where A is the earth's albedo and S the solar irradiance incident upon a normal plane surface at the top of the atmosphere (1380 W m^{-2}).

7.1.3 How the balance is achieved

Figure 7.1 shows one of the estimates of the global energy balance for the earth–atmosphere system, calculated on the basis of actual data, with slight adjustments to insure that (7.6), (7.10), and (7.11) are satisfied. The 100 units of solar radiation incident on the top of the atmosphere represent an irradiance of $S/4 = 345$ W m^{-2}, as calculated in Problem 6.5.

Of the 100 units of incident solar radiation, 19 are absorbed during passage through the atmosphere: 16 in cloud-free air and 3 in clouds. A total of 30 units are reflected back to space: 20 from clouds, 6 from cloud-free air, and 4 from

† The flux of energy by the solar wind and cosmic rays is more than five orders of magnitude smaller than the radiative flux.

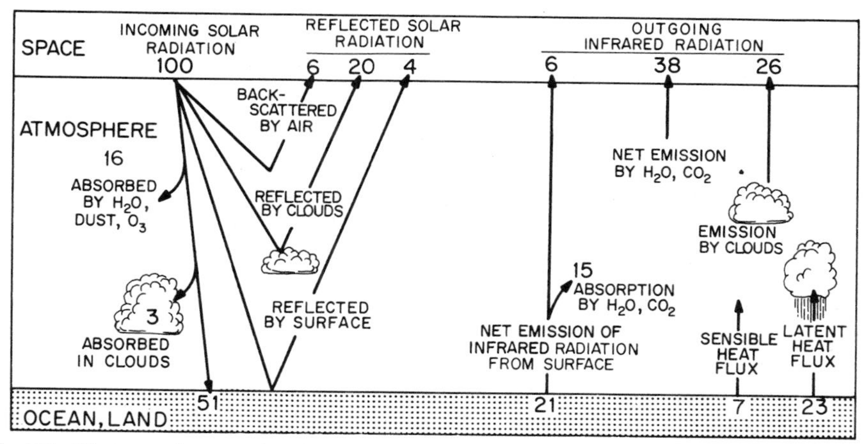

Fig. 7.1 The annual mean global energy balance for the earth–atmosphere system. (Numbers are given as percentages of the globally averaged solar irradiance incident upon the top of the atmosphere.) See text for further explanation. [Adapted from "Understanding Climatic Change," U.S. National Academy of Sciences, Washington, D.C. (1975), p. 14, and used with permission.]

the earth's surface. The remaining 51 units are absorbed at the earth's surface. The earth disposes of this energy by a combination of infrared radiation and sensible and latent heat flux, as summarized on the right-hand side of Fig. 7.1. The net infrared emission, which represents the upward emission from the earth's surface, minus the downward emission from the atmosphere, amounts to 21[†] units, 15 of which are absorbed in passing through the atmosphere and 6 of which reach space. The remaining 30 units are transferred from the earth's surface to the atmosphere by a combination of latent and sensible heat flux, as indicated in (7.10).

From the viewpoint of the atmosphere alone, there is a net loss of 49 units of infrared radiation (70 units emitted to space from the top of the atmosphere minus 21 units of net upward flux from the earth's surface) which exceeds, by 30 units, the energy gained as a result of the absorption of solar radiation. This deficit is balanced by an influx of 30 units of latent and sensible heat from the earth's surface, as indicated in (7.6). Thus, in the global average, the atmosphere experiences a net radiative cooling which is balanced by the latent heat of condensation that is released in regions of precipitation, and by the conduction of sensible heat from the underlying surface. Were it not for the fluxes of latent and sensible heat, the earth's surface would have to be considerably hotter (on the order of 340°K as compared with the observed value of 288°K) in order to emit enough infrared radiation to satisfy the balance requirements for thermal equilibrium.

[†] It should be noted that the net radiation through the earth's surface represents the difference between a large (greater than 100 units) upward irradiance from the earth's surface minus a slightly smaller downward irradiance from the atmosphere.

7.1.4 The roles of conduction and convection

Together with radiative transfer, exchanges of latent and sensible heat play important roles as mechanisms for redistributing energy within the earth's atmosphere. Such a redistribution can take place as a result either of *conduction* (the exchange of individual molecules) or of *convection* (the exchange of macroscale air parcels or currents containing differing amounts of enthalpy or moisture per unit mass).† Throughout most of the atmosphere convection is by far the more effective mechanism. Conduction is important only in two regions:

- within the molecular boundary layer, immediately adjacent to the earth's surface, where fluid motions are strongly suppressed by friction (this layer is generally less than 1 cm in thickness);
- above the turbopause ($\simeq 100$ km) where the mean free path between molecular collisions is comparable to, or larger than, the dimensions of the fluid motions.

Despite its limited region of influence within the atmosphere itself, molecular conduction plays an indispensible role as a mechanism for the exchange of energy between the atmosphere and the underlying surface. The consequences of this exchange of energy are discussed in Section 7.4.

7.2 THE ENERGY BALANCE OF THE UPPER ATMOSPHERE‡

Above the tropopause level the absorption of solar radiation plays a dominant role in the energy balance. Most of the absorption occurs in association with photoionization and photodissociation of various gaseous constituents of the upper atmosphere by ultraviolet and X-ray radiation. We will begin with a description of the radiation itself, and its sources in the sun's atmosphere.

7.2.1 The spectrum of solar radiation

Nearly all the radiant flux from the sun is emitted from the *photosphere*, a layer a few hundred kilometers in thickness which marks the outermost limit of its visible disk. This layer is centered near a radius of 7×10^5 km, which corresponds to the level of unit optical depth for a wide range of the electromagnetic spectrum. To a first approximation, the sun radiates to space as a blackbody with an effective temperature of about 5780°K, which is characteristic of the level of unit optical depth.

The solar spectrum exhibits some marked departures from an idealized blackbody spectrum. For example, the reader may have noticed that there is a substantial difference between the sun's "color temperature" and its "effective temperature," as computed in Problems 6.3 and 6.4. The departures are partic-

† Note that this definition of *convection* is somewhat different from the one given in Chapter 2. The same word has two distinct meanings, depending upon the context in which it is used.

‡ In an introductory survey course which stresses the lower atmosphere this section may be omitted without serious loss of continuity.

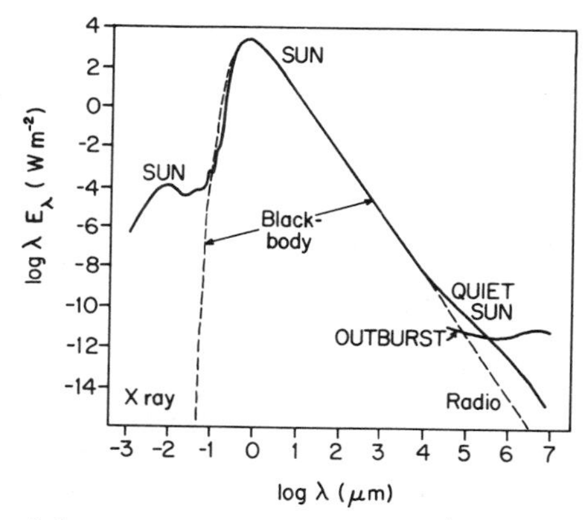

Fig. 7.2 The spectrum of the sun (——) and a black body at 5780°K (– – –). Note that in this representation the area under the curve is not proportional to irradiance. [Adapted from *Quart. J. Roy. Meteor. Soc.* **84**, 311 (1958).]

ularly large in the X-ray and far-ultraviolet ($\lambda < 0.1$ μm) regions and in the microwave regions of the spectrum where the solar emission is many orders of magnitude larger than that of a blackbody at 5780°K (see Fig. 7.2). Within these enhanced, outer portions of the spectrum, the solar emission is highly variable. However, the total solar output, which is dominated by wavelengths between 0.3 and 10 μm, appears to be constant, at least to within a few percent. In order to account for these peculiar properties of the solar radiation, it is necessary to consider the vertical structure of the sun's outer atmosphere and the mechanisms by which energy generated by thermonuclear reactions deep within the interior of the sun is transferred outward and finally disposed of as radiation to space.

The photosphere is actually the thin outer "skin" of a much deeper layer, at least 10^5 km in thickness, in which the lapse rate of temperature is nearly adiabatic and convection is the primary mechanism for the outward flux of energy. The visible "surface" of the photosphere (shown in Fig. 7.3) is marked by a rather uniform pattern of convective cells called *granules*, which have an average size of about 2000 km and a typical lifetime on the order of 10 min. The bright centers of the cells have temperatures about 100 deg higher than the outer edges. The buoyant gas in each hot convective "bubble" rises to the surface where it quickly radiates away its excess heat. As it cools, it loses its buoyancy and begins to drop back into the interior of the sun along the outer edge of the bubble. Photospheric convection also shows evidence of a larger scale of organization called supergranulation, which has a characteristic scale on the order of 30,000 km.

Because of the steep lapse rate within the photosphere, the gas molecules that emit radiation to space exhibit temperatures ranging over at least a few hundred degrees. The absorption coefficient of these gases displays some wavelength dependence, which accounts for much of the distortion in the shape of the solar spectrum relative to the blackbody curve.

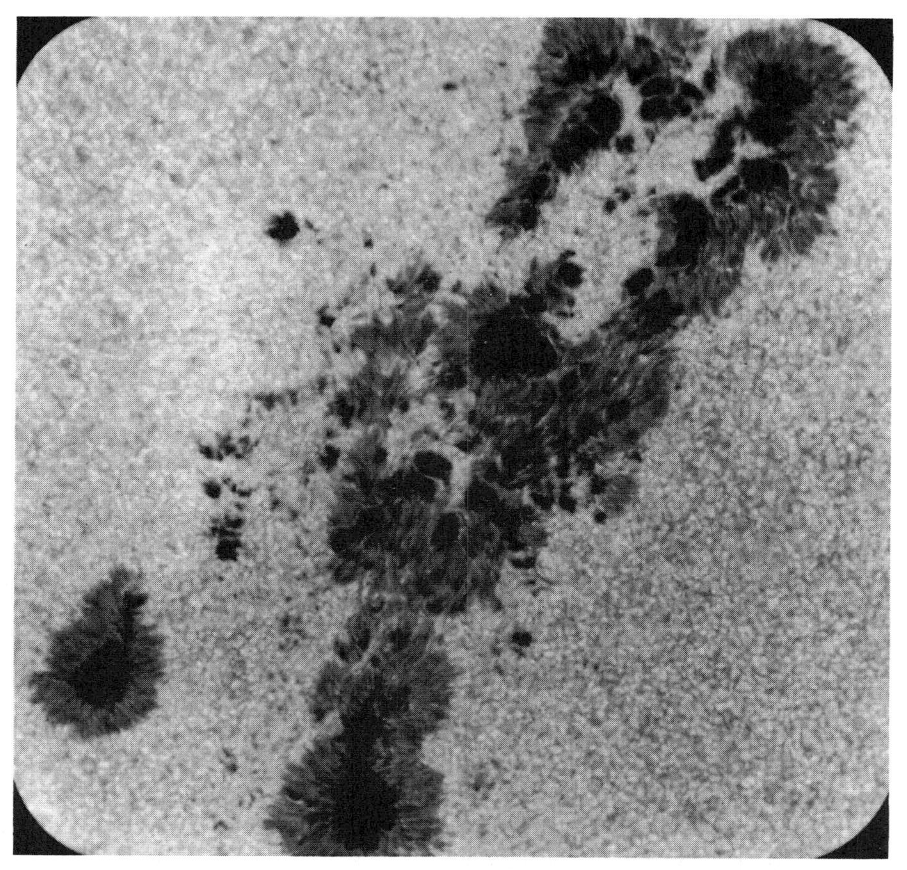

Fig. 7.3 High-resolution photograph of the sun showing granulation pattern interrupted by a complex group of sunspots. (Courtesy of R. Dunn, Sacramento Peak Observatory.)

Extensive observations of the sun have revealed the existence of a number of interrelated phenomena that appear intermittently in *active regions* of the sun:

- *Sunspots* are dark (cool) areas that interrupt the regular pattern of photospheric granulation as shown in Fig. 7.3. Sunspot groups are accompanied by strong magnetic fields which have a preferred orientation that reverses between solar hemispheres. Sunspots have lifetimes ranging from days up to a few months.
- *Faculae* are bright (hot) spots in the supergranulation pattern that often appear in conjunction with sunspots. They have a typical time scale comparable to that of sunspots and are also accompanied by strong magnetic fields.
- *Flares* are intense bursts of radiation and high-energy particles emanating from the sun's outer atmosphere within active regions. They are characterized

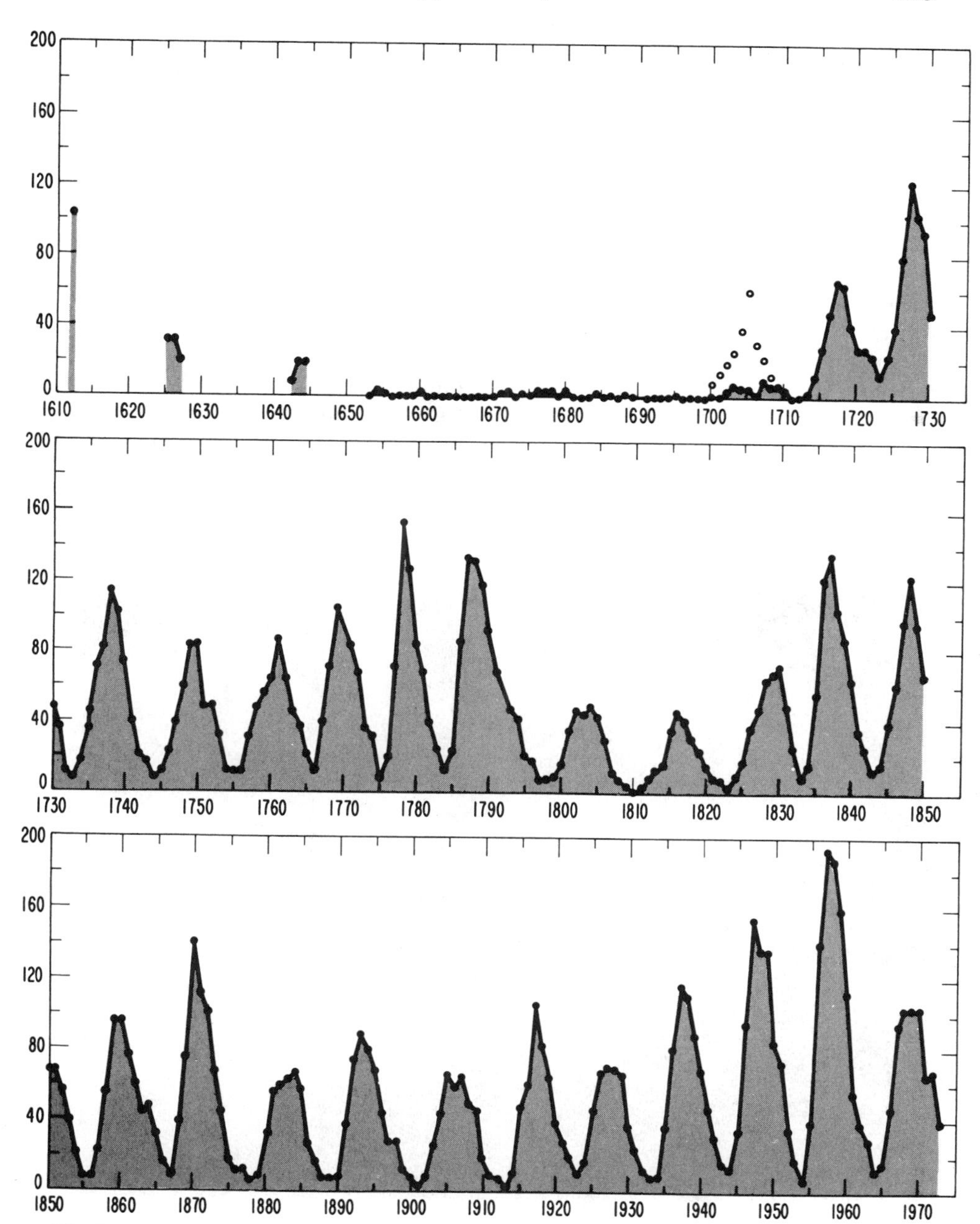

Fig. 7.4 Time series of annual average sunspot number. The lack of sunspot activity during the 17th century is believed to be real. (Courtesy of J. A. Eddy.)

by strong magnetic fields and violent motions. A typical flare lasts on the order of an hour.

The frequency of solar activity is modulated by the 11-yr *solar cycle* which is illustrated in Fig. 7.4. During the years corresponding to maxima in the cycle (such as the International Geophysical Year, 1958), sunspots, faculae, and flares are numerous, while near the times of the minima (such as the International Year of the Quiet Sun, 1964), solar activity practically disappears. Between successive cycles there is a reversal in the orientation of the magnetic fields associated with sunspot groups; hence, a full solar cycle is sometimes regarded as 22 yr.

Above the photosphere lies the *chromosphere*, a layer about 2500 km thick in which the temperature increases from a minimum of about 4300°K near the bottom to values on the order of 10^5 °K at its outer limit. Emanating outward from the chromosphere in all directions is an irregular stream of extremely hot, ionized particles known as the *solar wind*. In the earth's orbit, outward velocities in the solar wind are on the order of 500 km s^{-1} and temperatures occasionally reach values close to 10^6 °K. Photospheric radiation scattered by the free

Fig. 7.5 The solar corona observed during the 12 November 1966 eclipse, in red light. A filter which had a radial gradient covering a range of 10^4 in transmission was used to compensate for the rapid decrease of coronal brightness with increasing radius, thus allowing faint features in the outer corona to be observed without overexposing the chromosphere. (Courtesy of G. Newkirk, Jr., High Altitude Observatory, Boulder, Colorado.)

electrons of the denser, inner portion of the solar wind produces a diffuse luminosity called the *corona*,† which is observable during total eclipses, as shown in Fig. 7.5. In accordance with Wien's displacement law, the gases in the outer chromosphere and corona emit radiation with peak monochromatic irradiance in the far-ultraviolet and X-ray regions of the spectrum. At wavelengths shorter than 0.1 μm, these extremely hot gases account for virtually all the solar emission.

In contrast to the apparent steadiness of the emission from the photosphere, the radiation from the outer chromosphere and the corona increases in response to solar activity and is greatly enhanced at the times when flares are in progress. The enhancement in response to flares is particularly strong in the X-ray region of the spectrum. For example, radiation with wavelengths less than 0.02 μm is more than an order of magnitude stronger during a flare than during quiet periods. It should be noted that the total flux of energy from the sun's outer chromosphere and corona accounts for less than 1 part in 10^5 of the total solar output; hence, even the strongest flares have no detectable effect upon the total amount of solar radiation reaching the earth.‡

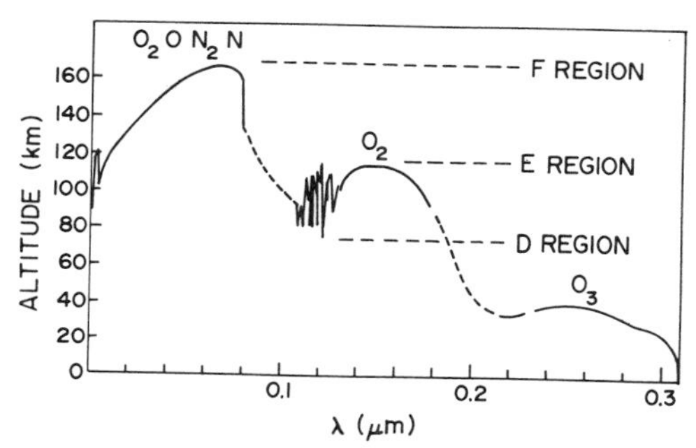

Fig. 7.6 Penetration of solar radiation into the atmosphere. The curve indicates the level at which the monochromatic irradiance is reduced by a factor of e, which corresponds to the level of unit optical depth. [Adapted from H. Friedman, The Sun's Ionizing Radiations, in "Physics of the Upper Atmosphere" (J. A. Ratcliffe, ed.), Academic Press, New York (1960), p. 134.]

As solar radiation in a particular wavelength band penetrates downward into the earth's atmosphere, most of it is absorbed within one or two scale heights of the level of unit optical depth (as defined in Section 6.6.3) for that range of wavelengths. The level of unit optical depth for the earth's atmosphere is displayed as a function of wavelength in Fig. 7.6. The curve can be divided into four broad "steps" on the basis of wavelength, as indicated in Table 7.1. Each of these four wavelength bands affects a different layer within the earth's atmosphere, and each is associated with a different absorption mechanism.

† Not to be confused with the corona described in Section 5.1.4.

‡ For an elementary description of solar phenomena the reader is referred to R. Jastrow and M. H. Thompson, "Astronomy: Fundamentals and Frontiers," 2nd ed., John Wiley and Sons, Inc., New York, 1974.

Table 7.1

Absorption of solar radiation in the earth's atmosphere

Wavelength band $\lambda(\mu m)$	Fraction of total solar energy	Source in solar atmosphere	Layer where absorbed in earth's atmosphere (km)	Primary absorption mechanism	Fraction absorbed in atmosphere
$\lambda < 0.1$	3 parts in 10^6	Mainly chromosphere and corona	90–200	Photoionization	All
0.1–0.2	1 part in 10^4	Photosphere	50–110	Photodissociation of O_2 (7.12)	All
0.2–0.31	1.75%	Photosphere	30–60	Photodissociation of O_3 (7.14)	All
$\lambda > 0.31$	98%+	Photosphere	0–10	Absorption by water vapor	~17%

7.2.2 Photoionization and the thermosphere

Solar radiation with wavelengths shorter than 0.1 μm contains enough energy to ionize the predominant atmospheric constituents. Most of this radiation is absorbed at levels above 90 km, where the photoionization of N_2, O_2, and O gives rise to the E- and F-layers of the ionosphere† (see Section 1.4.1). Like the ionized gases in the sun's outer atmosphere, the thin air at these high levels is very inefficient at disposing of energy by radiative transfer. Therefore, in order to maintain thermal equilibrium, the temperature must increase with height rapidly enough so that random molecular motions can conduct heat downward (from hotter air above to cooler air below) at the same rate as solar energy is being absorbed at the higher levels. Since molecular conduction is a relatively inefficient mechanism for transferring heat, a rather large vertical temperature gradient is needed in order to satisfy the energy balance requirement. This region of increasing temperature with height corresponds to the thermosphere.

The existence of a thermosphere in the earth's atmosphere and in the atmospheres of the other planets is mainly a consequence of radiation emitted by gases in the sun's outer chromosphere and corona. Although this radiation constitutes only a minute fraction of the sun's total output, it is absorbed by such a small fraction of the atmospheric mass that the actual input of energy per unit mass is extremely large.

As noted in Sections 1.4.1 and 1.5, the structure of the ionosphere and the temperature of the upper thermosphere exhibit strong fluctuations in response

† The D-layer is believed to be a consequence of the narrow spectral "window" near 0.12 μm (see Fig. 7.6) which permits a small amount of solar radiation to penetrate to the 60–90 km layer where it ionizes nitric oxide.

to solar activity. Some of these effects can be traced to the temporary enhancement of the radiation from the sun's outer atmosphere, in response to disturbances in the photosphere. Other effects such as auroras are a consequence of the enhanced flux of high-energy particles emanating from the sun in the solar wind. Disturbances transmitted by radiation affect the earth's upper atmosphere within a matter of minutes, whereas those transmitted by the solar wind usually require several days to reach the earth.

7.2.3 Photodissociation of oxygen

Solar radiation in the second wavelength band ($0.1 < \lambda < 0.2\ \mu m$) is virtually all absorbed in the photodissociation reaction

$$O_2 + h\nu = 2O \tag{7.12}$$

The atomic oxygen produced in this reaction is a major atmospheric constituent at levels above 100 km (see Section 1.3.2). Moreover, it is highly reactive so that, even at lower levels, where it is a trace constituent, it plays a vital role in a large number of chemical reactions which determine the concentrations of other trace substances. Of particular importance is the reaction

$$O_2 + O + M = O_3 + M \tag{7.13}$$

which is the dominant mechanism for the production of ozone in the atmosphere. (Here M is a third molecule which is required to carry away the excess energy released in the reaction.) The probability of the three-body collision (7.13) increases in proportion to the square of the density of the gas. At very low density a free oxygen atom can exist almost indefinitely, whereas at high density its probable lifetime is relatively short. Because of this density dependence, atomic oxygen is a stable species in the upper mesosphere and thermosphere, while in the stratosphere it combines very rapidly to form ozone.

The layer from 65 to 90 km corresponds to the minimum in the vertical temperature profile known as the mesopause (see Fig. 1.8). Within this layer the input of energy due to the absorption of solar radiation is less than the energy emitted to space in the long wavelength part of the spectrum. The deficit in the radiation balance is canceled by the convection of heat upward from the stratopause region.

7.2.4 The ozone layer

Ultraviolet radiation with wavelengths longer than $0.2\ \mu m$ is not strongly absorbed by the photodissociation of oxygen and thus is able to penetrate deeper into the atmosphere where it encounters ozone and is absorbed in the photodissociation reaction

$$O_3 + h\nu = O_2 + O \tag{7.14}$$

The free oxygen atom left in the wake of this reaction quickly recombines by (7.13) to form another ozone molecule. Hence, when (7.13) and (7.14) take place in sequence, there is no net chemical change but only an absorption of radiation and a resulting input of heat. Through this sequence of reactions, repeated many times, the generation of a single oxygen atom in (7.12) ultimately leads to the absorption of many photons of radiation. Because of the high absorptivity of ozone at $\lambda \lesssim 0.31\ \mu m$, and the rapidity with which dissociated ozone molecules are replaced through (7.13), the trace amounts of ozone present in the stratosphere are capable of absorbing virtually all the solar radiation in the wavelength band $0.2 < \lambda < 0.31\ \mu m$.

As seen in Fig. 7.6 and Table 7.1, the maximum absorption of solar radiation throughout most of this wavelength band takes place in a layer centered near 50 km, where it results in a relatively large input of energy per unit mass. This absorption is responsible for the temperature maximum which defines the stratopause level in the earth's atmosphere (see Fig. 1.8). The level of maximum ozone concentration is located somewhat lower, near 25 km.

The photochemistry of ozone has attracted considerable interest because of the vital importance of the ozone layer in shielding terrestrial life from the harmful effects of ultraviolet radiation. Recently, there has been concern that the introduction of pollutants into the stratosphere either directly, as in the case of the exhaust from supersonic aircraft, or indirectly by way of the troposphere, might reduce the equilibrium concentration of ozone, thereby allowing more ultraviolet radiation with $\lambda \simeq 0.31\ \mu m$ to reach the earth's surface. Water vapor, nitric oxide, and Freon have all been mentioned as trace constituents that might conceivably interfere with the normal ozone photochemistry. At the time of this writing these questions have not been resolved.

7.3 THE TROPOSPHERIC ENERGY BALANCE

The structure and dynamics of the troposphere are profoundly influenced by the underlying surface through the fluxes of latent and sensible heat which penetrate upward all the way to the tropopause. These fluxes are the primary energy source for the troposphere and for the atmosphere as a whole (see Fig. 7.1). Radiative transfer also plays an important role, primarily as an energy sink but also as a source of about a third of the energy put into the troposphere. In this section we will consider the relation between these various energy transfer mechanisms in the tropospheric energy balance.

7.3.1 The role of radiative transfer

The various absorption mechanisms considered in the discussion of the upper atmosphere involve less than 2% of the solar radiation incident upon the earth's atmosphere. The remaining 98% that is associated with wavelengths longer than 0.31 μm suffers very little depletion until it reaches the tropopause.

The subsequent fate of this radiation depends very much upon the local distribution of clouds, aerosol, and water vapor within the troposphere. Absorption may range from only a few percent in a clear dry atmosphere with an overhead sun up to a substantial fraction of the incident radiation in moist air with clouds and/or a low sun angle. Likewise, the backscattering to space may range from a few percent in clear air to over 50% in regions of dense, layered clouds. In view of the complexity of the short-wave radiation budget, the global average estimates given in Fig. 7.1 are subject to considerable uncertainty.

Figure 7.7 shows the spectrum of solar radiation reaching the earth's surface for the case of an overhead sun (the lower curve) together with the spectrum of solar radiation incident upon the top of the atmosphere (the upper curve). The area between the two curves represents the depletion of the incident radiation during its passage through the atmosphere. The depletion is divided into two parts: the unshaded area represents the combined effects of backscattering and absorption by clouds and aerosol, and backscattering by air molecules, while the shaded area represents the absorption by air molecules. Nearly all the shaded area can be identified with discrete absorption bands, the most important of these being the water vapor bands in the near infrared. More than half the solar radiation reaching the ground penetrates through the broad spectral "window" which corresponds roughly to visible wavelengths.

The banded structure of the absorption spectrum of a clear atmosphere can be seen more clearly in Fig. 7.8 which spans the wavelength range of both solar and terrestrial radiation. By comparing the absorption spectrum for the whole atmosphere (Fig. 7.8c) with that for the upper atmosphere (Fig. 7.8b) it is

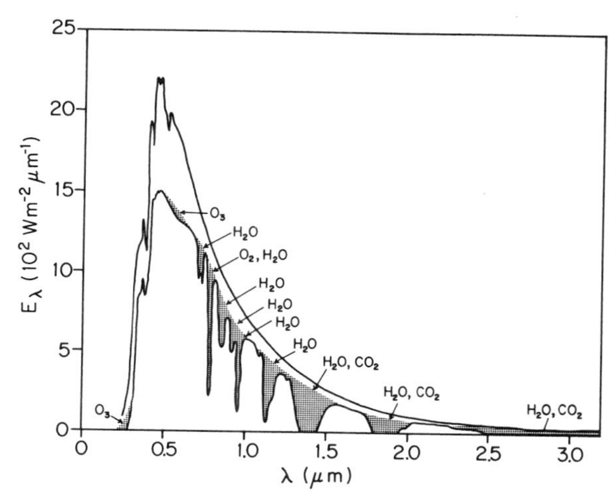

Fig. 7.7 Spectrum of solar radiation at the "top" of the atmosphere (upper curve) and at sea level (lower curve) for average atmospheric conditions and an overhead sun. The shaded area represents absorption by gaseous constituents, as indicated. [Adapted from "Handbook of Geophysics and Space Environments," McGraw-Hill, New York (1965), p. 16–20.]

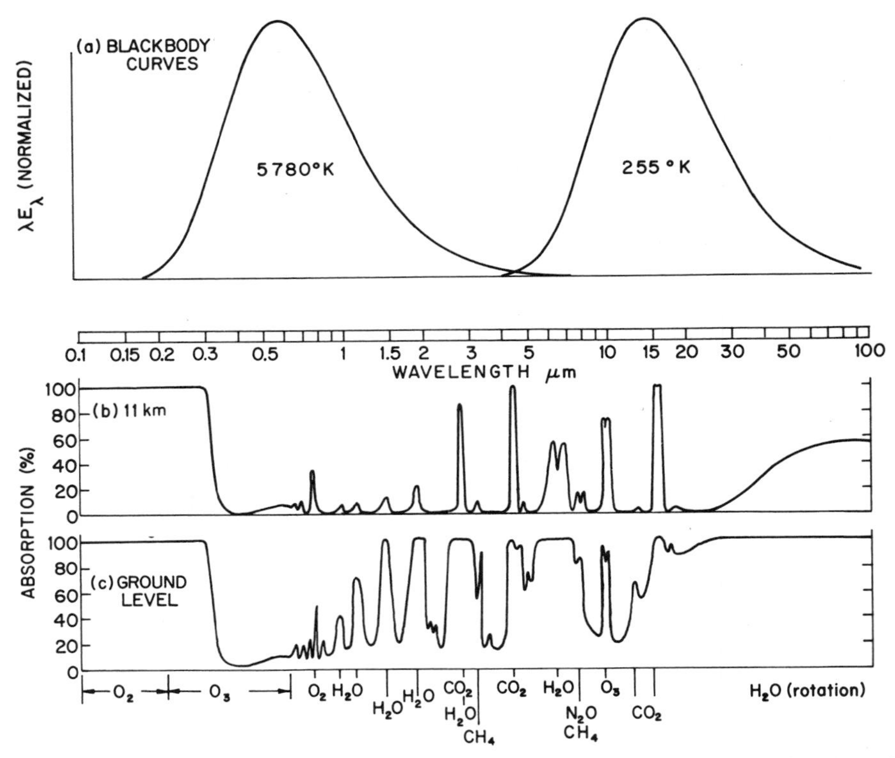

Fig. 7.8 (a) Normalized blackbody curves for 5780°K and 255°K, plotted so that irradiance is proportional to the areas under the curves. (c) Atmospheric absorption in clear air for solar radiation with a zenith angle of 50° and for diffuse terrestrial radiation. (b) Same as (c) but for the portion of the atmosphere lying above the 11-km level, near the middle-latitude tropopause. [Adapted from R. M. Goody, "Atmospheric Radiation," Oxford Univ. Press (1964), p. 4.]

possible to bring together a number of points that have already been mentioned in the previous discussion:

- Virtually all solar radiation with $\lambda < 0.31\ \mu m$ is absorbed before it reaches the tropopause level. The upper atmosphere is nearly transparent to solar radiation with $\lambda > 0.35\ \mu m$.
- There is very little absorption of solar radiation at visible wavelengths by the gaseous constituents of the atmosphere. This remarkable window in the absorption spectrum coincides with the wavelengths of maximum solar emission.
- The absorption of solar radiation at infrared wavelengths takes place mostly within the troposphere, where most of the atmospheric water vapor is located.

From a further inspection of Fig. 7.8c it is evident that the average atmospheric absorptivity is considerably larger at wavelengths of terrestrial radiation than at wavelengths of solar radiation, just as in the case of the hypothetical atmosphere in Problem 6.8. Radiation emitted by the earth's surface can penetrate through the atmosphere only at the wavelengths of the two narrow windows near 10 μm. At most other wavelengths, the terrestrial radiation that reaches outer space is emitted by gases in the middle and upper troposphere, where the ambient temperatures are considerably lower than those at the earth's surface. Hence the "effective temperature" of the earth (as calculated in Problem 6.5) is considerably lower than the average temperature of the surface of the planet. The insulating effect of tropospheric water vapor and carbon dioxide is enhanced by the presence of extensive cloud layers which effectively block emission through the windows over large areas of the earth. High cirroform cloud layers are the most effective insulators (that is to say, they contribute the most to the "greenhouse effect") because they radiate to space at the lowest temperatures.

It is instructive to consider what the vertical temperature profile would look like if the troposphere were in pure radiative equilibrium. In this hypothetical situation (7.10) would reduce to

$$[\overline{E_s}] = [\overline{E_L\uparrow}] - [\overline{E_L\downarrow}]$$

so that the net upward transfer of infrared radiation would have to be roughly 51% of the solar irradiance instead of the 21% given in Fig. 7.1 as an estimate for the real atmosphere. In order to facilitate such a large net upward flux of infrared radiation the earth's surface would have to be very warm and the middle and upper troposphere would have to be relatively cold in comparison to the actual atmosphere. The resulting vertical temperature profile (calculated on the basis of a multilayer atmosphere as illustrated in Problems 6.27 and 6.28) is compared with the "standard atmosphere" in Fig. 7.9. The inversion above 10 km in the radiative equilibrium temperature profile is due to absorption of solar radiation by ozone in the stratosphere. Note that throughout most of the troposphere the radiative equilibrium lapse rate exceeds the dry adiabatic

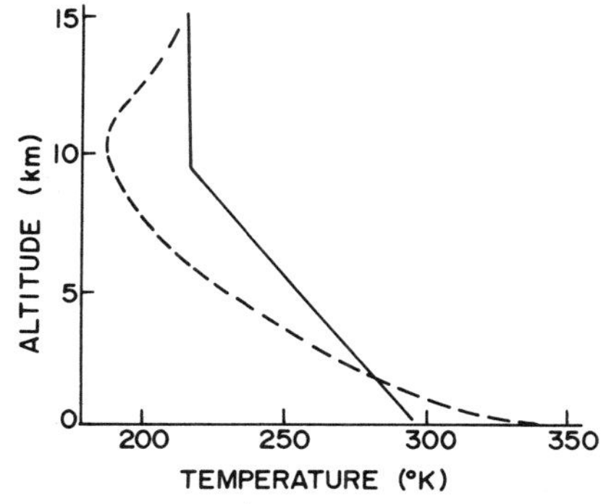

Fig. 7.9 Comparison between radiative equilibrium temperature profile (---) as calculated by Manabe and Strickler [*J. Atmos. Sci.* **21**, 370 (1964)] (——) and the U.S. Standard Atmosphere.

value of 9.8 deg km^{-1}, which is the criterion for the onset of convective overturning. A similar situation exists in the tropospheres of Mars and Venus, and in the solar photosphere.

7.3.2 The role of latent and sensible heat flux

The role of the fluxes of latent and sensible heat in the tropospheric energy balance can be described in terms of the changes in moist static energy that an idealized "typical" air parcel experiences as it moves through the atmosphere. For such a parcel, the time rate of change of moist static energy is governed by (7.4). During the periods that the parcel is isolated from the earth's surface its moist static energy is almost always decreasing as a result of the net radiative cooling that prevails throughout most of the troposphere. The parcel's supply of moist static energy is replenished during the brief periods when it is in contact with the earth's surface so that, in the long-term mean, it neither gains nor loses moist static energy.

Alternatively, one can view the fluxes of latent and sensible heat from the point of view of the energy budget of the middle and upper troposphere. This region is continually losing moist static energy as a result of radiative cooling. However, its supply is continually being replenished by the exchange of air with the lower troposphere. On the average, the air that is entering the region from below is richer in moist static energy than the air that is descending out of the region. (The situation is somewhat analogous to the money supply in Las Vegas!)

There exists a hierarchy of scales of motion that contribute to the vertical flux of sensible heat and water vapor. Each of these scales is dominant within a distinct layer of the troposphere as summarized in Table 7.2.

Table 7.2

Scales of motion that contribute to the vertical transport of latent and sensible heat

Layer	Vertical extent	Types of motion
Molecular boundary layer	<1 mm adjacent to surface	Molecular conduction and diffusion
Surface layer	Lowest few tens of meters	Microscale turbulence
Mixed layer	~ 1 km	Thermals, rolls
Troposphere	~ 10 km middle latitudes ~ 17 km tropics	Deep convection and large-scale, thermally driven circulations

Within the *molecular boundary layer* fluid motions are strongly suppressed so that molecular conduction and diffusion are the principal mechanisms for the vertical transport of sensible heat and water vapor.

In the layer immediately above the molecular boundary layer, thermal convection is still constrained by the presence of a lower boundary, and the dominant type of fluid motion is microscale turbulence induced mechanically by flow over irregularities in the underlying surface. Within this so-called *surface layer* sensible heat and water vapor are transferred upward by microscale "eddies" which perform the same role as individual molecules in the case of molecular conduction. The surface layer is characterized by strong vertical wind shear with wind speed proportional to the logarithm of the height above the surface. Under conditions of strong upward heat flux from the earth's surface the lapse rate within this layer may be strongly superadiabatic, with typical temperature decreases of 5 deg within the first few meters above the earth's surface. Toward the top of the surface layer, the vertical wind shear grows rather small, the lapse rate approaches the dry adiabatic value, and the microscale turbulence becomes weaker and more intermittent, with short "bursts" of a few seconds duration, separated by longer, quiescent periods as illustrated in Fig. 7.10. The intermittency of the turbulence reflects the influence of larger-scale motions, which modulate the distributions of static stability and moisture. The depth of the surface layer ranges from less than a meter under conditions of light winds and very strong surface heating to several tens of meters or more under conditions of strong winds and weak surface heating.

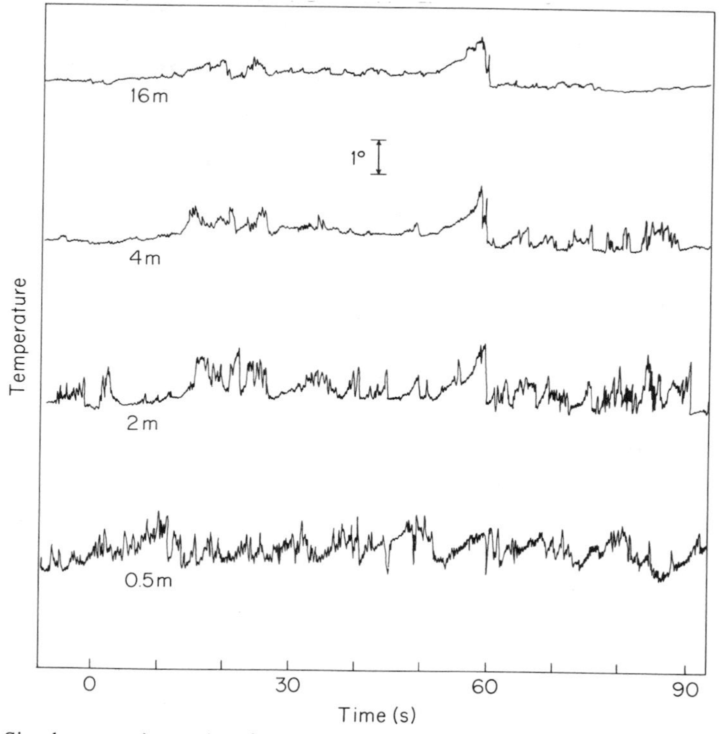

Fig. 7.10 Simultaneous time series of temperature at four heights above the ground, as indicated, showing the transition from the surface layer (lower levels) to the mixed layer (upper levels). Observations were taken over flat, plowed ground on a clear day with moderate winds. The top three temperature sensors were aligned in the vertical; the 0.5-m sensor was located 50 m away from the others. (Courtesy of J. E. Tillman.)

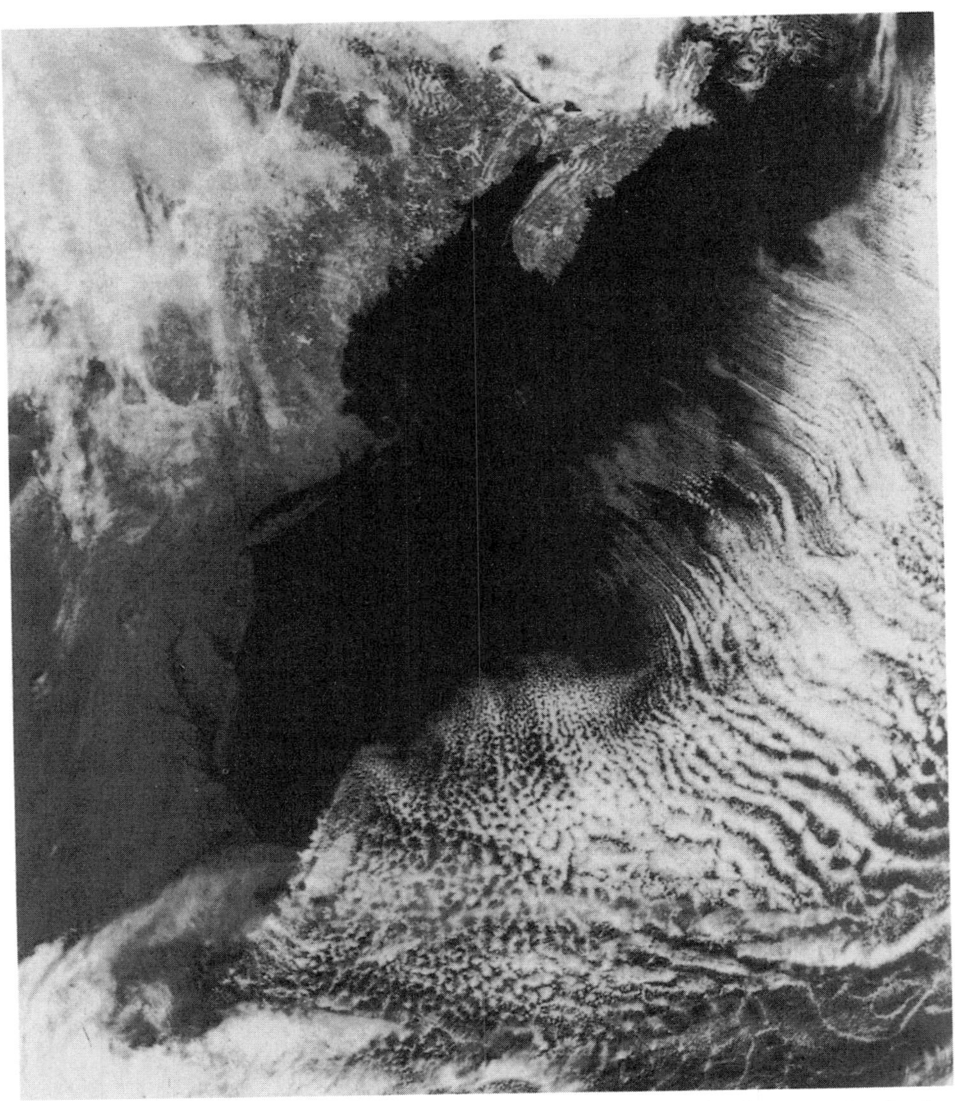

Fig. 7.11a Visible image from the NOAA-2 satellite 22 February 1975 showing convective clouds in the mixed layer in northwesterly flow off the east coast of the United States. Note also the lee-wave clouds over Nova Scotia, New Brunswick, and Maine. (NOAA Photo.)

Fig. 7.11b The corresponding infrared image. The coastline is less pronounced than in 7.11a but is still clearly visible. Bright areas are cold, dark areas warm. The light-gray areas immediately off the coast are associated with cold surface water. Note how the clouds form as the air flows out over the warm (dark) waters of the Gulf Stream. (NOAA Photo.)

As one moves upward out of the surface layer, convectively driven thermals or rolls assume an increasing share of the vertical transports of latent and sensible heat until they eventually become dominant, as seen in Fig. 7.10. As mentioned previously in Section 5.1, thermals tend to be dominant in regions of weak surface winds and rolls predominate in regions of stronger surface winds. These circulations are so efficient as a mechanism for the vertical transport of sensible heat, moisture, and momentum that potential temperature, mixing ratio, and wind show relatively little change with height within the layer in which they prevail;† hence, the term *mixed layer*. The upward penetration of thermals and roll circulations is often blocked by an inversion at the top of the mixed layer.

From the top of the mixed layer ($\sim 1-2$ km) up to the tropopause the upward transport of latent and sensible heat is accomplished by deep cumulus convection and synoptic- or planetary-scale circulations. In addition, these same two motion regimes control virtually all the conversion of latent to sensible heat in irreversible condensation processes, and they modulate the thermals and rolls within the mixed layer, just as the thermals and rolls modulate the microscale turbulence within the surface layer.

In regions where cold air flows over a warm surface, the fluxes of latent and/or sensible heat are generally large and the layered structure described above is likely to be quite pronounced. For example, the situation in Fig. 5.16 corresponds to the flow of warm ocean air over a hotter land surface. In another example shown in Fig. 7.11, cold, dry continental air is flowing out over the warm Gulf Stream. In such situations, the lapse rate within the surface layer tends to be convectively unstable. The magnitude of the fluxes of latent and sensible heat increases with the degree of convective instability and with the wind speed in the surface layer. The latent heat flux also depends upon the humidity of the air in the surface layer; the drier the air the faster the evaporation from the underlying surface.

In the opposite situation corresponding to the flow of warm air over a colder surface, the lapse rate within the surface layer is stable and the fluxes of latent and sensible heat are very small. In such situations there is usually no well-defined mixed layer, as such. The structure of the motion regime within the lowest kilometer of the atmosphere is discussed further in Section 9.6.2.

7.4 THE ENERGY BALANCE AT THE EARTH'S SURFACE‡

The regions of land, water, and ice that make up the surface of the earth fulfill their respective energy balance requirements in widely differing ways.

The land masses have a rather small capacity for storing heat. The specific

† Above the lifting condensation level (when one is present), equivalent potential temperature and wind tend to be nearly constant with height.

‡ For a more comprehensive treatment of the topics covered in Sections 7.4 and 7.5, the reader is referred to W. D. Sellers, "Physical Climatology," University of Chicago Press, 1965.

heats of soil, rock, sand, and clay are rather low; roughly one fourth that of water. More importantly, the only mechanism for the vertical transport of energy within the solid earth is conduction, and the conductivity of the solid earth is very small. The transfer of energy downward into the earth is so slow that the diurnal cycle in soil temperature penetrates to a depth of less than a meter and even the annual cycle is not felt below a depth of a few meters.

Because of their low heat capacity, land masses respond quickly to changes in the energy balance so that locally, even in an average over a few weeks, there is a very close balance between incoming and outgoing energy, as defined (in a global context) by (7.10). For example, an abrupt increase in the absorption of solar energy would give rise to a rapid rise in surface temperature, with consequent increase in outgoing long-wave radiation, sensible heat flux, and possibly also latent heat flux that would be just enough to compensate for the increased absorption. Thus, the temperature of land surfaces is governed directly by the energy balance requirements and the characteristic "adjustment time" is rather short.

In contrast to the land masses, the oceans have an enormous capacity for storing heat. The solar energy absorbed near the sea surface is distributed by convection through a "mixed layer" (in some respects, the counterpart of the mixed layer in the atmosphere) which has a typical thickness on the order of 100 m or less (see Fig. 7.18). The mixed layer is separated from the colder waters of the deep oceans by the *thermocline* (or, in higher latitudes, *the seasonal thermocline*), a layer characterized by strong static stability which is analogous to an inversion in the atmosphere. Unlike the land masses, the oceans can transport large amounts of energy laterally so that locally there is no requirement for a balance between the incoming and outgoing fluxes of energy at the sea surface.

The global distribution of sea surface temperature shown in Fig. 7.12 is not determined in any direct or simple way by the energy balance requirements. Wind-driven ocean circulations are capable of creating and sustaining large local imbalances in the energy exchanges at the sea surface. For example, upwelling of water from below the thermocline is responsible for the low sea surface temperatures observed along the coasts of California, Peru, and Chile, and West Africa north of 20°N, and along the equator in the eastern Atlantic and Pacific. Within these regions of cold water, the surface layers of the ocean absorb much more energy than they release to the atmosphere; hence, the oceanic heat budget (Fig. 7.13) shows a large surplus in these regions. Much of the excess energy absorbed in regions of upwelling is given back to the atmosphere in the western oceans where boundary currents such as the Gulf Stream carry warm tropical surface waters into high latitudes. The oceans dispose of massive amounts of energy in the winter hemisphere in regions where cold, dry continental air flows over these warm surface waters (for example, as in the situation depicted in Fig. 7.11). Hence the western boundary currents correspond to regions where the oceanic heat balance (Fig. 7.13) shows a large deficit.

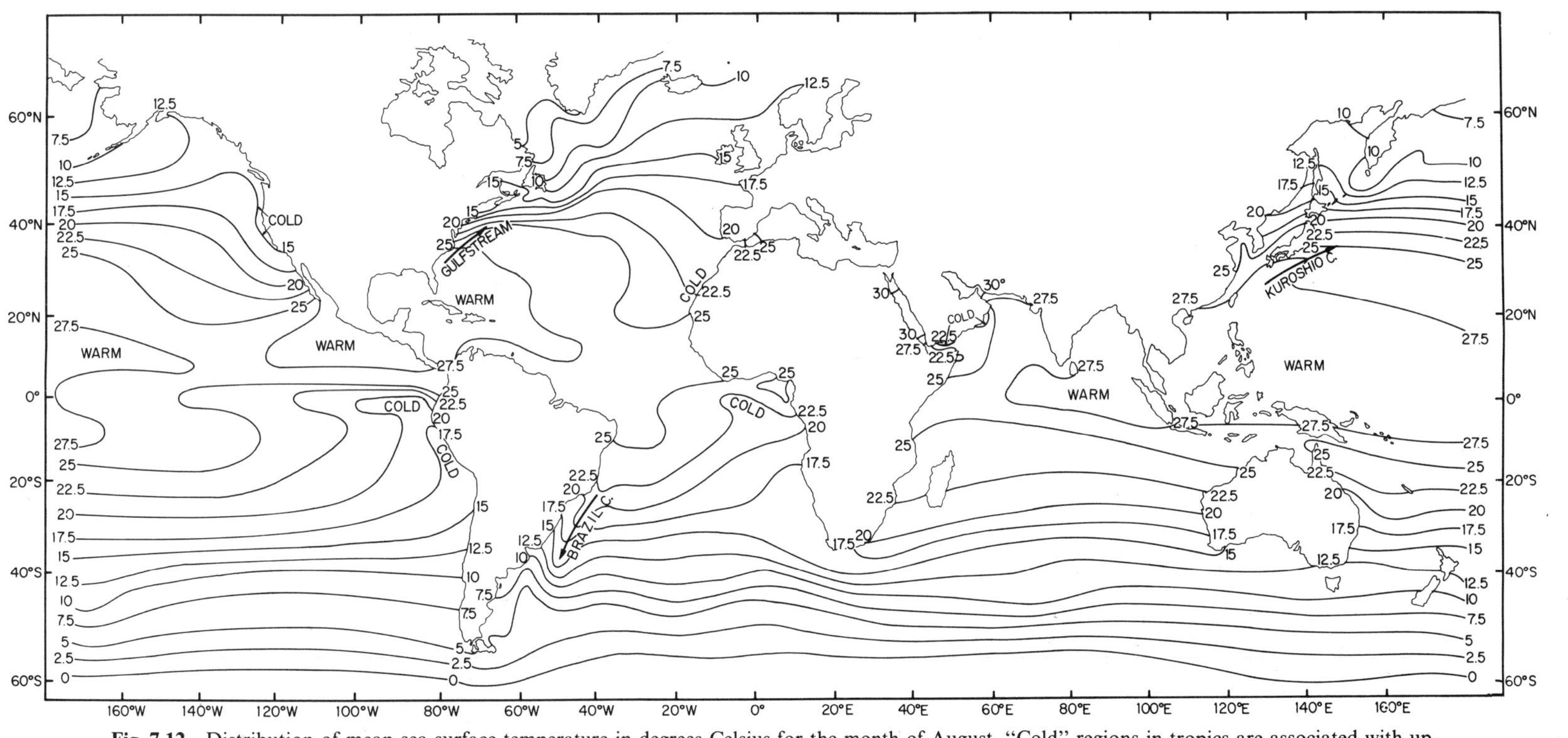

Fig. 7.12 Distribution of mean sea surface temperature in degrees Celsius for the month of August. "Cold" regions in tropics are associated with upwelling. Major western boundary currents are labeled. (Adapted from H. U. Sverdrup *et al.*, "The Oceans," Prentice-Hall, Englewood Cliffs, New Jersey, 1942, Chart III.)

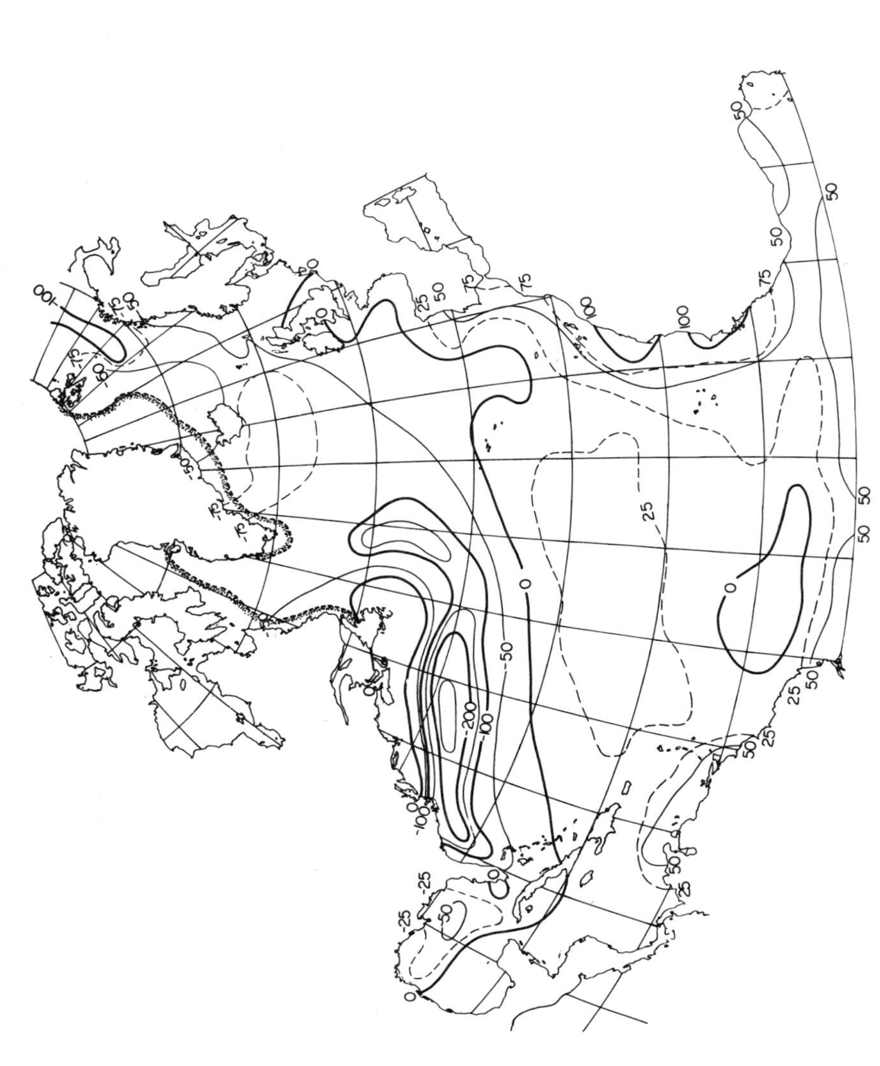

Fig. 7.13 Average annual energy flux through the sea surface in units of watts per square meter. Positive values indicate a downward flux. [From *Bull. Amer. Met. Soc.* **57**, 670 (1976).]

Land and sea surfaces differ also with respect to the relative importance of the fluxes of latent and sensible heat.† On the average over all land masses, the two fluxes are of roughly comparable importance. The latent heat flux predominates in regions covered by vegetation, while the sensible heat flux predominates in desert regions and cities. Over the oceans the flux of latent heat (as inferred from statistics on the average rate of precipitation over the globe) is about an order of magnitude larger than the sensible heat flux.

The presence of snow and ice influences the local energy balance at the earth's surface in a number of ways:

- The absorption of solar radiation is greatly reduced because of the high reflectivity.
- During warm periods the surface temperature remains near 0°C, regardless of the air temperature. Any excess incoming over outgoing radiation goes into melting or sublimation.
- During cold periods, the loss of heat from the underlying surface is greatly reduced. Fresh snow is particularly effective as an insulator.

7.5 TIME VARIATIONS IN THE ENERGY BALANCE

Some of the time variability observed in the structure of the atmosphere may be viewed as a direct response to the changing distribution of energy sources and sinks. In this section, we will consider three specific examples with widely differing time scales: the diurnal cycle, seasonal variations, and climatic change.

7.5.1 The diurnal cycle

The diurnal periodicity in solar heating affects the atmospheric energy balance in two ways: directly through the various absorption mechanisms discussed earlier in this chapter, and indirectly through changes in the energy balance at the earth's surface. The direct effects are felt most strongly in the upper atmosphere where they give rise to large diurnal and semidiurnal variations in temperature, pressure, and wind. These *thermally driven atmospheric tides* are several orders of magnitude larger than gravitationally induced tidal motions in the atmosphere, in marked contrast to the situation in the oceans where the gravitational tide predominates. In this subsection we will be concerned primarily with indirect effects of the diurnal cycle in solar heating, which account for most of the diurnal temperature variation in the mixed layer, over land.

On a calm, clear evening, the emission of infrared radiation is usually strong enough to cool land surfaces to below the ambient air temperature within the first few hours after sunset. Thus, throughout most of the night, the air just above the surface loses energy by emitting radiation both upward and down-

† The fraction F_h/F_m is sometimes called the *Bowen ratio*.

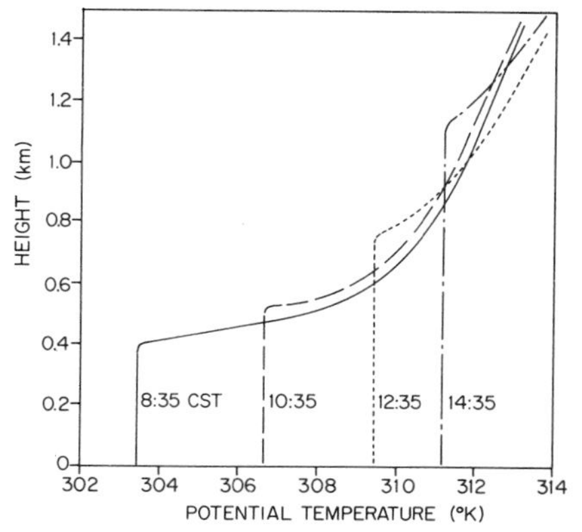

Fig. 7.14 Growth of the mixed layer on a typical summer day over land. Approximate observed data at O'Neill, Nebraska, 25 August 1953. [*J. Atmos. Sci.* **30**, 1097 (1973).]

ward. This radiative cooling gives rise to a low-level inversion which reaches its maximum strength during the late night hours.

Within a very short time after sunrise the temperature of the land surface rises to the level of the ambient air temperature. With further absorption of solar radiation at the ground, the lapse rate within the surface layer becomes convectively unstable and microscale turbulence breaks out. Soon buoyant thermals of heated air begin to rise out of the surface layer, creating a well-defined mixed layer, as shown in Fig. 7.14. The first thermals are confined to a thin layer adjacent to the earth's surface, since they quickly lose their buoyancy as they rise above the base of the inversion. Gradually the potential temperature of the air in the mixed layer rises in response to the steady input of sensible heat carried upward from the surface. The heating of the mixed layer is accompanied by a gradual increase in its depth as increasingly buoyant thermals from the surface layer are able to penetrate deeper and deeper into the inversion.† The bombardment by ever more buoyant thermals continues either until the inversion is completely eaten away from below or until the sun begins to sink in the west. If the lapse rate above the inversion is conditionally unstable, and if the lifting condensation level is low enough, the first thermals that succeed in breaking through the top of the inversion may rise to great heights, signaling the onset of deep cumulus convection.

Problem 7.1 On a summer morning over land the mixed layer extends upward to a height of 400 m above the ground and above it there is an inversion layer in which the temperature increases at a rate of 10 deg km^{-1}. The sensible heat flux from the earth's

† A similar growth of the mixed layer is apparent in Fig. 7.11 as dry, cold continental air is warmed by its passage over the Gulf Stream. Clouds first begin to form at the point where the top of the mixed layer reaches the lifting condensation level. The depth of the mixed layer increases in the downstream direction, and with it the height and size of the clouds.

surface is 500 W m^{-2} and the vertical flux of infrared radiation is negligible. Calculate the rate of rise of the inversion base.

Solution We begin by applying (7.4) to a unit column extending from the earth's surface upward to the top of the mixed layer:

$$\frac{d}{dt}\int_0^{400\text{m}} \rho(c_p T + \Phi + Lw)\,dz = \int_0^{400\text{m}} \rho(\dot{H}_R + S_h + S_m)\,dz$$

In the context of this problem, Φ and w are not changing with time and $\dot{H}_R = S_m = 0$. Therefore, the above equation reduces to

$$\frac{d}{dt}\int_0^{400\text{m}} \rho c_p T\,dz = \int_0^{400\text{m}} \rho S_h\,dz$$

Now since the sensible heat flux from the ground is completely absorbed by the air within the mixed layer, the integral on the right-hand side of the equation is equal to 500 W m^{-2}. For this thin layer, the density ρ can be regarded as a constant equal to 1.25 kg m^{-3}, and, since the layer is at constant potential temperature, dT/dt can also be treated as being independent of height. Therefore,

$$\frac{dT}{dt} = \frac{500\ \text{W m}^{-2} \times 3600\ \text{s h}^{-1}}{1.25\ \text{kg m}^{-3} \times 400\ \text{m} \times 1004\ \text{J kg}^{-1}\ \text{deg}^{-1}} = 3.58\ \text{deg h}^{-1}$$

From Fig. 7.14 it is clear that the rate of rise of the inversion base z_B is given by

$$\frac{dz_B}{dt} = \frac{d\theta/dt}{(d\theta/dz)_{\text{inv}}} \tag{7.15}$$

From (2.72), at constant pressure,

$$\frac{d\theta}{dt} = \frac{\theta}{T}\frac{dT}{dt}$$

and, from (2.77),

$$\frac{d\theta}{dz} = \frac{\theta}{T}\left(\frac{dT}{dz} + \frac{g}{c_p}\right)$$

Substituting into (7.15) we obtain

$$\frac{dz_B}{dt} = \frac{dT/dt}{(dT/dz) + (g/c_p)} = \frac{3.58\ \text{deg h}^{-1}}{(10 + 9.8)\ \text{deg km}^{-1}} = 181\ \text{m h}^{-1}$$

The diurnal cycle in the energy balance at the ground is shown in Fig. 7.15 for a region of grassland (a) and a dry lake (b). Symbols are as defined in (7.7). The curves for E_n and F have been reversed in order to facilitate comparisons between the magnitudes of the various fluxes. The following points are worthy of note:

- Even on the time scale of a day the net flux F is small in comparison to the leading terms in the energy balance.
- The ratio F_m/F_h is quite different at the two locations.
- Both F_m and F_h are very small during the night. The convective boundary layer is present only during the daylight hours.

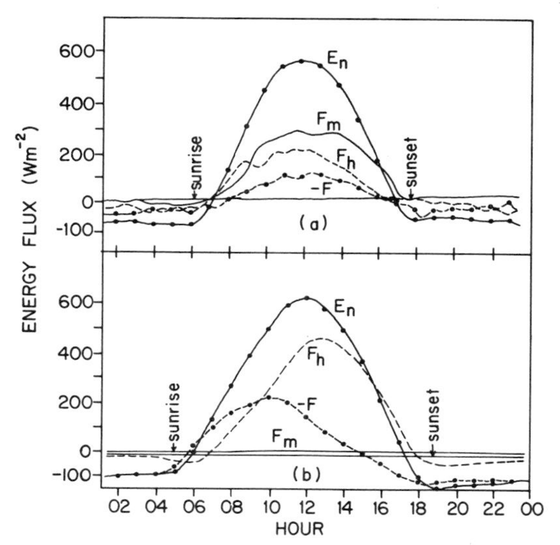

Fig. 7.15 Diurnal cycle in the energy balance for (a) a grassland site and (b) a dry lake. [After W. D. Sellers, "Physical Climatology," © University of Chicago Press, Chicago (1965), p. 112.]

The heat capacity of the mixed layer of the oceans is so large that the response to the diurnal cycle in solar heating is almost negligible. Since the sea surface temperature is nearly constant over the course of the day, there is very little diurnal variation in the sensible heat flux from the sea surface. The observed diurnal cycle in air temperature over the sea is only on the order of 1 deg.

As a result of the widely differing amplitudes of the diurnal temperature cycle over land and sea, there is a tendency for the air over land masses to be warmer than the air over neighboring oceans during the day and colder during the night. These reversible thermal contrasts are largely confined to the lowest kilometer of the atmosphere, where they drive mesoscale and synoptic-scale circulations that have a diurnally periodic component. During the afternoon there is a tendency for onshore flow from the oceans onto the heated land masses with rising air over the land and a return flow aloft (a sea-breeze circulation); while at night the circulation tends to be in the opposite sense, with offshore flow at low levels (a land breeze). In some regions the incursions of cool marine air onto land during the day have an identifiable leading edge that resembles a mesoscale cold front. Some of the diurnal variability in precipitation shown in Fig. 1.23 is due to synoptic-scale circulations induced by diurnally reversible land-sea heating contrasts.

7.5.2 Seasonal variations

The solar radiation incident upon a horizontal plane at the top of the atmosphere may be regarded as a parallel beam with irradiance

$$E = E_s \cos \phi_s (R/R_m)^{-2} \tag{7.16}$$

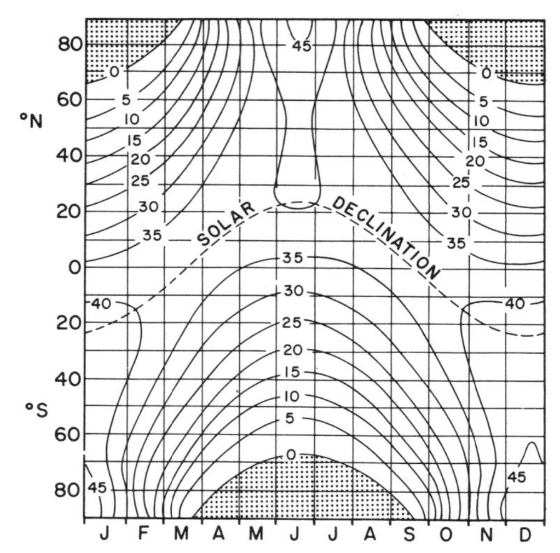

Fig. 7.16 Solar radiation incident on a unit horizontal surface at the top of the atmosphere as a function of latitude and date, expressed in units of megajoules per square meter per day. [Adapted from "Meteorological Tables" (R. J. List, ed.), 6th ed., Smithsonian Institute, Washington, D.C. (1951), p. 417.]

where R and R_m are the actual and mean distances between the earth and sun, respectively, E_s the solar irradiance upon a perpendicular plane at distance R_m from the sun (1380 W m^{-2}), and ϕ_s the solar zenith angle which is a function of latitude, time of day, and time of year. This instantaneous value of the irradiance can be integrated over the length of the daylight hours to obtain the daily incoming solar energy (sometimes called the *insolation*) as a function of latitude and time of year, as shown in Fig. 7.16.

From the figure, it is evident that, at the times of the equinoxes, when the length of daylight is the same at all latitudes, the insolation is proportional to the cosine of the latitude angle, with a maximum at the equator, where the sun is overhead at noon, and a value of zero at the poles, where the sun is on the horizon all day. In the summer hemisphere, the lengthening of the daylight hours with increasing latitude compensates for the increasing solar zenith angle, so that insolation is nearly constant over a wide range of latitude, and there is actually a slight maximum at the pole. In the winter hemisphere, the decreasing length of the day and the increasing solar zenith angle both contribute to the rapid decrease of insolation with increasing latitude until it reaches zero at the outer edge of the region of the polar night. The ratio of insolation at the summer solstice to that at the winter solstice is about 2:1 at 30°, 3:1 at 40°, and 5:1 at 50° latitude. The minor asymmetry between the distribution of insolation in the Northern and Southern Hemispheres is due to the slight eccentricity of the earth's orbit (see Problem 6.12).

The annual march of insolation produces a varied response at the earth's surface:

- Land surfaces exhibit large temperature fluctuations with peak-to-peak amplitudes as large as 50 deg in high-latitude continental regions. The response to the annual cycle damps out rapidly with increasing depth below the surface, as illustrated in Fig. 7.17.

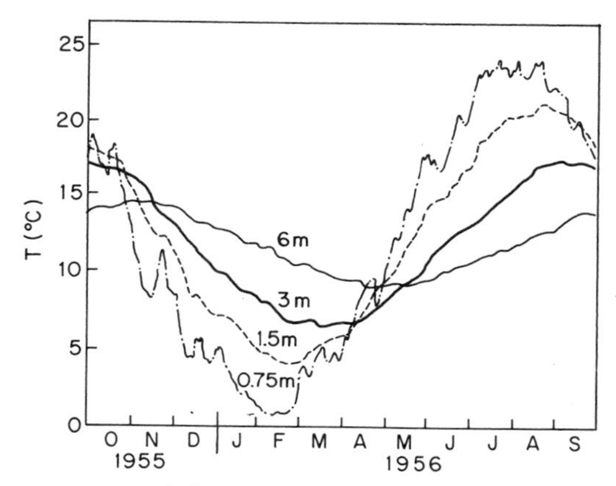

Fig. 7.17 Soil temperatures recorded at an exposed site at levels below the ground, as indicated. [Adapted from I. A. Singer, *Trans. Amer. Geophys. Union* **37**, 746 (1956), copyrighted by the American Geophysical Union.]

- Sea surface temperatures display much smaller but still significant temperature fluctuations with maxima and minima lagging about six weeks behind the solstices. The absorption of solar radiation during summer gives rise to a warm, shallow mixed layer which gradually deepens with time, as shown in Fig. 7.18 The bottom of this layer is called the seasonal thermocline.
- There are substantial seasonal changes in the fractional areas covered by snow and ice in polar regions and in the major mountain ranges. As an example, the seasonal variations in the limit of the Antarctic pack ice are shown in Fig. 7.19.

The differing responses of land and sea surfaces to the annual march of insolation has a profound influence upon global climate. During summer, the continents tend to be considerably warmer than the adjacent oceanic regions, while during winter they tend to be colder, with the land–sea contrasts being greatest just after the times of the solstices. The massive monsoon circulations in tropical and subtropical latitudes are driven by the seasonally reversing horizontal temperature gradients imposed upon the troposphere by the distribution of continents and oceans.† At all latitudes poleward of about 15° the

† A more extensive discussion of the monsoon circulations is given in Section 9.4.1.

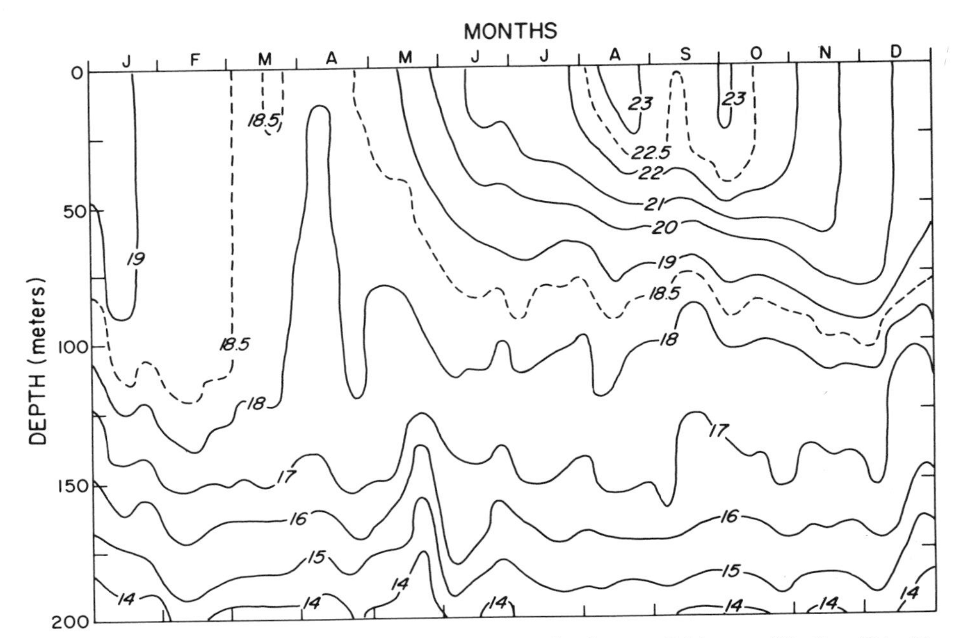

Fig. 7.18 Time–height section of water temperature, in degrees Celsius, at Weather Ship N (30°N, 140°W). [From *J. Phys. Oceanogr.* **4**, 645 (1974).]

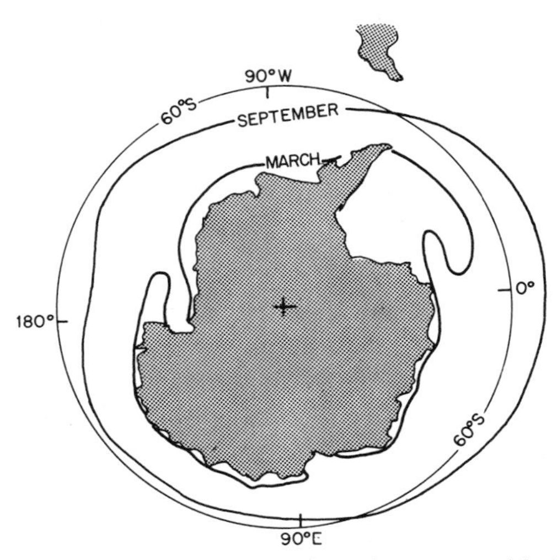

Fig. 7.19 Limits of Antarctic pack ice. [Adapted from data presented in "Atlas of Antarctica" USSR Main Directorate for Geodesy and Cartography, Moscow and Leningrad (1966).]

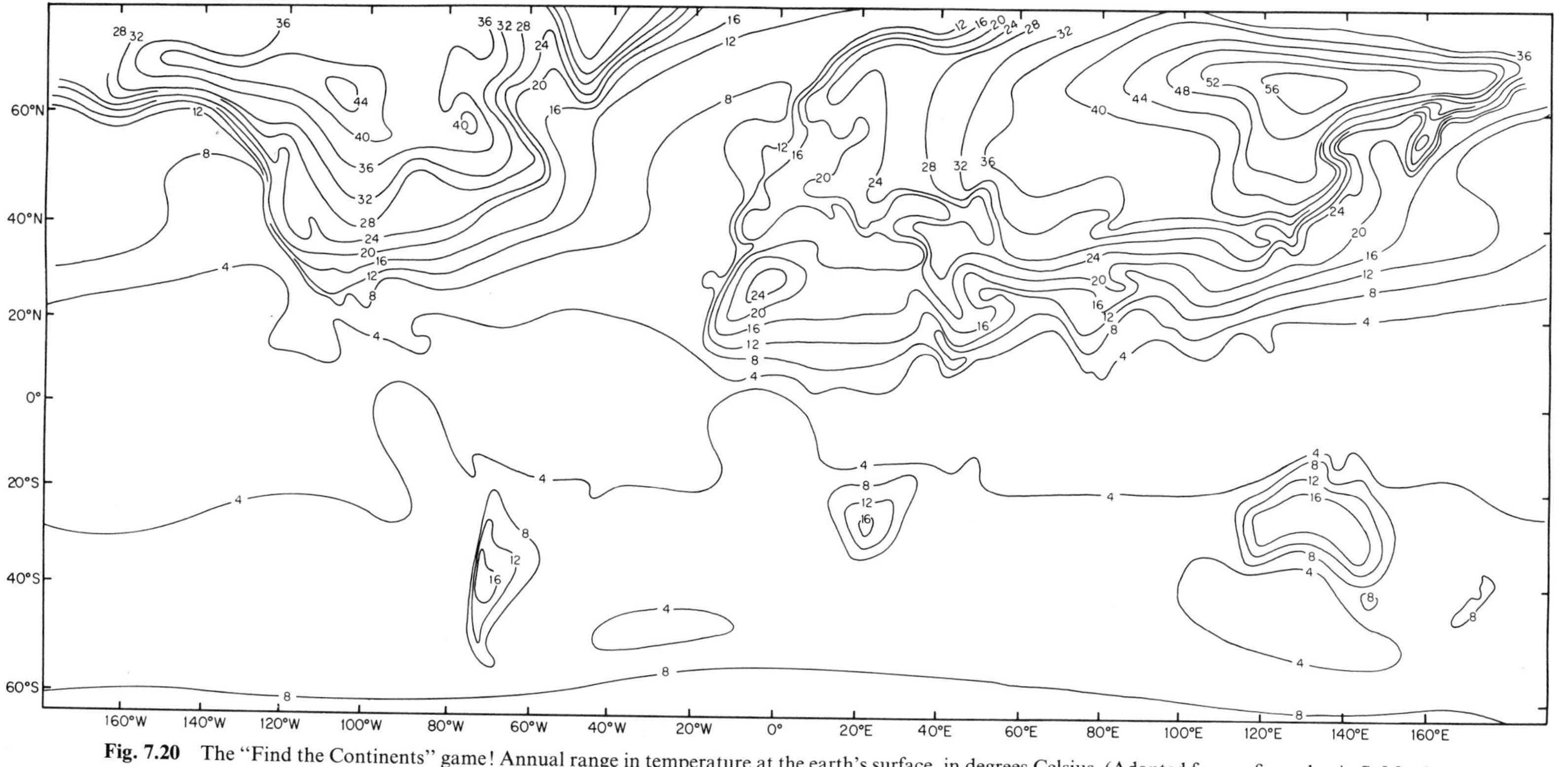

Fig. 7.20 The "Find the Continents" game! Annual range in temperature at the earth's surface, in degrees Celsius. (Adapted from a figure by A. S. Monin and P. P. Shirshov which appeared in Report No. 16, *GARP Publications Series*, World Meteorological Organization—International Council Scientific Unions, p. 203.)

character of the annual cycle is strongly affected by geographical position relative to land and sea, as shown in Fig. 7.20. Because of the asymmetry in the distribution of land and sea with respect to the equator, the amplitude of the annual temperature cycle at most Southern Hemisphere locations is smaller than at the corresponding locations in the Northern Hemisphere, despite the fact that the insolation undergoes a slightly larger seasonal variation in the Southern Hemisphere.

Large seasonal anomalies (deviations from climatological mean values) in precipitation and temperature can have serious consequences on world food production and energy requirements. The physical mechanisms responsible for these anomalies are not completely understood. However, it is clear that, at least in some situations, the coupling between the atmospheric and oceanic circulations plays a role in sustaining such anomalies over long time periods.

One suggested mechanism for the mutual interaction between atmospheric and oceanic circulations can be described as follows. Regardless of its cause, any sustained departure from the normal pattern of surface winds is likely to have some effect upon the wind-driven ocean circulation. Under certain circumstances even rather modest changes in the ocean circulation can produce dramatic changes in sea surface temperature patterns, particularly along the equator and in coastal areas that experience occasional upwelling.† Anomalies in the distribution of sea surface temperature in turn influence the atmospheric circulation through the energy fluxes at the sea–air interface, thus completing the feedback loop. In certain cases the characteristic time scale of these coupled fluctuations may be quite long because of slowness of the ocean circulation in adjusting to changes at the interface. Thus, once a large sea surface temperature anomaly is established it can sometimes persist for a season or longer.

There is at least one well-documented example of the kind of interaction described above: the so-called "Southern Oscillation," an intermittent 2–4 yr quasi-periodicity observed in sea level pressure, surface wind, sea surface temperature, cloudiness, and rainfall over a wide area of the Pacific Ocean and adjacent coastal areas, south of the equator. Fisheries in the region off Peru are adversely affected during the periods in which coastal upwelling is suppressed.

7.5.3 Climatic change

The field of *paleoclimatology* is concerned with reconstruction of the earth's climatic history, based on existing meteorological records together with various types of indirect evidence such as historical documents, archaeological findings, the widths of tree rings, the structure and chemical makeup of various types of sediment deposits and layers of ice that have been laid down gradually over

† Upwelling plays a crucial role in providing nutrients for plankton production in the oceans. Upwelling regions, totaling no more than 0.1% of the ocean surface, account for about half the world fish production.

long time periods, and so on. From such diverse pieces of evidence emerges a rather detailed and remarkably consistent picture of climatic variability during the past 10,000 yr, and a less detailed and more speculative picture that extends back much further in time.

About ten major ice ages have come and gone since mankind first appeared on earth. It is well established that the most recent of a series of major ice ages was at its peak about 18,000 yr ago. At that time the mass of the *cryosphere*† was a little more than twice as large as it is today. Massive continental ice sheets approximately 2 km thick covered the northern parts of Europe and North America, as shown in Fig. 7.21. Most major mountain ranges, including the Andes in the Southern Hemisphere, were much more heavily glaciated than they are now, and the sea level was about 100 m lower (hence, the distorted land outlines in Fig. 7.21). In the global average, the surface temperature of the earth was about 5 deg lower than at present; only a few degrees lower in the tropics, but as much as 20 deg lower in the Arctic. These differences are apparent from a comparison of the sea surface temperatures in Figs. 7.12 and 7.21.

Around 15,000 yr ago, for some unknown reason, the climate of the earth began to grow warmer, signaling the onset of the present *interglacial period.* The retreat of the continental ice sheets was slow and irregular; the last remnants of the North American ice sheet disappeared about 4000 B.C.

The interglacial period has been marked by continuous climatic variability on a wide range of time scales. The so-called "climatic optimum" occurred in the millenium around 4500 B.C., when temperatures over much of the earth were several degrees higher than at present. Since that time, there is evidence of alternating cooler periods marked by some increase in glaciation and milder periods marked by some glacial retreat. The period from 900 to 1300 A.D., which corresponds to the Viking colonization of Iceland and Greenland, was relatively mild compared to the present, and the period from 1500 to 1850 was relatively cold.

More detailed data available during the past century show evidence of more subtle climatic fluctuations on shorter time scales. It is well established that the climate of the Northern Hemisphere grew warmer during the period from 1880 to 1940, with temperature rises of about 0.5 deg over the Arctic and a few tenths of a degree in middle latitudes. The temperature rises were accompanied by a pronounced retreat of many glaciers. Many theories have been proposed to explain observed climatic variability in the past and to provide a basis for predicting future climatic trends. Generally speaking, these theories can be grouped into two categories:

† The cryosphere is the entire mass of frozen water substance on the earth's surface, including ice sheets, glaciers, pack ice, and snow cover. At present more than 90% of the mass of the cryosphere is contained in the Antarctic ice sheet, and about 8% is contained in the Greenland ice sheet. The cryosphere covers about 5% of the area of the Northern Hemisphere during summer and about 25% during winter.

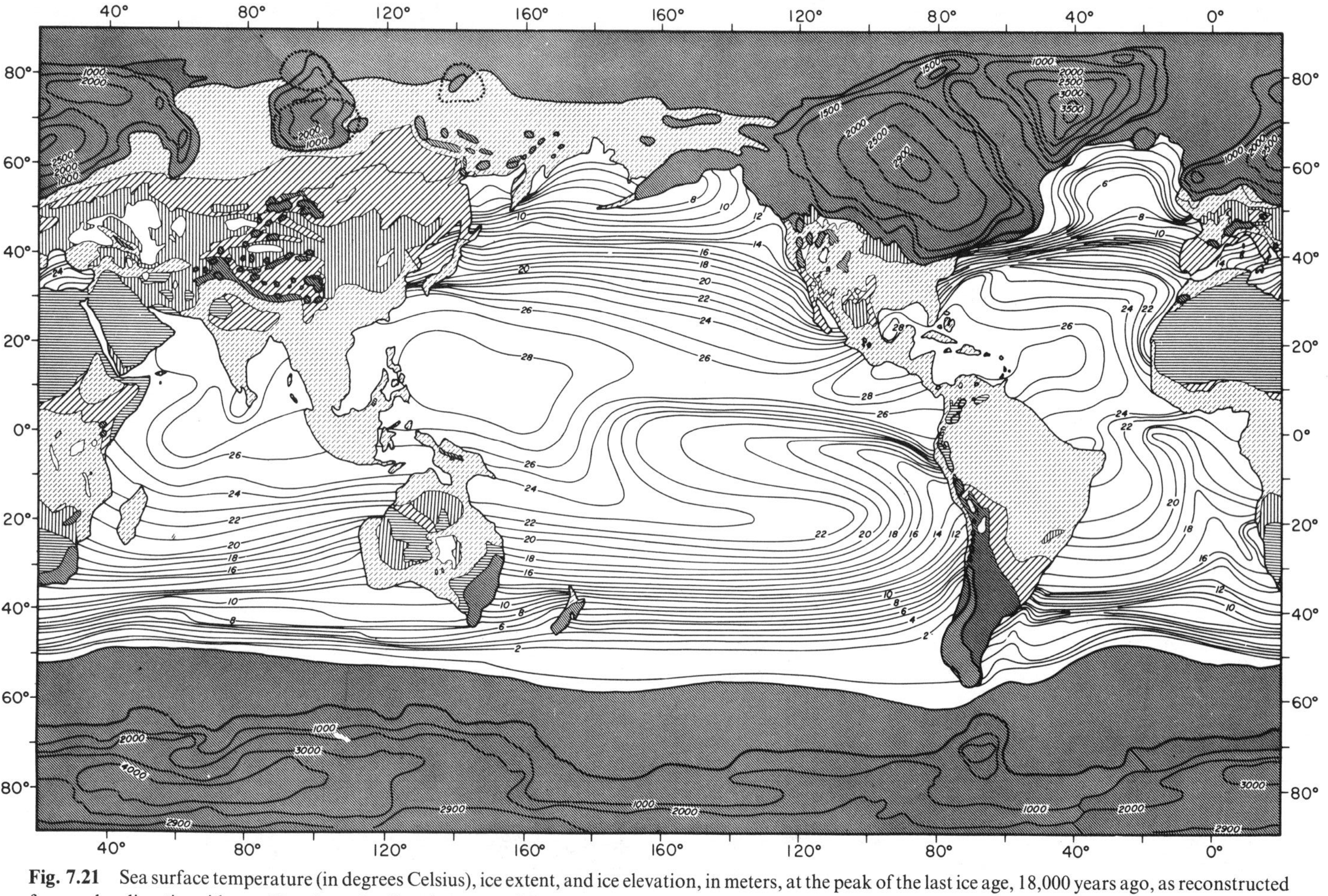

Fig. 7.21 Sea surface temperature (in degrees Celsius), ice extent, and ice elevation, in meters, at the peak of the last ice age, 18,000 years ago, as reconstructed from paleoclimatic evidence. Heaviest stippling denotes regions covered by snow and ice; various types of lighter stippling are used to distinguish between various types of vegetation and ground cover; white areas are covered by water. Continental outlines are based on an estimated 85 m lowering of sea level, relative to present value. [From *Science* **191**, 1132 (1976), copyright 1976 by the American Association for the Advancement of Science.] For a comparison of sea surface temperatures with current values, see Fig. 7.12.

- those that view climatic variability as the response of the coupled *atmosphere–ocean–cryosphere system*[†] to externally imposed changes in the global radiation balance,
- those based on the premise that the atmosphere–ocean–cryosphere system contains within itself physical processes capable of producing long-term variability, even in the absence of external forcing.

There is no shortage of possible external forcing mechanisms. Variability in solar emission, slow changes in the ellipticity of the earth's orbit or the tilt of its axis, and variations in atmospheric turbidity associated with volcanic activity have all been cited as possible causes of past climatic change. In addition, there are numerous mechanisms by which man's activities might conceivably be producing inadvertent climate modification, through their effect upon the radiative transfer characteristics of the atmosphere. Possible effects include the increase in carbon dioxide due to the burning of fossil fuels (see Fig. 1.3), increases in turbidity due to industrial pollution and burning of waste, changes in ozone concentrations due to stratospheric pollution, and albedo changes due to destruction of forests and grasslands.

Most of the external forcing mechanisms listed above would be too small to account for the large climatic changes that have occurred in the past, or to be of much concern with respect to possible future climatic deterioration, were it not for the interaction between the global radiation balance and the mass balance of the cryosphere. Any long-term decrease in solar energy absorbed at the earth's surface probably contributes to an increase in the fractional area of the earth covered by snow and ice, which raises the planetary albedo and further decreases the solar energy absorbed at the ground. Through this positive feedback mechanism, a small change in the input of solar radiation could conceivably produce a substantial change in global climate.

Theories which hold that climate change is generated internally within the atmosphere–ocean–cryosphere system are based upon the implicit assumption that climate is *intransitive*; that is to say, for a given spectrum of incident solar radiation, the atmosphere–ocean–cryosphere system can assume two or more widely differing states, corresponding to different climates, which are internally consistent and satisfy all the global energy balance requirements.[‡] The various proposed climate change mechanisms that fall within this category employ a wide range of physical processes to bring about transitions from one state to another (for example, from the present state back into another ice age.)

Regardless of the underlying cause it is apparent that important climatic changes in the past have taken place rather slowly. The stability of climate with

[†] In some proposed mechanisms the biosphere is involved as well.

[‡] Global climate appears to be intransitive at least in one sense: in contrast to the present climate, one can conceive of a completely ice-covered earth which, because of its high albedo, would have such a low radiative equilibrium temperature that ice could be sustained, even in the tropics. There is no evidence that such a "white earth" climate ever existed in the past; even during the ice ages the ice sheets covered much less than half the area of the earth.

respect to sudden changes is due, at least in part, to the vast "inertia" of the cryosphere in responding to changes in the global energy balance. In this respect, the cryosphere provides a long-term "memory" for the coupled system, just as the mixed layer of the oceans provides the shorter-term memory for the air–sea interactions that lead to climate fluctuations on the scale of individual seasons or years.

Many of the fundamental questions concerning the nature and causes of climate change are still largely unresolved because of our incomplete quantitative understanding of many of the physical processes that enter into the global energy balance and for lack of definitive observational data on which to test various theories. Under such circumstances it is far easier to propose new climate change hypotheses than it is to substantiate or disprove old ones.

PROBLEMS

7.2 Explain or interpret the following:

(a) The release of latent heat does not affect the moist static energy.

(b) During spring the net radiation at the top of the atmosphere is downward, in the hemispheric average.

(c) The globally averaged net radiation at the earth's surface is downward.

(d) Equation (7.10) is not valid for a seasonal average or for an average over only part of the earth's surface.

(e) The atmosphere emits more infrared radiation downward, to the earth's surface, than upward to space.

(f) The outline of the sun's visible disk is rather sharp.

(g) The photospheric radiation reaching the earth from near the center of the sun's visible disk is characteristic of a slightly hotter source than the radiation from the outer part of the disk.

(h) The solar corona is much hotter than the photosphere, yet it emits much less radiation.

(i) In photographs made with visible light, convective cells in the solar photosphere have bright centers and dark outlines.

(j) From an energetic point of view, it would be surprising if solar activity had any influence upon the troposphere.

(k) The vertical temperature profiles of Mars and Venus do not exhibit a warm stratopause like that of Earth.

(l) Very little ozone is produced at levels above 50 km.

(m) Even on a clear day, at a given zenith angle, the fraction of incoming solar radiation that reaches the ground may vary by more than 10%.

(n) On clear days, airplanes frequently encounter light turbulence immediately after takeoff and prior to landing. This turbulence is usually absent at night.

(o) Visibility in urban areas is generally better in the afternoon than in the morning.

(p) There is often an abrupt improvement in visibility in going from the mixed layer into the inversion above it.

(q) Serious air pollution rarely occurs in the presence of a deep mixed layer.

(r) Thermals whose momentum carries them beyond the level at which they lose their buoyancy produce a downward flux of sensible heat above that level.

(s) A breeze feels good on a hot day, especially when the air is humid.

(t) During summer, alpine snowfields disappear faster on a calm, humid day than on a dry, windy day with the same air temperature and insolation.

(u) On a clear, still night over land, the surface temperature usually rises if a low cloud moves overhead.

(v) The presence of smog usually contributes to the maintenance of low-level inversions during the daytime.

(w) The diurnal cycle in surface temperature is larger over deserts than over vegetated terrain.

(x) The presence of a snow cover tends to lower the air temperature near the surface, particularly on a clear night.

(y) Urban areas in the Arctic have potentially serious air pollution problems during winter.

(z) The continents tend to be colder than the neighboring ocean during winter and warmer during summer.

(aa) Sea surface temperature exhibits a far greater response to the annual cycle than to the diurnal cycle.

(bb) At high latitudes the latent and sensible heat flux from the oceans is much larger during winter than during summer.

(cc) On some days at locations on land, the surface temperature rises steadily until middle or late morning, then it levels off rather abruptly and remains steady for the remainder of the middle part of the day. Sometimes, but not always, this leveling off is accompanied by a sudden development of cumulus or cumulonimbus clouds.

(dd) Locally heavy snowstorms sometimes occur on the lee shores of the Great Lakes during early winter. (Suggest under what conditions.)

(ee) The lowest winter temperatures in the Northern Hemisphere are found over Siberia, rather than at the pole.

(ff) Summers are cooler in the polar regions than in middle latitudes, despite the fact that insolation is larger there.

(gg) An increase in the concentration of atmospheric aerosol could either raise or lower the planetary albedo.

(hh) Despite its effect in increasing planetary albedo, an increase in cloudiness, in the global average, would not necessarily lead to a decrease in the surface temperature of the earth.

7.3 (a) Prove that in a stably stratified atmosphere static energy increases with height.

(b) Show that (2.69) is equivalent to the statement that moist static energy is conserved in a pseudoadiabatic process.

7.4 Express the energy balance depicted in Fig. 7.1 in terms of globally averaged energy fluxes (in watts per square meter).

7.5 Rainfall, averaged over the earth's surface, is about 1 m y^{-1}, or 0.275 cm day^{-1}. On the basis of this information, calculate the following:

(a) The average upward flux of water vapor at the earth's surface.

Answer 3.17×10^{-5} kg m^{-2} s^{-1}

(b) The resulting energy flux.

Answer 79.4 W m^{-2}

(c) How does this flux compare with the globally averaged incoming solar radiation?

Answer 23% of it

(d) If this heat were uniformly distributed within the layer from 900 to 300 mb, what would be the resulting heating rate?

Answer 1.14 deg day^{-1}

7.6 A large nuclear power plant discharges waste heat into a lake at a rate of 10 GW (10^{10} W). The lake has a surface area of 10^3 km^2 (equivalent to a square about 32 km on a side) and a mean depth of 10 m. Assume that the heat is distributed uniformly throughout the lake.

(a) At what rate would the lake warm up if it had no means of disposing of the heat?

Answer 7.55 deg yr^{-1}

(b) Assume that the mean temperature of the lake is 10°C in the absence of the power plant. By how much would the lake have to warm in order to dispose of the waste heat entirely by radiation? (Assume that the lake radiates as a blackbody.)

Answer 1.93 deg

(c) How much would the rate of evaporation from the lake have to increase in order to dispose of all the waste heat?

Answer 12.6 cm yr^{-1}

7.7 (a) Verify that for typical ranges of values of atmospheric variables in the troposphere, the kinetic energy is several orders of magnitude smaller than the static energy. (It will be convenient to consider the energy per unit mass.)

(b) Show that for a parcel of air moving at the speed of sound c_s [$c_s = (\gamma RT)^{1/2}$, where $\gamma = c_p/c_v$, R is the gas constant for 1 kg and T is absolute temperature], the kinetic energy is equal to exactly one fifth of the enthalpy.

7.8 The effective temperature of Jupiter, as measured recently by the Pioneer spacecraft, is 125°K. It has been suggested that the difference between this value and the value given in Table 6.2 is due to a flux of energy from the interior of the planet. How large a flux would be required to account for this discrepancy?

Answer 7 W m^{-2}

7.9 Compute the daily insolation on the top of the atmosphere at the North Pole at the time of the summer solstice. The distance between earth and sun at this time of year is approximately 1.52×10^8 km, and the tilt of the earth's axis is 23.5°. Compare your result with the value in Fig. 7.16.

Answer 46.4 MJ m^{-2} day^{-1}

7.10 Compute the daily insolation on the top of the atmosphere at the equator at the time of the equinox (a) by integrating the flux over a 24-h period and (b) by purely geometrical considerations. Compare your result with the value in Fig. 7.16, and with the value computed in Problem 7.9.

Answer 38.0 MJ m^{-2} day^{-1}

7.11 As cold, continental air passes over the Gulf Stream on a certain winter day, the temperature of the air in the mixed layer rises by 10 deg over a distance of 300 km. Within this interval the mean depth of the mixed layer is 1 km and the average wind speed is 15 m s^{-1}. No condensation is taking place within the lowest kilometer and radiative fluxes are negligible. Calculate the sensible heat flux from the sea surface.

Answer $\simeq$600 W m^{-2}

7.12 Consider the simplified model of the short-wave energy balance shown in Fig. P7.12. The model atmosphere consists of an upper layer with transmissivity τ_1, a partial cloud layer with the fractional area f_c covered by clouds with reflectivity r_c in either direction, and a lower layer with transmissivity τ_2. The earth's surface has an average reflectivity r_s. For the sake of simplicity, assume that (1) no absorption takes place within the cloud layer, and (2) no scattering takes place within layers 1 and 2.

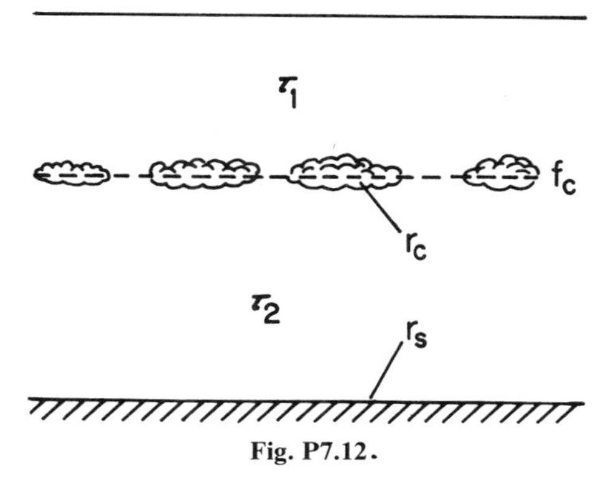

Fig. P7.12.

(a) Show that the total short-wave radiation reaching the surface of the planet divided by the solar radiation incident upon the top of the atmosphere is given by

$$F_s = \frac{[(1 - f_c) + f_c(1 - r_c)]\tau_1\tau_2}{1 - \tau_2^2 f_c r_c r_s}$$

(b) Show that the planetary albedo is given by

$$A = f_c r_c \tau_1^2 + F_s r_s[(1 - f_c) + f_c(1 - r_c)]\tau_1\tau_2$$

(c) For the following values of the model parameters, calculate the planetary albedo: $f_c = 0.5$, $r_c = 0.5$, $\tau_1 = 0.95$, $\tau_2 = 0.90$, $r_s = 0.125$.

Answer 0.278

(d) Use the model to estimate the albedo of a cloud free earth.

Answer 0.0914

(e) Use the model to estimate the albedo of a completely cloud-covered earth.

Answer 0.478

7.13 Suppose that in the model described in Problem 7.12, the surface reflectivity is of the form

$$r_s = f_w r_w + f_l r_l + f_i r_i$$

where f_w, f_l and f_i are the fractional areas of the earth covered by water, land, and ice, respectively, and the r's denote the corresponding reflectivities, here taken to be 0.05, 0.15, and 0.70, respectively. At present, the values of f_w, f_l and f_i are roughly 0.70, 0.22, and 0.08. (It may be verified that when these values are substituted into the above expression $r_s \simeq 0.125$, as assumed in Problem 7.12c.)

(a) Calculate r_s corresponding to ice age conditions: $f_w = 0.70$, $f_l = 0.12$, $f_i = 0.18$, roughly.

Answer 0.187

(b) Use the model in Problem 7.12 to calculate the planetary albedo during ice age conditions. Assume $f_c = 0.5$.

Answer 0.306

(c) Calculate the difference in the effective temperature of the earth between present conditions as calculated in Problem 7.12c, and ice age conditions. [Hint: first show that $\delta T_E/T_E = \frac{1}{4}/[\delta(1 - A)/(1 - A)]$.

Answer 6.8 deg

7.14 (a) How much energy would it take to raise the average temperature of a unit ($1\ m^2$) column of the earth's atmosphere by 1 deg?

Answer 1.03×10^7 J

(b) If all the sources of energy were instantly "turned off" and the atmosphere were allowed to radiate to space as a blackbody with an effective temperature of 250°K, compute the rate at which its average temperature would begin to drop.

Answer 1.86 deg day^{-1}

(c) In terms of thermal capacity, the atmosphere is equivalent to a layer of water how many meters deep?

Answer 2.46 m

Chapter

8

Atmospheric Dynamics

Until relatively recently, day-to-day weather forecasting was largely based upon subjective interpretation of synoptic charts. From the time of the earliest synoptic networks, it has been possible to make forecasts by extrapolating the past movement of the major features on the charts. The development of empirical models such as the polar front cyclone model described in Chapter 3 made it possible to predict the time evolution of weather systems and to anticipate the development of new disturbances. As forecasters began to acquire experience from a backlog of past weather situations they were able to refine their forecasts still further by making use of historical analogs of the current weather situation. However, it soon became apparent that the effectiveness of this so-called "analog technique" is inherently limited by the unavailability of very close analogs in the historical records. On a global basis the three-dimensional structure of the atmospheric motion field is so complicated that a virtually limitless number of distinctly different flow patterns are possible, and hence the probability of two very similar patterns being observed, say, during the same century, is extremely low. By the 1950s the more advanced weather forecasting services were already beginning to approach the limit of the skill that could be obtained strictly by the use of empirical techniques.

It is only since the development of high-speed digital computers that it has become possible to make extensive use of forecasting techniques based upon the numerical solutions of the basic equations that govern the time evolution of the large-scale atmospheric motion field—the so-called primitive equations.[†] In the United States, John von Neumann,[‡] one of the pioneers in the development of high-speed digital computers, was among the first to recognize numerical weather prediction as a fertile area for the application of computer technology. Shortly after the end of World War II he invited a group of meteorologists to come to the Institute for Advanced Studies at Princeton to develop a dynamical model of the synoptic-scale motion field that could be used as a basis for day-to-day weather forecasting. Early in the 1950s this group produced a highly simplified numerical model that showed some skill at making one-day forecasts. Parallel but independent efforts by Kibel[§] and collaborators in the Soviet Union produced successful numerical forecasts at about the same time.

In comparison to the prototype numerical weather prediction models of the early 1950s, today's models contain a much more accurate and detailed representation of the physical processes that enter into the primitive equations. The increase in the physical complexity of the models has been made possible by rapid advances in computer technology. As an example of the changes that have taken place during the past two decades, we note that if one were to make a single forecast for one day in advance by running the current operational NMC[¶] model on a computer of the type available during the early 1950s, several months of computer time would be required!

The current numerical prediction models are far superior to any of the empirical techniques in terms of their performance in forecasting the time evolution of upper-level flow patterns. However, in making weather forecasts most weather services find it advantageous to employ a combination of dynamical and empirical techniques. For example,

[†] Here the term *primitive* is used not in the sense of outmoded or archaic but rather in the sense of fundamental or basic.

[‡] **John von Neumann** (1903–1957) Born in Budapest, Hungary. Professor in the Institute for Advanced Study, Princeton University, 1933–1954. Known for his contributions to point-set theory, theory of continuous groups, quantum mechanics, mathematical logic, application of game theory to problems in the social sciences, and design of computers.

[§] **I'lya A. Kibel** (1904–1970) Regarded by many as the founder of the Soviet school of modern dynamical meteorology. Made important contributions to the theory of gas dynamics, non homogeneous turbulence in a compressible fluid, the atmospheric boundary layer, cloud dynamics, global climate, flow over irregular terrain, and mesoscale wind systems. In 1940 he developed a simplified set of equations (based upon the primitive equations) which could be used to produce short-range (1–2-day) forecasts without the aid of computers. Later, when high-speed digital computers became available in the Soviet Union, he and his colleagues applied them to the forecasting problem, first using the simplified equations and later using the full primitive equations.

[¶] The United States National Meteorological Center (NMC) in Suitland, Maryland, a facility of the National Oceanographic and Atmospheric Administration (NOAA) is charged with the responsibility for providing numerical guidance for forecasts issued by National Weather Service offices throughout the country. Similar facilities exist in a number of other countries.

- surface maps generated by the dynamical models are modified, on the basis of past experience, to correct for systematic errors in the model that are known to occur in certain synoptic situations;
- statistical relationships between model output and surface weather conditions are routinely used to forecast such quantities as maximum and minimum temperature, precipitation probability, and so on;
- output from two or more dynamical models are subjectively compared in making certain forecasting decisions.

In view of the importance of the primitive equations in weather forecasting, it is appropriate that we devote this chapter to deriving and interpreting them.

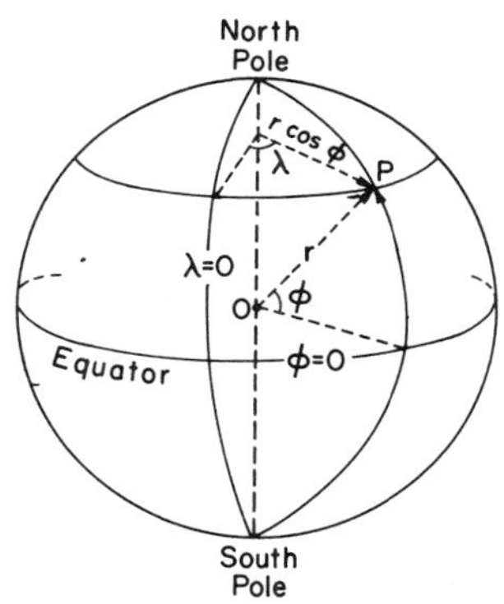

Fig. 8.1 Coordinates used to define atmospheric motions: r is the distance from the center of the earth O to a point P at latitude ϕ and longitude λ.

8.1 COORDINATE SYSTEMS

Large-scale atmospheric motions can be conveniently and accurately represented in terms of a modified spherical coordinate system, rotating with the earth, as illustrated in Fig. 8.1. The horizontal coordinates are latitude† ϕ and longitude λ. The angles are often replaced by the geometric coordinates

$$dx \equiv r\, d\lambda \cos\phi \qquad \text{and} \qquad dy \equiv r\, d\phi$$

where x and y are distance east of the Greenwich meridian along a latitude circle, and distance north of the equator, respectively, and r is the distance from the center of the earth. In dealing with motions below 50 km, r is nearly always replaced by R_E, the mean radius of the earth (6.37×10^6 m). The error introduced by this approximation is less than 1% and the resulting simplification of the equations is well worth it.

† Oceanographers and applied mathematicians often use the colatitude $\theta = \pi/2 - \phi$ instead of ϕ.

Equations of motion for the atmosphere assume a particularly simple form if geopotential height

$$Z \equiv \frac{\Phi - \Phi_0}{g_0}$$

is used as a vertical coordinate. In the above expression Φ_0 refers to the geopotential surface that coincides with mean sea level (see Section 2.2.1) and g_0 is the average gravitational acceleration at sea level (9.8 m s^{-2}). With this choice of vertical coordinate, "horizontal planes" coincide with surfaces of constant geopotential, and therefore the force of gravity does not have any horizontal component. It is clear from Table 2.1 that below 50 km there is very little difference between geopotential height and geometric height above sea level.

In this x, y, Z coordinate system, the three velocity components are defined as

$$u \equiv \frac{dx}{dt} = R_E \cos\phi \frac{d\lambda}{dt} \quad \text{(the \textit{zonal velocity component})}$$

$$v \equiv \frac{dy}{dt} = R_E \frac{d\phi}{dt} \quad \text{(the \textit{meridional velocity component})}$$

and

$$w \equiv \frac{dZ}{dt} \simeq \frac{dr}{dt} \quad \text{(the \textit{vertical velocity component})}$$

The *horizontal velocity vector* $\mathbf{V}$ is given by $\mathbf{V} \equiv u\mathbf{i} + v\mathbf{j}$, where $\mathbf{i}$ and $\mathbf{j}$ are the unit vectors in the zonal and meridional directions, respectively. Positive and negative zonal velocities are called *westerly* (from the west) and *easterly* (from the east) winds,[†] respectively; positive and negative meridional velocities are called *southerly* and *northerly* winds, respectively.

8.1.1 Pressure as a vertical coordinate

Some of the equations to be discussed in subsequent sections take on simpler forms if pressure p is used as the vertical coordinate in place of geopotential height. The transformation from height (x, y, Z) to pressure (x, y, p) coordinates is particularly convenient and straightforward because

- large-scale atmospheric motions are hydrostatic so that there is a monotonic and very simple relationship between pressure and height (see Section 2.2), and
- surfaces of constant pressure are so flat that the horizontal distributions of wind and temperature are virtually the same on a constant pressure surface as on a nearby constant height surface.

[†] The student who wishes to avoid confusion is advised not to consult the dictionary for meanings of *westerly*, *easterly*, and so forth.

In the (x, y, p) coordinate system, the horizontal and vertical wind components are not orthogonal. Strictly speaking, the horizontal wind vector $\mathbf{V}$ refers to motions in planes of constant Z, as defined above, and not to motions in the planes of constant pressure surfaces. The vertical velocity is expressed in terms of

$$\omega \equiv \frac{dp}{dt} \quad \begin{pmatrix}\text{the time rate of change of pressure} \\ \text{following an individual air parcel}\end{pmatrix}$$

ω is often expressed in units of millibars per day. To convert to SI units one can use the relation

$$1 \text{ mb day}^{-1} = 1.16 \times 10^{-3} \text{ Pa s}^{-1}$$

It should be noted that positive values of ω correspond to sinking motion and vice versa. It can be shown that to within about 5%

$$\omega \simeq -\rho g_0 w = -w\frac{p}{H}$$

where ρ is the air density and H is the scale height as defined in (2.27).† A rough conversion can also be made by referring to the standard atmosphere geometric height scale that accompanies the pseudoadiabatic chart (back end papers). For example, in the lower troposphere

$$100 \text{ mb day}^{-1} \simeq 1 \text{ km day}^{-1} \simeq 1 \text{ cm s}^{-1}$$

8.1.2 Other vertical coordinates

Both constant height surfaces and constant pressure surfaces sometimes intersect the ground, particularly in mountainous regions. In order to eliminate this problem, meteorologists have devised the (x, y, σ) coordinate system, where $\sigma \equiv p/p_s$ (p_s being the surface pressure) is a nondimensional, scaled, pressure coordinate which ranges from a value of unity at the earth's surface (regardless of the terrain height) to zero at the "top" of the atmosphere. The transformation from pressure to σ coordinates simplifies the geometrical problems associated with the treatment of the lower boundary of the atmosphere, but it introduces additional terms into the equations of motion that do not arise in height or pressure coordinates.

For certain applications, the potential temperature θ is very well suited for use as a vertical coordinate; (x, y, θ) are referred to as *isentropic coordinates*, since θ surfaces are surfaces of constant entropy. As we have seen in Chapter 3, narrow, sloping frontal zones contain much of the important vertical structure of large-scale atmospheric disturbances. When such zones are mapped into isentropic coordinates the strong horizontal contrasts are spread out somewhat

† See Problem 8.25 and J. R. Holton, "An Introduction to Dynamic Meteorology," Academic Press, New York, 1972, p. 59.

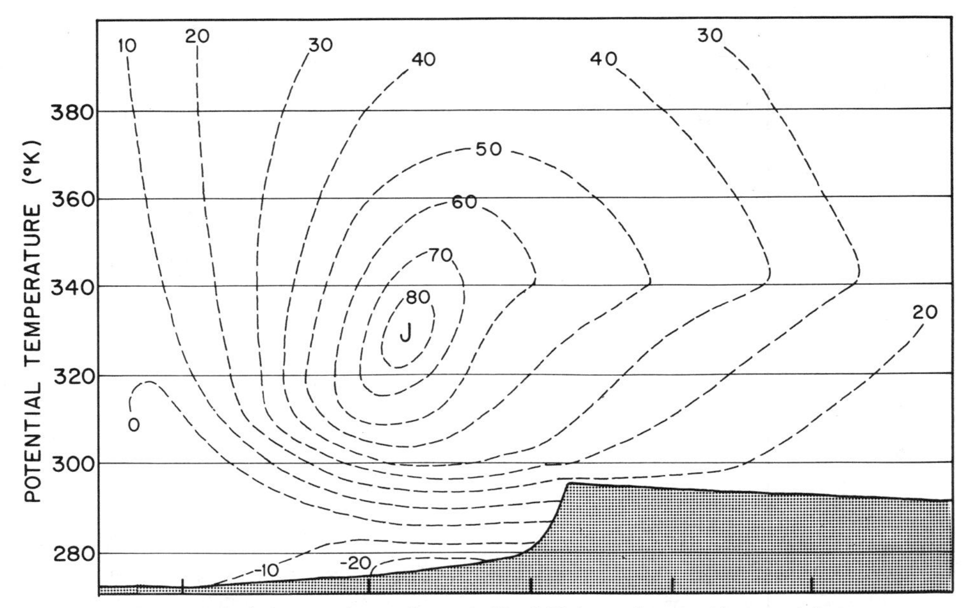

Fig. 8.2 Same vertical cross section as shown in Fig. 3.20, but redrawn with potential temperature as a vertical coordinate. The shaded area is below the ground.

and the configuration of isotachs takes on a simpler form. An illustration is provided in Fig. 8.2. Note how the zone of strong vertical wind shear loses its slope as a result of the transformation.

Isentropic coordinates have the additional advantage that the vertical motion component $d\theta/dt$ can be related directly to diabatic processes taking place within the atmosphere. In the absence of such processes, the flow is two dimensional within layers bounded by isentropic surfaces.

Unfortunately, isentropic coordinates have certain inherent disadvantages. Regions with adiabatic lapse rates, such as the mixed layer, cannot be represented properly because θ does not increase with height. Treatment of the lower boundary is awkward because the earth's surface exhibits a large apparent slope when mapped in isentropic coordinates, especially in the vicinity of frontal zones, as seen in Fig. 8.2. In the remainder of this chapter we will confine our attention to the (x, y, p) and (x, y, Z) coordinate systems.

8.1.3 Natural coordinates

On any "horizontal" surface (that is, a surface of constant Z, p, θ, and so on, depending upon the coordinate system) it is possible to define a set of *streamlines*: arbitrarily spaced lines whose orientation is such that they are everywhere parallel to the horizontal velocity vector **V** at a particular level and at a particular instant in time. At any point on the surface one can define *natural coordinates*:

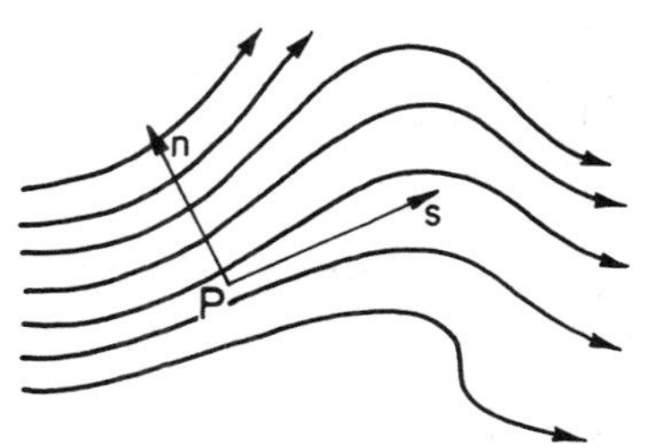

Fig. 8.3 Natural coordinates (s, n) defined at point P in a horizontal wind field. The curved arrows represent streamlines.

s the arc length directed downstream along the local streamline and n the distance directed normal to the streamline and toward the left, as shown in Fig. 8.3. It follows that, at any point in the flow,

$$V \equiv |\mathbf{V}| = \frac{ds}{dt} \quad \text{and} \quad \frac{dn}{dt} = 0$$

8.2 APPARENT FORCES IN A ROTATING COORDINATE SYSTEM

In applying Newton's† second law of motion to a coordinate system rotating with the earth it is convenient to introduce "apparent forces" that compensate for the acceleration of the frame of reference. As an illustration of how these forces arise let us consider the case of an air parcel moving from west to east relative to the earth, which is rotating with angular velocity‡

$$\Omega \equiv 2\pi \text{ rad day}^{-1} = 7.292 \times 10^{-5} \text{ s}^{-1}$$

Note that by definition the zonal velocity u is positive if the relative motion is in the same sense as the earth's rotation and negative if it is the opposite sense.

From the viewpoint of an omniscient observer external to the earth, the parcel is moving along a circular trajectory with a velocity ($\Omega R_A + u$) where R_A is the distance from the earth's axis of rotation and the term ΩR_A is the tangential velocity of the coordinate system at the location of the parcel. Because the parcel

† **Sir Isaac Newton** (1642–1727) Renowned English mathematician, physicist, and astronomer. A posthumous, premature ("I could have been fitted into a quart mug at birth") and only child. Discovered the laws of motion, the universal law of gravitation, calculus, the colored spectrum of white light, and constructed the first reflecting telescope. He said of himself: "I do not know what I may appear to the world, but to myself I seem to have been only like a boy playing on the sea-shore, and diverting myself in now and then finding a smoother pebble or a prettier shell than ordinary, while the great ocean of truth lay all undiscovered before me." Turned to government administration in his later years when he was appointed Master of the Mint!

‡ "Day" here refers to the sidereal day, which is 23 h 56 min.

is traveling in a circle of radius R_A with velocity $(\Omega R_A + u)$, it has an acceleration toward the center of the circle given by

$$(\Omega R_A + u)^2/R_A$$

as shown in Fig. 8.4a. Now, according to Newton's second law, the vectorial sum of the real forces (per unit mass) ΣF acting upon the parcel must be identical in magnitude and in direction to the above acceleration, as indicated in Fig. 8.4b.

From the viewpoint of an observer on the rotating earth the parcel appears to violate Newton's second law. The apparent acceleration toward the earth's axis of rotation is only u^2/R_A, as indicated in Fig. 8.4c, and yet the vectorial sum of the real forces (per unit mass) is exactly the same as that shown in Fig. 8.4b, namely

$$\sum F = \frac{(\Omega R_A + u)^2}{R_A} = \Omega^2 R_A + 2\Omega u + \frac{u^2}{R_A}$$

The apparent violation of Newton's second law can be eliminated by introducing the *apparent forces* per unit mass $\Omega^2 R_A$ and $2\Omega u$ directed outward from the axis of rotation, as indicated in Fig. 8.4d.

Problem 8.1 A 200-lb man steps onto a train moving westbound at 20 m s^{-1} along the equator. Calculate the resulting change in his apparent weight.

Solution The man's weight, before stepping onto the train, is given by

$$W = mg = m(g^* - \Omega^2 R_E)$$

where m is his mass, g the force per unit mass of effective gravity, g^* the force per unit mass of the true gravitational attraction, Ω the angular velocity of the earth and R_E the radius

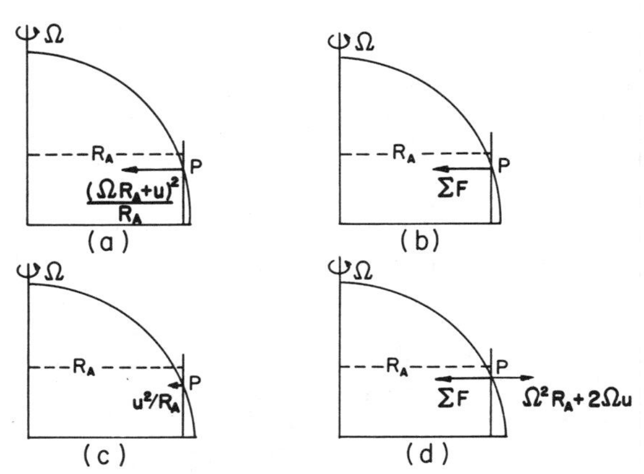

Fig. 8.4 (a) Acceleration of a parcel of air at point P with zonal velocity u as noted by an observer external to the earth. (b) Corresponding real forces (per unit mass) noted by an external observer. (c) Acceleration at P noted by an observer rotating with the earth. The latter observer has to introduce the apparent force $\Omega^2 R_A + 2\Omega u$ shown in (d) in order to satisfy Newton's second law. Boldface notion in this and following figures indicates vectors.

of the earth. After he steps onto the train, his apparent weight becomes

$$W' = m\left[g^* - \frac{(\Omega R_E - c)^2}{R_E}\right]$$

where c is the speed of the train. A minus sign appears in front of c because the train is moving in the direction opposite to the earth's rotation. The apparent change in the man's weight is given by

$$\delta W = W' - W = m(2\Omega c - c^2/R_E)$$

Substituting for m from the first equation above,

$$\delta W = \frac{W}{g}(2\Omega c - c^2/R_E)$$

Substituting in values

$$\delta W = \frac{200 \text{ lb}}{9.8 \text{ m s}^{-2}}\left[2 \times 7.29 \times 10^{-5} \text{ s}^{-1} \times 20 \text{ m s}^{-1} - \frac{(20 \text{ m s}^{-1})^2}{6.37 \times 10^6 \text{ m}}\right]$$

$$= \frac{200 \text{ lb}}{9.8}(2.92 - 0.06) \times 10^{-3}$$

$$= 0.058 \text{ lb increase}$$

8.2.1 Effective gravity

The force per unit mass, called *gravity* or *effective gravity* and denoted by the symbol g, is actually the vectorial sum of the true gravitational attraction g^* that draws all elements of mass toward the center of mass of the earth and the much smaller apparent force $\Omega^2 R_A$ that pulls all objects outward from the axis of planetary rotation, as indicated in Fig. 8.5. Surfaces of constant geopotential Φ, which are normal to g, assume the shape of oblate ellipsoids, as indicated by the dashed line in Fig. 8.5. Under the strong pull of effective gravity the equilibrium sea surface and the large-scale configuration of the earth's surface have assumed the shape of a surface of constant geopotential.

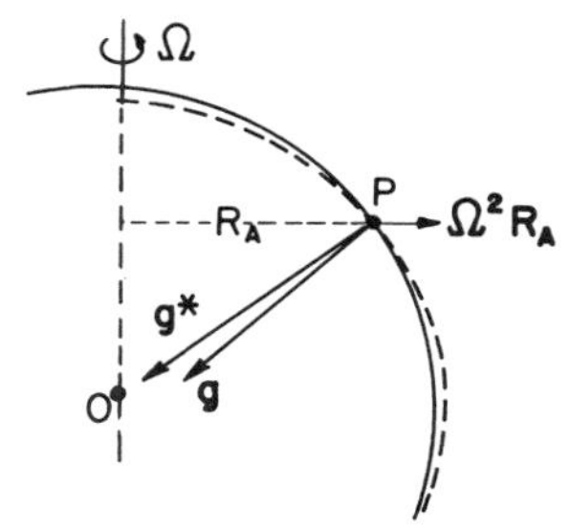

Fig. 8.5 Effective gravity g is the vectorial sum of the true gravitational acceleration g^* directed toward the center of the earth O and the apparent force $\Omega^2 R_A$. The acceleration g is normal to the dashed line, which represents a surface of constant geopotential. The solid line represents a true spherical surface.

Since a body rotating with the earth has no way of separately sensing the gravitational and centrifugal components of effective gravity, there is nothing to be gained by expressing the two as separate forces in the equations of motion. Therefore, the $\Omega^2 R_A$ term does not appear explicitly in the equations of motion; it is included implicitly as part of g.

8.2.2 The Coriolis force

The second of the apparent forces in Fig. 8.4d ($2\Omega u$ per unit mass) is fundamentally different from the $\Omega^2 R_A$ term in that it depends crucially upon the zonal velocity u. This so-called *Coriolis*[†] *force* is directed radially outward from the axis of rotation if the zonal motion is in the same sense as the planetary rotation ($u > 0$, westerly flow), and radially inward toward the axis of rotation if the zonal motion is in the opposite sense to the planetary rotation ($u < 0$, easterly flow).

A Coriolis force also arises in the case of radial motion toward or away from the axis of rotation. The existence of such a force can be deduced from the conservation of angular momentum. As a parcel with unit mass moves toward or away from the axis of rotation in the absence of real forces, angular momentum is conserved; that is,

$$\frac{d}{dt} R_A(\Omega R_A + u) = 0$$

where the two terms in parentheses represent the contributions of the planetary rotation and the zonal flow to the absolute tangential velocity of the parcel. Carrying out the differentiation, we obtain

$$2\Omega R_A \dot{R}_A + u\dot{R}_A + R_A \frac{du}{dt} = 0$$

where the overdots refer to time derivatives. If the zonal velocity u is zero, the above expression reduces to

$$\frac{du}{dt} = -2\Omega \dot{R}_A$$

Hence, if the parcel is moving radially outward from the earth's axis of rotation (for example, in association with horizontal motion toward the equator, as shown in Fig. 8.6), it experiences an easterly acceleration which induces a westward drift relative to fixed points on earth. Similarly, if a parcel is moving toward the axis of rotation, it experiences a westerly acceleration.[‡] An observer

[†] **G. G. de Coriolis** (1792–1843) French engineer, mathematician, and physicist. Gave the first modern definitions of kinetic energy and work. Studied motions in rotating systems.

[‡] A quantitative example is given in Problem 9.2 which is worked in the text on p. 420.

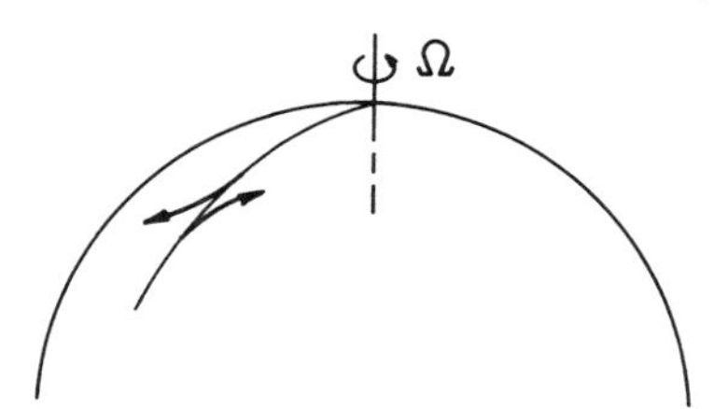

Fig. 8.6 Horizontal trajectories of air parcels that develop a zonal motion component as a consequence of the conservation of angular momentum.

on earth ascribes these accelerations and the resulting curvature of air trajectories to the action of an apparent force (the Coriolis force). Just as in the case of zonal motion discussed previously, the magnitude of the force (per unit mass) is proportional to the relative velocity of the parcel, and the constant of proportionality is the factor 2Ω.

The above results can be generalized in terms of the following statements:

(a) A parcel with velocity $\mathbf{c} \equiv \mathbf{u} + \dot{\mathbf{R}}$ in the plane perpendicular to the axis of rotation experiences a Coriolis force (per unit mass) of magnitude $2\Omega|\mathbf{c}|$.

(b) The Coriolis force is directed in the plane perpendicular to the axis of rotation, at right angles to $\mathbf{c}$.

(c) If the planetary rotation is clockwise (or counterclockwise) as viewed from space, the Coriolis force is directed toward the left (or right) of the velocity vector. Thus, when viewed from above, the Coriolis force is directed toward the right of the velocity vector in the Northern Hemisphere and toward the left in the Southern Hemisphere.

We have determined the magnitude and direction of the Coriolis force but we have yet to relate these results to the spherical coordinate system used for describing atmospheric motions. In Fig. 8.7 it is shown that the Coriolis force $2\Omega u$ associated with an eastward velocity u can be resolved into a horizontal force $2\Omega u \sin \phi$, directed southward in the Northern Hemisphere, and an upward force $2\Omega u \cos \phi$. A meridional velocity v can be resolved into a component

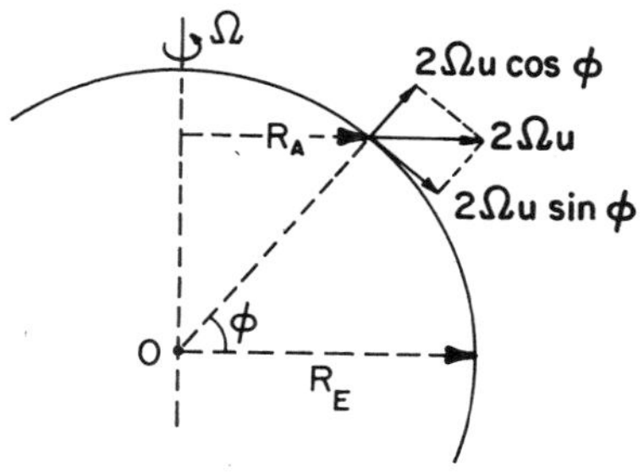

Fig. 8.7 Resolution of the Coriolis force $2\Omega u$ associated with zonal velocity u (directed into the page) into horizontal and vertical components.

$v \cos \phi$ parallel to the axis of rotation and $v \sin \phi$ in the plane perpendicular to the axis of rotation. The latter induces a force $2\Omega v \sin \phi$ in the zonal direction.

The above results can be generalized in terms of a few brief statements which describe the horizontal Coriolis forces induced by horizontal motions in a spherical coordinate system.

(d) A parcel with horizontal velocity V experiences a Coriolis force whose horizontal component has a magnitude $2\Omega V \sin \phi$.

(e) The horizontal component of the Coriolis force is directed perpendicular to the horizontal velocity vector; toward the right in the Northern Hemisphere and toward the left in the Southern Hemisphere.

Problem 8.2 Describe the horizontal motion of a hypothetical body which is constrained to move on a surface of constant geopotential in the absence of real horizontal forces. Assume that this so-called *inertia motion* takes place within a small region where $\sin \phi$ can be considered constant and the curvature of the earth can be neglected.

Solution Since the Coriolis force is always directed perpendicular to the velocity vector, it does not change the speed of motion; hence, the initial speed V_0 of the body will remain unchanged. Furthermore, since the Coriolis force $2\Omega V_0 \sin \phi$ is also independent of time, the curvature of the trajectory of the body must be constant; that is to say, the body must be moving in a circular arc. Equating the acceleration toward the center of curvature of the arc to the Coriolis force per unit mass, we obtain

$$V_0^2/R_T = 2\Omega \sin \phi \; V_0$$

where R_T is the radius of curvature of the trajectory. The time period required for the body to describe a complete *inertia circle* is given by

$$T = \frac{2\pi R_T}{V_0} = \frac{\pi}{\Omega \sin \phi}$$

and, since $\Omega = 2\pi \text{ day}^{-1}$, it follows that

$$T = 1 \text{ day}/2 \sin \phi$$

Hence, in the vicinity of the poles the so-called *inertial period* is 12 h; near 30° lat it is about one day; and so on. In the Northern Hemisphere inertia motion is clockwise while in the Southern Hemisphere it is counterclockwise. Inertia motion is sometimes observed quite clearly in the oceans.

The Coriolis force appears so frequently in the horizontal equations of motion that the factor $2\Omega \sin \phi$ is often designated by the symbol f (l in the Soviet Union) and called the *Coriolis parameter*. It is a slowly varying function of latitude with values ranging from zero at the equator to $\pm 1.46 \times 10^{-4} \text{ s}^{-1}$ at the poles. Throughout most of the middle latitudes $f \simeq 10^{-4} \text{ s}^{-1}$. Note that by definition f is positive in the Northern Hemisphere and negative in the Southern Hemisphere. (It could have just as easily been defined in the opposite sense.)

In vector notation the horizontal component of the Coriolis force can be expressed in the form

$$\mathbf{C} = -f\mathbf{k} \times \mathbf{V} \tag{8.1}$$

where $\mathbf{k}$ denotes the unit vector in the vertical direction. The operator $-(\mathbf{k}\times)$ indicates that the force lies in the horizontal plane, perpendicular to the velocity vector $\mathbf{V}$, directed toward the right of $\mathbf{V}$ in the Northern Hemisphere (where $f > 0$) and toward the left in the Southern Hemisphere (where $f < 0$). The magnitude of the force is given by the product of f and the wind speed.

Vertical motions also give rise to Coriolis forces that have a horizontal component. However, the vertical velocities in large-scale atmospheric motions are so small that these effects are usually ignored. Conversely, horizontal motions give rise to vertical Coriolis forces. However, it will be shown in Section 8.5 that these forces are negligible in comparison with the other forces that appear in the vertical equation of motion. Hence, it is only the horizontal component of the Coriolis force induced by horizontal motions in the atmosphere that is of importance in atmospheric dynamics.

8.3 REAL FORCES

In the previous section we indicated the real forces that act on a parcel of air by the symbol ΣF. These forces are gravitation (which has already been discussed), the pressure gradient force, and the frictional drag exerted by neighboring air parcels or by the underlying surface. (Free electrons and ions experience additional electrostatic and magnetic forces which are transmitted to the neutral particles by collisions, but the fraction of the total atmospheric mass that exists in an ionized state is so minute that these effects are negligible at levels below 100 km in the earth's atmosphere.)

8.3.1 The pressure gradient force

Consider a small block of fluid with dimensions δn, δs, and δz as shown in Fig. 8.8. The coordinate system is chosen so that the s axis is oriented parallel to the local isobars, the n axis points in the direction of higher pressure, and the z axis points upward, parallel to the line in which local gravity acts. The pressure

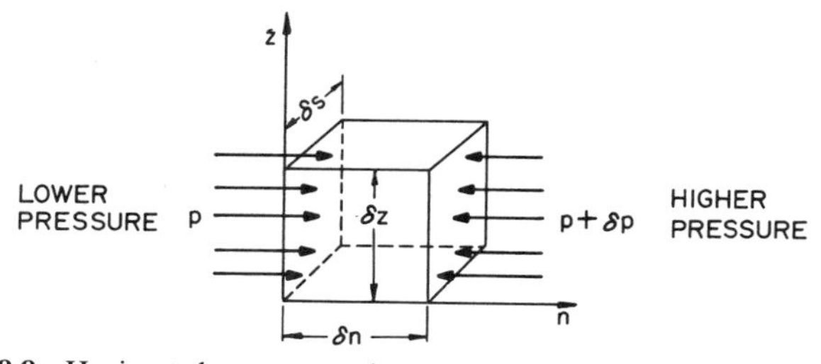

Fig. 8.8 Horizontal pressure acting upon an infinitesimal block of fluid.

force exerted by the surrounding air upon the left-hand n-face of the block is given by $p\ \delta s\ \delta z$, where p is the average pressure on this face of the block. There is an almost identical, but oppositely directed, force on the other side of the block given by $(p + \delta p)\ \delta s\ \delta z$, where the pressure increases by the small increment δp as n increases from the left to the right-hand side of the block. If the dimensions of the block are very small in comparison with the scale of the pressure fluctuations, we can write

$$\delta p \simeq \frac{\partial p}{\partial n}\ \delta n$$

where the partial derivative notation indicates that the derivative is taken at constant s and z. The net n component of the pressure force on the block is simply the vectorial sum of the forces on the two n faces, which is

$$-\frac{\partial p}{\partial n}\ \delta n\ \delta s\ \delta z$$

The negative sign indicates that the force is directed toward lower values of n; that is, from higher pressure toward lower pressure. Dividing by the mass of the block $(\rho\ \delta n\ \delta s\ \delta z)$, where ρ is the density of the air, we obtain the n component of the *pressure gradient force per unit mass*

$$P_n = -\frac{1}{\rho}\frac{\partial p}{\partial n} \tag{8.2}$$

Since the pressure forces on the s face of the block exactly cancel, (8.2) gives the magnitude of the total horizontal pressure gradient force. The force is directed perpendicular to the isobars on a horizontal surface and down the pressure gradient; that is, from higher pressure toward lower pressure.

In a similar manner it can be demonstrated that the vertical component of the pressure gradient force is given by

$$P_z = -\frac{1}{\rho}\frac{\partial p}{\partial z} \tag{8.3}$$

In (x, y, p) coordinates the horizontal pressure gradient force can be expressed in terms of the gradient of geopotential or geopotential height on pressure surfaces. Consider two points Q and R, separated by the vertical distance δz and the horizontal distance δn in the plane normal to the isobars, as indicated in Fig. 8.9. We will assume that the separation between the two points is small enough that the spatial gradients of p are uniform (which is equivalent to assuming that the pressure surfaces are parallel planes with equal spacing for equal increments of δp). The difference in pressure between the two points is given by

$$\delta p = \frac{\partial p}{\partial n}\ \delta n + \frac{\partial p}{\partial z}\ \delta z$$

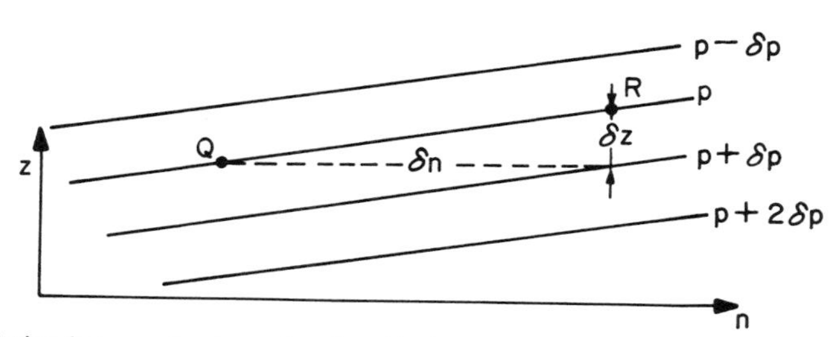

Fig. 8.9 Relation between horizontal and vertical pressure gradients and the slope of the surfaces of constant pressure.

Now if the two points lie on the same pressure surface $\delta p = 0$, and in the limit as the separation of the points approaches zero $\delta n \to dn$ and $\delta z \to dz$. Therefore, from the above expression,

$$\left(\frac{dz}{dn}\right)_p = -\frac{\partial p/\partial n}{\partial p/\partial z} \tag{8.4}$$

Thus, at any point, the ratio of the horizontal pressure gradient force to the vertical pressure gradient force is equal to the local slope of the pressure surface passing through that point. The slopes of pressure surfaces in large-scale atmospheric disturbances rarely exceed 1 part in 10^3. Therefore, it is evident that in these systems the horizontal pressure gradient force is typically at least three orders of magnitude smaller than the vertical pressure gradient force.

Introducing the hydrostatic equation (2.18) into the last expression, we obtain

$$\frac{1}{\rho}\frac{\partial p}{\partial n} = g\left(\frac{dz}{dn}\right)_p$$

In (x, y, p) coordinates $(dz/dn)_p$ and $\partial z/\partial n$ are identical. Incorporating this identity into the above expression and making use of (8.2), we have

$$P_n = -\frac{1}{\rho}\frac{\partial p}{\partial n} = -g\frac{\partial z}{\partial n}$$

Furthermore, noting from (2.20) and (2.22) that

$$g\,dz = d\Phi = g_0\,dZ$$

the above expression can be expanded in the form

$$P_n = -\frac{1}{\rho}\frac{\partial p}{\partial n} = -g\frac{\partial z}{\partial n} = -\frac{\partial \Phi}{\partial n} = -g_0\frac{\partial Z}{\partial n} \tag{8.5a}$$

Problem 8.3 At a certain point the horizontal gradient of sea level pressure is 5 mb per 100 km. Estimate the slope of the 1000 mb surface at this point. (Assume that the air density $\rho = 1.25$ kg m^{-3}.)

Solution

$$P_n = -\frac{1}{\rho}\frac{\partial p}{\partial n} = \frac{1}{1.25 \text{ kg m}^{-3}} \times \frac{500 \text{ Pa}}{10^5 \text{ m}}$$

$$= 4 \times 10^{-3} \text{ m s}^{-2}$$

Making use of (8.5a), we have

$$P_n = g_0 \frac{\partial Z}{\partial n} = 4 \times 10^{-3} \text{ m s}^{-2}$$

from which

$$\frac{\partial Z}{\partial n} = 4.08 \times 10^{-4} \quad \text{or} \quad 40.8 \text{ m per } 100 \text{ km}$$

In vector notation, the horizontal pressure gradient force assumes any of the forms

$$\mathbf{P}_n = -\frac{1}{\rho}\nabla p = -g\,\nabla z = -\nabla\Phi = -g_0\,\nabla Z \tag{8.5b}$$

where the so-called *gradient operator* $\nabla(\ )$ denotes a vector oriented normal to the isobars, height contours, or geopotential contours, pointing toward higher values and having a magnitude given by $\partial(\ )/\partial n$, as defined in the context of the previous discussions.† The minus signs indicate that the horizontal pressure gradient force points in the direction opposite to $\nabla(\ )$; that is, toward lower values of p, z, Φ, or Z, respectively. It is implicitly understood that ∇p is taken at constant z while ∇z, $\nabla\Phi$, and ∇Z are taken at constant p.

8.3.2 Friction

As pointed out in Section 2.5.1, the concept of an air parcel composed of a discrete set of molecules is artificial in some respects. In reality, there is a continual exchange of molecules between air parcels and their environment as they move through the atmosphere. Some of this exchange is due to the random motions of the molecules. However, below the turbopause the mean free path between collisions is so small that the exchanges due to random molecular motions are unimportant in comparison to the exchanges resulting from fluid motions on a scale smaller than the dimension of the parcel considered.

The frictional force in the equations of motion represents the collective effects of all scales of motion smaller than the one under consideration in exchanging momentum between air parcels and their environment. The vertical exchanges of momentum are the most important; they always act to smooth out the vertical profile of **V**. Other conditions being the same, the larger the vertical wind shear $\partial\mathbf{V}/\partial z$ the larger the vertical exchange of momentum between adjacent layers. The amount of mixing depends not only upon the wind shear but also upon the

† The gradient is usually defined as a three-dimensional operator, but when it appears in the horizontal equations of motion it is always understood in this narrower, two-dimensional context.

intensity of the smaller-scale motions that are responsible for the mixing. For example, for a given vertical wind shear the vertical exchanges of momentum tend to be much stronger within an adiabatic layer with vigorous convection than within a stably stratified layer. For the same reason, frictional effects in flow over rough terrain tend to be stronger and extend deeper into the atmosphere than those in flow over smooth terrain.

Throughout most of the atmosphere frictional forces are sufficiently small that, to a first approximation, they can be neglected. A notable exception is the so called *planetary boundary layer* (corresponding roughly to the lowest 1 km of the atmosphere) where the flow over a stationary underlying surface gives rise to a frictional drag force which is comparable in magnitude to the other terms in the horizontal equations of motion. For the purposes of this chapter it will be sufficient to represent this frictional drag in the highly simplified form

$$\mathbf{F} = -a\mathbf{V} \tag{8.6}$$

where a is a positive coefficient, the magnitude of which varies with wind speed, roughness of the underlying surface, static stability, and so on.

8.4 THE HORIZONTAL EQUATION OF MOTION

Having discussed the real and apparent forces that enter into Newton's second law when applied to atmospheric motions, we can now obtain the horizontal equation of motion simply by equating the horizontal acceleration to the sum of the real and apparent horizontal forces acting upon a parcel of unit mass. The horizontal equation of motion is therefore

$$\frac{d\mathbf{V}}{dt} = \mathbf{P}_n + \mathbf{C} + \mathbf{F}$$

or, substituting from (8.1), (8.5b), and (8.6) we obtain, in pressure coordinates,

$$\frac{d\mathbf{V}}{dt} = -\nabla\Phi - f\mathbf{k} \times \mathbf{V} - a\mathbf{V} \tag{8.7}$$

8.4.1 The geostrophic wind

For large-scale horizontal motions, typical horizontal velocities are on the order of 10–30 m s^{-1} and the time scale over which individual air parcels experience significant changes in velocity is on the order of a day or so ($\sim 10^5$ s).† Thus a typical parcel acceleration is on the order of, say 20 m s^{-1} in 10^5 s or 2×10^{-4} m s^{-2}. In middle latitudes, where $f \sim 10^{-4}$ s^{-1} an air parcel moving with a horizontal velocity of 20 m s^{-1} experiences a Coriolis force per unit mass on the order of 2×10^{-3} m s^{-2}, or about an order of magnitude larger

† "On the order of," designated by the symbol ($\sim$), may be taken to mean within a factor of two in either direction.

than the typical parcel accelerations. Note that the predominance of the Coriolis term is not dependent upon the particular choice of wind speed that we assume. It depends only upon the fact that the characteristic time scale over which air parcels experience large changes in their relative motion is long in comparison to $(1/f)$, which can be viewed as the time scale over which parcels experience large changes in their absolute motion as a result of the earth's rotation.

Since frictional drag is usually small except within the planetary boundary layer, the only term that is capable of balancing the Coriolis force is the pressure gradient force. Thus, to within about 15% in middle and high latitudes, (8.7) reduces to

$$f\mathbf{k} \times \mathbf{V} \simeq -\nabla\Phi$$

Making use of the vector identity

$$\mathbf{k} \times (\mathbf{k} \times \mathbf{V}) = -\mathbf{V}$$

this expression can be transformed into

$$\mathbf{V} \simeq \frac{1}{f}\mathbf{k} \times \nabla\Phi$$

For any given horizontal distribution of pressure (or geopotential height on pressure surfaces) it is possible to define a *geostrophic*† *wind field* for which the above relationship is exactly satisfied. In vector form, the geostrophic wind is given by

$$\mathbf{V}_g \equiv \frac{1}{f}(\mathbf{k} \times \nabla\Phi) = \frac{g_0}{f}(\mathbf{k} \times \nabla Z) = \frac{1}{\rho f}(\mathbf{k} \times \nabla p) \tag{8.8a}$$

where the last two identities follow from (8.5b). In scalar form, the geostrophic wind speed is given by

$$V_g \equiv \frac{1}{f}\frac{\partial\Phi}{\partial n} = \frac{g_0}{f}\frac{\partial Z}{\partial n} = \frac{1}{\rho f}\frac{\partial p}{\partial n} \tag{8.8b}$$

where n is the direction normal to the isobars (or geopotential height contours), pointing toward higher values. The balance of horizontal forces that results in the geostrophic wind field is illustrated in Fig. 8.10, which applies to the Northern Hemisphere. In order for the Coriolis and pressure gradient forces to balance, the geostrophic wind must blow parallel to the isobars leaving low pressure to the left, in the Northern Hemisphere. It may be noted that, in either hemisphere, the geostrophic wind around a center of low pressure is always in the same sense as the earth's rotation. Such flow is said to be *cyclonic*. Geostrophic flow around high pressure centers is said to be *anticyclonic*.

The tighter the spacing of the isobars or geopotential height contours, the stronger the Coriolis force that is required to balance the pressure gradient force, and hence, the stronger the geostrophic wind. As a device for remembering

† From the Greek: *geo* (earth) and *strephen* (to turn).

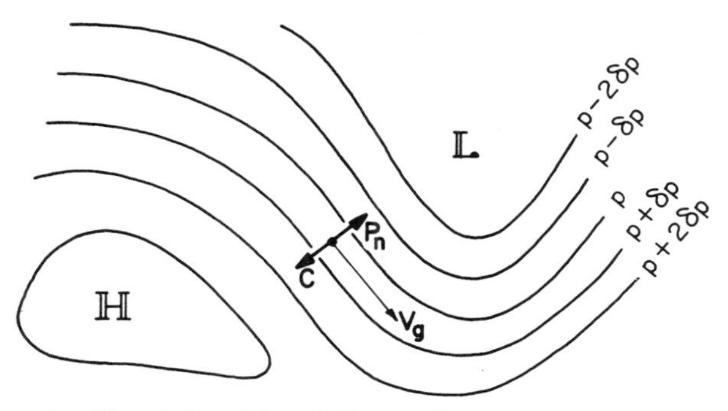

Fig. 8.10 The geostrophic wind and its relation to the horizontal pressure gradient force P_n and the Coriolis force C.

this relationship, it may be helpful to think of each of the spaces between isobars or height contours as a channel, through which a fixed amount of fluid must pass, irrespective of its width. The narrower the channel the higher the velocity required to pass the fluid. (For a given horizontal pressure gradient the geostrophic wind is stronger at low latitudes than at high latitudes and therefore the channel analogy is not strictly correct when large ranges of latitude are involved.)

Throughout middle and high latitudes the large-scale wind field tends to be *quasi-geostrophic*; that is to say, the wind direction closely parallels that of the isobars and the wind speed is within about 15% of the geostrophic value. Larger departures from geostrophic flow are observed near the earth's surface, where the frictional force is large enough to play a significant role in the balance of forces, and in flow with strongly curved trajectories. We will consider these exceptions in the next two subsections.

8.4.2 The effect of friction

The frictional retardation of the flow within the lowest kilometer is continually tending to make the wind speed *subgeostrophic*, so that the Coriolis force is not quite strong enough to balance the horizontal pressure gradient force, which is acting to push the air across the isobars from high toward low pressure (or from high toward low geopotential height). The effect is analogous to that experienced by a train coasting along a sloping surface. In the absence of friction, the train could maintain a constant speed on a track that follows along a constant height contour. However, in the real world, where friction is present, it is possible to coast at a constant speed only along a track that drifts steadily downhill at some angle relative to the contours. The stronger the friction, the larger the angle between the "constant speed" track and the height contours.

The three-way balance of forces required for a uniform flow ($d\mathbf{V}/dt = 0$) in the Northern Hemisphere in the presence of friction is illustrated in Fig. 8.11. In accordance with (8.7), **P** is directed normal to the isobars, **C** is directed

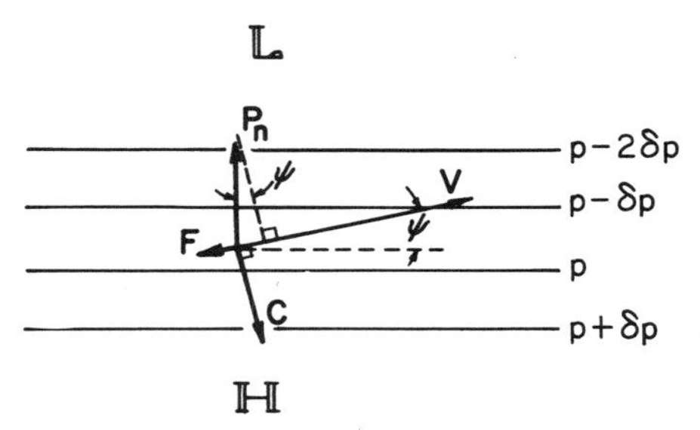

Fig. 8.11 The three-way balance of forces required for uniform flow in the presence of friction in the Northern Hemisphere. The solid lines represent isobars in a horizontal plane.

to the right of the velocity vector **V**, and **F** is directed opposite to the velocity vector. The angle between **V** and $\mathbf{V}_g$ (denoted here by ψ) is determined by the requirement that the component of **P** in the direction of the velocity vector be equal and opposite to **F** and the wind speed is determined by the requirement that **C** be just large enough to balance the component of **P** in the direction normal to the velocity vector. The larger the linear damping coefficient a the larger the angle ψ and the more subgeostrophic the wind speed. The observed tendency for the wind to blow parallel to the isobars above the lowest kilometer is evidence of the smallness of the frictional force at these levels.

Problem 8.4 Within the lowest kilometer in middle latitudes, the angle between the wind vector and the isobars is typically on the order of 15°. Estimate the time required for the linear damping term $-a\mathbf{V}$ to reduce the horizontal velocity by a factor of e in the absence of other forces.

Solution In the absence of other forces, (8.7) reduces to

$$\frac{dV}{dt} = -aV$$

which has a solution of the form

$$V = V_0 e^{-at}$$

where t is time and V_0 is the velocity at $t = 0$. The velocity will be reduced by a factor of e at $t = 1/a$.

From Fig. 8.11 it is evident that

$$|\mathbf{F}| = |\mathbf{P}| \sin \psi \qquad \text{and} \qquad |\mathbf{P}| \cos \psi = |\mathbf{C}| = f|\mathbf{V}|$$

Combining these relationships, we obtain

$$|\mathbf{F}| = f|\mathbf{V}| \tan \psi$$

It follows from (8.6) that $a = f \tan \psi$. Thus, in middle latitudes where $f \sim 10^{-4}\ \mathrm{s}^{-1}$, the typical damping time in the lowest kilometer is given by

$$t = \frac{1}{a} = \frac{1}{f \tan \psi} \simeq 3.7 \times 10^4 \quad \mathrm{s}$$

or about half a day.

8.4.3 The gradient wind†

Within sharp troughs in the middle-latitude westerlies the observed velocities are often subgeostrophic by as much as 50%, even though the streamlines still tend to be oriented parallel to the isobars. These large departures from geostrophic balance are a consequence of the large centripetal acceleration associated with the sharply curved flow within such regions. In order to illustrate how these accelerations alter the balance between the Coriolis force and the pressure gradient force it is convenient to represent them in terms of an apparent centrifugal force. The three-way balance between pressure gradient and Coriolis and centrifugal forces is shown in Fig. 8.12 for cyclonic and anticyclonically

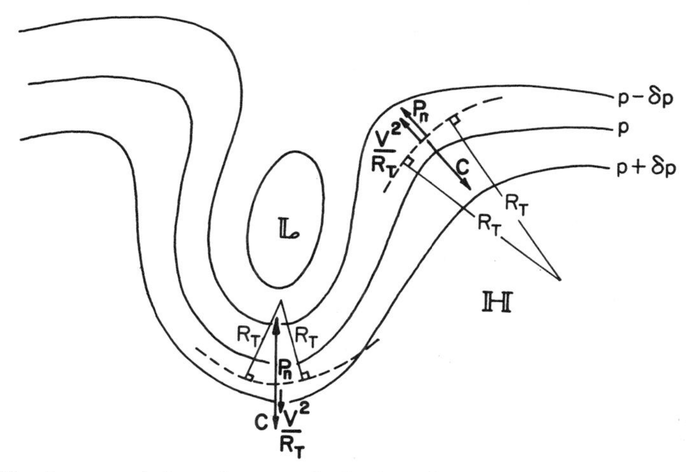

Fig. 8.12 The three-way balance between the horizontal pressure gradient force, the Coriolis force and the centrifugal force, in flow along curved trajectories (---) in the Northern Hemisphere.

curved trajectories. In both cases, the centrifugal force is directed outward from the center of curvature of the air trajectories (denoted by the dashed lines) and has a magnitude given by V^2/R_T, where R_T is the local radius of curvature. In the cyclonic case, the centrifugal force reinforces the Coriolis force so that, in effect, a balance of forces can be achieved with a wind velocity smaller than would be required if the Coriolis force were acting alone. Thus, in this case, it is possible to maintain a subgeostrophic flow parallel to the isobars. For the anticyclonically curved trajectory the situation is just the opposite: the centrifugal force opposes the Coriolis force and, in effect, necessitates a *supergeostrophic* wind velocity in order to bring about a three-way balance of forces.

† In an introductory survey course at the undergraduate level this subsection can be omitted without serious loss of continuity.

For the combination of sharp anticyclonic curvature and a strong pressure gradient it is not possible to achieve a three-way balance of forces; that is to say, for all possible values of V,

$$P_n > fV - \frac{V^2}{R_T}$$

It is no coincidence, then, that the combination of sharp troughs and tight pressure gradients is not uncommon, whereas sharp ridges are rarely if ever observed in conjunction with tight pressure gradients.

The wind associated with a three-way balance between the pressure gradient and Coriolis and centrifugal forces is called the *gradient wind.* To illustrate further the relationship between geostrophic and gradient wind balance, let us consider the following example.

Problem 8.5 In a region 50 km out from the center of an intense hurricane, a radial pressure gradient of 50 mb per 100 km is observed. The storm is located at 20°N. Calculate the geostrophic and gradient wind speeds.

Solution The geostrophic wind speed V_g is given by (8.8b); therefore,

$$V_g = \frac{1}{\rho f}\frac{\partial p}{\partial n} = \frac{50 \times 10^2 \text{ kg m}^{-1}\text{ s}^{-2}}{1.25 \text{ kg m}^{-3} \times 2(7.29 \times 10^{-5})\text{ s}^{-1} \sin 20^\circ \times 10^5 \text{ m}}$$

$$= 802 \text{ m s}^{-1}!!$$

If the air trajectories can be assumed to be circular about the storm center, the gradient wind can be determined from the three-way balance of forces in cyclonic flow, which is given by

$$\frac{V^2}{R_T} + fV = P_n$$

Solving for V by the quadratic formula, we obtain

$$V = \tfrac{1}{2}\left|-R_T f \pm (R_T{}^2 f^2 + 4P_n R_T)^{1/2}\right|$$
$$V = \tfrac{1}{2}\left|-2.49 \pm (6.21 + 8000)^{1/2}\right|$$
$$V = \left|-1.25 \pm 44.75\right| = 43.5 \quad \text{or} \quad 46.0 \quad \text{m s}^{-1}$$

By substituting the solutions into the original equation, it is easily verified that the larger of the two roots corresponds to the "anomalous" situation of anticyclonic flow around the hurricane, which also happens to satisfy the above equation. The smaller root corresponds to the correct solution (see Fig. 8.13).

In the above problem the centrifugal force was so much stronger than the Coriolis force that a reasonable accurate approximate solution could have been obtained by assuming *cyclostrophic balance*, namely that

$$\frac{V^2}{R_T} = P_n$$

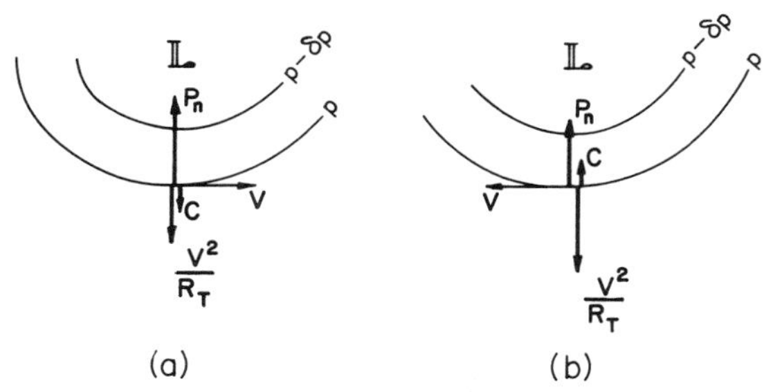

Fig. 8.13 The three-way balance of forces corresponding to the two solutions of the gradient wind equation in the Northern Hemisphere: (a) normal solution and (b) anomalous solution.

The flow near the center of hurricanes is usually close to cyclostrophic balance, whereas in typical middle-latitude disturbances it is usually closer to geostrophic balance. Both geostrophic and cyclostrophic balance may be viewed as special cases of gradient wind balance that obtain in the limit when one of the forces in the three-way force balance becomes negligible in comparison to the other forces.

In applying the gradient wind equation it is extremely important to recognize the distinction between horizontal *streamlines*, as defined in Section 8.1.3, and the horizontal projection of the air *trajectories* (the actual three-dimensional paths that air parcels follow as they move through space). Only under very specialized conditions are the two ever the same. Figures 8.14 and 8.15 show some examples that illustrate the importance of this distinction.

Figure 8.14 shows the flow in an idealized eastward-propagating wave disturbance. The solid lines represent horizontal streamlines at a particular instant in time, and the dashed lines represent the horizontal streamlines at the same level, a short time later, after the wave has moved some distance eastward. We will assume that there are no vertical motions. Let us consider the trajectory of an air parcel located at point A in the trough of the wave at the initial time.

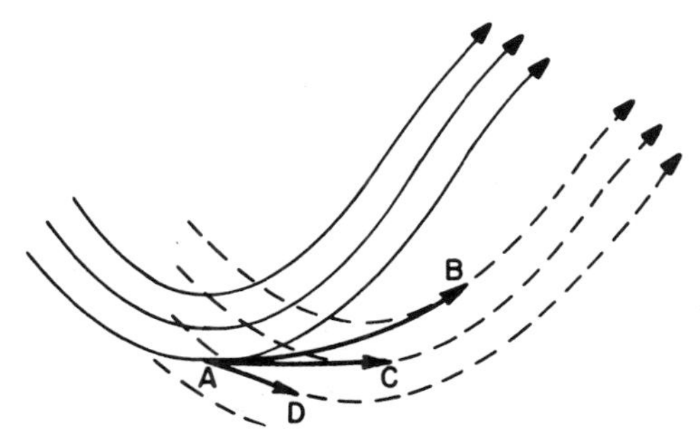

Fig. 8.14 Streamlines and trajectories in an eastward-moving wave. Heavy arrows denote horizontal projections of air trajectories, solid thin arrows denote earlier streamlines and dashed arrows denote later streamlines.

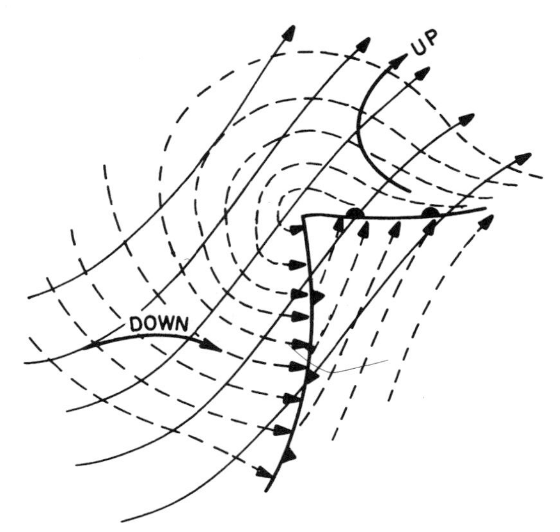

Fig. 8.15 Streamlines and trajectories in an idealized steady-state, middle-latitude disturbance. Heavy arrows represent horizontal projections of air trajectories. Thin solid arrows represent upper-level streamlines, and dashed arrows represent low-level streamlines.

Now if it happens that the speed of the wind at A is exactly equal to the rate of eastward propagation of the wave, the parcel will remain coincident with the wave trough as it moves eastward as indicated by the trajectory AC. If the wind speed at A is faster than the rate of propagation of the wave, the air parcel will gradually move ahead of the trough into the region of southwesterly flow, and will drift northward as indicated by the trajectory AB. On the other hand, if the wind speed at A is slower than the rate of propagation of the wave, the parcel will fall behind the trough and will move southward as indicated by the trajectory AD. Note that the three trajectories are parallel to the initial streamline passing through point A and to the later streamline passing through points B, C, or D, respectively. Trajectory AB is the longest, since it corresponds to the highest wind speed.

Figure 8.15 shows the three-dimensional air trajectories in an idealized steady-state middle-latitude disturbance. The solid lines indicate the horizontal streamlines at the 500-mb level and the dashed lines indicate the horizontal streamlines at sea level. Both sets of streamlines are assumed to be unchanging with time (an unrealistic assumption in this particular case). The heavy arrows represent the horizontal projections of the three-dimensional trajectories of two hypothetical air parcels. The parcel behind the surface cold front sinks from the 500-mb level at the beginning of the trajectory to sea level at the end of the trajectory, while the parcel ahead of the surface warm front does just the opposite. The sinking trajectory starts out parallel to the 500-mb streamlines and ends up parallel to the sea level streamlines, and vice versa.

In both Figs. 8.14 and 8.15, the curvature of the horizontal projections of the air trajectories is markedly different from the curvature of the horizontal streamlines. Trajectory AD in Fig. 8.14 and both trajectories in Fig. 8.15 curve much more anticyclonically than the related streamlines; a situation commonly observed in the real atmosphere. In applying the gradient wind equation it is necessary to estimate the curvature of the air trajectories in order to estimate the centrifugal force term. If streamlines are mistakenly used in place of trajectories the resulting centrifugal force term is almost certain to be incorrect, and it may even be of the wrong sign!

8.4.4 Smaller-scale motions

It should be emphasized that the concepts of geostrophic and gradient wind have meaning only in the context of large-scale atmospheric motions. In mesoscale and convective-scale motions the relative importance of the terms in (8.7) is entirely different.

For example, in the case of convective-scale motions the characteristic time scale over which significant parcel accelerations take place is on the order of minutes (10^2–10^3 s), which is short in comparison to $(1/f)$. It is easily verified that, for any assumed characteristic velocity, the typical horizontal accelerations are between one and two orders of magnitude larger than the Coriolis force. Hence, the earth's rotation plays only a very minor role in the dynamics of convective-scale motions.

8.5 THE VERTICAL EQUATION OF MOTION

For motion in the vertical, Newton's second law assumes the form

$$\frac{dw}{dt} = -\frac{1}{\rho}\frac{\partial p}{\partial z} - g + C_z + F_z \tag{8.9}$$

expressed in (x, y, z) coordinates, where C_z and F_z are the vertical components of the Coriolis and frictional forces, respectively.

The analysis of the magnitude of the various terms in (8.9) is somewhat more complicated and subtle than the one we just made for the horizontal equation of motion, and so we will only quote the results here.

For the large-scale motions, in which virtually all the kinetic energy is contained in the horizontal wind component, the vertical accelerations are so small that it is not practically feasible to determine them as a residual in (8.9). To within about 1 part in 100, there is a balance between the pressure gradient force and gravity so that the hydrostatic equation (2.18) is satisfied, not only for the mean atmosphere but also for the large-scale deviations from the mean. Thus to a very satisfactory degree of approximation, the *prognostic* equation (8.9) can be replaced by the *diagnostic* equation (2.18).†

† Prognostic equations like (8.7) and (8.9) yield time rates of change of the dependent variables. Diagnostic equations express interrelationships between the dependent variables that are valid at any instant in time.

The pressure, temperature, and density fields in large-scale atmospheric motions also obey the ideal gas equation (2.2), and therefore the hypsometric equation is also valid as a diagnostic relation between the temperature field and the distribution of geopotential (or geopotential height) on pressure surfaces. For the purposes of this chapter it is convenient to deal with the hypsometric equation in the form

$$\frac{\partial \Phi}{\partial p} = -\frac{RT}{p} \tag{8.10}$$

which is adapted from (2.23).

Convective-scale motions are not in hydrostatic balance, and therefore the vertical acceleration and the frictional drag terms need to be retained in (8.9).

8.6 THE THERMAL WIND†

By combining the hypsometric equation (2.29) and the geostrophic equation, it is possible to obtain the so-called *thermal wind equation* which relates the vertical shear of the geostrophic wind to the horizontal temperature gradient.

Let us begin by considering the special case of an atmosphere which is characterized by a total absence of horizontal temperature (thickness) gradients. In such a *barotropic*‡ atmosphere, the height contours on all pressure surfaces can be neatly stacked on top of one another in any order, like a set of matched dinner plates. It follows that the direction and speed of the geostrophic wind must be independent of height; that is to say, in a barotropic atmosphere there is no vertical shear of the geostrophic wind.

Now let us proceed to the somewhat more general case of an *equivalent barotropic* atmosphere, in which there exist horizontal temperature gradients but with the constraint that the thickness contours are everywhere parallel to the height contours. In such an atmosphere any given height contour is also a line of constant thickness, and so the shape of the height contours and consequently the direction of the geostrophic wind must be independent of height, just as in the pure barotropic case. However, because of thickness variations in the direction normal to height contours, the slope of the pressure surfaces and hence the speed of the geostrophic wind varies from level to level, as illustrated in Fig. 8.16.

In an equivalent barotropic atmosphere, centers of high and low pressure are also centers of positive or negative temperature anomalies. In warm highs and cold lows, the amplitude of the pressure anomalies and the speed of the

† In a one-semester introductory survey course at the undergraduate level it may be necessary to omit this section. If this is done, Sections 9.3.3 and 9.5.1–2 should also be omitted.

‡ The term barotropic is derived from the Greek *baro*, relating to pressure, and *tropic*, changing in a specific manner: that is, in such a way that surfaces of constant pressure are coincident with surfaces of constant temperature (or density), so that $\nabla T = 0$ on constant pressure surfaces.

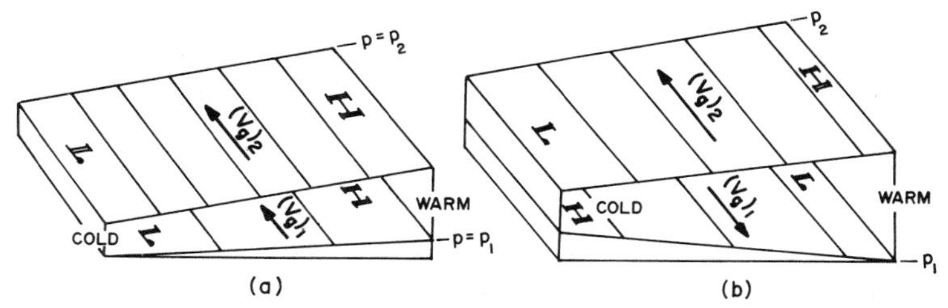

Fig. 8.16 The change of the geostrophic wind with height in an equivalent barotropic flow in the Northern Hemisphere: (a) V_g increasing with height within the layer and (b) V_g reversing direction within the layer.

geostrophic wind increase with height. In cold highs and warm lows the situation is just the opposite; the geostrophic wind speeds decrease with height and the wind direction may even reverse if the temperature anomalies persist through a deep enough layer, as indicated in Fig. 8.16b.

The relationship between temperature and geostrophic wind speed in an equivalent barotropic atmosphere can be expressed quantitatively by writing (8.8b) for any two pressure levels [denoted here by a lower level (1) and an upper level (2)] and subtracting to obtain

$$(V_g)_2 - (V_g)_1 = \frac{g_0}{f}\frac{\partial}{\partial n}(Z_2 - Z_1)$$

provided that the gradient of geopotential height does not reverse within the layer. Substituting for the thickness $(Z_2 - Z_1)$ from the hypsometric equation (2.29) we obtain the wind difference in the form

$$(V_g)_2 - (V_g)_1 = \frac{R}{f}\frac{\partial \bar{T}}{\partial n}\ln\frac{p_1}{p_2} \simeq \frac{\text{const}}{f}\frac{\partial \bar{T}}{\partial n} \tag{8.11}$$

where $\bar{T}$ is the mean temperature of the intervening layer. Thus the difference in wind speed between level (1) and level (2), a measure of the vertical shear of the geostrophic wind, is proportional to the horizontal gradient of the mean temperature of the intervening layer.

To a first approximation many of the disturbances in the earth's atmosphere are equivalent barotropic. For example, the isotherms and height contours in the vicinity of the frontal zone in Fig. 3.19 are oriented in roughly the same direction (southwest to northeast) through a deep layer of the atmosphere. The cross section is oriented normal to the isotherms and height contours, and to the winds. Hence the speed of the wind component normal to the section is not much different from the speed of the wind itself. The isotachs in the figure actually refer to geostrophic wind speeds which were inferred from the slopes of the pressure surfaces by measuring the horizontal distances between adjacent height contours in Figs. 3.13–3.17, and similar maps for intermediate levels.

In agreement with (8.11) there is a strong correlation between the strength of the horizontal temperature gradient and the vertical shear of the geostrophic wind. For example, both quantities have large values in the frontal zone and much smaller values in the warm air which is nearly pure barotropic. In passing upward through the jet stream level, the horizontal temperature gradient and the vertical wind shear simultaneously undergo a reversal in sign. Hence, just above the jet stream level the temperature is usually higher on the poleward side of the jet stream than on the equatorward side.

Problem 8.6 During winter in the middle-latitude troposphere, the zonally averaged temperature gradient is about 1° per degree of latitude (see Fig. 1.11) and the mean (averaged around a latitude circle) zonal component of the geostrophic wind at the ground is close to zero. Calculate the mean zonal wind at the jet stream level, near 250 mb.

Solution Taking the zonal component of (8.8a) and averaging around a latitude circle, we have

$$[u_g] = -\frac{g}{f}\frac{\partial[Z]}{\partial y}$$

where u_g is the zonal component of the geostrophic wind and the brackets [ ] operator denotes an average around latitude circles. Applying the above expression to the 250- and 1000-mb pressure levels and subtracting we obtain

$$[u_g]_{250} - [u_g]_{1000} = -\frac{g}{f}\frac{\partial}{\partial y}\{[Z]_{250} - [Z]_{1000}\}$$

where the subscripts refer to the pressure levels. Noting that $[u_g]_{1000} \simeq 0$ and making use of (2.29), the above expression can be transformed to

$$[u_g]_{250} = -\frac{R}{f}\frac{\partial[T]}{\partial y}\ln\frac{1000}{250} \qquad (\text{where } R \simeq R_d)$$

$$= \frac{-287\ \text{J deg}^{-1}\ \text{kg}^{-1}}{10^{-4}\ \text{s}^{-1}} \times \ln 4 \times \frac{-1.00\ \text{deg}}{1.11 \times 10^5\ \text{m}}$$

$$= 35.8 \quad \text{m s}^{-1}$$

Note that this result is in close agreement with Fig. 1.11.

Finally, let us consider the most general case of a *baroclinic*† atmosphere in which horizontal temperature gradients exist without any restrictions regarding their relationship to the height contours at the same level. In such an atmosphere the shape as well as the spacing of the height contours varies from one pressure level to another so that the geostrophic wind varies with height in direction as well as in speed.

† The term *baroclinic* is derived from the Greek *baro*, relating to pressure, and *klines*, inclining or intersecting. In a baroclinic atmosphere surfaces of constant pressure intersect surfaces of constant temperature (or density) so that, in general, $\nabla T \neq 0$ on constant pressure surfaces. An equivalent barotropic atmosphere may be regarded as a special kind of baroclinic atmosphere in which ∇T and $\nabla \Phi$ on pressure surfaces are linearly related.

The distinguishing characteristic of a baroclinic atmosphere, as just defined, is the existence of regions in which the height and thickness contours intersect one another at some angle so that the geostrophic wind has a component normal to the isotherms (or thickness contours). This geostrophic flow across the isotherms is associated with *geostrophic temperature advection.* The concept of advection is developed in a more quantitative way in 8.7.1. For the present discussion it will be sufficient to define *cold advection* as flow across the isotherms from a colder region into a warmer one, and *warm advection* as just the opposite. Now let us consider how the geostrophic wind changes with height within a layer in which temperature advection is occurring. Typical situations corresponding to cold and warm advection in the Northern Hemisphere are illustrated in Fig. 8.17. On the pressure level at the bottom of the layer the height contours are oriented from west to east, with lower heights toward the north, as indicated by the light solid lines in the figure. The height at point O on the middle contour is Z_1 and the geostrophic wind vector at this level is denoted by V_1. The distribution of vertically averaged temperature within the layer is represented in terms of thickness contours (dashed lines) which are drawn at the same interval ΔZ as the height contours. The thickness along the middle contour is denoted by Z_T, and higher thickness lies toward the east in Fig. 8.17a and toward the west in Fig. 8.17b. The height of the pressure surface at the top of the layer can be found at each intersection simply by adding the thickness of the layer to the height of the bottom pressure surface. For example, the height at point O is $Z_1 + Z_T$, and at point P it is $Z_1 + (Z_T + \Delta Z)$, and so forth. The upper-level height contours (heavy solid lines) are drawn by connecting intersections with equal values of this sum. The upper-level geostrophic wind vector $\mathbf{V}_2$ blows parallel to these upper-level contours.

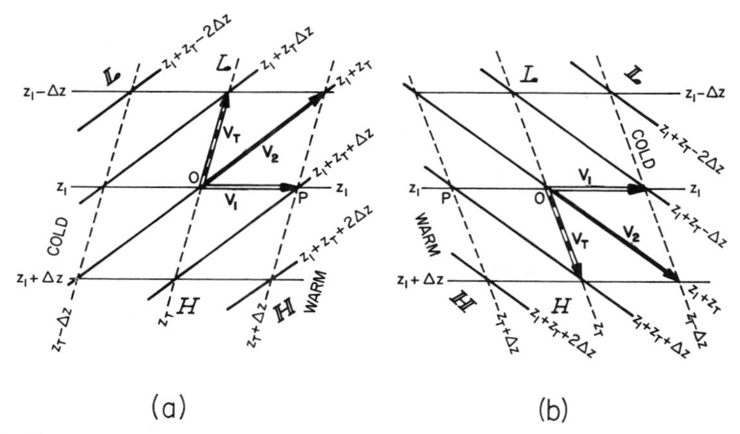

Fig. 8.17 The relation between isotherms, geopotential height contours, and geostrophic wind in layers with (a) cold and (b) warm advection. The light solid lines denote the geopotential height contours at the bottom of the layer and heavier solid lines denote the geopotential height contours at the top of the layer. The dashed lines represent the isotherms or thickness contours within the layer.

From the figure it is apparent that cold advection is characterized by *backing* (counterclockwise rotation) of the geostrophic wind vector with height and warm advection is characterized by *veering* (clockwise rotation). By experimenting with other configurations of height and thickness contours, it is easy to convince oneself that this relationship holds, regardless of the direction of the geostrophic wind at the bottom of the layer, or the orientation of the isotherms. The same results can be extended to the Southern Hemisphere; the only modification required is that veering and backing have to be redefined in more general terms, so that in either hemisphere backing means turning in the same sense as the planetary rotation and veering means turning in the opposite sense.

In order to derive the thermal wind equation, we repeat the derivation of (8.11), only instead of using (8.8b) as a starting point we use the more general vector form (8.8a). It is easily verified that the final result is

$$(\mathbf{V}_g)_2 - (\mathbf{V}_g)_1 = \frac{R}{f} \ln \frac{p_1}{p_2} (\mathbf{k} \times \nabla \bar{T}) \simeq \frac{\text{const}}{f} (\mathbf{k} \times \nabla \bar{T}) \tag{8.12}$$

This form of the thermal wind equation is analogous to that of the geostrophic equation (8.8a) except for the fact that the geostrophic wind vector is replaced by the thermal wind vector for the layer in question, and the geopotential height is replaced by the mean temperature of the layer. Thus the thermal wind bears the same relation to the horizontal distribution of temperature that the geostrophic wind bears to the horizontal distribution of geopotential height. For example, in the Northern Hemisphere the thermal wind "blows" parallel to the isotherms, leaving low temperature to the left.

As an illustration of the thermal wind relation we show in Fig. 8.18 the vertical wind shear in the vicinity of the developing cyclone discussed in Chapter 3. The light solid arrows represent geostrophic wind vectors at the 1000-mb level and the heavy solid arrows represent geostrophic wind vectors at the 500-mb level. The difference between these two vectors ($\mathbf{V}_{500} - \mathbf{V}_{1000}$) is the vertical wind shear in the 1000–500-mb layer, represented by the dashed arrows. By comparing Figs. 3.21 and 8.18, it is readily verified that the vertical wind shear vectors are parallel to the thickness contours for the 1000–500-mb layer. Note that the frontal zones behind cold fronts, which are regions of strong cold advection, are characterized by backing of the geostrophic wind with height, while the frontal zones ahead of warm fronts, which are regions of strong warm advection, are characterized by veering of the geostrophic wind with height. There is little or no turning of the geostrophic wind with height in the warm air masses which tend to be barotropic at low levels and equivalent barotropic in the middle troposphere. On the cold side of surface lows the structure is nearly equivalent barotropic and resembles the example shown in Fig. 8.16b.

Figure 8.19 shows vertical wind soundings for selected stations, taken at the same time as the data for the previous figure. The sounding for Nashville, which is in the cold frontal zone at the ground, exhibits strong backing of the wind

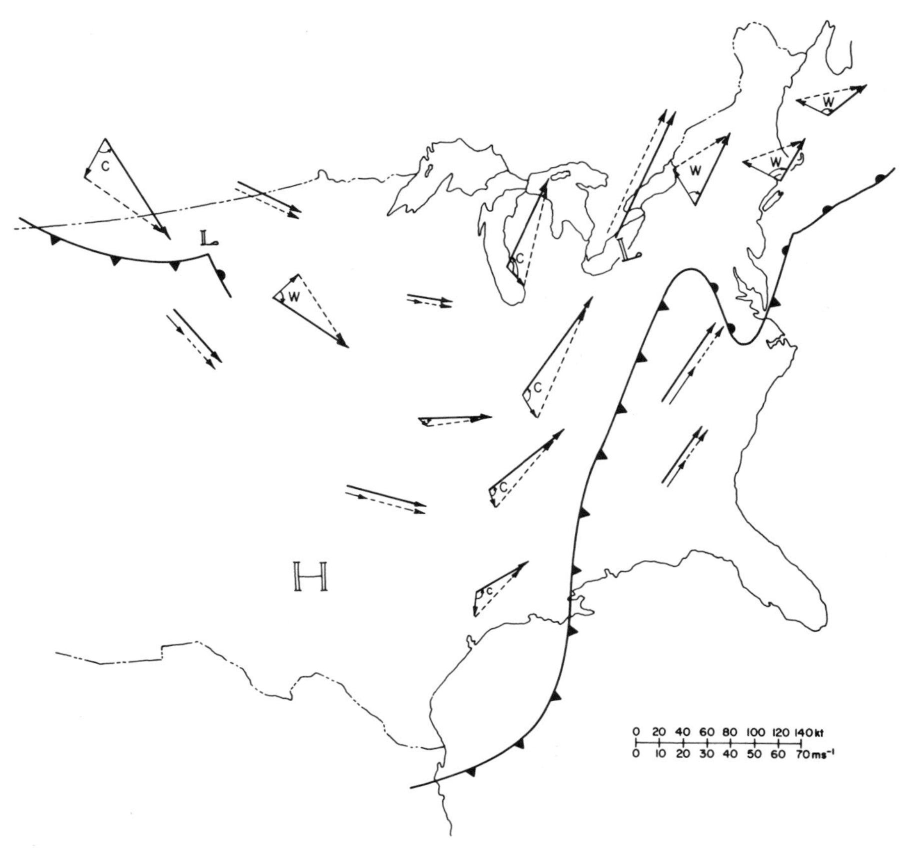

Fig. 8.18 Vertical shear of the geostrophic wind in the 1000–500-mb layer at selected locations at 00 GCT 20 November 1964. Light solid arrows represent the geostrophic wind at the 1000-mb level, heavy solid arrows represent the geostrophic wind at the 500-mb level, and dashed arrows represent the thermal wind vector for the 1000–500-mb layer. The letter C denotes cold advection and W denotes warm advection. These vectors were derived from the three sets of contours in Fig. 3.21.

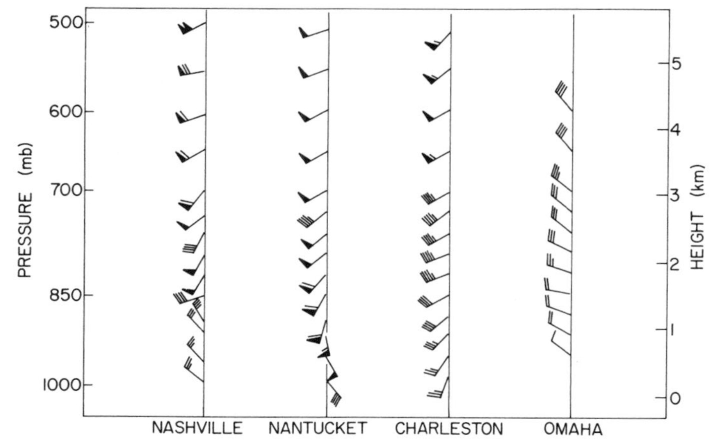

Fig. 8.19 Vertical soundings of wind at 00 GCT 20 November 1964 at Nashville, Tennessee; Nantucket Island, Massachusetts; Charleston, South Carolina; and Omaha, Nebraska.

with height up to the 850-mb level, which corresponds to the position of the cold front. Similarly, the sounding for Nantucket, Massachusetts, which is located in the warm frontal zone at the ground, shows strong veering up to the 850-mb level, which corresponds to the position of the warm front at that station. The soundings for Charleston, located on the warm side of the front, and Omaha, located deep within the cold air, show very little turning of the wind with height, other than some slight frictional veering within the lowest kilometer at Charleston.

Through the use of the thermal wind equation (8.12) it is possible to define completely the geostrophic wind field on the basis of a knowledge of the distribution of $T(x, y, p)$ together with boundary conditions for either $p(x, y)$ or $\mathbf{V}_g(x, y)$ at the earth's surface or at some other "reference level." Thus, for example, a set of sea level pressure observations together with a grid of infrared temperature soundings from satellites constitutes a sufficient observing system for determining the three-dimensional distribution of $\mathbf{V}_g$.

8.7 THE THERMODYNAMIC ENERGY EQUATION

The equations that we have dealt with in the previous sections provide only a partial description of the physical processes that govern the time evolution of the large-scale motion field. Notably absent so far is any prognostic relationship for determining how the large-scale pressure distribution is changing with time.

The time evolution of the pressure field is governed not only by dynamical processes, as embodied in Newton's second law, but also by thermodynamical processes as embodied in the First Law of Thermodynamics. The First Law can be expressed in the form of a prognostic equation, which yields the time rate of change of temperature of an air parcel as it moves through the atmosphere. These changes in temperature are reflected in the thickness pattern, which is closely related to the distribution of Φ on pressure surfaces. Thus any physical processes that change the temperature field ultimately change the distribution of the horizontal pressure gradient force $(-\nabla\Phi)$ in (8.7).

The First Law of Thermodynamics (2.49) can be written in the form

$$\dot{H}\, dt = c_p\, dT - \alpha\, dp$$

where $\dot{H}$ represents the diabatic heating rate expressed in units of energy per unit mass per unit time (joules per kilogram per second), and dt is an infinitesimal time interval. Now we divide through by dt and rearrange the order of the terms as follows:

$$c_p \frac{dT}{dt} = \alpha \frac{dp}{dt} + \dot{H}$$

Finally, we substitute for α, using the ideal gas equation (2.3) and divide through by c_p to obtain the *thermodynamic energy equation*

$$\frac{dT}{dt} = \frac{\kappa T}{p}\omega + \frac{\dot{H}}{c_p} \tag{8.13}$$

where $\kappa = R/c_p$. The first term on the right-hand side represents the rate of adiabatic temperature change due to expansion or compression and the second term represents the diabatic heating rate. Let us compare the relative magnitude of these two terms. The adiabatic temperature change in degrees per day is given by $\kappa T\ \delta p/p_{\mathrm{m}}$ where δp is a typical pressure change over the course of a day, following an air parcel,† and p_{m} is the mean pressure level along the trajectory. In a typical middle-latitude disturbance, air parcels in the middle troposphere ($p_{\mathrm{m}} \simeq 500$ mb) frequently undergo vertical displacements on the order of 200 mb day^{-1}. Assuming $T \simeq 250$°K, the resulting adiabatic temperature change is on the order of 30° day^{-1}.

The second term on the right-hand side of (8.13) represents the effects of diabatic heat sources and sinks: absorption of solar radiation, absorption and emission of infrared radiation, latent heat release, and, in the upper atmosphere, heat absorbed or liberated in chemical and photochemical reactions. In addition, it is customary to include, as a diabatic process, the heat added to or removed from the parcel through the exchange of mass between the parcel and its environment due to scales of motion smaller than the dimension of the parcel. For example, from the standpoint of large-scale atmospheric motion, convective-scale circulation systems act as a diabatic heat source within the mixed layer.

Throughout most of the lower atmosphere there tends to be a considerable amount of compensation between the various radiative heat sources and sinks so that the net radiative heating rates are only on the order of 1° day^{-1}. Latent heat release is zero except in regions of precipitation where it sometimes produces effects comparable to the adiabatic temperature changes discussed above. Under certain conditions the convective heat flux within the mixed layer is also important. However, throughout most of the troposphere, the sum of the diabatic heating terms is small in comparison to the adiabatic temperature change term in (8.13).

8.7.1 The local time rate of change of temperature

All the time-dependent equations that have been dealt with thus far in this chapter yield information on the time rate of change (d/dt) of various quantities following individual air parcels as they move along their three-dimensional trajectories through space. This form of the equations is convenient for dealing

† $\delta p = \omega\,\delta t = dp/dt\,\delta t$, where $\delta t = 1$ day.

with problems in which it is necessary to follow particular air parcels that have some special property (for example, emissions from a smokestack). However, for most problems in atmospheric dynamics it is not necessary to keep track of individual air parcels; it is sufficient to know the distribution of certain dependent variables ($\mathbf{V}$, p, T, and so on) as continuous functions of the three spatial coordinates and time. For such problems it is convenient to transform the equations so that the time derivative ($\partial/\partial t$) refers to changes taking place at fixed points in space.

The desired transformation can be accomplished by expanding $d\psi$ in the form

$$d\psi = \frac{\partial \psi}{\partial x}\, dx + \frac{\partial \psi}{\partial y}\, dy + \frac{\partial \psi}{\partial p}\, dp + \frac{\partial \psi}{\partial t}\, dt$$

or, alternatively, in natural coordinates,

$$d\psi = \frac{\partial \psi}{\partial s}\, ds + \frac{\partial \psi}{\partial p}\, dp + \frac{\partial \psi}{\partial t}\, dt$$

where ψ is any dependent variable. Dividing through by dt, making use of the definitions of u, v, ω, and V, and rearranging the terms, we obtain

$$\frac{\partial \psi}{\partial t} = -u \frac{\partial \psi}{\partial x} - v \frac{\partial \psi}{\partial y} - \omega \frac{\partial \psi}{\partial p} + \frac{d\psi}{dt} \tag{8.14a}$$

or, in natural coordinates,

$$\frac{\partial \psi}{\partial t} = -V \frac{\partial \psi}{\partial s} - \omega \frac{\partial \psi}{\partial p} + \frac{d\psi}{dt} \tag{8.14b}$$

or, in vector notation,

$$\frac{\partial \psi}{\partial t} = -\mathbf{V} \cdot \nabla \psi - \omega \frac{\partial \psi}{\partial p} + \frac{d\psi}{dt} \tag{8.14c}$$

It should be noted that

$$-V \frac{\partial \psi}{\partial s} = -V|\nabla \psi| \cos \beta = -\mathbf{V} \cdot \nabla \psi$$

where β is the angle between $\mathbf{V}$ and $\nabla \psi$.

In all three versions of (8.14) the time derivative term on the left-hand side refers to the *local time rate of change* of ψ, apparent to an observer located at a fixed point in space, whereas the time derivative on the right refers to the time rate of change of ψ apparent to an observer instantaneously at the same point in space but moving with the three-dimensional flow. The two time derivatives may differ, because the air at any fixed point in space is continually being replaced by air from farther and farther upstream, which may have different values of ψ. For example, let us suppose that ψ increases upstream along the horizontal streamline. Then, since by definition s increases downstream, $\partial\psi/\partial s < 0$, and the term $(-V\,\partial\psi/\partial s)$ contributes to increasing ψ at the point in question. The magnitude of this effect depends upon the horizontal velocity V and the rate at which ψ is increasing in the upstream direction. This term (including the minus sign) is called the *horizontal advection.* The minus sign is included within the definition so that the "advection of ψ" will be positive if it contributes to increasing the value of ψ at a fixed point in space, and vice versa.† The same sign convention is used for the vertical advection term $-\omega\,\partial\psi/\partial p$, which has an analogous physical interpretation.

Problem 8.7 A cold front has just passed a station and the temperature is 10°C and falling at a uniform rate of $3°\ \mathrm{h}^{-1}$. The wind is blowing straight out of the north at $40\ \mathrm{km\ h}^{-1}$, and the vertical velocity is zero. At a station located 100 km to the north, the temperature is -2°C. Estimate the time rate of change of the temperature of air parcels as they move southward behind the front.

Solution We begin by rewriting (8.14) in component form for temperature, with the total derivative on the left-hand side and with $u = w = 0$:

$$\frac{dT}{dt} = \frac{\partial T}{\partial t} + v\frac{\partial T}{\partial y}$$

On the basis of the information given in the problem we have no choice but to assume that the meridional wind velocity and the temperature gradient are uniform in the vicinity of the two stations. Substituting these into the equation we obtain

$$\begin{aligned}\frac{dT}{dt} &= \frac{-3°}{\mathrm{h}} + \frac{-40\ \mathrm{km}}{\mathrm{h}} \times \frac{-(10+2)°}{100\ \mathrm{km}}\\ &= (-3 + 4.8)°\ \mathrm{h}^{-1}\\ &= 1.8°\ \mathrm{h}^{-1} \quad \text{(warming)}\end{aligned}$$

We conclude that the air is warming as it moves southward because the temperature at the station is not falling as fast as it would be if horizontal advection were the only process acting.

Problem 8.8 At a certain station located at 43°N the geostrophic wind at the 1000-mb level is blowing from the southwest (230°) at $15\ \mathrm{m\ s}^{-1}$ while at the 500-mb level it is blowing from the west-northwest (300°) at $30\ \mathrm{m\ s}^{-1}$. Calculate the geostrophic temperature advection.

Solution Let

$$\mathbf{V}_a \equiv \tfrac{1}{2}[(\mathbf{V}_g)_{500} + (\mathbf{V}_g)_{1000}] \quad \text{(the mean wind in the layer)}$$

and

$$\mathbf{V}_T \equiv (\mathbf{V}_g)_{500} - (\mathbf{V}_g)_{1000} \quad \text{(the thermal wind in the layer)}$$

The relationship between the various wind vectors are shown in Fig. 8.20. The thickness contours (or isotherms of mean temperature in the layer) are drawn parallel to $\mathbf{V}_T$ with

† Great difficulties in communication arise when the minus sign is omitted from the definition.

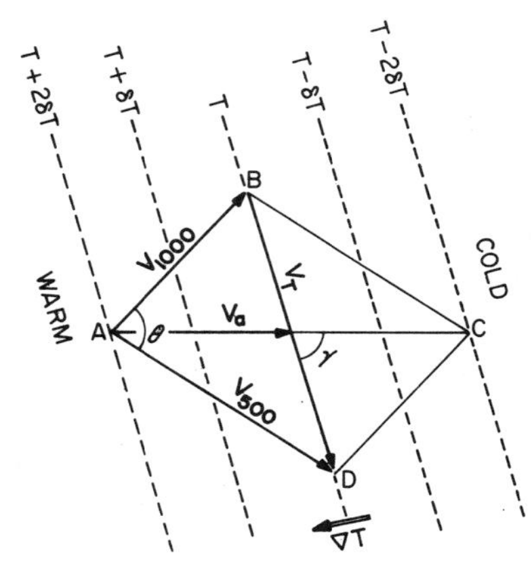

Fig. 8.20 Relationships between various wind vectors in Problem 8.8

low temperatures lying to the left, as required by the thermal wind equation (8.12) which states that

$$\mathbf{V}_\mathrm{T} = \frac{R}{f} \ln \frac{1000}{500} (\mathbf{k} \times \nabla \bar{T})$$

where $\bar{T}$ is the layer mean temperature. Solving this expression for the absolute value of the horizontal temperature gradient, we obtain

$$|\nabla \bar{T}| = A V_\mathrm{T}$$

where

$$A = \frac{f}{R \ln \frac{1000}{500}} = \frac{10^{-4}\ \mathrm{s}^{-1}}{287\ \mathrm{J\ kg}^{-1}\ \mathrm{deg}^{-1} \times 0.694} = 5.01 \times 10^{-7}\ \mathrm{m}^{-2}\ \mathrm{s\ deg}$$

The geostrophic temperature advection (GTA) is given by

$$\mathrm{GTA} = -V_\mathrm{a} \frac{\partial \bar{T}}{\partial s}$$

where s is in the direction of $\mathbf{V}_\mathrm{a}$. From Fig. 8.20 it is evident that

$$\frac{\partial \bar{T}}{\partial s} = -|\nabla \bar{T}| \sin \gamma$$

where γ is the angle between $\mathbf{V}_\mathrm{T}$ and $\mathbf{V}_\mathrm{a}$. Substituting for $|\nabla \bar{T}|$ from the thermal wind equation above we obtain

$$\frac{\partial \bar{T}}{\partial s} = -A V_\mathrm{T} \sin \gamma$$

Therefore,

$$\mathrm{GTA} = A(V_\mathrm{a} V_\mathrm{T} \sin \gamma)$$

It may be noted that the expression in parentheses is equivalent to the area of the parallelogram ABCD in Fig. 8.20. The same area is represented by the expression $V_{500} V_{1000} \sin \theta$

where θ is the angle between the 1000- and 500-mb geostrophic wind vectors. It follows that

$$\begin{aligned} \text{GTA} &= AV_{500}V_{1000} \sin\theta \\ &= 5.01 \times 10^{-7}\ \text{m}^{-2}\ \text{s}^{-1}\ \text{deg} \times 30\ \text{m s}^{-1} \times 15\ \text{m s}^{-1} \times \sin 70^\circ \\ &= 2.12 \times 10^{-4}\ \text{deg s}^{-1} \qquad \text{or} \qquad 18.3\ \text{deg day}^{-1} \end{aligned}$$

To express the thermodynamic energy equation in terms of the local time derivative, we combine (8.13) and (8.14b) to obtain

$$\frac{\partial T}{\partial t} = -V\frac{\partial T}{\partial s} + \omega\left(\frac{\kappa T}{p} - \frac{\partial T}{\partial p}\right) + \frac{\dot{H}}{c_p} \tag{8.15}$$

Referring back to Section 2.5.3 it is easily shown that in the case of an adiabatic lapse rate the term in parentheses vanishes. In a stably stratified atmosphere, $\partial T/\partial p$ must be smaller than in the adiabatic case, and thus the term in parentheses must be positive, so that sinking motion always contributes to local temperature rises and vice versa. In general, the more stable the lapse rate the larger the local rate of temperature rise that results from a given rate of downward motion ω.

In order to give a physical interpretation of the various terms in (8.15), it is instructive to begin by considering the artificial situation of adiabatic motion ($\dot{H} = 0$) in a stably stratified atmosphere in which the configuration of the potential temperature surfaces is not changing with time. At any fixed point in (x, y, p) coordinates the temperature must remain constant so that (8.15) reduces to

$$\frac{\omega}{V} = \frac{\partial T/\partial s}{(\kappa T/p - \partial T/\partial p)}$$

which is simply a statement of the fact that in this artificial situation the slopes of the three-dimensional trajectories are everywhere identical to the slopes of the local isentropic (constant potential temperature) surfaces.

Now let us retain the assumptions of adiabatic flow and stable stratification while relaxing the constraint that the isentropic surfaces remain fixed. Since air parcels conserve potential temperature, the isentropic surfaces behave like material surfaces that are carried along with and distorted by the three-dimensional motion field. At any fixed point in (x, y, p) space the time rate of change of temperature depends upon the imbalance between the first two terms on the right-hand side of (8.15). This imbalance, in turn, is proportional to the difference in slope between the three-dimensional parcel trajectories and the isentropic surfaces.

The typical relationship between the three-dimensional parcel trajectories and the isentropic surfaces at tropospheric levels in synoptic-scale middle-latitude disturbances is illustrated in Fig. 8.21. The trajectories usually slope in the same sense as the isentropic surfaces, and their inclination is typically about half as large. Thus, in a statistical sense, the motion along these sloping trajectories is acting to flatten out the isentropic surfaces, thereby reducing the horizontal temperature gradients. If the wind is blowing from left to right in the figure, the horizontal wind component will produce warm advection while

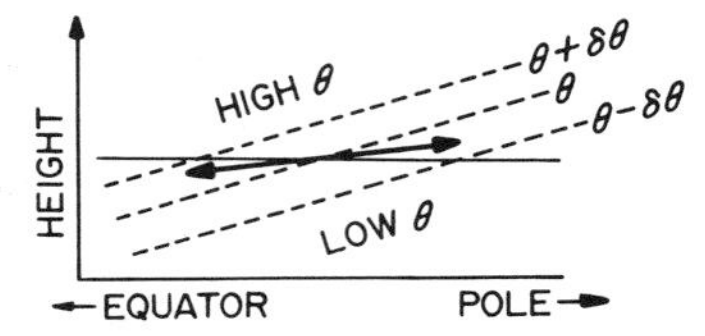

Fig. 8.21 Relation between parcel trajectories (→) and isentropes (---) in a meridional cross section through a typical synoptic-scale disturbance in middle latitudes.

the associated lifting contributes to cooling. The warm advection term predominates because motion along the sloping trajectory is replacing the air at O with air that has a higher value of potential temperature. If the wind is blowing from right to left the horizontal advection term contributes to cooling at O, while the associated sinking contributes to warming, with the former effect being predominant. Note that in either case the horizontal advection term and the vertical motion term in (8.15) tend to be of opposite sign, and the horizontal advection term tends to be dominant. In the absence of diabatic effects the local time rate of change of temperature tends to be, on the average, about half as large as that which would be inferred from the horizontal advection term alone. Typical local time rates of change are on the order of 10° day^{-1}, still about an order of magnitude larger than the diabatic heating rates (with the exceptions noted above) so that, to a first approximation, middle-latitude synoptic-scale disturbances can be regarded as adiabatic.

Within the tropical troposphere the relative sizes of the terms in (8.15) differ greatly from those in middle latitudes. Horizontal temperature gradients are very weak and temperature at a fixed point changes little from day to day. Observations indicate that the local time rate of change term and the horizontal advection term are only on the order of tenths of a degree per day. Thus, to a good approximation, (8.15) reduces to a balance between the vertical motion term and the diabatic heating term. Virtually all the upward motion in the tropics is concentrated in relatively small areas of precipitation where the release of latent heat of condensation almost exactly compensates for the adiabatic cooling. The prevailing lapse rate in the tropics is very nearly moist adiabatic so that the lifting of saturated air does not result in large temperature changes, even when the rate of ascent is very large, as in a hurricane. Within the much larger regions of slow sinking motion, warming due to adiabatic compression is balanced by radiative cooling. Because of the smallness of the time derivative term, (8.15) in its present form is not useful as a prognostic equation for the tropics.

8.8 THE CONTINUITY EQUATION

Even though the vertical wind component accounts for only a minute fraction of the kinetic energy of large-scale motions, it plays a crucial role in atmospheric dynamics as a mechanism for coupling between the wind and temperature fields. The vertical velocity appears as a factor in one of the leading terms in the thermodynamic energy equation. For the range of static stability observed throughout most of the earth's atmosphere even very small vertical motions ($\sim 1 \text{ cm s}^{-1}$) have a pronounced effect upon the time evolution of the temperature field. In addition, the distribution of precipitation and latent heat release is largely controlled by vertical motions.

For all scales of motion the time rate of change of the vertical wind component is governed by dynamical processes as embodied in Newton's second law. However, as shown in Section 8.5, Newton's second law provides no prognostic information when applied to vertical motions that are close to a state of hydrostatic balance.

The vertical motion field is also subject to a *kinematic*† constraint, based upon the conservation of mass. This constraint can be expressed in terms of the *continuity equation*, which relates the time rates of change of density and volume of an air parcel as it moves through the atmosphere. For the special case of hydrostatic motions, the continuity equation takes the form of a diagnostic relationship which can be used to obtain the vertical velocity field.

A thick, soggy pancake (of the type served in many college dormitories) provides an excellent medium for demonstrating the continuity equation. Flattening or squashing in the vertical dimension (for example, when the pancake is squeezed between two flat plates) is accompanied by spreading or divergence in the horizontal dimension as the pancake attempts to conserve its original volume. When air parcels are deformed by the large-scale motion field, they behave in qualitatively the same manner, with one notable exception: unlike the ideal, soggy pancake, an air parcel is compressible; that is to say, it may experience changes in volume. In general, these volume changes are of two types:

(a) nonhydrostatic fluctuations associated with vertically propagating sound waves. These fluctuations are of extremely small amplitude and high frequency and are of absolutely no importance as far as large-scale atmospheric motions are concerned.

(b) more gradual, hydrostatic changes due to the expansion and compression which accompanies hydrostatic pressure changes. These changes may be quite large and are therefore relevant to the dynamics of large-scale motions.

When the continuity equation is formulated in (x, y, p) coordinates the irrelevant nonhydrostatic effects are automatically eliminated, while the effects of hydrostatic expansion and compression are implicitly taken into account without adding to the complexity of the relation.

Let us consider an air parcel shaped like a block with dimensions δx, δy, δp, as indicated in Fig. 8.22. If we assume that the atmosphere is in hydrostatic balance, the mass of the block is given by

$$\delta M = \rho\, \delta x\, \delta y\, \delta z = \frac{-\delta x\, \delta y\, \delta p}{g}$$

† "*Kinematic*" refers to properties of the flow that can be deduced without reference to Newton's laws.

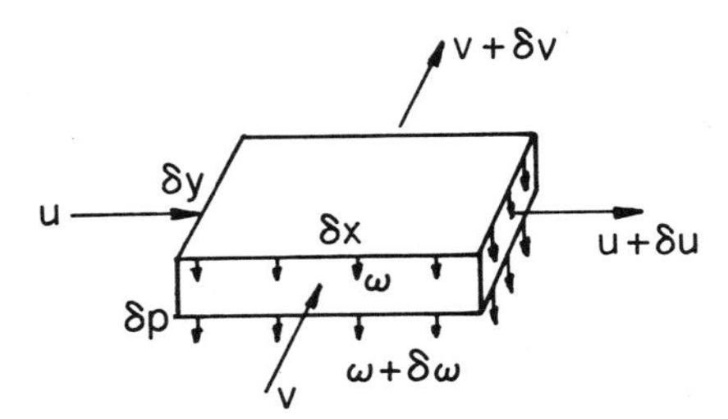

Fig. 8.22 Relationships used in the derivation of the continuity equation.

With the passage of time, the shape of the block will be twisted and distorted beyond recognition by the shears and deformations in the motion field. However, we will only be concerned about the rates of change of the dimensions of the parcel at the initial moment when it is shaped like a parallelepiped, as pictured in Fig. 8.22. Since the mass of the block is not changing with time, we can write

$$\frac{d}{dt}(\delta x\, \delta y\, \delta p) = 0 \tag{8.16}$$

or, in expanded form,

$$\delta y\, \delta p \frac{d}{dt} \delta x + \delta x\, \delta p \frac{d}{dt} \delta y + \delta x\, \delta y \frac{d}{dt} \delta p = 0$$

The time rate of change of the zonal dimension δx is equal to δu, the difference between the zonal velocities averaged over the two y–p faces of the block. If the dimensions of the block are very small in comparison to the space scale of the velocity fluctuations in the atmosphere, we may write

$$\frac{d}{dt}(\delta x) = \delta u = \frac{\partial u}{\partial x} \delta x$$

where the partial notation indicates that the derivative is taken at constant y and p. The time rates of change of δy and δp can be expressed in an analogous form. Substituting for the time derivatives in the expanded form of (8.16), and dividing through by $\delta x\, \delta y\, \delta p$, we obtain the *continuity equation* in pressure coordinates

$$\frac{\partial u}{\partial x} + \frac{\partial v}{\partial y} + \frac{\partial \omega}{\partial p} = 0 \tag{8.17}$$

In order to gain a clearer physical insight into this relationship, let us go back and expand (8.16) in the form

$$\delta p \frac{dA}{dt} + A \frac{d}{dt} \delta p = 0$$

where $A = \delta x\, \delta y$ is the area of the block in the horizontal plane. Substituting for δp as before, and dividing through by $A\, \delta p$, we obtain the continuity equation in the alternate form

$$\frac{1}{A}\frac{dA}{dt} + \frac{\partial \omega}{\partial p} = 0 \tag{8.18}$$

Combining (8.17) and (8.18) we obtain

$$\frac{\partial u}{\partial x} + \frac{\partial v}{\partial y} = \frac{1}{A}\frac{dA}{dt} \tag{8.19}$$

The expression on the left of (8.19) may be recognized as the Cartesian form of the horizontal *divergence* of the horizontal wind vector **V**, as abbreviated by the symbol $\nabla \cdot \mathbf{V}$. The reader unfamiliar with vector calculus can identify divergence with changes in the horizontal area of air parcels, as indicated by the right-hand side of (8.19). The continuity equation states that horizontal divergence ($\nabla \cdot \mathbf{V} > 0$) is accompanied by a squeezing together in the vertical ($\partial\omega/\partial p < 0$), as indicated in Fig. 8.22, and horizontal convergence is accompanied by vertical stretching.

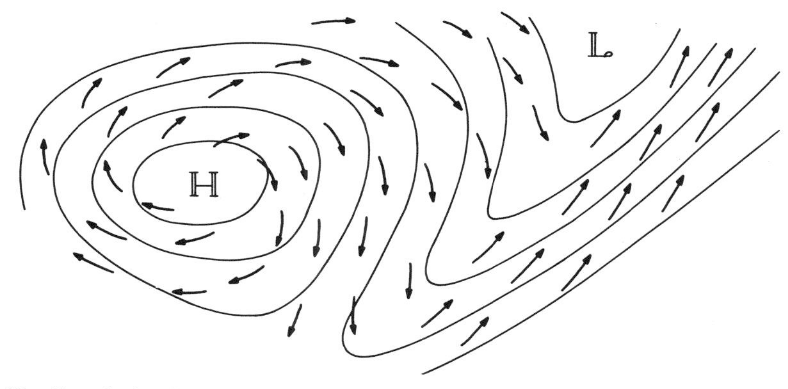

Fig. 8.23 Cross isobar flow induced by frictional dissipation within the planetary boundary layer. Solid lines represent isobars of sea level pressure.

As a qualitative example, let us consider a region of convergence in the lower troposphere. The continuity equation requires that, within such a region, $\partial\omega/\partial p > 0$. This relationship requires that ω increase with pressure from negative values (indicative of upward motion) in the middle troposphere to near zero† at the ground. Thus, within regions of horizontal convergence, the geometric vertical velocity w becomes more positive with increasing height, in agreement with intuitive reasoning.

† Even in the absence of topography, $\omega = dp/dt$ is not generally equal to zero at the ground, since air parcels may experience pressure changes even in the absence of vertical motions. However, it may be shown that in regions of reasonably flat terrain, in middle and high latitudes $\omega \sim 10$ mb day^{-1} at the ground, which is about an order of magnitude smaller than in the middle troposphere.

Within the planetary boundary layer we have seen that there is a systematic tendency for the large-scale wind field to have a component across the isobars toward lower pressure, because of the frictional drag force. This cross isobaric flow gives rise to horizontal convergence in regions of low pressure and divergence in regions of high pressure, as illustrated in Fig. 8.23. The convergence within low-pressure areas is associated with rising motion, enhanced cloudiness, and precipitation, while the divergence within high-pressure areas is associated with large-scale subsidence, suppression of convective clouds, and precipitation, and the frequent formation of low-level inversions. (As will be shown in the next chapter, there are other dynamical processes that influence the distribution of cloudiness and precipitation.)

8.8.1 The kinematic vertical velocity

It is possible to estimate the vertical velocity $\omega(x, y, p)$ at any desired pressure level by the kinematic method, which involves integrating (8.17) or (8.18) vertically from some reference level p^*, where ω is known, to level p. At a given point (x, y)

$$d\omega = -(\nabla \cdot \mathbf{V})\, dp$$

Therefore,

$$\omega(p) = \omega(p^*) - \int_{p^*}^{p} (\nabla \cdot \mathbf{V})\, dp \tag{8.20}$$

In this form, the continuity equation serves as a crucial link between the vertical velocity field and the other dependent variables in the system of equations that governs large-scale atmospheric motions. In principle, (8.20) is an exact relationship that defines a vertical velocity field that is everywhere consistent with the horizontal motion field. However, when computed from actual observational data, the *kinematic vertical velocity field* given by (8.20) is subject to a number of uncertainties which stem from an imperfect knowledge of ω at the ground or any other reference level, and observational errors in the vertical profile of the horizontal divergence. From (8.20) it is clear that the uncertainty in ω due to errors in divergence increases in proportion to the thickness of the layer over which the integration is carried out. The determination of vertical velocity by the kinematic method is demonstrated in the following example:

Problem 8.9 The area of a large cumulonimbus anvil is observed (by geosynchronous satellite) to increase by 20% over a 10-min period. Assuming that this increase in area is representative of the average divergence within the 300–100-mb layer, and that the vertical velocity at 100 mb is zero, calculate the vertical velocity at the 300-mb level (see Fig. 8.24).

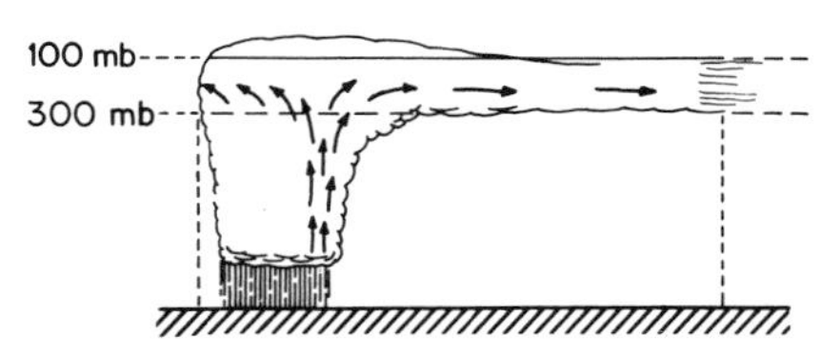

Fig. 8.24 Physical situation in Problem 8.9.

Solution From (8.18),

$$\nabla \cdot \mathbf{V} = \frac{1}{A}\frac{dA}{dt} = \frac{0.20}{600\ \text{s}} = 3.33 \times 10^{-4}\ \text{s}^{-1}$$

Making use of (8.20), we have

$$\begin{aligned}\omega_{300} &= \omega_{100} - (\nabla \cdot \mathbf{V})(300\ \text{mb} - 100\ \text{mb}) \\ &= 0 - 3.33 \times 10^{-4}\ \text{s}^{-1} \times 200\ \text{mb} \\ &= -6.66 \times 10^{-2}\ \ \text{mb s}^{-1}\end{aligned}$$

$$w_{300} \simeq -\omega\frac{H}{p} = 6.66 \times 10^{-2}\ \text{mb s}^{-1} \times \frac{7\ \text{km}}{300\ \text{mb}} \simeq 1.5\ \ \text{m s}^{-1}$$

Note that this is the 300-mb vertical velocity averaged over the area of the anvil. The vertical velocity within the updraft may be much larger.

Problem 8.10 Figure 8.25 shows the vertical velocity profile in a certain rain area in the tropics. The horizontal convergence into the rain area in the 1000–800-mb layer is $10^{-5}\ \text{s}^{-1}$, and the average water vapor content of this converging air is 16 g kg^{-1}. (a) Calculate the divergence in the 200–100-mb layer. (b) Estimate the rainfall rate, using the assumption that all the water vapor is condensed out during the ascent of the air.

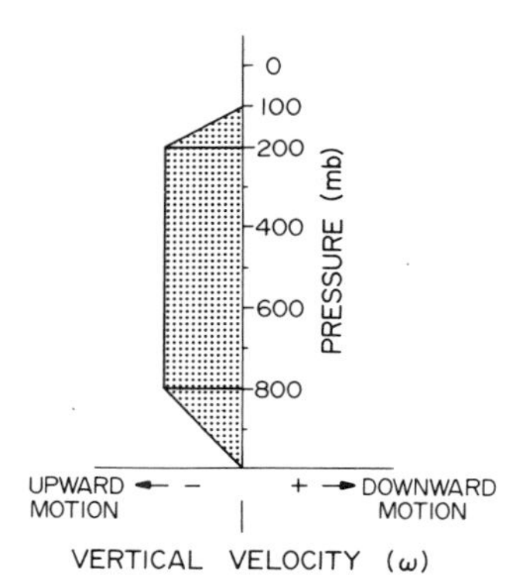

Fig. 8.25 Vertical velocity profile in Problem 8.10.

Solution (a) It is clear that $|\partial\omega/\partial p|$ is twice as large in the 200–100-mb layer as in the 1000–800-mb layer, and therefore the absolute magnitude of the divergence is twice as large; that is, $2 \times 10^{-5}\ \text{s}^{-1}$.

(b)

$$\begin{aligned}\omega_{800} &= -\int_{1000}^{800} (\nabla \cdot \mathbf{V})\, dp + \omega_{1000} \\ &= (\nabla \cdot \mathbf{V})(1000\ \text{mb} - 800\ \text{mb}) + \omega_{1000} \\ &= (-10^{-5}\ \text{s}^{-1})(200\ \text{mb}) + 0 \\ &= -2 \times 10^{-3}\ \ \text{mb s}^{-1}\end{aligned}$$

Making use of the relationship $\omega \approx -\rho g w$, the vertical mass flux ρw can be obtained by dividing ω (in SI units) by g. Therefore, the rate at which liquid water is being condensed out is

$$\frac{2 \times 10^{-1}\ \text{Pa s}^{-1}}{9.8\ \text{m s}^{-2}} \times 0.016 = 3.27 \times 10^{-4} \quad \text{kg m}^{-2}\,\text{s}^{-1}$$

We can express the rainfall rate in more conventional terms by noting that 1 kg m^{-2} of liquid water is equivalent to a depth of 1 mm. Thus the rainfall rate is 3.27×10^{-4} mm s^{-1} or

$$3.27 \times 10^{-4}\ \text{mm s}^{-1} \times 8.65 \times 10^{4}\ \text{s day}^{-1} = 28.2 \quad \text{mm day}^{-1}$$

which is a typical rate for moderate rain.

8.8.2 The horizontal divergence in natural coordinates

Let us consider an incremental element of area $\delta s\, \delta n$ as it moves along in the flow, as pictured in Fig. 8.26. The horizontal divergence is given by

$$\frac{1}{A}\frac{dA}{dt} = \frac{1}{\delta s\, \delta n}\frac{d}{dt}\delta s\, \delta n = \frac{1}{\delta s\, \delta n}\left(\delta s \frac{d}{dt}\delta n + \delta n \frac{d}{dt}\delta s\right)$$

We will assume that the dimensions of the block are very small in comparison to the scale of the motion field so that the block is essentially trapezoidal in shape with an incremental angle $\delta\psi$ between the two s sides. The rate of stretching in the n dimension is given by

$$\frac{d}{dt}\delta n = V\,\delta\psi = V\frac{\partial\psi}{\partial n}\delta n$$

while the rate of stretching in the s dimension is

$$\frac{d}{dt}\delta s = \delta V = \frac{\partial V}{\partial s}\delta s$$

Substituting into this expression for the horizontal divergence, and simplifying, we obtain

$$\frac{1}{A}\frac{dA}{dt} = V\frac{\partial\psi}{\partial n} + \frac{\partial V}{\partial s} \tag{8.21}$$

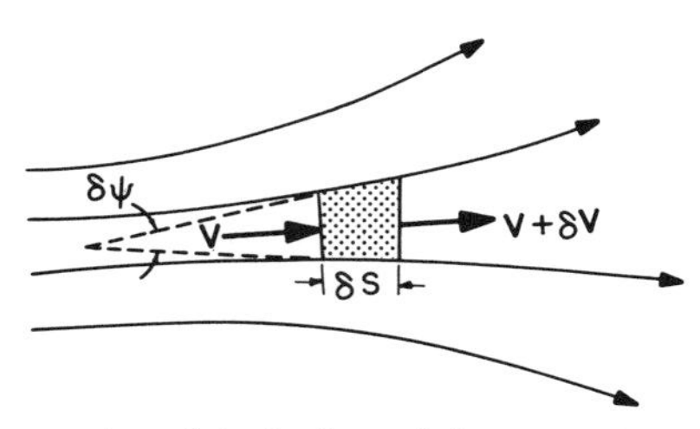

Fig. 8.26 Representation of the horizontal divergence in natural coordinates.

The first term on the right-hand side represents the *diffluence* (or, if negative, the *confluence*) of the horizontal flow; literally the flowing apart (or flowing together) of the streamlines. The second term can be interpreted in terms of a stretching or squashing due to changes in velocity along a given streamline.

For an analogous interpretation of the terms in (8.21) let us consider the example of the flow of traffic on a multilane highway at a point where the number of usable lanes abruptly decreases in the direction of travel. Usually in these situations, both terms on the right-hand side of (8.21) are negative. The first term represents the convergence of automobiles due to the narrowing of the roadway. The second term represents an additional convergence due to the inevitable decrease in speed that accompanies any abrupt increase in traffic density.

The large-scale atmospheric motion field behaves quite differently from traffic on a highway. Confluence is usually accompanied by increasing speeds and vice versa so that the two terms on the right-hand side of (8.21) are normally of opposite sign. Because of the tendency for compensation between the two terms, the horizontal divergence is usually smaller, by almost a full order of magnitude, than the diffluence or the stretching terms taken individually. A similar compensation is observed between the two terms on the left-hand side of (8.19).

In both (8.19) and (8.21) the horizontal divergence assumes the form of a small difference between two much larger terms involving spatial derivatives of the horizontal wind field. In either equation, even rather modest errors in the estimation of these derivatives can lead to serious errors in the computed values of divergence. A number of special analysis techniques have been developed to minimize the impact of these errors upon the reliability of the kinematic vertical velocity field.

8.8.3 The magnitude of the vertically averaged divergence

Within the earth's atmosphere there exists a high degree of compensation in the vertical profile of divergence; that is to say, strong convergence at one level is almost always accompanied by strong divergence at some other level, so that a typical value of the vertically averaged divergence

$$\{\nabla \cdot \mathbf{V}\} \equiv \frac{1}{p_0} \int_0^{p_0} (\nabla \cdot \mathbf{V})\, dp \ll |\nabla \cdot \mathbf{V}|$$

where p_0 is the pressure at the earth's surface and $|\nabla \cdot \mathbf{V}|$ is the absolute value of the typical convergences or divergences observed at a particular level in the atmosphere. The tendency for low-level convergence to be accompanied by upper-level divergence and vice versa was first noted by Dines† on the basis of

† **William H. Dines** (1855–1927) English meteorologist. Leading exponent of experimental meteorology in England during the late 19th and early 20th centuries. Carried out research on anemometry (Dines anemometer), upper air observations (with kites attached to several miles of wire), and radiation.

observational evidence. That such a compensation does, in fact, exist can be demonstrated on the basis of the following argument.

Applying (8.14b) to the pressure at the earth's surface, we obtain

$$\frac{\partial p_0}{\partial t} = -V_0 \frac{\partial p_0}{\partial s} - w \frac{\partial p_0}{\partial z} + \omega_0 \tag{8.22}$$

Combining the identity

$$\omega_0 \equiv \int_0^{p_0} \frac{\partial \omega}{\partial p} \, dp$$

with (8.17) we arrive at the expression

$$\omega_0 = -\int_0^{p_0} (\nabla \cdot \mathbf{V}) \, dp$$

Thus, in a hydrostatic atmosphere, ω at any level is completely determined by the mass convergence into the vertical column extending upward from the level in question to the top of the atmosphere. Incorporating this result into (8.22), we obtain Margules'[†] *pressure tendency equation*:

$$\frac{\partial p_0}{\partial t} = -V_0 \frac{\partial p}{\partial s} - w \frac{\partial p}{\partial z} - \int_0^{p_0} (\nabla \cdot \mathbf{V}) \, dp \tag{8.23}$$

Near the earth's surface, the local time rate of change of pressure is on the order of 10 mb day^{-1} and the horizontal advection term is even smaller (see Problem 8.25). Furthermore, at the earth's surface, $w \simeq 0$ except in regions of rough terrain, so that the vertical advection term usually vanishes. It follows that

$$\int_0^{p_0} (\nabla \cdot \mathbf{V}) \, dp = p_0 \{\nabla \cdot \mathbf{V}\} \sim 10 \quad \text{mb day}^{-1}$$

Since $p_0 \simeq 1000$ mb and 1 day $\sim 10^5$ s,

$$\{\nabla \cdot \mathbf{V}\} \sim 10^{-7} \quad \text{s}^{-1}$$

which is about two orders of magnitude smaller than the typical divergences observed at a particular level in the atmosphere (see Problem 8.26).

A physical interpretation of the preceding result can be given in terms of Fig. 8.27, which shows the two-dimensional flow that accompanies a wave that is propagating uniformly toward the right in a stably stratified liquid (that is, a liquid with density decreasing with height). Figure 8.27a shows a pure "external wave" in which divergence is independent of height and vertical velocity increases linearly with height to a maximum at the free surface of the liquid.

[†] **Max Margules** (1856–1920) Meteorologist, physicist, and chemist, born in Brody, Galicia (now Ukranian USSR). Worked in intellectual isolation on atmospheric dynamics from 1882 to 1906, during which time he made many fundamental contributions to the subject. Thereafter, he returned to his first love, chemistry. Died of starvation while trying to survive on a government pension equivalent to $2 a month in the austere post-World War I period. Many of the current ideas concerning the kinetic energy cycle in the atmosphere (see Sections 9.1–9.3) stem from Margules' work.

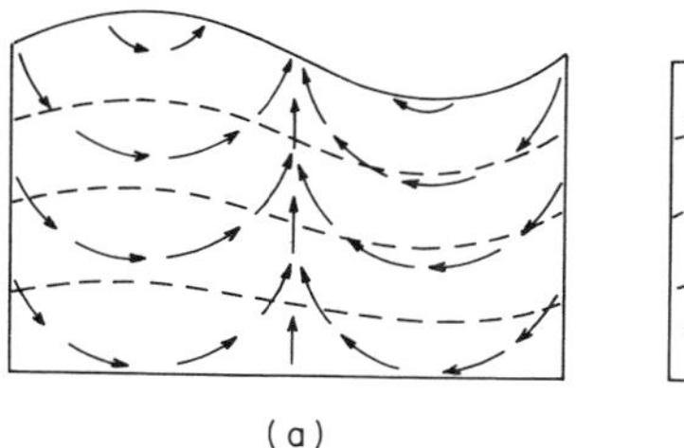

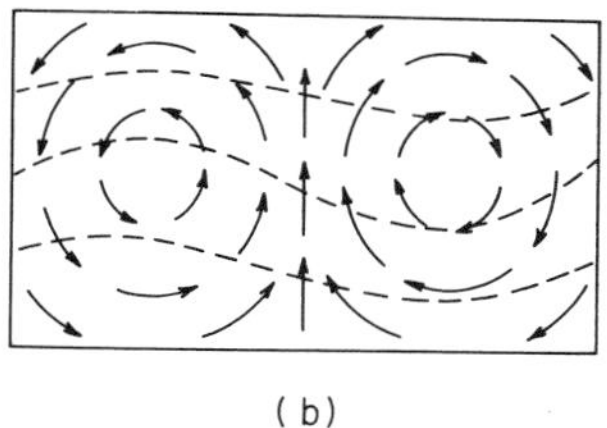

Fig. 8.27 Motion field in a vertical cross section through a two-dimensional wave propagating from left to right in a stably stratified liquid. The dashed lines represent surfaces of constant density. (a) An "external wave" in which the maximum vertical motions occur at the free surface of the liquid and (b) an "internal wave" in which the vertical velocity vanishes at the top of the fluid because of the presence of a rigid lid.

Figure 8.27b shows a pure "internal wave" in a fluid which is bounded by a rigid upper lid, so that $w = 0$ at both top and bottom. Note the compensation between low-level convergence and upper-level divergence and vice versa. In the earth's atmosphere, motions that resemble the internal wave contain several orders of magnitude more kinetic energy than those that resemble the external wave. In some respects, the stratosphere plays the role of a "lid" over the troposphere, as evidenced by the fact that w is typically almost an order of magnitude smaller in the lower stratosphere than in the middle troposphere.

8.9 THE PRIMITIVE EQUATIONS

In order to organize and summarize the diverse topics covered in the preceding sections of this chapter, we will begin by writing down the system of equations that governs large-scale atmospheric motions. Here the equations appear in component form, where all time derivatives refer to changes at a fixed point in space, in the (x, y, p) coordinate system:

$$\frac{\partial u}{\partial t} = -u\frac{\partial u}{\partial x} - v\frac{\partial u}{\partial y} - \omega\frac{\partial u}{\partial p} - \frac{\partial \Phi}{\partial x} + fv + F_x \qquad (8.24\text{a})^\dagger$$

$$\frac{\partial v}{\partial t} = -u\frac{\partial v}{\partial x} - v\frac{\partial v}{\partial y} - \omega\frac{\partial v}{\partial p} - \frac{\partial \Phi}{\partial y} - fu + F_y \qquad (8.24\text{b})^\dagger$$

$$\frac{\partial \Phi}{\partial p} = -\frac{RT}{p} \qquad (8.24\text{c})$$

$$\frac{\partial T}{\partial t} = -u\frac{\partial T}{\partial x} - v\frac{\partial T}{\partial y} + \omega\left(\frac{\kappa T}{p} - \frac{\partial T}{\partial p}\right) + \frac{\dot{H}}{c_p} \qquad (8.24\text{d})$$

$$\frac{\partial \omega}{\partial p} = -\left(\frac{\partial u}{\partial x} + \frac{\partial v}{\partial y}\right) \qquad (8.24\text{e})$$

† In the full primitive equations (8.24a) and (8.24b) contain an additional small inertial term on the right-hand side, which arises because of the curvature of the spherical coordinate system.

This system of so-called *primitive equations* represents a synthesis of

- Newton's Second Law for the horizontal component of the motion,
- the Ideal Gas Equation, which is implicit in (8.24c) and (8.24d),
- the First Law of Thermodynamics, and
- the Conservation of Mass

as applied to motions that are in a state of hydrostatic balance. Together with the surface pressure tendency equation (8.23) and suitable boundary conditions for ω or w at the earth's surface and the "top" of the atmosphere, it constitutes a fully determined system of five equations involving five dependent variables (u, v, T, Φ, and ω) which are all continuous functions of x, y, and p. Also appearing in the equations are diabatic heating $\dot{H}$ and friction F, which may be either specified as external parameters, or represented as functions of the dependent variables. In situations where the latent heat of condensation is important as a heat source, an additional equation, based upon the conservation of mass of water substance, can be incorporated into the system. Analogous systems of primitive equations can be formulated in each of the other coordinate systems described in Section 8.1.

The primitive equations, together with (8.23) and additional equations specifying the boundary conditions and the distributions of $\dot{H}$ and F, can be solved in various ways to predict how the state of the atmosphere (as defined by the fields of the five dependent variables) will evolve starting from some particular set of initial conditions. Such "time integrations" of the primitive equations are the basis for numerical weather prediction and for general circulation models of the type that will be described in Section 9.5. In addition, the primitive equations are the starting point for constructing simpler and more specialized systems of equations which elucidate specific phenomena or dynamical processes.

From a cursory analysis of the magnitudes and signs of the various terms in the primitive equations, it has been possible to infer a number of simple, approximate relationships between the dependent variables: namely, the geostrophic, gradient wind, and thermal wind equations which stem from the horizontal equations of motion. It has also been possible to give some indication of how a balance is achieved between the various terms in the thermodynamic energy equation and the continuity equation.[†]

[†] For a more rigorous derivation of the primitive equations and a more extensive discussion of some of their applications, the reader is referred to J. R. Holton, "An Introduction to Dynamical Meteorology," Academic Press, New York, 1972. Holton's treatment emphasizes the quasi-geostrophic model: a consistent, simplified set of equations that can be obtained by performing a number of simple operations on the primitive equations and neglecting certain small terms that appear in the resulting relationships. The quasi-geostrophic model is extremely valuable as a heuristic tool for elucidating the more important physical processes that control the development and evolution of large-scale atmospheric motions. It is the basis for much of the discussion in the following chapter of this book.

It may have been noticed that in the physical interpretation of the primitive equations we have purposely steered clear of such questions as: Why do the large-scale wind and temperature fields remain close to a state of geostrophic or gradient wind balance? Why do the air trajectories in the middle-latitude troposphere usually slope in the same sense as the isentropic surfaces, but not as steeply? Why does the horizontal divergence tend to be almost an order of magnitude smaller than the horizontal gradients of the wind field? In fact, we have even avoided the fundamental question of why atmospheric motions exist in the first place. We will attempt to come to grips with these questions in the next chapter which deals with the general circulation of the atmosphere.

PROBLEMS

8.11 Explain or interpret the following on the basis of the principles discussed in this chapter.

(a) In the definitions of u and v, the distance from the center of the earth is treated as a constant (that is, r is replaced by R_E).

(b) ω and w are nearly always of opposite sign.

(c) The primitive equations assume a simpler form in pressure coordinates than in height coordinates.

(d) The thermodynamic energy equation assumes a particularly simple form in isentropic coordinates.

(e) A person of fixed mass experiences a slight loss in weight by moving toward the equator.

(f) A person of fixed mass experiences a slight weight gain when stepping off an east-bound train onto a west-bound one.

(g) A geostationary satellite can remain in orbit at a fixed point above the equator.

(h) The existence of the Coriolis force is not apparent to an observer located at a fixed point on earth.

(i) The vertical component of the Coriolis force is not important in atmospheric dynamics.

(j) The Coriolis forces induced by vertical motions are not important in atmospheric dynamics.

(k) The Coriolis force is directed in the opposite sense in the Northern and Southern Hemispheres.

(l) At the equator, horizontal motions do not induce any horizontal Coriolis force.

(m) The vertical component of the pressure gradient force is much larger than the horizontal component. Nevertheless, the horizontal component plays an important role in the equations of motion.

(n) The geostrophic wind cannot be defined at the equator.

(o) The winds in the tropics are not quasi-geostrophic.

(p) The geostrophic wind speed corresponding to a given horizontal pressure gradient decreases with latitude.

(q) It would be counterproductive for a meteorologist in Australia to memorize the rule that the geostrophic wind *l*eaves *l*ow pressure to the *l*eft.

(r) The strong winds in the vicinity of hurricanes are usually highly subgeostrophic.

(s) For a given horizontal pressure gradient the actual wind tends to be stronger in ridges than in troughs.

(t) The horizontal projections of streamlines and trajectories often curve in the opposite sense.

(u) The flow in mountain valleys is usually far from geostrophic.

(v) Surface winds are usually closer to geostrophic balance over ocean areas than over land areas.

(w) The flow around hurricanes is always cyclonic, whereas in small rotating systems like dust devils it may be in either direction.

(x) Synoptic-scale motions are close to hydrostatic balance.

(y) Jet streams tend to occur near levels at which the horizontal temperature gradient is zero, and reversing with height.

(z) In the Northern Hemisphere, northerly winds tend to be associated with cold advection.

(aa) It is often possible to tell whether cold or warm advection is taking place by comparing the direction of the surface wind with the direction of motion of cloud layers aloft.

(bb) Veering of the wind with height in the planetary boundary layer is not necessarily an indication of warm advection.

(cc) In an equivalent barotropic atmosphere there is no horizontal temperature advection.

(dd) There is no advection of pressure by the geostrophic wind.

(ee) During winter the atmosphere above the Gulf Stream experiences cold advection most of the time, yet the air temperature at any fixed point usually remains quite warm.

(ff) Pressure falls in the vicinity of a center of low pressure are an indication that the low is deepening.

(gg) The terms $\partial w/\partial z$ and $\partial \omega/\partial p$ are usually of the same sign.

(hh) Rain areas tend to be associated with convergence in the lower troposphere and divergence in the upper troposphere.

(ii) It is difficult to measure divergence accurately.

(jj) Confluence doesn't necessarily imply convergence.

8.12 An air parcel is moving westward at 20 m s^{-1} along the equator. Compute

(a) the apparent acceleration toward the center of the earth from the point of view of an observer external to the earth, and in a coordinate system rotating with the earth, and

(b) the apparent Coriolis force in the rotating coordinate system.

Answer (a) 311×10^{-4} and 0.627×10^{-4} m s^{-2}; (b) 29.1×10^{-4} m s^{-2} downward

8.13 (a) A projectile is fired vertically upward with velocity w_0 from a point on earth. Show that in the absence of friction the projectile will land a distance

$$\frac{4{w_0}^3\Omega}{3g^2}\cos\phi$$

to the west of the point from which it was fired.

(b) Calculate the displacement for a projectile fired upward on the equator with a velocity of 500 m s^{-1}.

Answer 126 m

8.14 (a) Show that the free surface of a cylindrical tank of liquid in solid-body rotation has a shape given by

$$Z_0 = Z_{00} + \frac{\Omega^2 r^2}{2g}$$

where Z_{00} is the height of the free surface at the axis of rotation, Ω the rotation rate, and r the distance from the axis of rotation. [Hint: Make use of the fact that the surface is everywhere normal to the direction of apparent gravity.]

(b) Can you suggest a system of coordinates in which "gravity" would appear as a force only in the vertical equation of motion?

8.15 A locomotive with a mass of 2×10^4 kg is moving along a straight track at 40 m s^{-1} at 43° lat. Calculate the magnitude and direction of the transverse horizontal force on the track.

Answer 80 N to the right of the direction of movement

8.16 Within a local region near 40N the 500-mb geopotential height contours are oriented east–west and the spacing between adjacent (60 m) contours is 300 km, with geopotential height decreasing toward the north. What is the direction and speed of the geostrophic wind?

Answer $V_g = 21.0$ m s^{-1} from the west

8.17 Two moving ships pass close to a fixed weather ship within a few minutes of each other. The first ship is steaming eastward at a rate of 5 m s^{-1} and the second is steaming northward at 10 m s^{-1}. During the 3-h period that the ships were in the same vicinity, the first recorded a pressure rise of 3 mb while the second recorded no pressure change at all. During the same 3-h period, the pressure rose 3 mb at the location of the weather ship (50N, 140W). On the basis of the above data, calculate the geostrophic wind speed and direction at the location of the weather ship. (Use $\rho = 1.25$ kg m^{-3}.)

Answer $V_g = 19.8$ m s^{-1} from the west

8.18 At a certain station located 43N, the surface wind has a speed of 10 m s^{-1} and is directed across the isobars from high toward low pressure at an angle $\psi = 20°$. Calculate the magnitude of the frictional drag force and the horizontal pressure gradient force (per unit mass).

Answer $|\mathbf{F}| = 3.63 \times 10^{-4}$ m s^{-2} and $|\mathbf{P}_n| = 1.06 \times 10^{-3}$ m s^{-2}

8.19 Figure P8.19 shows the pressure and horizontal wind fields in an atmospheric Kelvin wave which propagates zonally along the equator. Pressure and zonal wind oscillate sinusoidally with longitude and time, while $v = 0$ everywhere. Such waves are observed in the stratosphere where frictional drag is negligible in comparison with the other terms in the horizontal equations of motion.

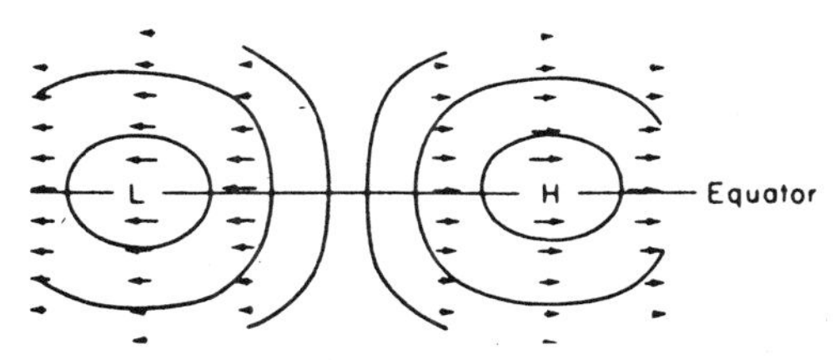

Fig. P8.19.

(a) Prove that the waves propagate eastward.

(b) Prove that the zonal wind component is in geostrophic equilibrium with the pressure field.

8.20 Show that in the case of anticyclonic air trajectories, gradient wind balance is possible only when

$$P_n \leq f^2 R_T/4$$

8.21 At a certain station the 1000-mb geostrophic wind is blowing from the northeast (050°) at 10 m s^{-1} while the 700-mb geostrophic wind is blowing from the west (270°) at 30 m s^{-1}. Subsidence is producing adiabatic warming at a rate of 3°C day^{-1} in the 1000–700-mb layer, and diabatic heating is negligible. Calculate the time rate of change of the thickness of the 1000–700-mb layer. The station is located at 43N.

Answer Decreasing at a rate of 139 m day^{-1}

8.22 During winter in middle latitudes, the meridional temperature gradient is typically on the order of 1° per degree of latitude, while potential temperature increases with height at a rate of roughly 5 deg km^{-1}. What is a typical slope of the potential temperature surfaces in the meridional plane? Compare this result with the slope of the 500-mb surface in Problem 8.16.

Answer 1.8×10^{-3} versus 0.2×10^{-3}

8.23 Making use of the approximate relationship $\omega \simeq -\rho g w$ show that the vertical motion term in (8.15) is approximately equal to

$$-w(\Gamma_d - \Gamma)$$

where Γ_d is the dry adiabatic lapse rate and Γ is the actual lapse rate.

8.24 A tank is filled with an incompressible fluid with constant density ρ_0. The height of the free surface of the fluid is given by $\zeta(x, y)$. Neglect atmospheric pressure on the free surface.

(a) Show that the horizontal component of the pressure gradient force (per unit mass) is independent of height and given by the expression

$$P_n = -g\frac{\partial \zeta}{\partial n}$$

where n is directed toward higher levels of the free surface. [Hint: First show that, at any point in the tank,

$$p(x, y, z_0) = \int_{z_0}^{\zeta(x,y)} \rho_0 g \, dz]$$

(b) Show that the continuity equation takes the form

$$\frac{\partial u}{\partial x} + \frac{\partial v}{\partial y} + \frac{\partial w}{\partial z} = 0$$

(c) If u and v are independent of height, show that

$$\frac{d\zeta}{dt} = -\zeta\left(\frac{\partial u}{\partial x} + \frac{\partial v}{\partial y}\right) \quad \text{and} \quad \frac{\partial \zeta}{\partial t} = -\left(\frac{\partial}{\partial x}u\zeta + \frac{\partial}{\partial y}v\zeta\right)$$

8.25 (a) Prove that in large-scale (hydrostatic) motions

$$\omega = \frac{\partial p}{\partial t} - \rho g w + V\frac{\partial p}{\partial s}$$

where s is defined as distance downstream along the local streamline.

(b) Prove that the absolute magnitude of the last term in the preceding expression is given by

$$|f\rho V V_g \sin \psi|$$

where ψ is the angle between the isobars and the streamlines on a horizontal surface passing through the point in question. Estimate the magnitude of this term on the basis of the following data, which represent typical conditions in middle latitudes; in the midtroposphere, $\rho = 0.7$ kg m^{-3}; V, $V_g = 25$ m s^{-1}; $f = 10^{-4}$ s^{-1}; and $\psi = 10°$ (an upper limit).

Answer 6.55 mb day^{-1}

(c) In the middle-latitude troposphere, typical values of ω are on the order of 200 mb day^{-1} in synoptic-scale atmospheric motions, while local pressure changes are usually on the order of 10 mb day^{-1} at a fixed point in (x, y, z) coordinates. On the basis of these observations, together with the results of (a) and (b), verify that, to within about 5%,

$$\omega \simeq -\rho g w$$

8.26 In middle-latitude winter storms, rainfall (or melted snowfall) rates on the order of 20 mm day^{-1} are not uncommon. Most of the convergence into these storms takes place within the lowest 1–2 km of the atmosphere (say, below 850 mb) where the mixing ratios are on the order of 5 g kg^{-1}. Estimate the magnitude of the convergence into such storms.

Answer 3×10^{-5} s^{-1}

8.27 (a) Prove that the divergence of the geostrophic wind is given by

$$\nabla \cdot \mathbf{V}_g \equiv \frac{\partial u_g}{\partial x} + \frac{\partial v_g}{\partial y} = -\frac{v_g}{f}\frac{\partial f}{\partial y} = -v_g \frac{\cot \phi}{R_E}$$

Give a physical interpretation of this result.

(b) Calculate the divergence of the geostrophic wind at 45N at a point where $v_g =$ 15 m s^{-1}.

Answer 1.67×10^{-6} s^{-1}

8.28 (a) In Fig. 1.10, shade the regions in which temperature decreases with latitude.

(b) In Fig. 1.12, shade the regions in which westerly winds are increasing with height.

(c) Are the regions of shading consistent with the thermal wind equation?

Chapter

9

The General Circulation

9.1 INTRODUCTION

Atmospheric fluid motions may be divided into two broad classes, both of which owe their existence to the uneven distribution of diabatic heating in the atmosphere:

(a) Motions driven either directly or indirectly by horizontal heating gradients in a stably stratified atmosphere account for more than 98% of the atmospheric kinetic energy. Nearly all this kinetic energy is associated with the synoptic- and planetary-scale horizontal wind field, which has a globally averaged root mean square velocity of about 17 m s^{-1}. Some of the small-scale motions observed in the atmosphere derive their energy from this large-scale horizontal wind field through mechanisms discussed below.

(b) Motions driven by convective instability account for the remainder of the atmospheric kinetic energy. Convection is continually breaking out within discrete regions of the atmosphere as a consequence of the vertical gradient of diabatic heating. The resulting motions have space scales ranging from about 30 km in the largest thunderstorms down to less than 1 mm in microscale motions within the surface layer. Despite their small contribution to the atmospheric kinetic energy, convectively driven motions play an important role in the upward transport of latent and sensible heat, as indicated in Section 7.3.2.

The term *general circulation* is used by some meteorologists to denote the totality of atmospheric fluid motions, while others in the field use the term in a more restrictive sense, to denote motions described under (a).

Defined either way, the general circulation can be viewed in the context of a *kinetic energy cycle* in which atmospheric fluid motions continually draw on the reservoir of potential energy inherent in the spatial distribution of atmospheric mass in order to maintain themselves against frictional dissipation, which is continually transforming the kinetic energy of fluid motions into the internal energy of random molecular motions. In the presence of this continual drain, the potential energy of the mass field is maintained by the spatial gradients of diabatic heating which, in a statistical sense, are always acting to lift the center of gravity of the atmosphere.

From a dynamical point of view, large-scale horizontal motions owe their existence to the pressure gradient force which drives a slow horizontal flow across the isobars from higher toward lower pressure. In the presence of the earth's rotation this flow across the isobars induces a circulation parallel to the isobars whose speed tends toward a state of geostrophic balance[†] with the horizontal pressure gradient (and thermal wind balance with the temperature gradient). From an energetic point of view, the same cross-isobar flow, together with its attendant vertical motions, is responsible for the conversion from potential to kinetic energy.

In lower latitudes, most of the atmospheric kinetic energy is contained in quasi-steady, *thermally driven circulations* which are directly related to the geographical distribution of sources and sinks of heat; as a result, the observed weather over most of the tropics varies relatively little from day to day (apart from diurnal variations) at a fixed location, although it may vary greatly from one location to another. These thermally driven circulations include the seasonally varying *monsoons*, which are the atmospheric response to land–sea heating contrasts, and a large-scale meridional overturning over the mid-Atlantic and Pacific Oceans which gives rise to the *intertropical convergence zone*, a narrow east–west band of heavy cloudiness and precipitation.

In middle and high latitudes much of the kinetic energy is associated with moving disturbances called *baroclinic waves*, which develop spontaneously within zones of strong horizontal temperature gradients. Most of the significant day-to-day weather changes at these latitudes can be attributed to the passage of these systems with their attendant mesoscale frontal zones.

As mentioned previously, planetary- and synoptic-scale atmospheric disturbances are subject to frictional dissipation, which causes them to gradually lose their kinetic energy. The energy is not transferred directly from the large-scale motions into random molecular motions. It is first siphoned off by small-scale fluid motions which interact among themselves to transfer energy to smaller and smaller scales and ultimately down to the random molecular motions.

[†] Gradient wind balance when the flow is strongly curved.

Roughly half the frictional dissipation takes place within the lowest kilometer of the atmosphere, as a result of turbulent motions generated mechanically by flow over irregularities in the underlying surface. The other half takes place higher in the atmosphere in discrete patches where small-scale disturbances are generated as a result of shear instability of the vertical wind profile.

In a gross sense the atmosphere may be viewed as a vast but rather inefficient heat engine to which heat is added at a high temperature and removed at a somewhat lower temperature. The mechanical output of the heat engine is the supply of kinetic energy required to maintain the general circulation against frictional dissipation.

9.2 THERMALLY DRIVEN CIRCULATIONS IN THE ABSENCE OF ROTATION

The physical processes responsible for the generation and maintenance of large-scale atmospheric motions can be demonstrated by means of two simple laboratory analogs.

9.2.1 A "before and after" analog

Figure 9.1a shows a tank filled with equal volumes of two homogeneous immiscible liquids of differing densities, placed side by side and separated by a movable partition. The shaded liquid on the right is more dense than the unshaded liquid. In (b) the partition has been removed and fluid motion has developed. We will not be concerned about the details of the motion, but only the final state after friction has brought the liquids to rest. In the new equilibrium configuration shown in (c) the heavy fluid occupies the bottom half of the tank.

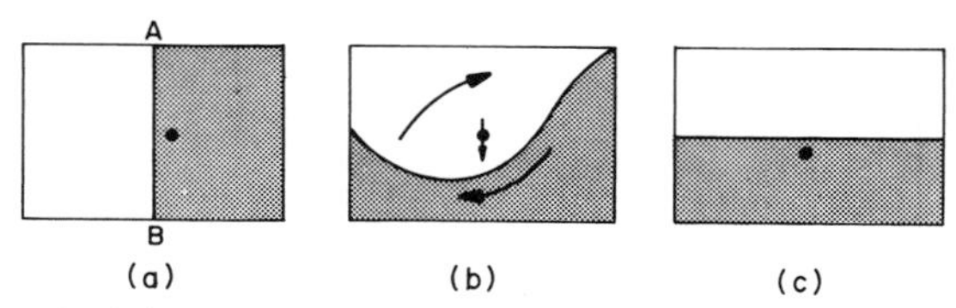

Fig. 9.1 (a) Heavier (shaded) and lighter fluids separated by a movable partition AB. The dot represents the center of gravity of the two-fluid system. (b) Fluids in motion following the removal of the partition. (c) Equilibrium configuration of the fluids after the motion has been dissipated by friction.

Now let us consider this sequence of events from the point of view of the energetics. In the initial configuration, the center of gravity of the fluid, denoted by the dot, is exactly halfway between the top and bottom. After the partition is removed, the center of gravity drops as the denser liquid slides under the lighter one. Through the sinking of denser liquid and the rising of lighter liquid [indicated in (b)] potential energy is converted into the kinetic energy of fluid motions. Frictional dissipation eventually converts all the fluid motions to random molecular motions so that, in the final state (c), the only evidence of the conversion that took place is the drop in the center of gravity of the system and a very slight increase in the temperatures (or internal energy) of the liquids. The energy cycle is summarized in Fig. 9.2. Here we have used the term *available potential energy* in recognition of the fact that only a small fraction of the potential energy of the initial state is really available for conversion to kinetic energy, since no matter what kind of motions develop the center of gravity cannot possibly drop below the level shown in Fig. 9.1c.

Fig. 9.2 The kinetic energy cycle.

Problem 9.1 In the experiment described above the depth of the tank is 1 m and two fluids have specific gravities of 1.1 and 0.9, and the same specific heat of 4×10^3 J kg^{-1} deg^{-1}. What is the maximum possible root mean square (rms) velocity of fluid motion that can be realized in the experiment. How much does the temperature increase as a result of frictional dissipation? (Assume that the rms velocity and the temperature rise are the same for the two fluids.)

Solution Let ρ_1 and ρ_2 be the densities of the heavier and lighter fluids, respectively, D the depth of the tank, m the total mass of fluid in the tank, c the specific heat of the fluids, V the rms fluid velocity, δT the temperature increase, and δz the drop in the center of gravity. From the principle of the conservation of energy, we can write

$$c\,\delta T = V^2/2 = g\,\delta z$$

Therefore,

$$V = (2g\,\delta z)^{1/2}$$

and

$$\delta T = g\,\delta z/c$$

Taking moments† about the mid-depth of the tank ($D/2$) and noting that the two fluids occupy the same volume, we have

$$(\rho_1 + \rho_2)\,\delta z = \left(\rho_1 \frac{D}{4} - \rho_2 \frac{D}{4}\right)$$

from which

$$\delta z = \frac{D}{4} \times \left[\frac{\rho_1 - \rho_2}{\rho_1 + \rho_2}\right] = \frac{1}{4} \times \frac{0.2}{2.0} = 0.025 \quad \text{m}$$

† A moment, in this context, is the product of mass times the vertical distance between the center of mass and the mid-depth of the tank. The mass of the whole fluid times its "moment arm" δz is equal to the mass of the lighter fluid times its moment arm $D/4$ plus the mass of the heavier fluid times its moment arm $-D/4$.

Therefore,

$$V = (2 \times 9.8 \times 0.025)^{1/2} = 0.7 \quad \text{m s}^{-1}$$

and

$$\delta T = 9.8 \times 0.025/4 \times 10^3 = 6.11 \times 10^{-5} \quad \text{deg}$$

If the two liquids are miscible, the final state is characterized by a continuous density gradient between heavier fluid at the bottom and lighter fluid at the top. The location of the center of gravity in the final configuration depends upon how much mixing takes place between the liquids after the removal of the partition. One possibility corresponds to the state of no mixing, which we have already examined. This situation represents the maximum possible drop in the center of gravity and the fullest realization of the potential energy inherent in the original configuration. The other limit corresponds to the case of complete mixing and no release of potential energy. A transition to the latter state can be brought about if the partition is replaced by a permeable membrane which allows the liquids to mix by molecular diffusion without any fluid motions whatsoever. Thus the amount of kinetic energy that is actually realized depends not only upon the amount of potential energy inherent in the original state but also upon the nature of the motions through which the conversion takes place. In general, the more reversible the motions, the larger the fraction of the original potential energy that is actually converted into kinetic energy.

9.2.2 A "steady-state" analog

In the real atmosphere, available potential energy is constantly being replenished by diabatic heating so that there is a continuous flow of energy through the cycle indicated in Fig. 9.2. The following "steady-state" laboratory analog will help to illustrate this situation. Fig. 9.3 shows a tank full of liquid, which has internal heat sources along the bottom and left wall and matching heat sinks along the top and right wall. The liquid expands with increasing temperature.

The gradient of heating drives a slow, clockwise circulation cell. As a parcel of fluid is carried around the tank in this cell it grows warmer as it passes along the bottom of the tank and as it ascends along the left wall. Then it grows cooler as it moves across the top of the tank and down along the right wall to

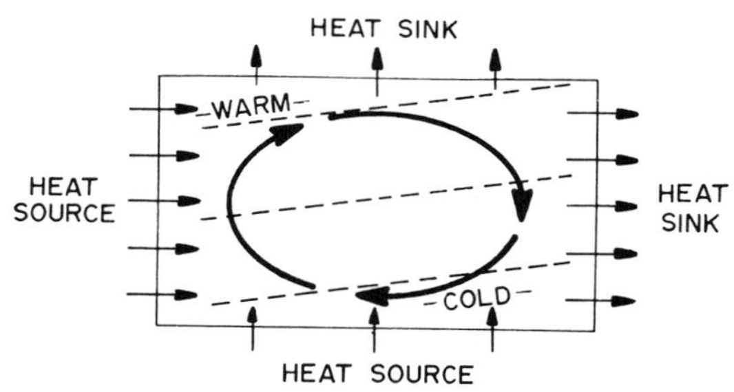

Fig. 9.3 Steady state circulation (heavy arrows) in a liquid, driven by the distribution of heat sources and sinks as indicated. Dashed lines denote isotherms.

complete the circuit. In order to be consistent with these temperature changes, the isotherms must slope from lower left to upper right, as indicated by the dashed lines in the figure. Since temperature increases with height, the fluid in the tank is stably stratified.

In applying the above arguments to the atmosphere, the compressibility of air must be taken into account by considering changes in potential temperature (rather than temperature) as an air parcel moves around the circulation cell. The isotherms in the figure are thus indicative of isentropes in the atmosphere, where an increase of potential temperature with height is the criterion for stable stratification (see Section 2.7.1).

At any given level in the tank, lighter fluid is rising along the left-hand side of the tank and an equal volume of heavier fluid is sinking along the right-hand side. This vertical exchange of equal volumes of fluid with different densities produces a net downward flux of mass which should tend to lower the center of gravity of the fluid, just as in the previous analog, converting available potential energy to kinetic energy. In this steady-state model, the conversion proceeds at exactly the same rate as the kinetic energy of the fluid motions is being destroyed by frictional dissipation. The lowering of the center of gravity of the fluid is opposed by diabatic heating which is always warming and expanding the fluid near the bottom of the tank, and cooling and compressing the fluid near the top. In order to accommodate this expansion and compression, a very small mean upward motion is required at intermediate levels. This mean upward mass flux exactly cancels the downward mass flux due to the circulation cell so that the center of gravity of the fluid remains at a constant level.

In the laboratory analog, the heat source and sink adjacent to the side walls of the tank represent the horizontal gradients of diabatic heating in the earth's atmosphere. The heat source near the bottom of the tank represents the combined effects of the absorption of solar radiation, the exchange of infrared radiation with the earth's surface, and, more importantly, the input of sensible heat associated with convectively driven motions in the mixed layer and the release of latent heat in clouds. Collectively, these processes result in a strong input of energy into the lower troposphere. The heat sink near the top of the tank represents the emission of infrared radiation to space, which is strongest in the upper troposphere.

The importance of atmospheric water vapor in these large-scale thermally driven circulations is worth emphasizing. Latent heat added to the atmosphere in the lower branch of the cell is converted into sensible heat when the water vapor condenses in the rising branch. This additional source of heating serves to enhance the horizontal heating contrasts that would have existed in the absence of the cell. As a result, thermally driven circulations tend to be much stronger than they would be in an atmosphere without water vapor.

The maintenance of large-scale thermally driven circulations requires both horizontal and vertical gradients of diabatic heating. In the absence of horizontal heating contrasts, the heat source and sink at the bottom and top of the tank destroy the stable stratification and initiate convectively driven motions on a scale much smaller than the dimensions of the tank. In these convection cells, bubbles of warm light fluid rise, and are replaced by cooler, denser fluid from above. In the equilibrium situation the downward mass flux in the convection cells is just enough to maintain the center of gravity of the fluid at a constant level.†

9.2.3 The generation of kinetic energy

In hydrostatic motions characterized predominantly by the rising of warm light air and the sinking of cold dense air, kinetic energy is being generated as a result of mechanical work done by the horizontal pressure gradient force. To demonstrate this relationship between the release of potential energy and the generation of kinetic energy, let us consider the two simple circulation cells shown in Fig. 9.4a,b, in which warm air is rising on the left-hand side and cold air is sinking on the right-hand side. The solid lines in the figure depict two possible configurations of pressure surfaces. The vertical spacing between the surfaces (the thickness) is larger on the warm side than on the cold side, in accordance with (2.29).

Fig. 9.4 Possible distributions of motion (heavy arrows) and pressure surfaces (solid lines) in vertical cross sections through circulation cells that convert potential energy to kinetic energy.

For the distribution of pressure shown in (a), the horizontal motion is directed across the isobars from high to low pressure in both the upper and lower branches of the cell. At both levels the pressure gradient force is tending to strengthen the horizontal motions. In the absence of other horizontal forces, each air parcel is experiencing an acceleration

$$\frac{d\mathbf{V}}{dt} = \mathbf{P}_n$$

† The steady-state situation that corresponds to the absence of vertical heating gradients has no counterpart in the atmosphere. If the bottom and top of the tank were perfectly insulated, the circulation cell pictured in Fig. 9.3 would produce a lowering of the center of gravity or, equivalently, an increase in the static stability of the fluid. The greater the vertical density gradient the smaller the vertical motions required to counteract the density changes induced by the horizontal heating contrasts. Thus the cell would grow progressively weaker. Eventually, the vertical temperature gradient would become so strong that molecular diffusion would begin to transfer heat downward. The ensuing density changes would tend to oppose the drop in the center of gravity resulting from the cell until ultimately a state of equilibrium would be reached.

directed down the horizontal pressure gradient. Here $\mathbf{V}$ is the horizontal wind vector, t the time, and $\mathbf{P}_n$ the horizontal pressure gradient force. Since in this example the acceleration and the velocity are in the same direction, the rate of change of kinetic energy (per unit mass) of the parcel is given by

$$\frac{d}{dt}\left(\frac{V^2}{2}\right) = V\frac{dV}{dt} = P_n V$$

It may be noted that the term $P_n V$ has the units of work per unit time (per unit mass). Thus the above equation states that the rate of increase of kinetic energy of each air parcel is equal to the rate at which work is being done upon it by the horizontal pressure gradient force.

The situation depicted in Fig. 9.4b is a little more ambiguous. The horizontal pressure gradient force is accelerating the air at the upper levels but it is decelerating the air at the lower levels. However, from an inspection of the slope of the pressure surfaces, it is evident that the horizontal pressure gradient must be stronger at the upper levels. It follows that the kinetic energy imparted to air parcels in the upper branch of the cell is greater than the kinetic energy lost within the lower branch, so that the cell as a whole is experiencing a net gain of kinetic energy. It can be proven that, for hydrostatically balanced motions, the net rate of generation of kinetic energy by the horizontal pressure gradient force is exactly equal to the rate of release of available potential energy by the sinking of dense air and the rising of lighter air.

Circulations characterized by rising of warm air and sinking of cold air, horizontal flow across the isobars toward lower pressure, and a conversion from available potential energy to kinetic energy are said to be *thermally direct*, and motions with the opposite characteristics are said to be *thermally indirect*. Both types of circulations are found within the atmosphere but the thermally direct type predominates.

9.3 THE INFLUENCE OF PLANETARY ROTATION UPON THERMALLY DRIVEN CIRCULATIONS

The cross-isobar component of the horizontal wind field accounts for only a few percent of the kinetic energy of the general circulation of the earth's atmosphere. The remainder of the kinetic energy is associated with the component of the wind that blows parallel to the isobars, which is close to a state of geostrophic balance with the pressure field. In order to understand how quasi-geostrophic motions develop and persist in the earth's atmosphere, we will need to consider how the large-scale thermally driven circulations described in the previous section are modified by the planetary rotation.

9.3.1 Conservation of angular momentum

Let us consider a hypothetical motion field in which pressure and velocity are independent of longitude; consequently, the zonal pressure gradient $\partial p/\partial\lambda$

is identically equal to zero. In the absence of frictional forces, any zonal ring of air conserves angular momentum as it moves about in the meridional plane. The angular momentum (per unit mass) M of the air in such a zonal ring is given by

$$M = R_{\mathrm{E}} \cos \phi \, (\Omega R_{\mathrm{E}} \cos \phi + u)$$

where R_{E} is the average radius of the earth, ϕ the latitude, Ω the angular velocity of the earth's rotation, and u the zonal velocity of the air relative to the rotating earth. In the above expression, we have neglected the thickness of the atmosphere and the small deviations from spherical geometry associated with the ellipsoidal shape of the earth. The term in parentheses represents the absolute tangential velocity of the ring, which is the sum of the tangential velocity associated with the planetary rotation (for example, 465 m s^{-1} at the equator on earth), plus the tangential velocity u of the ring relative to the solid planet. The term outside the parentheses represents the distance from the planetary axis of rotation. Now let us see what happens to the zonal velocity of the ring as it moves poleward conserving angular momentum.

Problem 9.2 A ring of air parcels originally at rest on the earth's equator moves to 30° latitude, conserving angular momentum. What is its zonal velocity when it reaches 30°?

Solution $M(\phi) = M\,(\text{equator}) = \Omega R_{\mathrm{E}}^{2}$

$$R_{\mathrm{E}} \cos \phi \, (\Omega R_{\mathrm{E}} \cos \phi + u) = \Omega R_{\mathrm{E}}^{2}$$

Therefore,

$$u = \frac{\Omega R_{\mathrm{E}}(1 - \cos^2 \phi)}{\cos \phi} = \Omega R_{\mathrm{E}} \sin \phi \tan \phi$$

$$u = 465 \text{ m s}^{-1}(0.500)(0.576) = 134 \text{ m s}^{-1}$$

Now if the poleward motion in the above problem is directed down the horizontal pressure gradient, the resulting westerly acceleration is in the same sense as the geostrophic wind. In a similar manner an equatorward flow down the horizontal pressure gradient will give rise to an easterly acceleration that is in the same sense as the geostrophic wind. The two situations are pictured in Fig. 9.5a,b, for a Northern Hemisphere location.

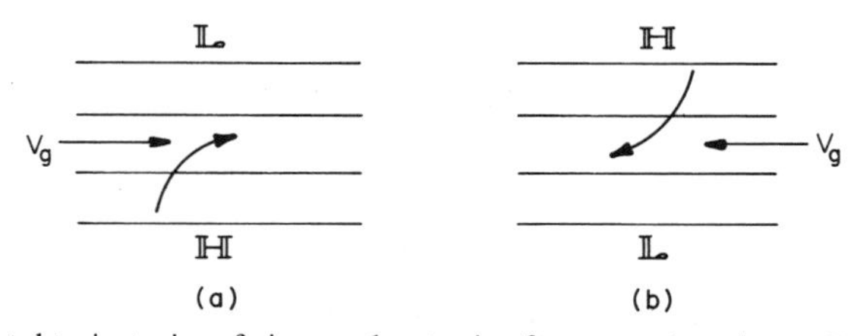

Fig. 9.5 Horizontal trajectories of air parcels, starting from rest, in a thermally direct circulation, showing how the conservation of angular momentum gives rise to motion in the same sense as the geostrophic wind.

To treat the more general case of motions that are not symmetric about the axis of planetary rotation, let us begin by considering the cross-isobaric flow converging into a region of low pressure, centered on some latitude ϕ_c. For simplicity, we will retain the assumption of circular symmetry of the pressure and motion fields, only now the axis of symmetry is located not at the pole but in the middle of the low-pressure area. In addition, we will assume that the horizontal dimension of the low is very small in comparison to the radius of the earth so that we may treat all the air parcels in the ring as if they were located at latitude ϕ_c. This simplification is equivalent to treating the horizontal motions as if they lay in a plane tangent to the earth at latitude ϕ_c. The angular momentum per unit mass of a circular ring of air parcels about a vertical axis located at the center of the low is given by

$$M_c = r^2(\Omega \sin \phi_c + \hat{\omega})$$

where r is the distance from the axis, $\Omega \sin \phi_c$ the component of planetary rotation about the axis, and $\hat{\omega}$ the rate of rotation of the ring, as viewed in the rotating frame of reference, defined as positive in the same direction as Ω. We recall from Section 8.4.1 that a circular relative motion in the same sense as the planetary rotation ($\hat{\omega} > 0$) is defined as a cyclonic circulation, while motion in the opposite sense is defined as an anticyclonic circulation.

Now let us assume that the air at some radius far from the center of this low-pressure area has no relative rotation about the axis of symmetry. As this air converges toward the low center, conserving angular momentum, $\hat{\omega}$ must become positive and increase rapidly in order to compensate for the decrease in r. Thus the air in a region of convergence tends to develop a cyclonic circulation. This mechanism is responsible for synoptic-scale cyclonic circulations in the earth's atmosphere, including those observed in the most intense hurricanes, and possibly even tornadoes.

Problem 9.3 A circular ring of air is converging into the center of a hurricane located at 20° latitude. The air in the ring originated at a radius of 500 km from the center where there was no net cyclonic rotation about the storm. Calculate the tangential wind speed at the time when the ring reaches a radius of 100 km, assuming that angular momentum has been conserved.

Solution

$$M_c \text{ (final)} = M_c \text{ (initial)}$$

Therefore, if the subscripts i and f refer to initial and final conditions, respectively,

$$r_f^{\,2}\left(\Omega \sin \phi_c + \frac{V_f}{r_f}\right) = r_i^{\,2}(\Omega \sin \phi_c) \qquad \text{from which} \qquad V_f = \Omega \sin \phi_c \left(\frac{r_i^{\,2} - r_f^{\,2}}{r_f}\right)$$

Hence,

$$V_f = 7.29 \times 10^{-5}\ \text{s}^{-1} \times 0.342 \times \left(\frac{500^2 - 100^2}{100}\right) \text{km} \times 10^3\ \text{m km}^{-1}$$

$$= 59.8 \quad \text{m s}^{-1}.$$

The opposite effect is felt when air with little or no circulation to start with diverges out of a region of high pressure. In this case, $\hat{\omega}$ becomes negative and decreases rapidly in order to compensate for the increase in r. Thus in regions where divergence is taking place, the air tends to develop an anticyclonic circulation.†

9.3.2 The maintenance of geostrophic balance

It is easily verified that, regardless of the curvature of the isobars, the Coriolis force associated with cross-isobar flow from higher toward lower pressure is always in the same sense as the geostrophic wind, and vice versa. Thus, cross-isobar flow toward lower pressure always tends to strengthen the wind and cross-isobar flow toward higher pressure always tends to weaken it.

In rotating fluids the cross-isobar component of the horizontal flow plays a crucial role in keeping the flow parallel to the isobars close to a state of geostrophic balance. As an illustration, let us consider the situation, depicted in Fig. 9.6, in which the speed of the wind component parallel to the isobars is

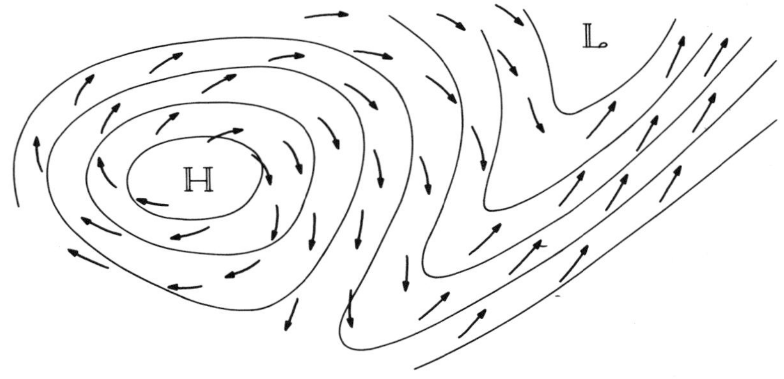

Fig. 9.6 Horizontal flow (arrows) in relation to isobars (solid lines) when the wind speeds are slightly subgeostrophic.

slightly subgeostrophic. In this case the Coriolis force, directed across the isobars from low toward high pressure, is not quite strong enough to balance the pressure gradient force. This imbalance of forces drives a weak cross-isobar flow which causes the air to spiral out of centers of high pressure and into centers of low pressure as indicated in the figure. The Coriolis force, acting on this cross-isobaric component, produces an acceleration to the right, which intensifies the flow parallel to the isobars and brings it closer to geostrophic

† The relationships between cyclonic (anticyclonic) rotation and convergence (divergence) are valid even in the absence of circular symmetry. A more general discussion, based upon the "circulation theorem," is given in J. R. Holton, "An Introduction to Dynamic Meteorology," Academic Press, New York, 1972, pp. 61–65.

balance with the pressure field. By a similar line of reasoning it is readily shown that supergeostrophic wind speeds tend to induce a weak flow across the isobars from low toward high pressure, which has the effect of decelerating the wind parallel to the isobars and bringing it back into geostrophic balance.

9.3.3 The maintenance of thermal wind equilibrium

The cross-isobaric flow, together with its associated rising and sinking motions, is always at work to maintain a state of thermal wind equilibrium between the wind component parallel to the isobars and the temperature field. To illustrate how this adjustment mechanism operates let us consider the special case of the equivalent barotropic atmosphere pictured in Fig. 9.7. Let us assume that the speed of the wind component parallel to the isobars is subgeostrophic at both levels in the figure so that the vertical wind shear between the two levels is not large enough to be in thermal wind equilibrium with the horizontal temperature gradient in the intervening layer. The imbalance between the pressure gradient force and the Coriolis force at the two levels induces the thermally direct circulation shown in the figure.

The cross-isobar flow toward lower pressure at both levels accelerates the flow parallel to the isobars, thus increasing the vertical wind shear. At the same time the associated rising and sinking motion produces adiabatic cooling in the region of high-temperature anomalies and adiabatic warming in the region of low-temperature anomalies, thus reducing the horizontal temperature contrasts. Thus the thermally direct circulation converts potential energy associated with the horizontal temperature gradients directly into kinetic energy of the horizontal wind component parallel to the isobars in order to restore thermal wind balance. In a similar manner, a thermally indirect circulation develops to restore the balance in regions where the vertical wind shear is too large to be in thermal wind balance with the horizontal temperature gradient.

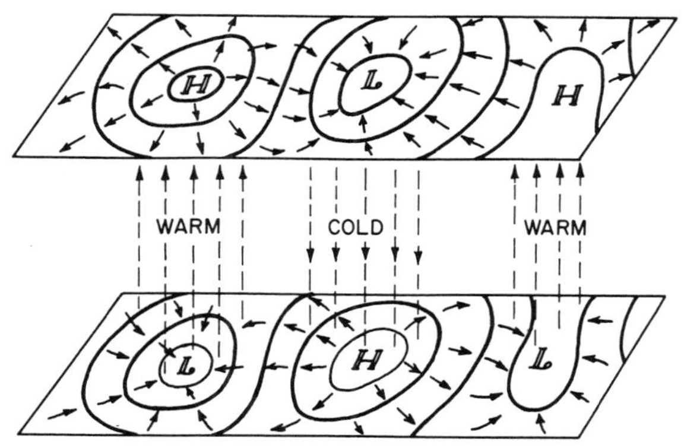

Fig. 9.7 Thermally direct circulation induced by subgeostrophic flow in equivalent-barotropic disturbances. Solid lines represent isobars or geopotential height contours. Short, solid arrows indicate cross-isobar flow. Dashed arrows indicate the sense of the vertical motions.

The adjustment process described above is so effective that the temperature field and the component of the wind field parallel to the isobars rarely get very far out of thermal wind balance. The cross-isobar flow that is responsible for the adjustment cannot become very large itself without reversing the imbalance of forces that brought it into existence in the first place. This explains why the direction of the actual wind closely parallels the isobars except within the lowest few hundred meters of the atmosphere, where the wind speeds are highly subgeostrophic because of the influence of friction.

9.4 THERMALLY DRIVEN CIRCULATIONS IN THE TROPICS

We recall from Fig. 1.22 that the regions of persistent cloudiness and heavy rainfall in the tropics correspond to the intertropical convergence zone and most of the land masses in the summer hemisphere, while the major dry regions correspond to the subtropical oceans, the equatorial dry zones in the Atlantic and Pacific, the land masses in the winter hemisphere, and the desert regions in the summer hemisphere. For the most part, the wet regions are warm relative to their surroundings at the earth's surface and, with the notable exception of the deserts, the dry regions are cool. (For example, we recall from Section 7.4 that the intertropical convergence zone corresponds to a belt of warm ocean water and the equatorial dry zone corresponds to a narrow belt of upwelling of cold water from below the thermocline. The land masses in the summer hemisphere are warmer than the surrounding ocean while the land masses in the winter hemisphere are colder.) Through the convective processes described in Section 7.3, the horizontal gradients in temperature at the earth's surface are imposed upon the overlying atmosphere. In fact, the lapse rate throughout the tropical troposphere is uniformly so close to the moist adiabatic lapse rate that the horizontal temperature distribution in the middle troposphere closely resembles that at the earth's surface. Thus, in a gross sense, the major climatological rain areas in the tropics are characterized by positive temperature anomalies while the dry regions are characterized by negative temperature anomalies.

Furthermore, it is evident that there must be a close relationship between the climatological distributions of precipitation and vertical motion. In the large-scale and long-term average, wet regions are characterized by rising motion and dry regions by sinking motion. Thus, in the climatological mean circulations in the tropics there is an overall tendency for the rising of warm air and the sinking of colder air. Hence, these circulations are predominantly of the thermally direct type.

Let us examine the climatological mean circulation patterns in the tropics in more detail. Regions of heavy precipitation are characterized by

(a) temperatures slightly higher than their surroundings throughout most of the depth of the troposphere,

(b) a slight depression of pressure surfaces in the lower troposphere and a slight upward bulge in pressure surfaces in the upper troposphere, relative to the surroundings.

(c) an upward mass-flux with maximum values in the middle troposphere,

(d) lower tropospheric convergence and upper tropospheric divergence, and

(e) cyclonic flow in the lower troposphere and anticyclonic flow in the upper troposphere.

The vertical structure defined by (a)–(e) is qualitatively consistent with respect to the governing equations; namely,

- (a) and (b) with respect to the hypsometric equation,
- (c) and (d) with respect to the continuity equation,
- (b) and (e) with respect to the geostrophic equation, and
- (a) and (e) with respect to the thermal wind equation.

9.4.1 The monsoon circulations

Figure 9.8 shows an idealized model of the three-dimensional structure of the summer monsoon circulations, based upon the relationships described above. "Wet monsoons" are associated with rising motion over the warm land masses. The available potential energy generated by the differential heating between land and sea is released by the rising of warm (light) air and the sinking of cold (dense) air. The vertical motions are accompanied by a horizontal motion

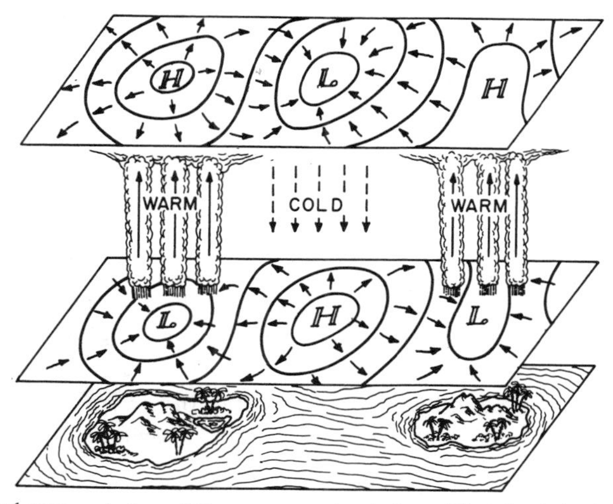

Fig. 9.8 Idealized representation of the monsoon circulations. The islands in the figure represent the tropical continents in the summer hemisphere. Solid lines represent isobars or geopotential height contours near sea level or 1000 mb (lower plane) and 14 km or 200 mb (upper plane). Short solid arrows indicate the cross-isobar flow. Vertical arrows indicate the sense of the vertical motions in the middle troposphere.

component directed from sea to land at low levels and from land to sea in the upper troposphere. At all levels, this motion is directed across the isobars from high to low pressure. Thus, the vertical motion field and the cross-isobar flow are components of a thermally direct circulation which converts potential energy to kinetic energy. Because of the Coriolis force associated with the earth's rotation, the flow across the isobars toward lower pressure induces motion parallel to the isobars.† This is the mechanism by which the monsoon circulation builds up during the early summer in response to the heating of the continents. After the monsoon circulation is established, the same thermally direct circulation maintains it by supplying kinetic energy to counteract the effects of frictional dissipation.

The latent heat of condensation greatly enhances the heating over the continents that drives the summer monsoon circulations. In a dry atmosphere monsoon circulations would exist, but they would not be nearly as strong as those observed on Earth.

9.4.2 Circulations over the tropical oceans

The same model, with appropriate modifications of the shapes of the isobars, can be applied to the circulation over the tropical oceans. Here most of the large-scale overturning takes place in the meridional plane, with warm air rising at the latitude of the intertropical convergence zone (ITCZ), and colder air sinking over the subtropical oceans. The winds in the lower troposphere have a component directed down the horizontal pressure gradient from the *subtropical high-pressure belts*, which prevail over both the Atlantic and Pacific in both the Northern and Southern Hemispheres, toward the *equatorial trough*, a belt of minimum pressure which lies just equatorward of the ITCZ. The Coriolis force induced by this equatorward flow generates lower tropospheric easterlies in both hemispheres. The *northeasterly tradewind belt*, which prevails over much of the tropical north Atlantic and Pacific Oceans, is simply a reflection of this quasi-geostrophic easterly flow around the southern flank of the subtropical high-pressure belts, with the equatorward, cross-isobar flow superimposed upon it. The southeasterly trades in the Southern Hemisphere can be interpreted in a similar manner. In the upper troposphere the circulation is just the reverse of that described above: the ITCZ corresponds to a belt of high pressure, and the cross-isobar flow toward higher latitudes generates westerly winds.

Figure 9.9 shows the wind field in the vicinity of the ITCZ over the tropical Atlantic on one particular day. The wind vectors shown in the figure are based

† It is interesting to note the resemblance between the monsoon circulation described above and the familiar seabreeze circulation which develops on sunny afternoons when the land surface is much warmer than the adjacent sea surface. From a dynamical point of view there is at least one important distinction between these two phenomena: the time scale. The short time interval during which the seabreeze circulation exists is not long enough to allow the Coriolis force to build up a strong flow parallel to the isobars. Hence the flow in the seabreeze circulation is mainly perpendicular to the coastline.

on the motions of clouds (tradewind cumulus at the lower level and filaments of cirrus from the tops of cumulonimbus clouds at the higher level) as viewed by a geosynchronous satellite in orbit at a fixed longitude over the equator. The shaded regions correspond to heavy cloudiness associated with deep cumulus convection along the ITCZ. The ITCZ, the subtropical high, and the northeast tradewind belt are all evident in the figure. Just south of the ITCZ there is a narrow belt of westerlies associated with the cross-equatorial flow from the Southern Hemisphere. This westerly motion is induced by the Coriolis force acting on this air as it flows poleward from the equator into the ITCZ. Note the reversal between the lower and upper tropospheric flow over most areas.

9.4.3 The Hadley circulation

When the wind and pressure fields in the tropics are averaged around latitude circles and over a period of a year we obtain a rather simple circulation that is almost symmetric about the equator and bears a strong qualitative resemblance to the circulations shown in Figs. 9.3 and 9.4a, where the equator is on the left-hand side of the figures and the subtropics are on the right. This meridional overturning is known as the *Hadley*† *circulation.* Because of the earth's rotation the Hadley circulation gives rise to lower tropospheric easterlies (the trade winds) and upper tropospheric westerlies in the zonal and time average circulation.

It should be emphasized that at individual longitudes and during individual seasons the meridional circulation in the tropics assumes a variety of configurations. Therefore the Hadley cell and its related zonal flow are useful not so much as a description of the actual motion field in the tropics but more as an indication of the idealized circulation that would exist in the absence of the complicating effect of inhomogeneities in the earth's surface and the annual cycle in solar declination angle.

9.4.4 The deserts

The extreme warmth and dryness of the desert regions during summer is the result of large-scale subsidence which heats the air by adiabatic compression and lowers its relative humidity to the point where cumulus convection is almost entirely suppressed. In light of the previous discussion, it may seem paradoxical that subsidence should be observed in a region that is warm relative to its surroundings. Indeed, the sinking of warm air and the rising of cooler air is contrary to the motions that should arise in a thermally direct circulation, driven by

† **George Hadley** (1685–1768) English meteorologist. Originally a barrister. Formulated a theory for the trade winds in 1735 but his explanation went unnoticed until 1793 when it was revived by John Dalton. Hadley clearly recognized the importance of what was much later called the Coriolis force.

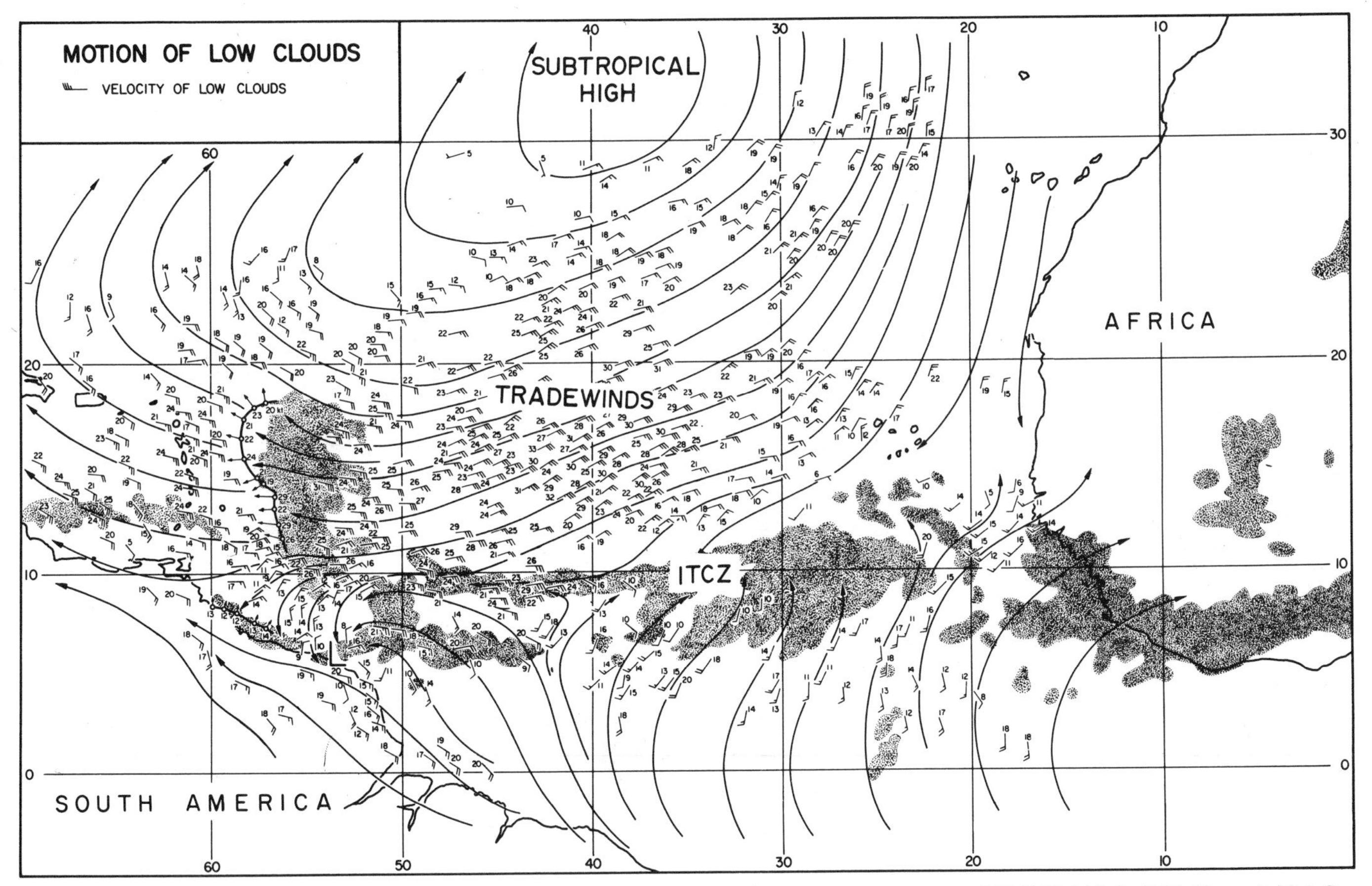

Fig. 9.9a Distribution of low-level cloud motions in the vicinity of the intertropical convergence zone, around 15 GCT 14 July 1969. [From *J. Met. Soc. Japan* **49**, 816 (1971).]

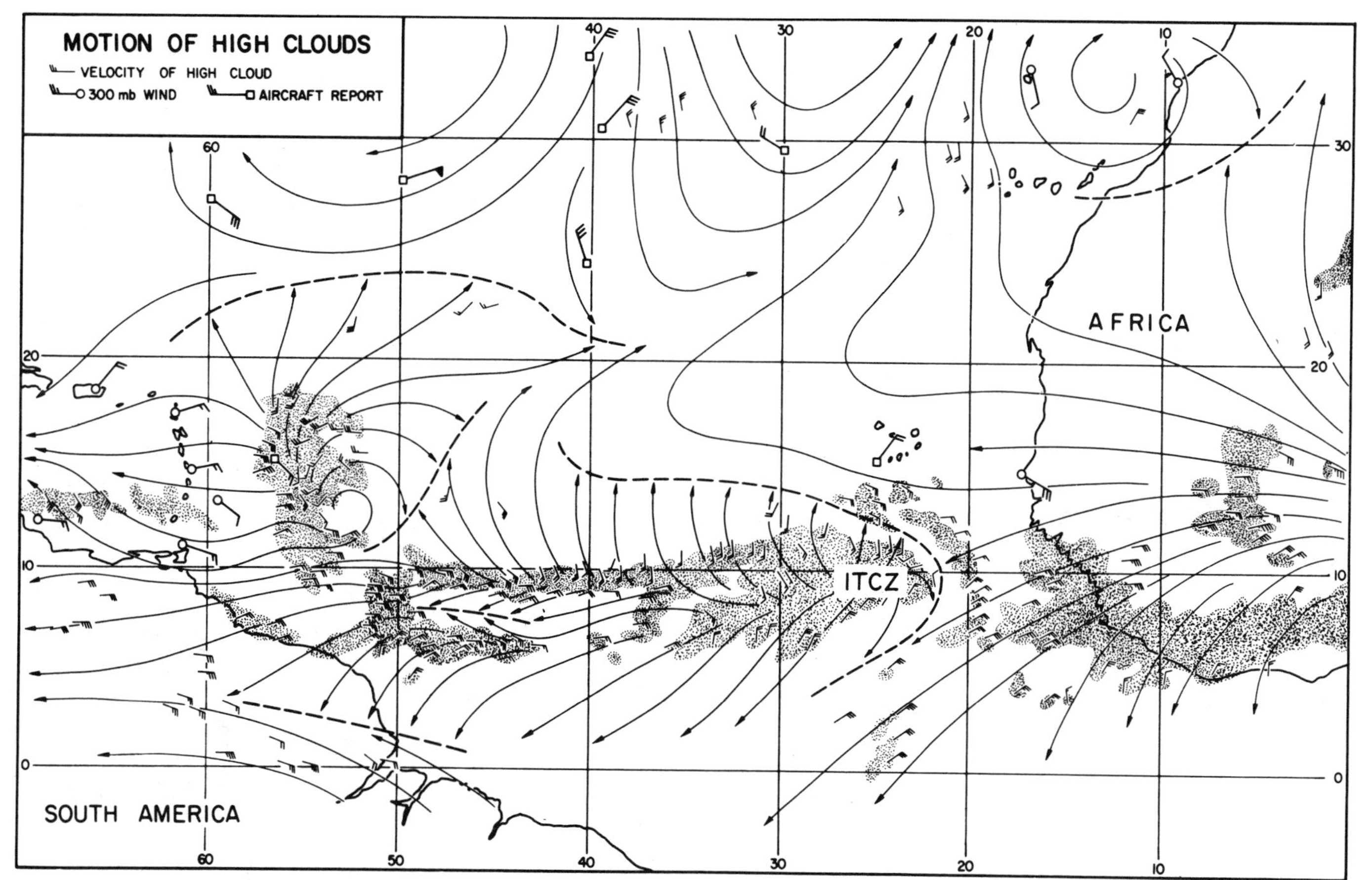

Fig. 9.9b Distribution of upper tropospheric cloud motions and wind observations in the vicinity of the intertropical convergence zone around 15 GCT 14 July 1969. [From *J. Met. Soc. Japan* **49**, 816 (1971).]

heating contrasts at the earth's surface. Thus, we are forced to conclude that the local circulations over desert regions during summer are of the thermally indirect type, driven, not by local heating contrasts, but by a continual input of kinetic energy from other regions of the tropics.

It is not at all obvious from intuitive reasoning why the atmosphere should choose to use up part of the kinetic energy generated in thermally direct circulations in one region to create a desert in another region, nor is it simple to explain why deserts should exist in the particular geographical regions where they are observed. However, in numerical models of the general circulation it has been possible to simulate the observed distribution of deserts with considerable accuracy.

It has been suggested that in semiarid regions large-scale changes in the characteristics of the earth's surface may have an important influence upon climate. For example, erosion due to overly intensive grazing of pasture land in regions of sandy soil tends to increase the local albedo (see Table 6.3). An increase in the albedo tends to reduce the solar energy absorbed at the earth's surface and given up to the atmosphere as sensible heat flux. In the absence of other changes, the reduced diabatic heating of the atmosphere over such a region might tend to favor increased subsidence (or reduced upward motion) and reduced precipitation. A decrease in precipitation would impose an additional stress upon the vegetation, thus hastening the transition from grassland to desert. Once a desert regime is established in such a region it is conceivable that it might persist even in the absence of grazing, because of the climate change brought about by the increase in surface albedo.

9.4.5 Hurricanes

The hurricane is characterized by a strong thermally direct circulation with the rising of warm air near the center of the storm and the sinking of cooler air outside (see Fig. 5.34). The low-level inflow is directed down the radial pressure gradient, from higher toward lower pressure, while the outflow takes place at much higher levels where the radial pressure gradient is very weak.

A distinctive characteristic of the hurricane is its ability to generate and maintain a "warm core" structure without any reliance upon preexisting heating gradients. As explained previously in Section 5.4, the positive temperature anomalies in the interior of the hurricane result from the heating and moistening of the low-level inflow air by the fluxes of water vapor and sensible heat from the highly disturbed sea surface, together with the release of latent heat by deep cumulus convection in the eye-wall cloud and spiral rainbands. The warm core of the hurricane serves as a reservoir of potential energy, which is continuously being converted into kinetic energy by the thermally direct circulation described above. This energy conversion is responsible for generating the high tangential wind speeds in the storm and maintaining them against frictional dissipation.

9.5 BAROCLINIC DISTURBANCES

Let us consider a hypothetical experiment involving the atmosphere on a mythical planet, rotating at the same rate as the earth.† The entire atmosphere is initially at rest relative to the surface of the planet, and there are no horizontal pressure gradients. At some instant in time diabatic heating is abruptly "turned on." We will assume that the meridional and vertical distribution of diabatic heating is identical to the zonally averaged distribution in the earth's atmosphere but, for the sake of simplicity, we will assume that there are no longitudinal variations in diabatic heating or in the surface characteristics of the planet.

In the early stages of the experiment the motion field is dominated by a thermally driven circulation with rising motion in the tropics and sinking motion in the polar regions. The cross-isobar flow associated with this equator to pole "Hadley circulation" gives rise to easterly winds at the surface of the planet and westerlies aloft, which are in thermal wind equilibrium with the meridional temperature gradient. With the passage of time the meridional temperature gradient steepens and the associated zonal wind field intensifies as the atmosphere moves toward an equilibrium with this newly imposed distribution of diabatic heating. In the process, potential energy generated by the differential heating is being converted into the kinetic energy of the zonal flow by the "Hadley circulation."

After the passage of several weeks a dramatic event occurs. The zonal symmetry of the flow is suddenly interrupted by large-scale wave disturbances which begin to develop in middle latitudes. Within a few days these disturbances amplify to the point where they account for roughly half the atmospheric kinetic energy. While the waves are amplifying in middle latitudes, the Hadley cell retreats to the tropics where the flow is still relatively undisturbed. Meanwhile, a weak reverse cell develops in middle latitudes. Gradually the kinetic energy of the atmospheric motions levels off and the atmosphere reaches some sort of equilibrium in this "wave regime." This state of equilibrium does not represent a steady, undisturbed flow; rather it is a continually changing mass and motion field in which the wave disturbances are forming and decaying at more or less the same rate, in a statistical sense. In this chaotic equilibrium state the hypothetical atmosphere resembles the earth's atmosphere.

The reader who is skeptical about numerical experiments involving mythical planetary atmospheres will be reassured to know that an experiment very similar to the one described above can be carried out in the laboratory using a rotating annulus similar to that shown in Fig. 9.10. The fluid in the annulus is heated along the outer wall and bottom, and cooled along the inner wall and top. If the radial heating gradient is gradually increased as the tank rotates at a constant rate about its axis of symmetry, a sequence of events similar to that

† This hypothetical experiment is suggested by numerical simulations of the atmosphere, based upon the primitive equations.

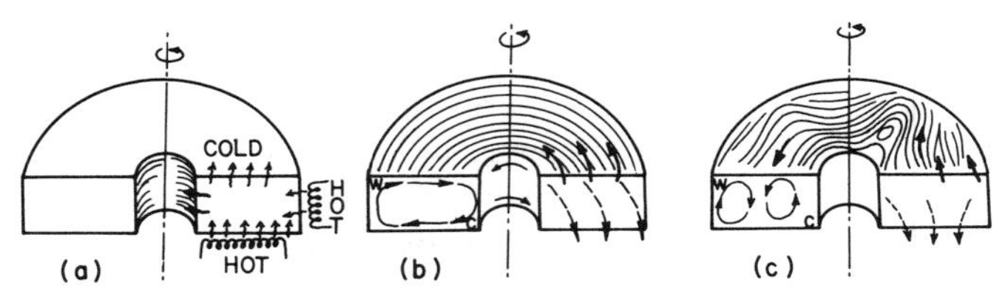

Fig. 9.10 A differentially heated rotating annulus experiment. (a) The distribution of heating and cooling. (b) Symmetric regime showing the "Hadley circulation" (cross section at left) and the azimuthal flow relative to the rotating annulus (inner wall and cross section at right). Note that the azimuthal flow varies with height. (c) The wave or "Rossby" regime. Cross section at left shows azimuthally averaged radial and vertical circulation. Arrows at top and right indicate flow relative to the rotating annulus. Note that the flow varies with azimuth angle and with height and that the radial component may be very large locally.

described above is observed. For a period of time a symmetric azimuthal circulation builds up under the influence of a thermally direct circulation which resembles the one shown in Fig. 9.3 and the Hadley circulation in the atmosphere. When the radial temperature gradient reaches some critical value, there is a sudden transition from the symmetric regime pictured in (b) to the wave or Rossby† regime pictured in (c).

The wave disturbances that develop in the atmosphere and in the tank are called *baroclinic waves* because horizontal temperature advection plays a crucial role in their development; the process by which they amplify is called *baroclinic instability*. This phenomenon was first investigated by Eady‡ and Charney.

9.5.1 Baroclinic instability

We will now describe the processes by which a baroclinic wave amplifies within a region of strong north–south temperature gradient. Let us assume that a weak wavelike perturbation already exists in a flow that is otherwise uniform and purely zonal at some particular level in the atmosphere. For the present, we will assume that this wavelike perturbation is moving at exactly the same speed as the zonal flow at the level under consideration, so that the air is not

† **Carl-Gustav Rossby** (1898–1957) Swedish-American meteorologist. Founded the first meteorology department in the U.S.A. at the Massachusetts Institute of Technology in 1928. He was among the first to recognize the important role of transient mid-latitude wave disturbances in the atmospheric general circulation. Some of the theoretical ideas which he proposed during the 1930s were instrumental in the development of numerical prediction models during the following decade. Many of the basic principles of modern dynamical meteorology were first elucidated in his papers written during this period. In 1950, Rossby returned to Sweden and established the Institute of Meteorology at the University of Stockholm. There he turned his attention to atmospheric chemistry and was largely responsible for the awakening of interest in that subject.

‡ **E. T. Eady** (1915–1966) English mathematician and meteorologist. Worked virtually alone in developing the theory of baroclinic instability while an officer in the Royal Air Force during World War II. One of us (P. V. H.) is grateful to have had him as a tutor.

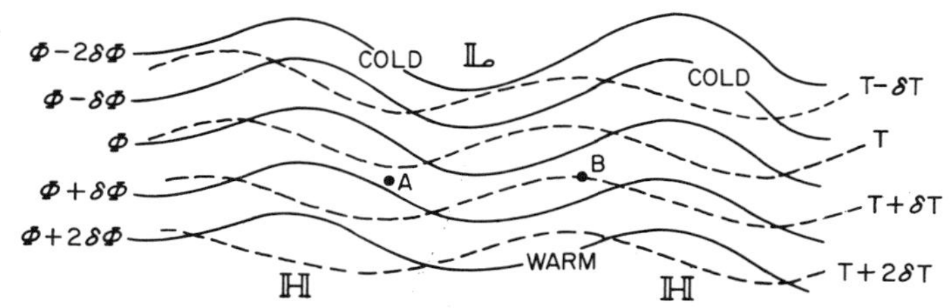

Fig. 9.11 Distribution of geopotential height (——) and temperature (– – –) on a constant pressure surface in a developing baroclinic wave in the Northern Hemisphere. The pressure surface is located near the level where the speed of the wave is the same as the speed of the mean zonal flow.

flowing through the wave. Then, if we view the motion in a coordinate system moving with the wave, the zonal flow vanishes, and the only motion we need consider is the meridional velocity associated with the wavelike perturbation. Let us consider how these meridional motions will distort the isotherms, which are assumed initially to be perfectly straight, and having an east–west orientation with cold air lying to the north. The situation is illustrated in Fig. 9.11.

It is apparent that the northerly wind in the vicinity of point A in the figure is carrying cold air southward, while the southerly wind in the vicinity of point B is carrying warm air northward, thus distorting the isotherms as indicated. From the point of view of (8.15) the local time rate of change of temperature (for example, the cooling at A and the warming at B) is a consequence of the horizontal advection term $(-v\,\partial T/\partial y)$. Note that if the flow is geostrophic the resulting "wave" in the temperature field is displaced one quarter wavelength to the west of the wave in the pressure or geopotential height field. Furthermore, it is clear that, in the absence of other influences, horizontal temperature advection associated with the geostrophic wind will further distort the isotherms from their original east–west configuration, thus causing the east–west temperature contrasts in the wave to further amplify.

In order for the wave to grow it is necessary not only that the temperature perturbations become larger but the kinetic energy associated with the wave disturbance must also increase. The mechanism for generating kinetic energy is a thermally direct circulation which keeps the wind and temperature fields close to a state of thermal wind equilibrium throughout the development process. In order that the wind fluctuations may amplify, at any given latitude cold air must be sinking and warm air must be rising. Applying this relationship to Fig. 9.11, we are led to conclude that the southward-moving air at A is sinking and the northward-moving air at B is rising. The slope of these air trajectories, as they appear when projected onto the meridional plane, is represented in Fig. 9.12 by the heavy arrow. Note that the isentropes and the parcel trajectories both slope upward toward the pole.

The slope of the air trajectories in the meridional plane must be less steeply inclined than that of the undisturbed isentropes, as indicated in Fig. 9.12. If

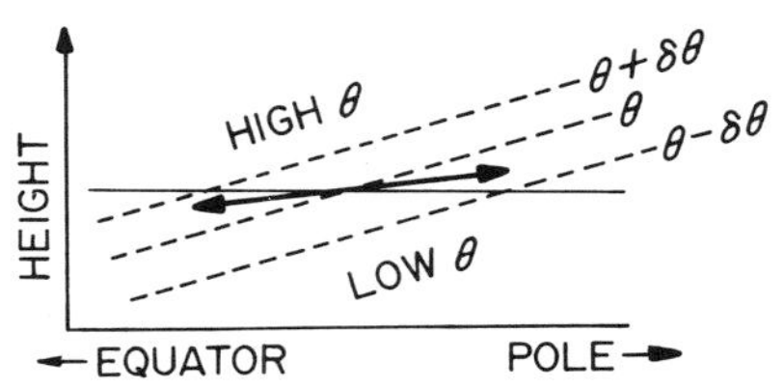

Fig. 9.12 Idealized meridional cross section through a developing baroclinic wave showing slopes of typical air parcel trajectories in relation to the slopes of the zonally averaged isentropes.

trajectories and isentropes had exactly the same slopes, the meridional motions in the waves would have no effect upon the temperature field, since, from (8.15), we would have

$$\frac{\partial T}{\partial t} = -v\frac{\partial T}{\partial y} + \omega\left(\frac{\kappa T}{p} - \frac{\partial T}{\partial p}\right) = 0$$

and so the temperature perturbations would not amplify. Furthermore, if the trajectories sloped more steeply than the isentropes, the meridional motions in the waves would cause the temperature perturbations to weaken rather than amplify.

It can be shown from theoretical arguments that, in order for baroclinic disturbances to amplify most rapidly, the slope of the air trajectories should lie about halfway between the two limiting values deduced in the above discussion; that is to say, it should be about half as steep as that of the isentropes. This theoretical result is consistent with the observed slope of air trajectories in the atmosphere (see pp. 395–396). Furthermore, it can be shown that, for a disturbance of any given horizontal scale, there exists a lower limit to the slope of the isentropes, below which amplification by baroclinic instability is not possible.

Thus far we have been considering the special case of a wave moving at exactly the same speed as the zonal flow. For the more general case in which there is zonal flow relative to the wave the analysis is more complicated and it is not possible, on the basis of simple qualitative reasoning, to deduce the exact phase of the temperature wave relative to the pressure wave. It is apparent that if the mean zonal wind speed $\bar{u}$ is faster than the rate of eastward propagation of the wave c, the wave in the isotherm pattern should be advected somewhat toward the east of its position in Fig. 9.11; just how far to the east depends upon the difference between $\bar{u}$ and c and the rate of amplification of the wave. In the limiting case of a wave which is not amplifying, it can be shown that the temperature pattern eventually assumes an equilibrium configuration with the warmest air in the ridges and the coldest air in the troughs. For the case of ($\bar{u} < c$) the temperature wave will lie to the west of its position in Fig. 9.11, the limiting case being an out-of-phase relationship between pressure and temperature; that is, cold ridges and warm troughs.

9.5.2 Three-dimensional structure of baroclinic waves

The structure of a developing baroclinic wave, as deduced from theoretical considerations, is remarkably consistent with the observed structure described in Chapter 3. We have summarized some of the important aspects of this structure in the longitude–height cross section shown in Fig. 9.13. In this diagram we have purposely omitted mesoscale features associated with fronts in order to emphasize the synoptic-scale structure. When interpreting the diagram it is important to bear in mind that baroclinic waves form in regions of strong vertical wind shear. Throughout the troposphere the mean westerly wind increases with height in agreement with the thermal wind equation. The waves themselves propagate eastward at a velocity comparable to the mean zonal wind speed in the vicinity of the 700-mb (3-km) level. At lower levels the air motion relative to the waves is from east to west, while at higher levels it is from west to east. The following relationships are evident in Fig. 9.13:

- In the vicinity of the 700-mb level, where $\bar{u} \approx c$, the coldest air lies about one quarter wavelength to the west of the trough in the pressure field and the warmest air lies about one quarter wavelength to the west of the ridge. In this range of levels, the wave structure closely resembles that described in Fig. 9.11.
- In the lower troposphere, where $\bar{u} < c$, the displacement between the pressure and temperature waves is greater than one quarter wavelength, while

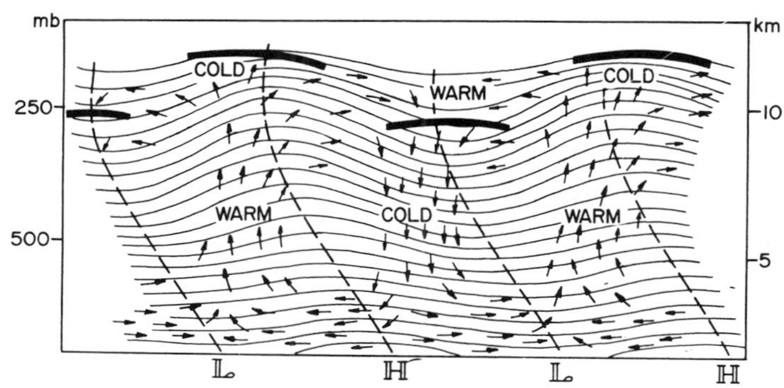

Fig. 9.13 Idealized sketch of the thermally direct circulation in the zonal plane in a developing baroclinic wave. Solid lines represent pressure surfaces with the slopes exaggerated by a ratio of about 5:1. The dashed lines show the vertical tilt of the wave axes in the geopotential height field. Heavy lines indicate the position of the tropopause. Arrows denote vertical motions and associated cross-isobar flow.

in the upper troposphere, where $\bar{u} > c$, it is less than one quarter wavelength, in agreement with the discussion at the end of the preceding subsection.

- In accordance with the hypsometric equation, the *wave axes* (the positions of ridge and trough in the pressure field) tilt westward with increasing height so that the wave trough in the upper troposphere lies over the cold air at lower levels and the ridge lies over the warm air. The typical phase shift is on the order of a quarter of a wavelength between the earth's surface and the 500-mb level, and another one eighth of a wavelength between the 500- and 250-mb levels.
- The maximum synoptic-scale upward motion (typically on the order of a few centimeters per second) is located just to the east of the surface low and just to the west of the ridge at the 500-mb level. These phase relationships are consistent with the cloud pattern shown in Figure 5.38.
- Throughout the troposphere there is a positive correlation between poleward motion, rising motion, and positive temperature anomalies, as required by the energetical considerations discussed previously in Section 9.5.1.
- At most levels there is cross-isobar flow toward lower pressure which (as a result of the earth's rotation) tends to strengthen the wind component parallel to the isobars as the wave intensifies.
- The temperature wave undergoes an abrupt phase reversal at the tropopause. This phase reversal can be viewed as a consequence of the abrupt increase in static stability in going from the troposphere to the lower stratosphere. This change in static stability affects the relative magnitude of the horizontal advection and the vertical motion terms in (8.15). In the lower stratosphere the vertical motion term controls the phase of the temperature wave. Note that in that region ascending air is cold and subsiding air is warm, and kinetic energy is locally being converted into potential energy.

When the structure of baroclinic waves in the earth's atmosphere is examined in greater detail (as in Chapter 3), we find that the strong horizontal temperature gradients tend to be concentrated into mesoscale baroclinic zones (or frontal zones), particularly in the vicinity of the earth's surface and the tropopause. Such zones are characterized by strong horizontal temperature advection (see Fig. 8.18) and vertical motion. When averaged over mesoscale areas, the vertical velocities within baroclinic zones are on the order of tens of centimeters per second, or about an order of magnitude larger than the synoptic-scale vertical motions in baroclinic waves. In fact, much of the vertical mass flux indicated in Fig. 9.13 actually takes place within mesoscale baroclinic zones. Regions of upward motion are particularly strongly concentrated because of the interaction with precipitation processes, as discussed in Chapter 5. Numerical models based upon the primitive equations are capable of simulating the development of baroclinic zones and fronts. However, the horizontal resolution of the current operational numerical weather prediction models is not fine enough to represent these features accurately.

9.6 THE DISSIPATION OF KINETIC ENERGY

The process of frictional dissipation of kinetic energy is described cogently in Richardson's† rhyme:

> Big whirls have little whirls which feed on their velocity.
> Little whirls have lesser whirls, and so on to viscosity.

The so-called *energy cascade* from larger scales of motion to the smaller scales and ultimately down to the scale of the molecular motions themselves is a consequence of the Second Law of Thermodynamics, which can be expressed as a statement to the effect that, in the absence of external forcing, any system must tend toward a state of greater and greater randomness and disorder. In the case in point the ultimate state of randomness (or maximum entropy) corresponds to the situation in which all the kinetic energy originally contained in the organized fluid motions has been imparted to the random molecular motions. The term *viscosity* in the rhyme describes the final stage in the dissipation process, in which the random molecular motions extract energy from the microscale fluid motions.

The energy cascade mechanism is capable of dissipating atmospheric motions with space scales ranging up to a few kilometers within a matter of minutes or, at most, a few hours. In contrast, the time required for frictional dissipation of large-scale atmospheric motions is on the order of a week. The relative ineffectiveness of the energy cascade mechanism in dissipating large-scale motions is due to the fact that, in a statistical sense, such motions accomplish most of their transfer of energy to smaller scales only within certain very limited regions of the atmosphere. The experienced pilot or airline passenger is likely to be familiar with the regions where the energy cascade is in progress: they are precisely those regions in which flying is bumpy due to the presence of vigorous small-scale motions; namely, within the lowest kilometer of the atmosphere and in discrete patches or layers of *clear air turbulence* (CAT).

9.6.1 Shear-induced turbulence

Most clear air turbulence and even some of the turbulent motions close to the ground are the result of small-scale wavelike undulations which develop spontaneously when the vertical shear of the horizontal flow exceeds some critical value. The process responsible for the development of these waves is called

† **Lewis F. Richardson** (1881–1953) English physicist and meteorologist. Youngest of seven children of a Quaker tanner. Served as an ambulance driver in France in World War I. Developed the use of finite differences for solving the differential equations for weather prediction, but at that time (1922) the computations could not be performed quickly enough to be of practical use. Pioneer in the mathematical investigation of the causes of war, which he described in his books: "Arms and Insecurity" and "Statistics of Deadly Quarrels," Boxward Press, Pittsburgh, 1960. Sir Ralph Richardson, the actor, is his nephew.

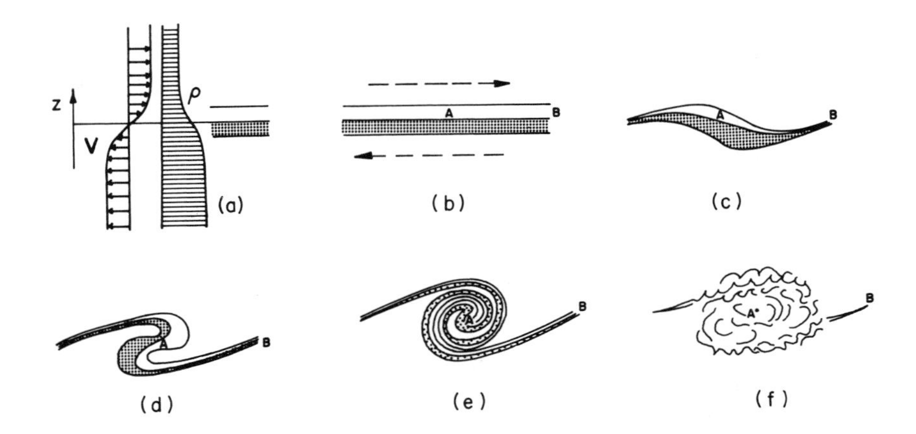

Fig. 9.14 Growth and breakdown of waves induced by shear instability in a stably stratified fluid. (a) Vertical profiles of velocity and density prior to wave development and their relation to the shaded and unshaded layers that appear in the subsequent panels (b–f). [Adapted from S. A. Thorpe, *Radio Sci.* 4, 1330 (1970), copyrighted by the American Geophysical Union.]

shear instability, and the resulting spectrum of small-scale motions is called *shear-induced turbulence.*

The development of shear-induced turbulence can be demonstrated by means of the laboratory experiment depicted in Fig. 9.14a–f. Initially there is straight horizontal flow along a channel with the velocity and density profiles as shown in (a). The fluid is stably stratified ($\partial\rho/\partial z < 0$, for a liquid) and there is a layer with vertical shear centered on the horizontal line in the figure. For any given density stratification, if the vertical shear is weak enough, the flow is stable and no disturbances develop, as shown in (b). However, if the shear exceeds some critical value, undulations spontaneously develop within the shear layer and evolve through the sequence of configurations shown in (c–f). Note the resemblance of the undulations in the wave-cloud formation shown in Fig. 9.15 to (c) and (d). The rolling up of the waves into billows in (e) produces strong velocity shears and density gradients on a very fine scale. These gradients give rise to further instability and, as a result, the flow within the billow breaks down into small-scale turbulent motions as shown in (f). Thus in the sequence (c–f) a cascade of energy to smaller scales has taken place by means of two discrete steps: the waves have amplified at the expense of the kinetic energy of the mean flow and they have subsequently broken down and given up their energy to smaller-scale turbulent motions.

While growing, the waves must do work in exchanging denser fluid from below with lighter fluid from above. In general, the more stable the stratification of the fluid the more work the waves must do against the force of gravity. In order to amplify, the waves must extract kinetic energy from the undisturbed shear flow at a rate faster than they are losing energy by doing work against

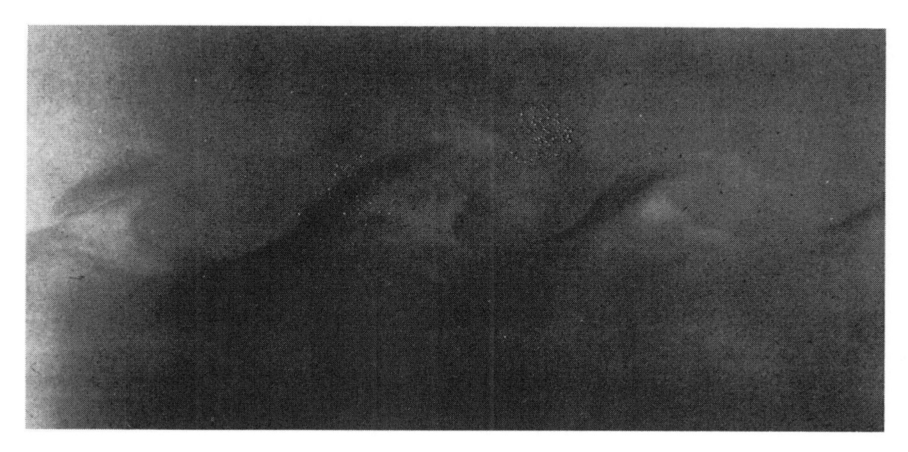

Fig. 9.15 Kelvin–Helmholtz billows as viewed on edge. (Photo: James E. Lovill.)

gravity. Thus the critical shear $\partial \mathbf{V}/\partial z$ required for the waves to amplify depends upon the static stability of the fluid: in general, the larger the static stability the larger the critical shear.

The stability criterion for the spontaneous growth of small-scale waves in a stably stratified atmosphere with vertical wind shear can be expressed conveniently in terms of the dimensionless *Richardson number*, defined as

$$\mathrm{Ri} \equiv \frac{(g/T)(\Gamma_d - \Gamma)}{(d\mathbf{V}/dz)^2} = \frac{N^2}{(\partial \mathbf{V}/\partial z)^2}$$

where N is the Brunt–Väisälä frequency defined in Problem 5.2. The Richardson number is a measure of the ratio of the work done against gravity by the vertical motions in the waves to the kinetic energy available in the shear flow. In general, the smaller the value of Ri the less stable the flow with respect to shear instability. The most commonly accepted value for the onset of shear instability is $\mathrm{Ri} = 0.25$. Once turbulence is established within a shear layer, energetical considerations indicate that it should be sustained so long as $\mathrm{Ri} \leq 1.0$.

The value of Ri within various atmospheric layers can readily be estimated on the basis of wind and temperature data obtained from radiosonde ascents. Within layers of active clear air turbulence, the value usually lies between the two theoretical limits cited above. There is a considerable body of observational evidence (based on cloud photographs, radar returns from clear air, and data from instrumented aircraft) which indicates that wave structures similar to those shown in Fig. 9.15 are often present within layers of moderate and severe clear air turbulence. The motions that produce the bumpiness are not the waves themselves but the smaller-scale motions that result from their breakdown.

In layers of strong static stability (for example, in frontal zones or in the stratosphere), the vertical wind shear must become very strong in order to reach the critical value for the onset of shear instability. In such situations, clear air turbulence outbreaks are likely to be rather severe, since there is, in effect, a large reservoir of kinetic energy for the small-scale motions to draw upon. On the other hand, in layers of near-neutral lapse rate even rather weak vertical wind shears are sufficient to trigger instability, so that small-scale motions are likely to be widespread, but not of sufficient intensity to be classified as clear air turbulence.

The frequency of aircraft encounters with clear air turbulence is about an order of magnitude higher over regions of mountainous terrain than it is over regions of flat terrain. Much, if not all, of the excessive turbulence over mountainous regions is associated with the breakdown of mesoscale waves induced by flow over the mountains, as described in Section 5.1.5.

9.6.2 The planetary boundary layer

Roughly half the frictional dissipation within the earth's atmosphere takes place within the *planetary boundary layer*, the region in which the influence of the lower boundary is directly felt. The horizontal wind speed increases from highly subgeostrophic values near the earth's surface to near geostrophic values at the top of the planetary boundary layer. Within the planetary boundary layer, small-scale turbulent motions, induced by flow over rough terrain features (trees, ocean waves, buildings, and so on) extract energy from the vertical shear of the horizontal flow and dissipate it through the energy cascade mechanism.

The structure and energetics of the planetary boundary layer can be described in terms of three regimes which correspond to different combinations of energy flux through the earth's surface and wind speed in the "free atmosphere" at the top of the planetary boundary layer.

- Under conditions of strong upward sensible heat flux through the earth's surface and low wind speeds the turbulent motions within the planetary boundary layer are closely associated with buoyant thermals rising out of the surface layer as described in Section 7.3.2. Since these motions are driven by dry convective instability within the surface layer this particular combination of conditions is often called the *unstable regime*. The depth of the unstable planetary boundary layer (or *mixed layer*, as it was called in Section 7.3.2) is determined purely by thermodynamical processes as illustrated in Problem 7.1. Typical depths are on the order of 1 km.
- When relatively warm air flows over a colder surface (as, for example, on a clear night over land or when warm air flows over cold water), diabatic cooling of the air just above the surface tends to create a low-level inversion. In this regime the sole source of energy for the turbulent motions in the planetary boundary layer is the kinetic energy of the horizontal motions. The turbulent eddies generated by flow over surface irregularities produce enough vertical mixing to maintain a thin planetary boundary layer with a nearly adiabatic lapse rate adjacent to the earth's surface. These turbulent

eddies decay rapidly with height because most of their kinetic energy is used up in doing work against gravity, lifting cold air from near the earth's surface, and mixing warmer lighter air downward. In this so-called *stable regime*, the depth of the planetary boundary layer increases with wind speed and surface roughness. The warmer the air relative to the underlying surface, the thinner the planetary boundary layer. For typical conditions in the Arctic during winter, the depth of the planetary boundary is on the order of 100 m, and even smaller depths have been observed under very stable conditions.

- When the sensible heat flux through the earth's surface is small and the wind speed relatively strong, the main source of energy for the turbulent motions within the planetary boundary layer is the kinetic energy of the wind field, just as in the stable regime already described. However, in this so-called *near-neutral regime*, gravity offers little or no resistance to the upward penetration of the turbulent eddies, and therefore the planetary boundary layer tends to be much deeper than in the stable regime. Roll circulations (as described in Sections 5.1.4 and 7.3.2) are often present under these conditions. When there is an appreciable upward heat flux from the earth's surface in the presence of strong winds, the planetary boundary layer exhibits characteristics of both unstable and near-neutral regimes.

As a result of the continual loss of momentum to the earth's surface. the flow within the planetary boundary layer generally tends to be subgeostrophic. The imbalance between the pressure gradient force and the Coriolis force induces a cross-isobar flow component from high toward low pressure. Continuity of mass requires a vertical mass flux at the top of the planetary boundary layer: downward in regions of high pressure and upward in regions of low pressure. Through the work done by these vertical motions, the free atmosphere is constantly transfusing kinetic energy into the planetary boundary layer, in a futile attempt to maintain a state of geostrophic balance in the presence of strong frictional dissipation.

9.7 THE KINETIC ENERGY CYCLE

Many of the fundamental concepts discussed in this chapter can be summarized and related to one another in terms of the kinetic energy cycle of the atmospheric general circulation, which is portrayed graphically in Fig. 9.16. The physical processes noted in the diagram are numbered in accordance with the points in the following discussion.

(1) In the global average, the vertical distribution of diabatic heating gives rise to a strong heat source at the earth's surface and a strong heat sink in the upper troposphere. The resulting lifting of the center of gravity of the atmosphere and the concomitant increase in potential energy is a necessary prerequisite for the maintenance of all types of atmospheric motions. (The effect of diabatic heating may also be viewed in terms of a destabilization of the globally averaged lapse rate of temperature.)

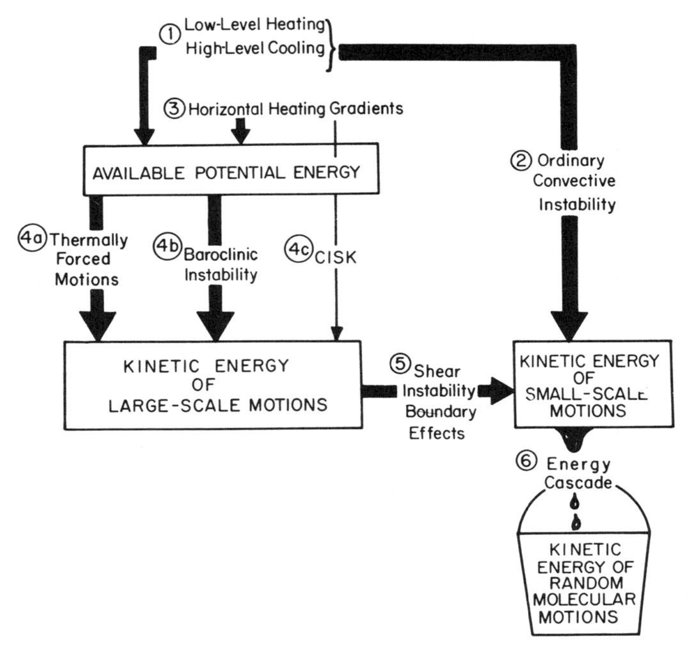

Fig. 9.16 Schematic flow diagram for the kinetic energy cycle in the atmospheric general circulation. Numbers in circles refer to paragraphs in text.

(2) Much of the potential energy generated by the vertical gradient of diabatic heating is released by convective instability which breaks out within limited regions of the atmosphere where conditions are favorable. In such regions spontaneous lifting of buoyant plumes or bubbles of warm air, and the compensating sinking of the colder surrounding air, gives rise to a spectrum of motions ranging from microscale turbulence to the largest thunderstorm cells.

(3) Within regions of stable stratification, which constitute most of the volume of the atmosphere, the presence of horizontal temperature contrasts generated by differential heating renders some of the potential energy available for conversion to kinetic energy. Most of this available potential energy is associated with planetary- and synoptic-scale horizontal temperature gradients.

(4) We have considered two† distinct physical mechanisms by which the conversion of available potential energy to kinetic energy is accomplished. Both

† A third mechanism called *Convective Instability of the Second Kind* (*CISK*), is responsible for hurricanes and other intense "warm core" disturbances which form over the sea. In contrast to baroclinic waves, which feed upon preexisting horizontal temperature gradients, these storms generate their own "warm cores" by absorbing latent and sensible heat from the sea surface and converting latent heat into sensible heat in the rain areas near their centers. Thus, in Fig. 9.16 the CISK process is represented as involving both the generation of available potential energy and the conversion of available potential energy to kinetic energy.

mechanisms involve thermally direct circulations characterized by the rising of warm air and sinking of cold air, and horizontal flow across the isobars from high to low pressure:

(a) *Thermally driven circulations* in which the wind field bears a direct relation to the horizontal heating pattern. The quasi-steady tropical circulations such as the monsoons and the intertropical convergence zone fall into this general category.

(b) *Baroclinic instability* in which the planetary-scale horizontal temperature gradient spontaneously breaks down into synoptic-scale wave disturbances of the ideal size to facilitate the conversion from potential to kinetic energy in middle and high latitudes.

(5) Shear instability and the presence of irregularities in the earth's surface give rise to mesoscale and small-scale disturbances such as shear-induced waves and rolls which extract kinetic energy from the large-scale wind field. These processes are largely confined to the planetary boundary layer and to layers of strong vertical wind shear which sometimes develop in the free atmosphere.

(6) Through the energy cascade mechanism the kinetic energy contained in the smaller scales of motion is ultimately transformed into the kinetic energy of the random molecular motions (that is, internal energy). However, the energy cascade is only a "drop in the bucket" in comparison to the other physical processes that enter into the budget of internal energy, and therefore we need not be concerned about the very small amount of "recycling" that may take place into (1) and (3) above as a result of vertical and horizontal gradients of frictional dissipation.

The amounts of kinetic energy flowing through the two parallel branches of the energy cycle are roughly the same order of magnitude. The relatively large storage of kinetic energy in synoptic-and planetary-scale motions as compared with that in convectively induced motions (a difference of roughly two orders of magnitude) is a consequence of the fact that the shear instability mechanism (5) requires rather large vertical wind shears in order to operate effectively as a brake on the large-scale general circulation. In this respect the comparison between the large-scale and convectively driven motions is analogous to two people coasting down a long hill, one on a racing bicycle and the other on a rusty pair of roller skates. From the same source of potential energy the former acquires a much larger velocity before the rate of frictional dissipation becomes equal to the rate of conversion from potential to kinetic energy. If the processes that generate kinetic energy in the earth's atmosphere were "turned off," the time required for frictional dissipation (operating at its present rate) to completely destroy all the kinetic energy of the large-scale motions is on the order of a week. The corresponding time scale for the convective scale motions is on the order of a few hours.

9.8 THE ROLE OF THE ATMOSPHERIC GENERAL CIRCULATION IN THE HYDROLOGIC CYCLE

The total mass of water substance stored in the atmosphere is equivalent to only about a week's precipitation over the globe, and it changes very little from day to day. Thus it is clear that, over a period of a week or more, there must be a very close balance between globally averaged precipitation (P) and evaporation (E). However, in localized regions there are large imbalances:

- Within the surface anticyclones that cover much of the subtropical oceans, the prevailing subsidence suppresses precipitation and favors cloud-free conditions which are conducive to large evaporation rates. Within such regions, $E \gg P$.
- Within the wet monsoons, the intertropical convergence zone, and the regions of cyclogenesis in middle latitudes, $P > E$.
- For the continents as a whole, $P > E$. The excess precipitation is the sole source of water for the world's great river systems. Conversely, for the oceans as a whole, $E > P$.
- Over land areas that have no drainage to the sea (for example, the Great Basin in the western United States), $P \simeq E$.

The zonally averaged distributions of precipitation and evaporation are shown in Fig. 9.17. The scale on the abscissa is linear in the sine of the latitude angle in order to make any interval of width proportional to the corresponding surface area on earth. Thus the global moisture balance requires that the total areas under the two curves be the same. In the subtropical belts of both hemispheres there is a surplus of evaporation over precipitation; a reflection of the dominance of oceanic anticyclones. Thus, from the point of view of the atmosphere, the subtropics serve as source regions for water vapor. In the equatorial belt and at latitudes above about 40°, precipitation exceeds evaporation. The equatorial region corresponds to the location of the ITCZ and the maximum

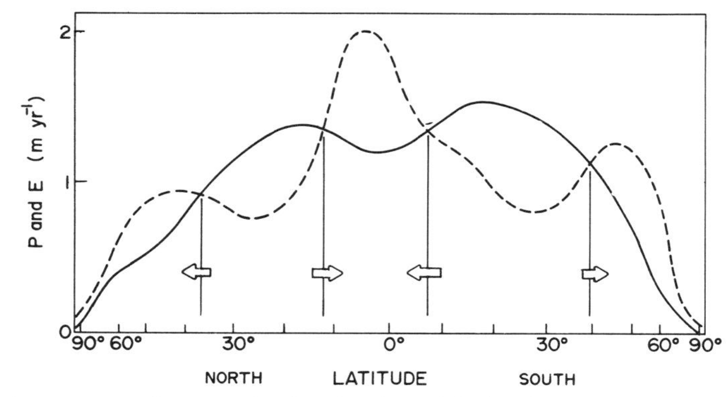

Fig. 9.17 Average annual evaporation (——) and precipitation (– – –) per unit area expressed in meters per year. Arrows represent the sense of the required water vapor flux in the atmosphere. [Adapted from W. D. Sellers, "Physical Climatology," © Univ. of Chicago Press, Chicago (1965), p. 84.]

monsoon rainfall in the annual average, and the precipitation peaks in high latitudes are associated with baroclinic wave activity. Both regions are moisture sinks for the atmosphere.

In the long-term average it is possible to define a closed *hydrologic cycle* which describes the transport of water substance in the earth–atmosphere system. The general circulation may be viewed as the atmospheric component of this cycle. Atmospheric motions transport water vapor† from source regions, where $E > P$, into sink regions, where $P > E$. On the ground, other processes such as river runoff, leakage through subsurface aquifers, and ocean currents transport liquid water in the opposite sense, from regions of excess precipitation to regions of excess evaporation, thus completing the cycle.

Within the tropics and subtropics most of the atmospheric transport of water vapor is due to the thermally driven circulations discussed in Section 9.4. The cross-isobar flow in the lower branch of these circulations carries moist air out of the surface anticyclones, where $E > P$, into the equatorial trough and monsoon depressions, where $P > E$. The return flow in the upper branch of these circulations is so dry that the net atmospheric transport is almost completely determined by the low-level flow.

Poleward of the subtropical anticyclones, the horizontal advection of water vapor in baroclinic waves is responsible for most of the required atmospheric transport. At these latitudes, poleward-moving air usually contains more moisture than equatorward-moving air. Thus, the exchange of equal masses of air containing differing amounts of moisture across a given latitude circle accomplishes a net poleward flux of water vapor.

9.9 THE ATMOSPHERIC TRANSPORT OF ENERGY

The *net* (downward) *radiation* at the top of the atmosphere is given by

$$E^{\downarrow}_{\text{net}} = [(1 - A)S/4] - E$$

where A is now the local reflectivity, S the solar irradiance, and E the infrared emission to space. According to (7.11) the globally averaged net radiation is zero in the long-term mean. However, in localized regions the net radiation may amount to a substantial fraction of the incoming radiation.

The zonally averaged distributions of $(1 - A)S/4$ and E at the top of the atmosphere are shown in Fig. 9.18. Again, the abscissa is scaled so that width is proportional to area on the earth's surface. In accordance with (7.11) the total areas under the two curves are identical. However the shapes of the two curves are considerably different. The net solar radiation $(1 - A)S/4$ drops off by about a factor of 4 between the equator and the poles. This strong latitudinal gradient is largely a reflection of the sharp decrease in insolation with latitude during the winter season (see Fig. 7.16) and the large local albedo that prevails in the

† The transport of solid and liquid water associated with cloud droplets is negligible in comparison to the water vapor transport.

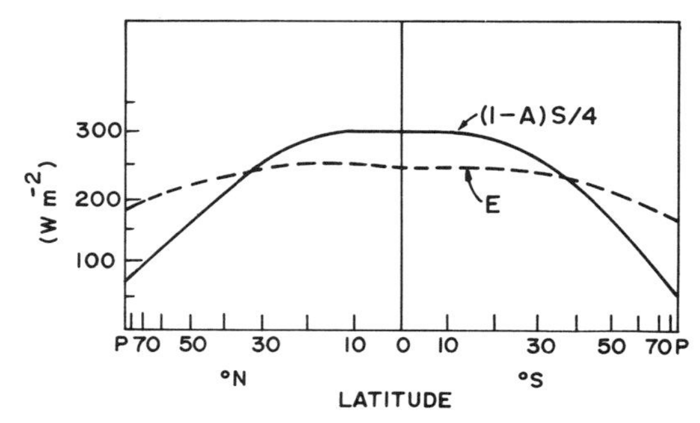

Fig. 9.18 Annual average net incoming solar radiation (——) and outgoing terrestrial radiation (– – –) as a function of latitude. [Adapted from *Space Res.* **11**, 645 (1970).]

polar regions because of the existence of icecaps and persistent summertime cloudiness. The curve corresponding to outgoing terrestrial radiation E is rather flat by comparison. The observed equator-to-pole difference in surface temperature is large enough to account for a factor of two difference in outgoing radiation according to the Stefan–Boltzmann law (6.7). However, because of the larger burden of atmospheric water vapor, and the higher colder cloud tops in the tropics, the "greenhouse effect" is stronger there, and so the equator-to-pole difference in outgoing radiation is considerably less than a factor of two. The two curves cross at a latitude of about 35°. Equatorward of this latitude the net radiation is positive (that is to say, the solar radiation absorbed by the earth–atmosphere system is larger than the infrared radiation emitted to space), and poleward of this latitude the net radiation is negative.

The atmosphere and oceans transport energy poleward, out of the tropics and subtropics, where there is a surplus of incoming solar energy (RN > 0), into higher latitudes, where there is a deficit, thus compensating for the large local imbalances in the radiation budget. Somewhat more than half the total transport is accomplished by the atmospheric general circulation.

The horizontal transport of energy within the atmosphere can be described in terms of (7.4) which relates the time rate of change of moist static energy ($c_pT + \Phi + Lw$) of individual air parcels to the distribution of energy sources and sinks. Because of the imbalances in the local radiation budget, air parcels located in the tropics and subtropics tend (in a statistical sense) to gain moist static energy, while parcels situated at high latitudes tend to lose it. It follows that parcels of air moving poleward across a given latitude circle tend to have higher values of moist static energy than parcels of air moving equatorward across the same latitude circle. Thus, by the exchange of equal masses of air containing differing amounts of moist static energy per unit mass, it is possible to have a net poleward transport of moist static energy. Thermally driven circulations and baroclinic waves both contribute to this transport.

Referring back to the laboratory analog of a thermally driven circulation cell in Fig. 9.3, we see that there is a net transport of internal energy from left to right due to the exchange of warm fluid in the upper branch of the cell with colder fluid in the lower branch. The analogy carries over to the atmosphere provided that internal energy in the model is replaced by moist static energy. The heat sources and sinks in the laboratory model then become sources and sinks of moist static energy in the atmosphere, and the vertical gradient of internal energy becomes the vertical gradient of moist static energy. In the troposphere, c_pT and Lw both usually decrease with height, but the decrease is more than offset by the increase in Φ from the lower troposphere to the upper troposphere. Thus thermally driven circulations responsible for the monsoons and the intertropical convergence zone transport energy horizontally from source regions to sink regions, in agreement with intuitive reasoning. Furthermore, it follows that the Hadley cell, which is nothing more than the zonally averaged thermally driven circulation in the tropics and subtropics, transports energy poleward from the equatorial belt into the subtropics of both hemispheres.

At higher latitudes, the poleward transport of energy is accomplished mainly by baroclinic waves, which produce a poleward flux of warm moist air and an equatorward flux of colder drier air at any given level. The exchanges are particularly large in the lower troposphere near the 850-mb level. During winter the poleward flux of sensible heat is responsible for most of the energy transport across middle latitudes, whereas during summer the poleward flux of latent heat is the dominant transport term.

9.10 THE ATMOSPHERE AS A HEAT ENGINE

In a gross, statistical sense, the sources of energy for the atmosphere are located at relatively low latitudes and low elevation, while the sinks of energy are located at somewhat higher latitudes and higher levels, where the temperatures are somewhat lower. Atmospheric motions transport energy poleward (as discussed in the previous section) and upward (as discussed in Section 7.3.2) from a warm source to a slightly colder sink.

Thus, the atmosphere may be viewed as a heat engine which receives heat at a high temperature and rejects it at a slightly lower temperature. The work done by the heat engine maintains the kinetic energy of the general circulation against the continual drain of frictional dissipation. Just as in the Carnot cycle discussed in Section 2.8, heat is added at a high temperature (in the lower troposphere) and rejected at a lower temperature (higher in the troposphere).

The thermal efficiency η of any heat engine is defined by (2.79):

$$\eta \equiv \frac{Q_1 - Q_2}{Q_1} = \frac{W}{Q_1}$$

where Q_1 is the heat added at the higher temperature, Q_2 the heat rejected at the lower temperature, and W the net mechanical work done by the cycle. By making use of (2.85), it is readily shown that the efficiency can also be expressed in the form

$$\eta = \frac{T_1 - T_2}{T_1}$$

where T_1 and T_2 refer to the (mean) temperatures at which heat is added and rejected, respectively.† The efficiency of the atmosphere is quite low because the difference between T_1 and T_2 is only a few degrees. A quantitative estimate of the efficiency can be obtained by calculating the globally averaged rate of production or dissipation of kinetic energy W and dividing it by the rate of absorption of solar radiation per unit area, averaged over the globe. (See Problem 9.9.)

While the atmosphere as a whole behaves as a heat engine, certain limited regions function in the opposite sense; accepting heat at low temperatures and rejecting it at higher temperatures. Such regions are dominated by thermally indirect circulations which convert kinetic energy, imported from other regions, into potential energy. From a thermodynamic point of view these localized thermally indirect circulations function as refrigerators, producing temperature gradients that are often completely at odds with the local distribution of diabatic heating. For example, two of the coldest regions of the atmosphere are the equatorial tropopause and the mesopause over the summer pole (Fig. 1.10). Adiabatic cooling induced by thermally indirect circulations maintains the temperatures of these regions far below their respective radiative equilibrium values. Just as a refrigerator in the home depends upon electricity generated elsewhere, these localized thermally indirect circulations could not be maintained were it not for the copious amounts of available potential energy generated by gradients of diabatic heating in other regions of the atmosphere.

PROBLEMS

9.4 Explain or interpret the following:

(a) Thermally direct circulations transport sensible heat upward.

(b) Despite the destabilizing influence of radiative transfer, the average lapse rate in the middle and upper troposphere tends to be considerably less than the moist adiabatic lapse rate.

(c) Most of the kinetic energy of atmospheric motions is associated with flow parallel to the isobars.

(d) In a dry atmosphere, the monsoon circulations would not be as strong as those observed in the earth's atmosphere.

(e) The "Hadley circulation" does not extend from pole to equator in the earth's atmosphere.

(f) Baroclinic waves do not often form within the tropics.

(g) The most favored regions for cyclogenesis and frontogenesis are along the east coasts of Asia and North America and along the coast of Antarctica, during the winter season.

† The expression $\eta = (T_1 - T_2)/T_1$ gives the maximum possible efficiency of a heat engine working between temperatures T_1 and T_2. Since the atmospheric heat engine is not completely reversible, its efficiency will be less than that given by this value (see Section 2.8).

(h) Frictional dissipation within the planetary boundary layer tends to enhance the vertical motion field in a developing baroclinic wave.

(i) Baroclinic waves slope westward with height in both Northern and Southern Hemispheres.

(j) The circulation in hurricanes is thermally direct.

(k) Clear air turbulence is rarely observed in the core of the jet stream, but often just above or below it.

(l) Shear-induced turbulence in a stably stratified atmosphere produces a downward heat flux.

(m) Over land the surface wind speed tends to be lowest at night and highest during the early afternoon, while the wind speed 1 km above the ground tends to be highest at night and lowest during the early afternoon.

(n) The combination of strong winds, smooth terrain, and very stable lapse rates near the ground can be hazardous for aviation.

(o) All the hydroelectric power produced by the damming of rivers may be viewed as a by-product of the general circulation.

(p) The equator-to-pole temperature gradient is much less than the radiative equilibrium value.

(q) If the earth had an atmosphere, but no oceans or water vapor, the equator-to-pole temperature gradient would be much stronger than the observed and the lapse rate in the troposphere would be steeper.

(r) If the earth had no mountains the kinetic energy of atmospheric motions would be somewhat larger than the observed.

(s) In considering the role of the general circulation in the global energy balance, it is permissible to ignore transports of kinetic energy.

(t) The Hadley cell transports sensible and latent heat equatorward, yet it transports total energy poleward.

(u) Over the Sahara Desert during summer there exists a large deficit in net radiation.

9.5 By how much would the center of gravity of the atmosphere have to drop in order to generate enough kinetic energy to increase the root mean square wind velocity from zero to 17 m s^{-1}.

Answer 14.7 m

9.6 For a homogeneous (constant density) fluid in a tank with a flat bottom, vertical walls, and no top:

(a) Show that the average potential energy per unit area is given by $\frac{1}{2}\rho g\overline{z^2}$, where z is the height of the free surface of the fluid above the bottom of the tank, and the overbar denotes an average over the area of the tank.

(b) Show that the average available potential energy per unit area is given by $\frac{1}{2}\rho g\overline{z'^2}$, where $z' \equiv z - \overline{z}$. [Hint: Begin by substituting $z = \overline{z} + z'$ in the expression for potential energy per unit area. Note that $\overline{\overline{z}z'} = \overline{z}\overline{z'} = 0$.]

(c) Show that the average rate of conversion from potential to kinetic energy is given by $-\rho g\overline{wz}$, where w is the vertical velocity of the free surface. Give a qualitative physical interpretation of this result in terms of horizontal flow across the isobars.

9.7 Prove that for horizontal flow governed by (8.7) the time rate of change of kinetic energy of any air parcel is given by

$$\frac{dK}{dt} = -\mathbf{V} \cdot \nabla\Phi - 2aK$$

where

$$K \equiv \frac{V^2}{2} = \frac{u^2 + v^2}{2}$$

Give a physical interpretation of this result.

9.8 Within an upper-level frontal zone the temperature is $\simeq 230°K$, the lapse rate is isothermal, and the vertical wind shear is 30 m s^{-1} km^{-1}

(a) Estimate the Richardson number (Ri).

Answer 0.465

(b) What does this value of Ri imply regarding the probability of the occurrence of CAT within this layer?

9.9 The globally averaged rate of kinetic energy generation by large-scale thermally direct circulations is about 2 W m^{-2}.

(a) If the root mean square velocity of large-scale atmospheric motions is 17 m s^{-1}, estimate the time required to completely replenish the kinetic energy of the general circulation.

Answer 8.5 days

(b) Estimate the thermal efficiency of the atmospheric heat engine.

Answer 0.84%

(c) Estimate the temperature difference between the atmospheric heat source and heat sink, as defined in the context of Section 9.10.

Answer $\sim 2°$

9.10 The world output of hydroelectric power is on the order of 2×10^{11} W. What fraction of the output of the atmospheric heat engine does this represent.

Answer Roughly 1 part in 10,000

9.11 Prove that in a dry atmosphere, steady-state, thermally direct circulations transport dry static energy $(c_p T + \Phi)$ horizontally from the heat source toward the heat sink, provided that the lapse rate is statically stable.

9.12 On the basis of the data given in Fig. 9.18, estimate the poleward transport of energy across 30N by the atmosphere and oceans. [Hint: Exactly half the surface area of the globe lies within 30° of the equator.]

Answer $\sim 5 \times 10^{15}$ W

9.13 Show that the general circulation is acting to increase the entropy of the universe. [Hint: Treat the atmosphere as a "black box" which receives heat at a high temperature and gives it back at a lower temperature.]

Index

B

U

ISBN 0-12-732950-1

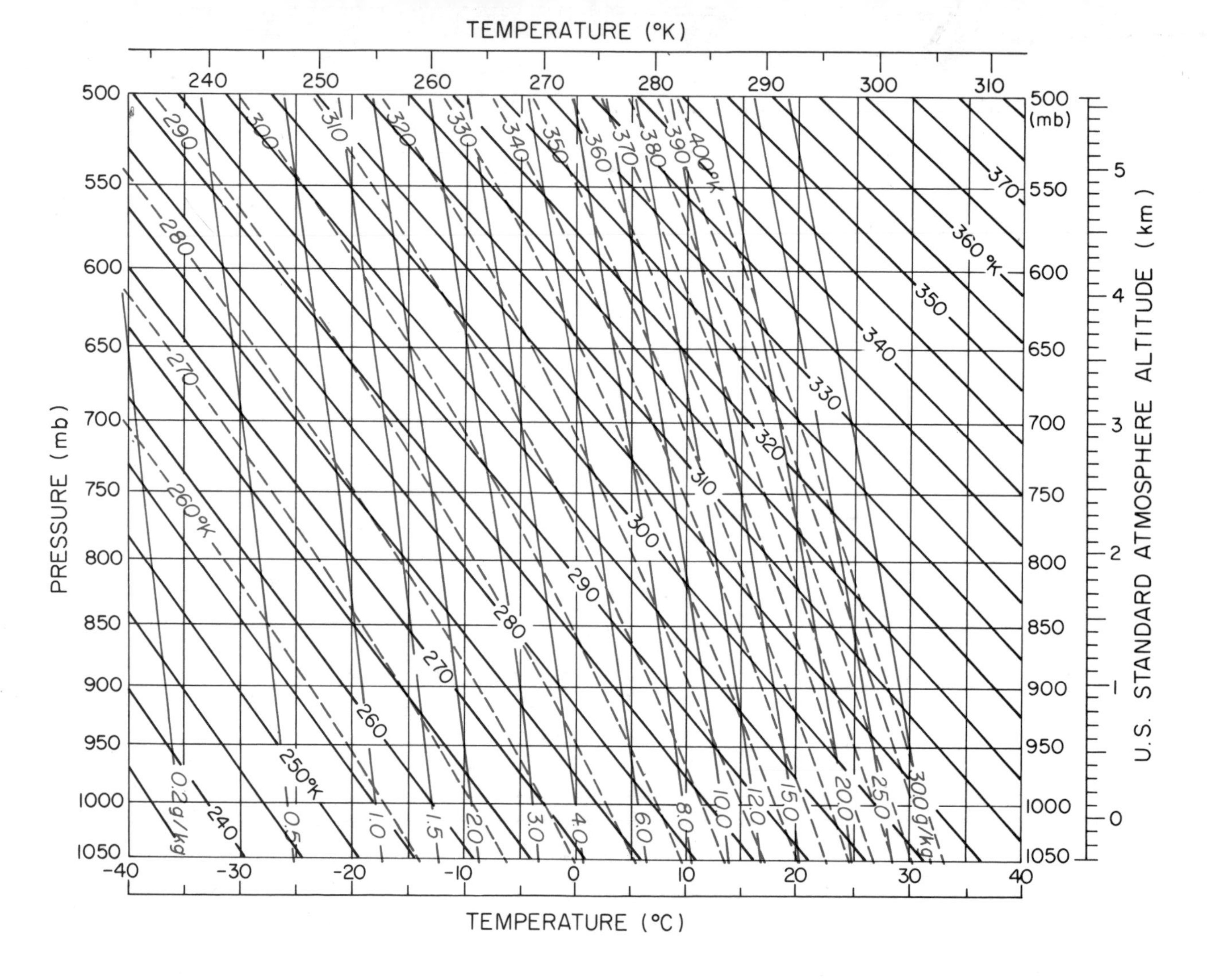

TEMPERATURE (°K)
240
250
260
270
280
290
300
310
PRESSURE (mb)
500
550
600
650
700
750
800
850
900
950
1000
1050
(mb)
U.S. STANDARD ATMOSPHERE ALTITUDE (km)
0
1
2
3
4
5
TEMPERATURE (°C)
-40
-30
-20
-10
0
10
20
30
40
0.2 g/kg
0.5
1.0
1.5
2.0
3.0
4.0
6.0
8.0
10.0
12.0
15.0
20.0
25.0
30.0 g/kg
240
250°K
260
270
280
290
300
310
320
330
340
350
360°K
370
260°K
400°K
390
380